Springer Proceedings in Mathematics & Statistics

This book series features volumes composed of selected contributions from workshops and conferences in all areas of current research in mathematics and statistics, including data science, operations research and optimization. In addition to an overall evaluation of the interest, scientific quality, and timeliness of each proposal at the hands of the publisher, individual contributions are all refereed to the high quality standards of leading journals in the field. Thus, this series provides the research community with well-edited, authoritative reports on developments in the most exciting areas of mathematical and statistical research today.

Jorge Vargas Florez • Irineu de Brito Junior
Adriana Leiras • Sandro Alberto Paz Collado
Miguel Domingo González Alvarez
Carlos Alberto González-Calderón
Sebastian Villa Betancur • Michelle Rodríguez
Diana Ramirez-Rios
Editors

Production and Operations Management

POMS Lima, Peru, December 2-4, 2021 (Virtual Edition)

Springer

Editors
Jorge Vargas Florez
Pontifical Catholic University of Peru
Lima, Peru

Irineu de Brito Junior
Sao Paulo State University
Sao Jose dos Campos, Sao Paulo, Brazil

Adriana Leiras
Pontifical Catholic University of Rio de Janeiro
Rio de Janeiro, Rio de Janeiro, Brazil

Sandro Alberto Paz Collado
Pontifical Catholic University of Peru
Lima, Peru

Miguel Domingo González Alvarez
Pontifical Catholic University of Peru
Lima, Peru

Carlos Alberto González-Calderón
Universidad Nacional de Colombia at Medellín
Medellín, Antioquia, Colombia

Sebastian Villa Betancur
Indiana University
Bloomington, IN, USA

Michelle Rodríguez
Universidad del Pacífico
Lima, Peru

Diana Ramirez-Rios
Rensselaer Polytechnic Institute
Troy, NY, USA

ISSN 2194-1009 ISSN 2194-1017 (electronic)
Springer Proceedings in Mathematics & Statistics
ISBN 978-3-031-06861-4 ISBN 978-3-031-06862-1 (eBook)
https://doi.org/10.1007/978-3-031-06862-1

This Springer imprint is published by the registered company Springer Nature Switzerland AG
The registered company address is: Gewerbestrasse 11, 6330 Cham, Switzerland

Preface

This book crowns the effort of the reconstructed Latin America & the Caribbean chapter of the Production and Operations Management Society, POMS (http://www.poms.org), to keep a biannual POMS International Conference in the region, even with the challenge imposed by restrictions due to the COVID-19 pandemic. POMS is an international professional organization congregating production and operations management professionals and academicians worldwide and aims at extending and integrating knowledge in the field of production and operations management (POM). POMS disseminates POM concepts to managers, scientists, educators, students, public and private organizations, national and local governments, and the public. POMS International Conferences provide an opportunity to faculty, doctoral students, and practitioners to share knowledge and insights.

The first POMS International Conference organized by the Latin America & the Caribbean chapter was held in December 2018 in Rio da Janeiro, Brazil, with 164 participants, many from outside Latin America. It was an in-person meeting and very successful. Particularly, the traditional Latin American warmth and hospitality had been very appreciated by the international audience.

This second edition of the POMS Latin America & the Caribbean International Conference was hosted by the Pontifical Catholic University of Peru in Lima (PUC-Lima) in December 2021, and held online due to the COVID-19 limitations. It is a tribute to the chapter board's efforts and dedication that it could be conducted during these troubled times.

The 2021 POMS International Conference in Lima was held from December 01 to 03, 2021, with the theme being "The operations management for an innovative and resilient society." The event had 2 plenary sessions, 2 tutorial sessions, and 71 papers presented in 12 sessions with 83 participants.

The full papers presented in the proceedings were organized in the following tracks:

- Artificial Intelligence and Data Analytics in Operations
- Defense, Tourism, and other Emerging OM Issues
- Healthcare Operations Management
- Humanitarian Operations and Crisis Management
- Logistics and Supply Chain Management
- Marketing and Operations Management
- Product Innovation and Technology in Operations Management
- Resilience and Risk in Operations
- Service Operations and Servitization
- Sustainable Operations

These papers offer a representative sample of POM research undertaken in the Caribbean and Latin America. And they are a tribute to the tenacity of the chapter board members, who were able to organize an academic meeting during such challenging period. Congratulations to them.

University of São Paulo, São Paulo, Brazil Hugo Yoshizaki

Contents

Part I
Business Operations Management

Socially Optimal Retail Return Strategies Under the Influence of Endowment Effect

Ali Shirzadeh and Ehsan Elahi

Abstract As a solution to bring security for customers in the retail markets, product return policies are widely utilized. As a result, many retailers take various return leniency measures to ease the applicability of product returns for customers. This, in turn, increases the frequency of returns in the market, which has huge economic impacts on retailers. Therefore, it is necessary to accurately understand the impact of return policy social welfare. To enable this, we present a novel, inclusive analytical model capable of capturing the impacts of major factors affecting the customers' behavior, including return leniency, customers' heterogeneity, and the endowment effect, in addition to the other commonly studied factors in the literature (product expenses, hassle cost, salvage value). In this model, we mathematize the probability of purchasing, keeping, and returning products. Utilizing these probabilities, in turn, help us determine the optimal price and refund levels to optimize the social welfare. We use a set of numerical experiments over a wide range of parameters' values that should cover almost all practical circumstances. Our analysis shows optimal return strategies under various circumstances to maximize the social welfare.

Key words Product pricing · Product return · Return leniency · Social welfare maximization · Endowment effect

---

A. Shirzadeh · E. Elahi (✉)
Department of Management Science and Information Systems, College of Management, University of Massachusetts Boston, Boston, MA, USA
e-mail: shirzadehchaleshtari@umb.edu; ehsan.elahi@umb.edu

J. Vargas Florez et al. (eds.), *Production and Operations Management*, Springer Proceedings in Mathematics & Statistics 391,
https://doi.org/10.1007/978-3-031-06862-1_1

1 Introduction

Product returns in retail markets have huge economic and environmental impacts. Optoro[1] estimates that returns cost retailers $400B each year in the United States alone.[2] In addition, every year, the United States generates 15 million tons of carbon emissions due to product returns.[3] The COVID-19 pandemic has further intensified the product return problems while further shifting the purchase behaviors toward e-commerce,[4] where at least 30% of all products are returned compared to 8.89% in brick-and-mortar stores.[5] On the other hand, many retailers have been implementing some measures of return leniency to ease the hassle of return for customers, which in turn increases the frequency and applicability of the product returns. This trend is being further intensified during the COVID-19 pandemic by adopting new policies such as a longer return window time that make product returns even easier [1, 2]. However, the return leniency can have varying degrees of time, money, effort, scope, and exchange on the return policy [3]. Therefore, the vast economic impact of the product returns necessitates a careful investigation of return policies.

Product return could happen due to the opportunistic manner of customers or the uncertainty in customers' valuation of a product (i.e., online purchases). Having opportunistic returns, the customer knows she wants to return the product from the moment of the purchase, but she makes a purchase to use the product for some while and then returns it. On the other hand, for customers with uncertain manners, the actual value of products is not clear before purchasing. In such cases, the actual value becomes apparent after purchase when she gets the chance to use the product. Upon the purchase, the customer knows only the range of values within which her true valuation would fall. If the customer is considering a product for the purchase, she knows in her mind up to what value she may be willing to pay for the product. This can happen more often in online shopping when customers cannot examine the product up-close before the purchase. In these situations, returning the product helps the customer make the purchase decision, knowing that she can return the product if she finds out the product is not worth the paid-price. There are, however, many factors that affect the customer's ultimate decision, such as the *product acquiring cost* and *salvage value* that we consider in our study.

---

[1] Leading retailers and brands use Optoro to improve how they process, manage, and sell returned and excess inventory. www.optoro.com.

[2] Creating Value from Returns this Holiday Season. https://rla.org/media/article/view?id=1179

[3] Research and Markets – The World's Largest Market Research Store. https://www.researchandmarkets.com/reports/4911530/the-environmental-impact-of-e-commerce-2020?utm_source=dynamic & utm_medium=GNOM & utm_code=xmpm26 & utm_campaign=1344501+-+2020+Report+on+the+Environmental+Impact+of+E-Commerce & utm_exec=joca220gnomd

[4] The pandemic has pushed more sales online, and that means more returns, too https://www.post-gazette.com/business/money/2020/09/14/retail-shopping-covid-19-pandemic-sales-online-in-store-returns/stories/202009130028

[5] E-commerce Product Return Rate – Statistics and Trends [Infographic] https://www.invespcro.com/blog/ecommerce-product-return-rate-statistics/

Furthermore, in this study, we consider customer-related factors, such as uncertainty about the product valuation and *endowment effect,* referring to the tendency of consumers to feel a higher value for a product once they own it [4–6]. This behavior increases the likelihood that people will keep possession of objects even when their actual product valuation is less than their expected value. In other words, the endowment effect makes a customer experience negative utility just by returning something that they have already owned [3, 7].

Finally, we consider the factors usually set by retailers, such as the selling price, refund level, and return hassle cost, which typically depends on how the retailer implements return policies. To ease this hassle, retailers may implement some measures of return leniency. A more lenient return policy can also increase *return efficiency*, which is another factor that we include in our model. Return efficiency refers to the occasions when a customer intends to return but fails to do so. Naturally, a more difficult return process results in a lower return efficiency. This can also correlate with other reasons, such as customers losing the purchase receipt or missing the return deadline. Our final goal is to show how the return leniency impacts the return strategies of the retailers and their profit and social welfare. In correspondence with the literature, the measurement of social welfare is done via the surplus metric (or expected surplus) in this study.

This research contributes to the literature in several ways. First, we develop an inclusive analytical model that captures all major factors affecting the purchase and returns decisions. Prior research papers either did not include these factors or simplified them to such an extent that their conclusions did not reflect these factors in their closed-form solutions for the optimal level of price and refund. (see, for example, [8–10]). In this project, however, we will examine the social welfare maximization strategies based on product return policy under the assumptions of monopolist retailers and include (among other more commonly studied factors) customers' heterogeneity and return efficiency, return leniency, randomness in customers' valuation of the product, and the endowment effect.

In order to consider different aspects of return leniency in our study, we included the hassle cost to audit the cost leniency, and we have the variable of refund level for monetary leniency. For all other aspects of leniency, we have assumed a single parameter return efficiency. This is because leniency factors are not fixed in the literature (in various references, different factors have been assumed to impact leniency, such as [3]). Besides, we present the first analytical research (to the best of our knowledge), which captures the return leniency (even though some may discuss simplifications). This is a considerable contribution compared to all previous analytical researches who have either neglected the return leniency (although the experimental research proved the high impact of this factor in this context [3]) or simplified it to only the hassle cost of return [8, 9].

We investigate the impact of various parameters on the retailer's optimal pricing and refund policies and the resulting social welfare by conducting numerical experiments over a wide range of problem parameters. Despite its limitations, numerical analysis allows us to examine the behavior of social welfare, and return levels with fewer simplifying assumptions and in a realistic setting. Specifically, our research contributes to the body of literature in this field by focusing on the following questions:

1. How can the price and refund be jointly optimized to maximize the social welfare under various market conditions?
2. What is the impact of various parameters on the resulting social welfare?
3. How the relative impact of return leniency on the purchase and return frequencies are compared to each other?

Our analyses show that return leniency increases social welfare, especially when the cost of acquiring product and salvage value are both high.

The remainder of this study is organized as follows: Sect. 2 reviews the related literature on product return management in revenue management. In Sect. 3, we describe the problem and its model setup. Section 4 presents the analytical framework. In Sect. 5, we present the numerical experiments, and Sect. 6 concludes the study with a few remarks.

2 Literature Review

To investigate the impact of various factors on the product return and social welfare, our research models the behavior of customers in a monopoly e-commerce market while capturing the impact of major related factors such as customers' heterogeneity, endowment effect, and customers' *return efficiency*, which refers to the occasions when a customer intends to return but fails to do so. Although the topic of product return has been studied in the literature, to the best of our knowledge, no research has studied all these factors together and addressed issues such as the impact of return leniency on the social welfare and retailers' optimal refund strategy and expected profit. Here we briefly review the related literature.

Product pricing in retail industry have been vastly studied, see for example [11–19]. When it comes to product return, there are researchers who study joint optimization of pricing and refund level in the context of supply chain management, such as [20–31]. In comparison, we focus on the joint optimization of price and refund level in the context of revenue management, excluding inventory management.

Research [8] studies the product return problem when target customers are homogeneous, return leniency solely consists of the impact of the *hassle* cost, and no endowment effect exists. In their model, every customer has the option of information acquisition to fully understand the value of the product before making a purchase, and in order to induce customers to stay informed or uninformed, the retailer has limited pricing and refund possible options. The authors characterize the optimal pricing and refund strategies for both informed and uninformed customers. They also characterize socially optimal pricing and refund strategies. The authors show the optimal refund is equal to the seller's salvage value for a returned unit unless customers are risk averse or if they can choose to acquire information on their own, or unless there exists a fully informed segment. In comparison, our research studies the problem in a market with heterogeneous customers (different levels of product valuations), and captures the impact of the return leniency, and endowment effect. Our findings indicate the value of return leniency for the retailer and social welfare under different circumstances and show, considering the additional factors, the optimal refund can be below the seller's salvage value.

Shulman et al. [9] conducts a similar study with their main focus on profit maximization objective. They model a monopolist retailer who sells two horizontally differentiated products. The customers have the option of exchanging their purchased products with the other one. The fit between consumers' preferences and product characteristics is a source of uncertainty in their model, which is revealed only after the purchase. They consider customers' utility for the product type to be a binary variable which is a second source of uncertainty for certain portion of customers. The authors jointly optimize the price and restocking fee and study the impact of full information provision which resolves all sources of uncertainty. Their model allows returns only for the customers who does not find any use for the product. They show that the marginal value of information to the seller is decreasing in the operational efficiency of the seller's forward and reverse logistics process as well as the level of product uncertainty. In comparison, our model allows returns whenever they generate a higher utility. Moreover, our framework is capable of capturing a full range of market heterogeneity in terms of different customers' initial product valuation. Besides it addresses the return leniency and endowment effect, and allows capturing the impact of those factors.

Becher et al. [10] conducts a similar research while considering the endowment effect and customer heterogeneity but not the return leniency. In their model, they consider the possibility of a fixed amount of reduction in the customers' valuation of the product after the purchase. They characterize jointly optimized retailer's price and return fee decisions. They compare their results under profit and social welfare maximization objectives in the presence and absence of endowment effect as well as when the customers underestimate the endowment effect. The authors characterize the conditions under which the legal upper limits on the return fees increase the social welfare. In comparison, the randomness in customers' valuation of products in our model let the actual valuation attain any value above or below the expected value within the defined range. Moreover, our research assesses the impact of return leniency on retailer's optimal pricing and refund policies.

Akturk et al. [32] presents a similar joint pricing and return optimization study for the monopoly market to see how technology-enabled countermeasures may benefit the retailer by mitigating return abuse. In their model, they consider opportunistic consumers who purchase product with the full intention of returning it. Akturk, Ketzenberg [32] assume an option of customer profiling system to identify such customers by using their personal identification and transaction history. In addition, they take into account fraudulent returns associated with the situations when a person engages in criminal activity such as shoplifting, price switching, and receipt fraud, among others. In this regard, they assume a product tracking system which identifies fraudulent returns by recording each transaction of a product through the use of unique identifiers. They demonstrate the conditions under which it is advantageous to adopt these technologies and when such investments should be avoided. In contrast, we consider a different type of customer heterogeneity where customers differ from each other by their needs and tastes regarding the product in question not their ethical behavior. We also incorporate major return influential factors such as return efficiency, endowment effect, and customer heterogeneity to capture their impact on the pricing and return strategies and the benefit of return leniency.

Most studies define return leniency based on the seller's return policy terms. See for example [33–37]. Suwelack and Krafft [38] note there are significant differences in retailers' return policy terms. [3, 39, 40] each classify the return policies in terms of different restriction factors, such as time, money, and scope. Analytical researches either simplify their modeling assumptions by neglecting return leniency, such as [10, 32], or restricting the leniency to the return hassle cost, such as [8, 9]. In our research, however, our model captures the impact of return leniency through a combination of hassle cost and return efficiency, which we define as the percentage of customers who intend to return the product and manage to do so. Lower return efficiency could be, in part, the result of higher hassle cost as well as all other factors that make the return more challenging (low return leniency).

Customers' hesitancy to return supports the endowment effect theory which considers a gap between valuation of a product before and after a purchase [4–6]. In other words, endowment effect happens when customers are reluctant to lose the ownership of products by returning them. Most of the experiments on the endowment effect have examined it immediately after an object is obtained. Strahilevitz and Loewenstein [41] show the endowment effect changes by the duration of ownership. The vast literature on the endowment effect demonstrates its influential impact on consumer's behavior and in particular on their returning behavior (for reviews see, for example, [42, 43]). Janakiraman and Ordóñez [44] find that longer deadlines for returning increase the attachment to the item. Kim and Wansink [34] identify an interaction between seller's recommendation and return policies, so that lenient return policies yield more positive post-purchase evaluations when the product was recommended. Our findings show that endowment effect increases the sensitivity of social welfare and retailers' profit respect to return leniency, while enabling the retailers to optimally offer higher refund levels.

3 Model Setup

A monopolist retailer sells a product online with unit price p. Each potential customer i has a unit demand for the product and has her own reservation price for the product, R_i, which reflects the value that the customer believes for the product. R_i is a uniformly distributed random variable over $\left[\overline{R_i} - \delta, \overline{R_i} + \delta\right]$, where $\overline{R_i}$ is the mean of the uniform distribution. From the retailer's perspective, $\overline{R_i}$ itself is also a uniformly distributed variable over the interval $[L, U]$, which reflects the variation in mean reservation prices of its pool of customers. Before buying the product, customer i is only aware of the distribution of his/her own reservation price for the product, not the exact amount of R_i. After buying the product, the actual amount of R_i is revealed to the customer, and hence the utility of buying and keeping the product, which would be R_i-p. Obviously, the utility can be negative when the revealed reservation price is less than the product price.

Having bought the product, a customer has the choice to return the product if he/she is not satisfied with it. Upon return, the customer has to endure some form of hassle cost h, as well as the disutility associated with the endowment effect, γR_i, where γ is the endowment effect factor. The case of $\gamma = 0$ represents the situation when customers experience no endowment effect. Modeling the endowment effect in this fashion is similar to the analytical model in Becher, Feess [10].

On the other hand, upon return, a customer receives the refund value of βp, where $0 \leq \beta \leq 1$. The cases of $\beta = 1$ and $\beta = 0$ refer to full refund and no refund policies, respectively. So, the utility of purchasing and returning the product is $-(1 - \beta) p - h - \gamma R_i$. To decide whether to return the product or not, the customer checks if he would get more utility by returning the product or by keeping it. Therefore, the return is preferable when $R_i - p < -(1 - \beta)p - h - \gamma R_i$, or equivalently $R_i < v_r$, where $v_r = (\beta p - h)/(1 + \gamma)$ is the *return value*. In this situation, the customer might fail to return the product due to different reasons, such as losing the receipt, missing the deadline, or simply forgetting. To take into account this possibility, we add to our model a return efficiency factor α $(0 \leq \alpha \leq 1)$, which reflects the probability of returning the product when return is preferable. Higher return leniency corresponds with less hassle cost and possibly higher return efficiency factor. Hence, the utility of a customer who has already bought the product can be stated as:

$$U_i = \begin{cases} R_i - p \quad \text{if } v_r \leq R_i (\text{when customer keeps the product}), \text{ or} \\ \qquad \text{if } R_i < v_r \text{ and customer fails to return the product} \\ \qquad \qquad \text{with the probability of} (1 - \alpha) \\ (\beta - 1)p - h - \gamma R_i \quad \text{if } R_i < v_r \text{ and customer returns the product} \\ \qquad \qquad \text{with the probability of } \alpha \end{cases} . \tag{1}$$

Considering the customers' uncertainty about the actual valuation of the product, a customer decides to buy the product if and only if the expected utility of the purchase is positive.

For the retailer, there exist a total expense of c to acquire and sell each unit of the product, and it sells each unit of the returned product with a salvage of s. To avoid selling and return arbitrage, we assume that $s<c$, which is a common assumption in the literature (see e.g., [8, 9]).

4 Problem Formulation

In this section we analytically determine the social welfare outcome of each pricing and refund strategy of the retailer, and show how the optimal price p^* and refund β^*can be specified.

Considering the stochastic utility of purchasing the product according to (1), *Proposition 1* characterizes the conditions under which customer *i* purchases the product. In other words, the proposition characterizes the conditions under which the customer's expected utility of purchase is positive.

Proposition 1 *Customer i purchases the product if and only if*

(i) $\overline{R}_i > \max(v_r + \delta, p)$, *or*

(ii) $v_r^e - \delta \leq \overline{R}_i \leq v_r^e + \delta$ *and one of the following two conditions is fulfilled:*

- $B^2 - 4AC < 0$
- $B^2 - 4AC \geq 0$ *and* $\overline{R}_i \notin [r_1, r_2]$

where

$A = \alpha + \gamma,$
$B = 2\delta - 2p + (2\delta + 2p)(1 - \alpha) + (-2(\beta - 1)p + 2h - 2\gamma_e\delta)\alpha,$
$C = (\delta + v_r - 2p)(\delta - v_r) + (-\delta + v_r - 2p)(v_r + \delta)(1 - \alpha)$
$+(2(\beta - 1)p - 2h - \gamma(-\delta + v_r))(v_r + \delta)\alpha,$
$r_1 = \frac{-B-\sqrt{B^2-4AC}}{2A}$ and $r_2 = \frac{-B+\sqrt{B^2-4AC}}{2A}.$

Part (*i*) of *Proposition 1* states a situation in which the customer's average reservation price is so high that guarantees a positive expected utility for the customer, which in turn results in a purchase. On the other end of the spectrum, when $\overline{R}_i < v_r - \delta$, the customer never buys the product since his expected utility of purchase is always negative. In the middle range of $\overline{R}_i$, as expressed by part (*ii*) of *Proposition 1*, the customer's expected utility is a convex quadratic function of $\overline{R}_i$ (see the proof of *Proposition 1* for more details). This means, depending on the sign of the determinant of this function, the customer's expected utility is positive only outside the range of the two roots of this function, r_1 and r_2, or it is always positive if the function does not have a root. In other words, the customer's expected utility as a function of $\overline{R}_i \in [L, U]$, crosses the line of zero utility either once or twice. This is quite a counter intuitive result. If a given customer's average reservation price, $\overline{R}_i$, results in positive expected utility, then one expects an increase in $\overline{R}_i$ should result in an increase in the expected utility and hence a more desirable purchase. This is not necessarily the case when the customer is subject to endowment effect and his expected utility as a function of $\overline{R}_i$ has two roots, r_1 and r_2. In this case customers with $\overline{R}_i < r_1$ might experience a positive expected utility of purchase while customers with larger average reservation price in the range of $r_1 < \overline{R}_i < r_2$ do not buy the product due to a negative expected utility of purchase (see Fig. 1).

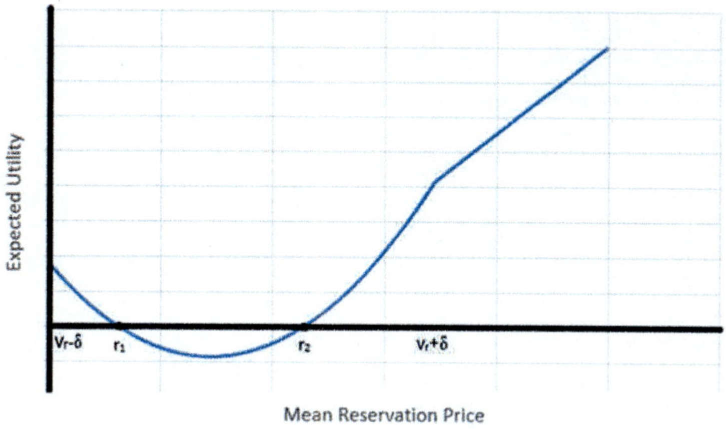

Fig. 1 Expected utility of purchase at different mean reservation prices

Let Φ be the set of average reservation prices which result in a positive expected utility for the customer, as characterized by proposition 1. That is:

$$\Phi = \left\{ \overline{R}_i \in [L, U] \middle| E\left[U_i\left(\overline{R}_i \right) \right] > 0 \right\}$$

Then, from the retailer's perspective, whose customers' average reservation prices are uniformly distributed over *[L, U]*, the probability of the purchase of a given customer *i* can be stated as:

$$P^{Purchase} = \frac{\int_{x \in \varphi} dx}{U - L} \tag{2}$$

As we indicated earlier, customer *i* keeps a purchased product if his actual reservation price, which is revealed after the purchase, turns out to be higher than the return value, $v_r = (\beta p - h)/(1 + \gamma)$. Even when this condition is not fulfilled, the customer might fail to return the product with a probability of $1 - \alpha$. *Proposition 2* characterizes the probability of a customer buys and keeps the product, $\Pr^{Keep}$, as well as the probability of a customer buys and returns the product, $\Pr^{Return}$. It is easy to verify that $\Pr^{Purchase} = \Pr^{Keep} + \Pr^{Return}$.

Proposition 2 *The probability of purchasing and keeping and the probability of purchasing and returning the product are determined by the following relations.*

$$\Pr^{Keep} = \frac{1}{2\delta(U-L)}\left(\frac{\overline{r_i}^2}{2} + (\delta - v_r)\overline{r_i} + (1-\alpha)\left(\frac{-\overline{r_i}^2}{2} + (\delta + v_r)\overline{r_i}\right)\right)\Big|_{v_r-\delta}^{v_r+\delta,\overline{r}_i\in\varphi} + \frac{U - \max(v_r+\delta, p)}{U-L} \tag{3}$$

$$\Pr^{Return} = \frac{1}{2\delta(U-L)}\left(\alpha\left(\frac{-\overline{r_i}^2}{2} + (\delta + v_r)\overline{r_i}\right)\right)\Bigg|_{v_r-\delta}^{v_r+\delta,\overline{r}_i\in\varphi} \tag{4}$$

Now we study the social welfare maximizing strategies of the retailer, i.e. the optimal selling price, p^*, and refund level, β^*, that maximize the social welfare. The efficiency requires the product to be delivered to precisely those customers whose product valuation exceed the acquisition cost (see e.g. [8, 10]). This simply infers that the price has to be set equal to cost of acquiring the product, c, in order to obtain the most overall efficiency. Assuming this optimal price, we proceed with specification of the refund. When a purchase happens and the given customer i keeps the product, the retailer and the customer surplus are R_i-c and none, respectively, so the overall surplus form the purchase is R_i-c, but when the customer returns the product after the purchase, the customer surplus turns $-h-(1-\beta)p-\gamma R_i$ and the retailer surplus changes to$s-c+(1-\beta)p$, so in this situation, the overall surplus from the purchase is$s-c-h-\gamma R_i$. Therefore, the overall surplus of purchasing and possibly returning the product for the retailer and customer i is

$$\Gamma_i = \begin{cases} R_i - c \\ \quad \begin{cases} f \;\; v_r \le R_i & customer\ keeps\ the\ product \\ f \;\; v_r > R_i & customer\ keeps\ the\ product\ with\ a\,probability\ of (1-\alpha) \end{cases} \\ s - c - h - \gamma R_i \\ \quad if \;\; v_r > R_i \quad customer\ returns\ the\ product\ with\ a\,probability\ of\ \alpha \end{cases} \tag{5}$$

Between the two keeping and return outcomes, the efficiency requires the one with the higher overall surplus happens. Based on this necessity, the refund level is simply inferred by the following proposition.

Proposition 3 *The efficient refund level, β^*, which maximizes the expected overall surplus is s/c.*

This specifies the absolute value of the refund equal to the salvage value of the product, which is again consistent with similar findings of the literature (see e.g. [8, 10]).

When the pricing and return strategies are in favor of surplus maximization, we denote the maximum surplus that is obtained as a result of each purchase decision by a given customer i as the maximal surplus, S_i. The following proposition specifies the maximal surplus expected.

Proposition 4 *The expected maximal surplus subsequent each purchase decision equals the following.*

$$E(S_i) = \frac{1}{U-L}\int_{L}^{\min\left(\max\left(L,\frac{s-h}{1+\gamma}-\delta\right),U\right),r\in\varphi} ((\overline{r}_i - c)(1-\alpha) + (s-c-h-\gamma\overline{r}_i).\alpha).$$

$$d\overline{r}_i + \frac{1}{U-L}\int_{\min\left(\max\left(L,\frac{s-h}{1+\gamma}-\delta\right),U\right)}^{\min\left(\max\left(L,\frac{s-h}{1+\gamma}+\delta\right),U\right),r\in\varphi}$$

$$\frac{1}{4\delta}\begin{pmatrix} (\alpha+\gamma\alpha)\overline{r}_i^{\,2} \\ +(4\delta + 2(-\delta - s + h - \gamma\delta)\alpha)\overline{r}_i \\ + \begin{pmatrix} -4c\delta \\ +\left(\delta - \dfrac{s-h}{1+\gamma} + 2s - 2h - \gamma\left(-\delta + \dfrac{s-h}{1+\gamma}\right)\right)\left(\dfrac{s-h}{1+\gamma} + \delta\right)\alpha \end{pmatrix} \end{pmatrix}.$$

$$d\overline{r}_i + \frac{1}{U-L}\int_{\min\left(\max\left(L,\frac{s-h}{1+\gamma}+\delta\right),U\right)}^{U,r\in\varphi} (\overline{r}_i - c).$$

$$d\overline{r}_i \tag{6}$$

5 Numerical Experiments

In this section we investigate the impact of various parameters on the refund strategies of the retailer, along the overall surplus as a measure of social welfare. In order to conduct these experiments, we normalize $[L,U]$ over [1, 2], and to avoid any negative values for the actual realization of the reservation prices, we confine the customers' reservation price uncertainty level, δ, to be always less than 1. This type of normalization of reservation prices is consistent with the literature (see, e.g., [8, 9]). To capture and demonstrate the impact of problem parameters, we consider a wide range of variations for each parameter from extremely low to extremely high levels in order to capture any likely realistic situation. Albeit for the sake of coherence and simplicity, we only present the results under the extreme cases, as the outcomes of the mid-levels are interpolation of the extreme cases. However, we present outcomes for the full range of return leniency variables. Moreover, to make our results more general, we define the two extreme levels of each parameter in terms of the normalized interval limits L and U; as shown in Table 1. We also associate the return efficiency with the hassle cost of return, since we expect a higher efficiency when returning the product is easier (low hassle cost) and vice versa. We consider the combination of the two as a measure of return leniency.

Return leniency clearly stimulates return in the market via reducing the negative impacts of returns for the customers. Subsequently, on one hand, this enhances

Table 1 Values of basic parameters for the numerical experiments

Parameter	Symbol	Low value	High value
Reservation price uncertainty	δ	$0.1 \times (U - L)$	$0.5 \times (U - L)$
Cost of acquiring the product	c	$\frac{1}{3} \times \frac{U+L}{2}$	$\frac{U+L}{2}$
Return Leniency (Hassle cost of return, Return efficiency)	(h, α)	$(0.1 \times c, 0.45)$	$(0.01 \times c, 0.9)$
Product salvage value	s	$0.1 \times c$	$0.9 \times c$
Endowment effect	γ	0.1	0.5

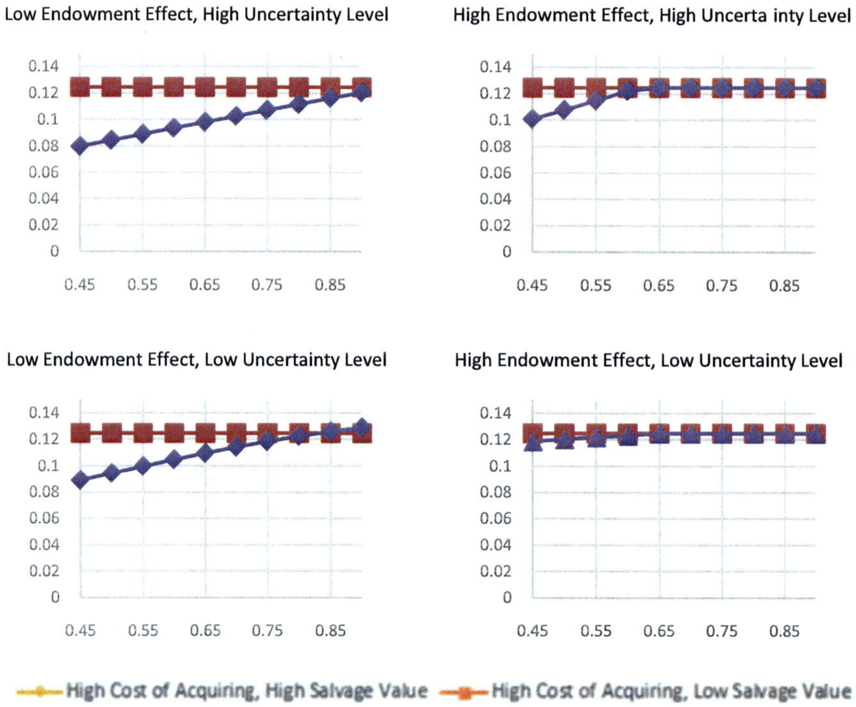

Fig. 2 Maximal surplus versus Return Efficiency (as a measure of return leniency)

customer's surplus by alleviating the negative impact of undesirable purchases, but on the other hand, increases negative surplus of the retailer while increasing the frequency of return. Due to these contradicting impacts, the overall result is not simply predictable. We study this topic in this section in terms of sensitivity of maximal surplus with respect to return leniency.

Figure 2 shows the sensitivity of maximal surplus respect to return leniency under different market conditions. The results of this part show that when the cost of acquiring is low, maximal surplus has no sensitivity respect to return leniency. Therefore, we are only presenting the results under high cost of acquiring in the figure. Based on the figure, there is no sensitivity when the salvage value is low either, since in this case, the actual value of the product for the customer hardly turns less than the price, therefore the return decision is not preferable for the customer, and with no return in the market, the return leniency plays no role in the outcomes. But when the cost of acquiring and salvage value are both high, maximal surplus is increasing respect to return leniency. This suggests that the positive impact of return leniency on the customers' surplus overweigh its negative impact on the retailer's surplus under the market conditions that we examined in here. In this situation, the sensitivity of maximal surplus respect to return leniency is less when the endowment effect is high or uncertainty is low, since under these conditions, fewer returns happen in the market.

Results of this part are also comparable with the [3] finding regarding the higher impact of the leniency on the purchases with respect to the returns. This finding is important as it economically justifies striving for more return leniency by retailers when the profit of higher purchase rates overweighs the expenses of more frequent returns. Figure 3 presents a set of graphs that visualize the results of our experiments in this regard. The first (left-hand-side) vertical axes in these graphs present the purchase probability of a given customer under the profit maximizing price and refund strategies of the retailer, while the second (right-hand-side) vertical axes stand for probability of return. Our results show that at low levels of endowment effect, the purchase and return probabilities hardly change while increasing the leniency level, unless the salvage value and uncertainty are both high where the return probability is decreasing respect to the return leniency. The trend is generally reverse when the problem is subject to high endowment effect. Under such condition, the return probability is usually increasing respect to the return leniency, while the purchase probability is increasing if the cost of acquiring is high, and if not, the purchase probability shows some decreases followed by increases unless the uncertainty is low and salvage value is high. Under the latter conditions, the return probability is decreasing respect to the return leniency.

6 Conclusion

In this paper we presented an analytical model to capture the impact of various parameters on the customers' purchase and return behavior in a monopoly online market. The model includes for the first time, to the best of our knowledge, the impact of several influential factors, among which customer heterogeneity and return leniency, are novel in the product return literature (revenue management context). We specified the conditions under which a given customer makes a purchase. This helped us determine the probabilities of purchasing, keeping, and returning the product, which in turn enabled us to calculate the optimal price and refund level aiming to optimize the overall surplus as a measure of social welfare.

We conducted extensive numerical experiments using normalized problem parameters to help us with generalization of the results. Through these numerical experiments, we examined the sensitivity of retailer's optimal refund and its

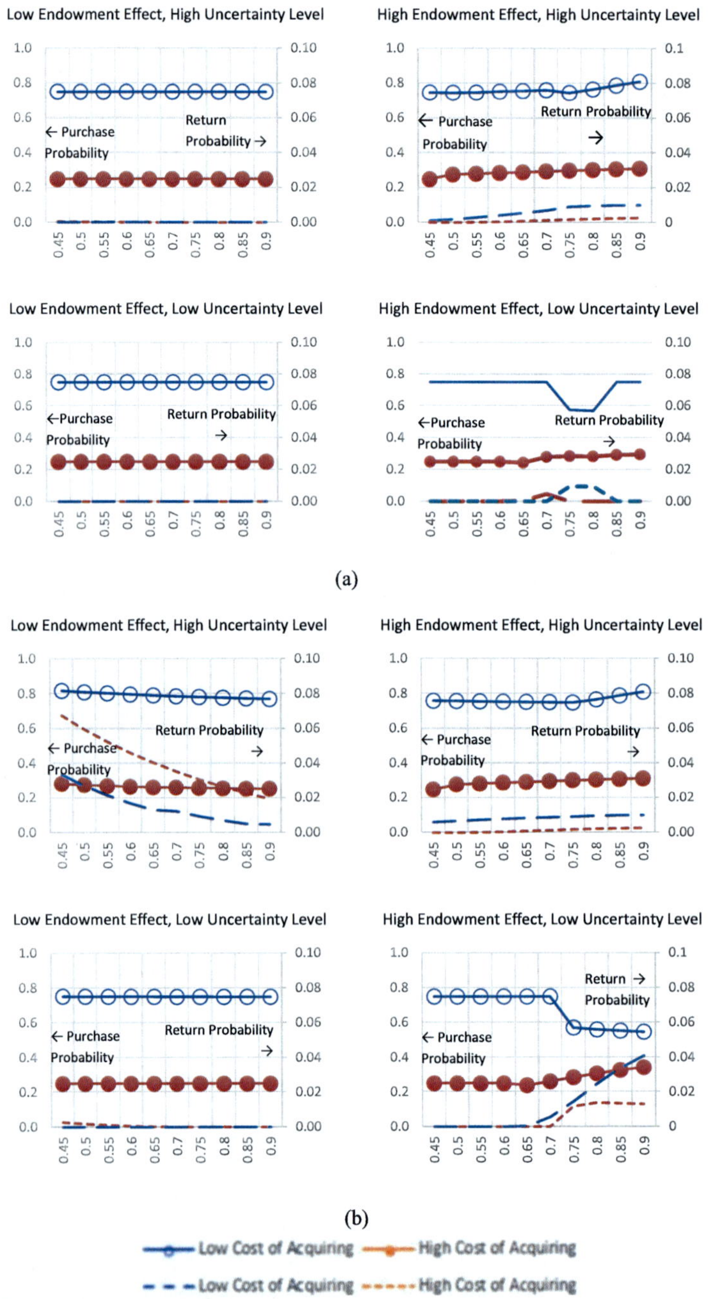

Fig. 3 Probabilities of purchase and refund versus Return Efficiency at (**a**) low salvage value, (**b**) high salvage value

expected profit and overall surplus with respect to the return leniency, whose results help retailers decide on taking measures toward providing return leniency to the customers.

Our results showed that has minor role when it comes to the impact of return leniency on the social welfare. There, the salvage value of product and cost of acquiring are the key factors, such that high values of those parameters result in increasing trend of overall surplus toward return leniency, unless the endowment effect and uncertainty are also high. Therefore, other than the specified conditions, higher return leniency does not lead to higher social welfare, thus is not justifiable, and this is more pronounced when the expenses of providing higher leniency are also considered.

We also examined the hypothesis of higher impact of return leniency on purchases compared to the return decisions. Our results showed that this is not always the case when the product price is optimized in accordance with the return leniency. There are cases where higher return leniency does not impact market demand, or even reduces it, such as when the salvage value and endowment effect are both high, and uncertainty and cost of acquiring are low. While under such conditions, the return leniency results in higher return frequency, there are cases that the negative impact of return leniency exists for both demand and returns. High salvage value and uncertainty, and low cost of acquiring, and negligible endowment effect is a clear condition of such.

The research of this paper can be extended in many different aspects. To mention a few, different market structures other than the monopoly can be studied, such as duopoly or oligopoly, the pricing and return strategies can be examined for the product bundles, and the impact of product quality, price adjustment strategy, and return insurance policies can be jointly studied with the current subject matter.

References

1. Roggio, A. *Pandemic Alters Retail Return Policies*. 2020 [cited 2020; Available from: https://www.practicalecommerce.com/pandemic-alters-retail-return-policies.
2. Thomas, L. *Retailers face another challenge during coronavirus: Handling returns*. 2020; Available from: https://www.cnbc.com/2020/04/14/coronavirus-dealing-with-returns-could-be-bigger-burden-for-retailers.html.
3. Janakiraman, N., H.A. Syrdal, and R. Freling, *The effect of return policy leniency on consumer purchase and return decisions: A meta-analytic review*. Journal of Retailing, 2016. **92**(2): p. 226–235.
4. Knetsch, J.L., *The endowment effect and evidence of nonreversible indifference curves*. American economic review, 1989. **79**(5): p. 1277–1284.
5. Kahneman, D., J.L. Knetsch, and R.H. Thaler, *Experimental tests of the endowment effect and the Coase theorem*. Journal of political Economy, 1990. **98**(6): p. 1325–1348.
6. Thaler, R., *Toward a positive theory of consumer choice*. Journal of Economic Behavior & Organization, 1980. **1**(1): p. 39–60.
7. Becher, S.I. and T.Z. Zarsky, *Open doors, trap doors, and the law*. Law & Contemp. Probs., 2011. **74**: p. 63.

8. Matthews, S.A. and N. Persico, *Information acquisition and refunds for returns.* 2007.
9. Shulman, J.D., A.T. Coughlan, and R.C. Savaskan, *Optimal restocking fees and information provision in an integrated demand-supply model of product returns.* Manufacturing & Service Operations Management, 2009. **11**(4): p. 577–594.
10. Becher, S.I., E. Feess, and R. Sarel, *The Impact of the Endowment Effect on the Law and Economics of Product Return Regulations.* SSRN Electronic Journal 2018.
11. Hartman, R.S., *An empirical model of product design and pricing strategy.* International Journal of Industrial Organization, 1989. **7**(4): p. 419–436.
12. Häckner, J., *Collusive pricing in markets for vertically differentiated products.* International Journal of Industrial Organization, 1994. **12**(2): p. 155–177.
13. Lindsey, R. and D.S. West, *Predatory pricing in differentiated products retail markets.* International Journal of Industrial Organization, 2003. **21**(4): p. 551–592.
14. Deltas, G. and E. Zacharias, *Entry order and pricing over the product cycle: The transition from the 486 to the Pentium processor.* International Journal of Industrial Organization, 2006. **24**(5): p. 1041–1069.
15. Peeters, R. and M. Strobel, *Pricing behavior in asymmetric markets with differentiated products.* International Journal of Industrial Organization, 2009. **27**(1): p. 24–32.
16. Johnson, J.P. and D.P. Myatt, *The properties of product line prices.* International Journal of Industrial Organization, 2015. **43**: p. 182–188.
17. González-Maestre, M. and L.M. Granero, *Competition with targeted product design: Price, variety, and welfare.* International Journal of Industrial Organization, 2018. **59**: p. 406–428.
18. Bauner, C. and E. Wang, *The effect of competition on pricing and product positioning: Evidence from wholesale club entry.* International Journal of Industrial Organization, 2019. **67**: p. 102525.
19. Garella, P.G., D. Laussel, and J. Resende, *Behavior based price personalization under vertical product differentiation.* International Journal of Industrial Organization, 2021. **76**: p. 102717.
20. Pasternack, B.A., *Optimal pricing and return policies for perishable commodities.* Marketing science, 1985. **4**(2): p. 166–176.
21. Lau, H.-S. and A.H.-L. Lau, *Manufacturer's pricing strategy and return policy for a single-period commodity.* European Journal of Operational Research, 1999. **116**(2): p. 291–304.
22. Pasternack, B.A., *Optimal pricing and return policies for perishable commodities.* Marketing science, 2008. **27**(1): p. 133–140.
23. Aviv, Y. and A. Pazgal, *Optimal pricing of seasonal products in the presence of forward-looking consumers.* Manufacturing & Service Operations Management, 2008. **10**(3): p. 339–359.
24. Ding, D. and J. Chen, *Coordinating a three level supply chain with flexible return policies.* Omega, 2008. **36**(5): p. 865–876.
25. Su, X., *Consumer returns policies and supply chain performance.* Manufacturing & Service Operations Management, 2009. **11**(4): p. 595–612.
26. Xiao, T., K. Shi, and D. Yang, *Coordination of a supply chain with consumer return under demand uncertainty.* International Journal of Production Economics, 2010. **124**(1): p. 171–180.
27. Samatli-Pac, G., W. Shen, and X. Hu, *The Impact of Consumer Loss Aversion on Returns Policies and Supply Chain Coordination.* International Journal of Operations Research and Information Systems (IJORIS), 2018. **9**(4): p. 1–20.
28. Li, G., et al., *Return strategy and pricing in a dual-channel supply chain.* International Journal of Production Economics, 2019. **215**: p. 153–164.
29. De Giovanni, P. and G. Zaccour, *Optimal quality improvements and pricing strategies with active and passive product returns.* Omega, 2019. **88**: p. 248–262.
30. Tian, X. and Z.-H. Zhang, *Capacitated disassembly scheduling and pricing of returned products with price-dependent yield.* Omega, 2019. **84**: p. 160–174.
31. Li, T., J. Xie, and X. Liu, *Consumer return policies in presence of a P2P market.* Omega, 2020. **97**: p. 102092.

32. Akturk, M.S., M. Ketzenberg, and B. Yıldız, *Managing consumer returns with technology-enabled countermeasures.* Omega, 2020: p. 102337.
33. Bower, A.B. and J.G. Maxham III, *Return shipping policies of online retailers: normative assumptions and the long-term consequences of fee and free returns.* Journal of Marketing, 2012. **76**(5): p. 110–124.
34. Kim, J. and B. Wansink, *How retailers' recommendation and return policies alter product evaluations.* Journal of retailing, 2012. **88**(4): p. 528–541.
35. Pei, Z., A. Paswan, and R. Yan, *E-tailer′ s return policy, consumer′ s perception of return policy fairness and purchase intention.* Journal of Retailing and Consumer Services, 2014. **21**(3): p. 249–257.
36. Khouja, M., H. Ajjan, and X. Liu, *The effect of return and price adjustment policies on a retailer's performance.* European Journal of Operational Research, 2019. **276**(2): p. 466–482.
37. Cui, H., S. Rajagopalan, and A.R. Ward, *Predicting product return volume using machine learning methods.* European Journal of Operational Research, 2020. **281**(3): p. 612–627.
38. Suwelack, T. and M. Krafft, *Effects of Money-Back and Low-Price Guarantees on Consumer Behavior*, in *Quantitative Marketing and Marketing Management.* 2012, Springer. p. 531–561.
39. Davis, S., M. Hagerty, and E. Gerstner, *Return policies and the optimal level of "hassle".* Journal of Economics and Business, 1998. **50**(5): p. 445–460.
40. Heiman, A., B. McWilliams, and D. Zilberman, *Demonstrations and money-back guarantees: Market mechanisms to reduce uncertainty.* Journal of Business Research, 2001. **54**(1): p. 71–84.
41. Strahilevitz, M.A. and G. Loewenstein, *The effect of ownership history on the valuation of objects.* Journal of consumer research, 1998. **25**(3): p. 276–289.
42. Marzilli Ericson, K.M. and A. Fuster, *The endowment effect.* Annu. Rev. Econ., 2014. **6**(1): p. 555–579.
43. Villanova, D., *The extended self, product valuation, and the endowment effect.* AMS Review, 2018: p. 1–15.
44. Janakiraman, N. and L. Ordóñez, *Effect of effort and deadlines on consumer product returns.* Journal of Consumer Psychology, 2012. **22**(2): p. 260–271.

Sales & Operations Planning a Practical Implementation Guide

Luciana de Oliveira Pedra Romão, Luiz Felipe Scavarda, and Eduardo Machado

Abstract Although the academic literature body in Sales and Operations Planning (S&OP) has grown significantly, as well as its implementations in the industry, companies are still struggling to implement successfully S&OP to obtain its expected benefits. This article addresses this research-practice gap with the purpose of offering a practical guide for S&OP implementation. This goal is achieved with a case study conducted at an audiovisual content production company, taking into account the evaluation of its S&OP implementation. The research builds upon a longitudinal study of one and a half years in the company with an exploratory approach. It contributes to theory by revealing barriers and enablers, as well as the main lessons learned. Practitioners can also take stock of the research findings, as the offered guide can help them to face the challenges of implementing S&OP in their real life settings and support on how the process can be implemented clearly and objectively.

Keywords S&OP · Integrated business planning · Business process

1 Introduction

Aligning supply and demand is a well-known challenge for Operations Management. As a result, discussions on how to introduce Sales and Operations Planning (S&OP) have been increasingly explored. S&OP unites different business plans within an organization in an integrated set of plans aiming to balance supplies and demand and to bridge strategic planning to the company's operational plans, within a country, a region or even the globe [1].

The literature highlights significant benefits of its application in companies, such as improvement in the accuracy of forecasts and in the service level [2], reduction in the stock level [2, 3], enhancement of the flow of information between demand and

L. de Oliveira Pedra Romão · L. F. Scavarda · E. Machado (✉)
Pontifícia Universidade Católica do Rio de Janeiro (PUC-Rio), Rio de Janeiro, Brazil
e-mail: emachado@eflix.com.br

J. Vargas Florez et al. (eds.), *Production and Operations Management*, Springer Proceedings in Mathematics & Statistics 391,
https://doi.org/10.1007/978-3-031-06862-1_2

supply [4], a decrease of deadlines, more stable production rates and a greater focus on the long-term horizon [5, 6]. Additionally, S&OP guarantees the balance among demand and all capacities, such as production, distribution, purchases and finances, aiming at the alignment with the strategic goals [7], being able to unite different goals of the organization in a single plan [1, 8]. However, even with the diversity of empirical studies being carried out and the benefits identified, the literature offers a research practice gap regarding how to apply S&OP practices to achieve such advantages [9, 10].

Although the concept of S&OP is easy to understand, it can be extremely challenging to implement, which may reveal the patterns of variable popularity over the years [11, 12]. Part of the difficulty is explained by the fact that S&OP requires companies to change not only a business process but also the company's culture [11]. Its implementation involves many organizational levels [13] and requires the linking of, often, contradictory departments in a company [14], being able to resort to the classic dilemma between marketing and manufacturing presented in [10]. It is also necessary that companies reverse the functional silos that make executives have very different incentives and, therefore, do not work towards a common goal [11, 15]. Furthermore, there is still a general lack of guidance in the literature on the implementation of the S&OP process [16], especially concerning the improvements and actions needed over the time to achieve horizontal and vertical integration [10, 17, 18].

The implementation of S&OP should be done in stages, starting simply, but in a way that executives perceive the effects on strategic plans and financial performance [11]. The so-called "maturity models" address these stages. These models consist of multiple successive stages that demonstrate the progress of S&OP implementation, each characterized by a precise set of dimensions [2, 11]. They aim to assess the effectiveness of S&OP processes and conduct on how to evolve to more advanced levels [19]. However, although they were designed to plan the transition between steps, they do not guide the dynamics of evolution from one stage to the next [19].

Within this context, the following research question (RQ) is posed: "how to implement S&OP successfully and address difficulties during this process?". With this RQ in mind, the goal of this article is to offer a practical guide for implementing S&OP based on the case study of the S&OP implementation in an audiovisual content production industry considering barriers and enablers, as well as the main lessons learned. This study complements [20] which focuses on the characterization of the S&OP process in the studied company.

The article is organized in six sections, being this first one the introduction. The second section presents the theoretical background. The third section describes the research methodology adopted. Results and discussions are offered in section four and the practical guide for S&OP implementation is presented in section five. Finally, the last section offers the authors conclusion and suggestions for future research.

2 Theoretical Background

Despite the growth of interest on S&OP, many companies have difficulty in its implementation and end up with limited success with the process [21]. The main factors reported for this lack of success are the lack of understanding of the process, disconnection with the organizational strategy, misalignment among stakeholders, lack of support from top management, without incentives and penalties, rigid and poorly integrated organizational culture, based on functional silos, lack of training with those involved and performance management metrics with failures and forecast errors, generating an absence of capacity to monitor and measure the progress of the process [2, 3, 10, 11, 18, 19]. Moreover, the challenges of measuring S&OP efficiency are related to how the activities that are part of the process are organized and managed [22]. Thus, for companies to realize the benefits of S&OP and reduce the risk of failure, it is important to develop a more extensive comprehension of how S&OP should be implemented [3, 10], bringing the need to have a better characterization of the process.

The literature suggests that the implementation of S&OP takes place in stages, starting simply, but in such a way that executives perceive the effects on strategic plans and financial performance [11]. Many studies have evolved to develop models to assess S&OP maturity. Understanding the degree of maturity of the S&OP process is essential to identify gaps and the possibility of evolution. The model by [11] demonstrates how a company operates in S&OP practices and develops its capabilities over time to move from one stage to another. The structure produced by [11] analyzes the following categories: meetings and collaboration, organizational structure, performance measures, information technology, and integrated planning. This maturity model shows a more detailed analysis using five levels of evolution - without S&OP, reactive S&OP, standard S&OP, advanced S&OP, and proactive S&OP. Several authors have referred to this study over the past few years as [2, 3, 9, 18, 23]. Danese et al. [19] used this model to study the key dimensions and the sequence of implementation of S&OP in companies.

3 Methodology

The methodology adopted in the article is the case study. According to [24], the case study is a type of research where the phenomenon (in this case, the S&OP process) can be studied in its natural environment, generating relevant theories from the understanding acquired through the observation of the real practice. For this article, an exploratory investigation is applied to a large Latin American industry, which produces audiovisual content. This study aims through the evaluation of the implementation of the S&OP process in an audiovisual content production industry, to understand barriers and benefits observed in each stage. For this, a longitudinal study was carried out over a year and a half (from January 2019 to July 2020). Based on the empirical analysis and outputs of this case study, it was possible to develop a practical guide for the application of the S&OP.

The choice of this company was given by the opportunity to evaluate, in a longitudinal way, the application of S&OP in a very specific industry with a production model that is still little explored in the literature. Besides, one of the authors of this paper was an integral part of the company's S&OP process, which allowed the collection of process information from unstructured interviews with those who were more involved, being they executives from the technical areas that support the production, demand, production, capacity and financial planning teams. Data was also collected in the company's information systems, in cyclical S&OP meetings in the firm, where the author of this study is part, and in direct observations. The study was also based on an observation guide that consisted of assessing the participation of these stakeholders in the stages of S&OP, including the commitment to metrics and the use of information systems throughout the process. Therefore, triangulation was conducted confronting the results from the interviews among themselves, and with the observed evidence obtain in the participatory research, observation in the S&OP meeting and data from the Company's internal reports.

4 S&OP Process in the Latin American Company

The Latin American Company under study is part of the entertainment content production sector. This sector has been changing in recent years. To be more efficient in its production process and improve its performance, the company started implementing S&OP in its production process in January 2019. After a year and a half cycle, it was possible to observe how the model was implemented, what particularities were found, and what the evolution needs were.

The S&OP implementation in the case company was not simple and it required different stages to arrive at the ideal model, as recommended in [11]. [1, 2, 18] highlight the need for some meaningful definitions such as roles and responsibilities of those involved, policies adopted and support tools, which were indeed essential to enable the implementation.

Among the challenges encountered by the company, it is possible to highlight the pillar of people. According to the literature, one of the failure factors of S&OP is the lack of support from top management and executive sponsorship, in addition to the rigid organizational culture [3, 18, 19, 25, 26]. When executives are engaged in the process, all functional areas and other stakeholders become more participatory, making the process happen, which was confirmed in the case study. The case study also confirms S&OP as a process that requires multifunctional collaboration and, because of that, echoes the literature on the necessity for involvement and engagement of the different areas of the company (e.g. [1–3, 19, 27]). To achieve the commitment of the areas in the process implemented in the company, it was necessary to involve top management from the beginning, showing the benefits that this process would bring to the areas themselves and the business as a whole.

From that point on, this commitment started to be cascaded for everyone in the structure. Furthermore, in the case of the analyzed company, having a team involved, even if not dedicated, was fundamental for the success of the implementation, corroborating [18]. This group, in addition to integrating the plans at first, ensures the alignment and checkpoints necessary for the completion of the cycle. They are responsible for collecting the lessons learned in the cycle so that there can be developed in the next cycle. It is primary to note that, as [11] discuss in their maturity model, the fact of not having a dedicated team for S&OP does not prevent the success of the process. The most important thing is that everyone involved is engaged and committed to the end result that the process can bring.

A second prominent pillar during implementation is related to the process itself. According to the literature, the difficulty of implementing S&OP may be linked to the lack of a structured and interactive process for building a consensus [2]. Because of this, it is necessary to define the stages of the cycle and the areas involved. The engagement of the sectors throughout the cycle is fundamental for the commitment to the final result. Before the implementation of S&OP, the company had barriers in having the areas committed in consolidated planning, as they did not feel active in the process. With the implementation of S&OP, the planning stages became clearer and the role of the areas, well defined. Thus, it was possible to involve and commit them to the integrated plan and the decisions made on it. This corroborates [22], where the efficiency of S&OP is related to how the activities that are part of the process are organized and managed. The S&OP of the analyzed company presents the stages with well-defined forums that count on the participation of all those involved in the process, consequently increasing everyone's engagement.

Finally, it is important to highlight the pillar of information technology. It is possible to realize that this is the pillar of the most difficult evolution. Despite the evolution in the way of structuring, consolidating, analyzing and reporting information, the process continues to be done manually and in spreadsheets. As a result, the process still requires significant time from the team involved to generate and consolidate information, which hinders any evolution towards automation of the information flow. However, it is important to note that this did not prevent the implementation of S&OP and the results already obtained, corroborating some studies that claim that in the early stages of S&OP spreadsheets and basic tools are already sufficient to implement the process [11, 18, 28]. At this stage, efforts need to be focused on strengthening the process and role of the S&OP team and not on investing in sophisticated technologies [11, 28].

For this company, S&OP has already been considered a fundamental process for managing resources and installed capacity. Prior to the implementation of S&OP, orders often arrived late, resulting in delays or extra costs; there was no visibility of demand for the coming months, making it impossible the proper management of the capacity. It was hard to plan the workforce, resulting in idleness or overload, impacting costs and quality. In these last 18 months, the period in which the S&OP was applied, it was possible to observe a significant improvement in the area plans' forecasts, portraying [2], who states that S&OP brings greater accuracy in predictions. Another point observed was the improvement in the indicators of usage of technical resources and workforce, directly impacting the better capacity management, corroborating [7] who state that S&OP guarantees the balance between demand and the company's supply capacity. With the implementation of S&OP, it was also possible to perceive a greater involvement of stakeholders, ensuring an improvement in the flow of information between the areas of demand and production, corroborating [4] who affirm that the application of S&OP brings an improvement in the flow of information between demand and supply. S&OP has also enabled, through more efficient resource management, better cost management. Although few studies address the integration of the financial plan in the S&OP process [1, 8], in the analyzed company, it was possible to notice significant gains from this integration. The S&OP process implemented in the studied company is focused on cost reduction, balancing demand and supply, and efficiently managing resources, which strengthens the benefits of integrating the financial plan. To evaluate S&OP performance, the company adopted the practice of comparing the S&OP result with what was, in fact, accomplished. That permits to identify deviations and map possible failures in the process. To date, most deviations are still related to changes in demand or lack of details of the project's scope.

Following the study by [1], S&OP can be applied in diverse contexts (for example, industries and production models) and different ways (for instance, cycle cycles, periodicity, functional areas of business involved). In the case of the analyzed company, even with the particularities of the audiovisual content production industry and the ETO production model, S&OP could be applied in a similar way to the models studied in the literature, with all the steps described in the model of [28] and with the results and benefits expected, as described by [1].

Table 1 summarizes how the failure factors mentioned in the literature were addressed in the implementation of S&OP in the analyzed company.

5 Practical Guide for S&OP Implementation

Based on the empirical evidence and the lessons learned within the case study, it was possible to synthesize the findings in a Practical Guide for S&OP implementation, as offered in Fig. 1 and explained next.

The S&OP implementation involves critical points that need to be address before its employment, as listed next.

- **Mapping of the application of S&OP in the business** to identify objectives and expected benefits to the business.
- **Alignment of the process objective and benefits with the executive team** to obtain the needed level of commitment to the process.
- **Definition of the areas to be involved for their alignment and engagement in the process** as the areas involved may also vary according to the form of implementation. These areas should be engaged since the beginning. Sharing discussions and ensuring that everyone is heard is essential to the success.

Table 1 Failure factors

	Literature approach	Company approach
People	Failure factors: lack of support from top management and executive sponsorship, in addition to the rigid organizational culture [3, 18, 19, 25, 26].	The implementation involved top management. A non-dedicated team was structured to integrate plans, alignments, check points and collect lessons learned in the cycle so that there would be developments in the next cycle.
Process	Failure factors: lack of a structured and iterative process for building a consensus [2].	Planning stages were defined with a clear role for the areas involved. Thus, it was possible to involve and commit them to the integrated plan and the decisions made on it.
Performance measures	Failure factors: evaluation of metrics not relevant to the business and the company's production model. According to [11, 22], measurement is essential, both for implementation and for continuous improvement. Wagner et al. [2] also highlights the importance of aligning goals between areas as a competitive advantage for companies.	The company worked on the most strategic indicators for the operation, always in an integrated view between the areas.
Information technology	In the initial stages, spreadsheets and basic tools are already sufficient to implement the process [11, 18, 28]. At this stage, efforts need to be focused on strengthening the process and role of the S&OP team and not on investing in sophisticated technologies [11, 28].	Although the process is still done manually and in spreadsheets, requiring significant time to generate and consolidate information, this did not prevent the implementation of the process and the results already obtained.

Source: developed by the authors

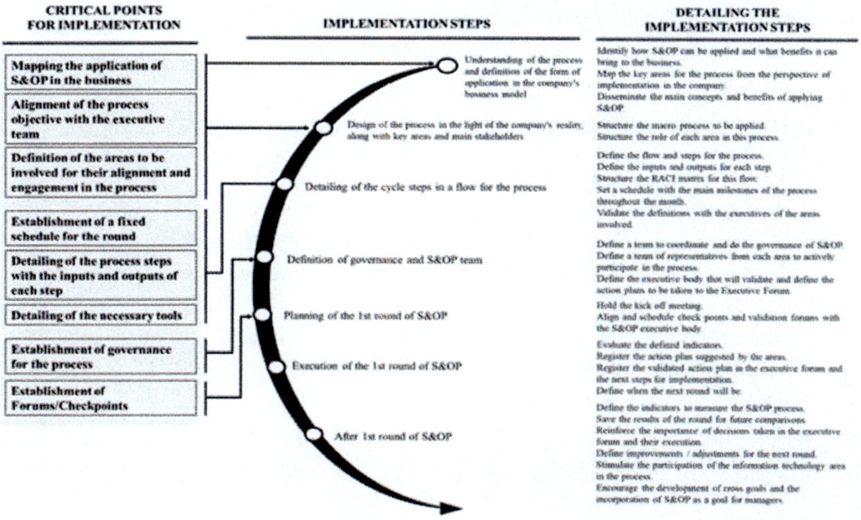

Fig. 1 Practical Guide for S&OP implementation. (Source: developed by the authors)

- **Establishment of a fixed schedule for the round,** which includes the input and output dates, agreed with those responsible. As it is a cyclical process, the demands coming from it cannot be treated as extra or occasional.
- **Detailing of the process steps with their inputs and outputs,** identifying the information needed to start and end each step, their sources and by whom and when can data be generated, detailing roles and responsibilities of each area.
- **Detailing of the necessary tools** to identify, with the areas, the need for tools to generate/process the information for the round. It is also important to map those responsible for developing and maintaining the tools and databases to be used.
- **Establishment of governance for the process** to establish a team that takes care of the process. This team does not need to be dedicated, but it sets up governance and ensures that the process happens.
- **Establishment of Forums/Checkpoints** to align with the executives involved in the process all information and premises used in S&OP.

With the understanding of the criticality of the points mentioned above, it is possible to implement S&OP aided by the following steps.

- **Understanding of the process and definition of the form of application in the company's business model**
  - Identify, together with the areas of demand and production planning, how S&OP can be applied and the main objectives and expected benefits.
  - Map the key areas for the process from the implementation´s perspective.
  - Disseminate to the executives of the areas involved and to the company leadership the main concepts and benefits of applying S&OP.
- **Design of the process in the light of the company's reality, along with key areas and main stakeholders**
  - Design the macro process to be applied, defining the stages of each phase in the light of the company's operation.
  - Draw, in a macro way and together with stakeholders, the role of each area in this process.
- **Detailing of the cycle steps in a flow for the process**
  - Define the flow for the process, detailing each step, as well as their main inputs and outputs. It is important to evaluate with the areas involved the need for new databases and reports and the possibilities of obtaining them.
  - Validate this flow with all stakeholders and define what, who and when for each step.
  - Set a schedule with the main milestones of the process throughout the month.
  - Validate with the executives of the areas involved: the RACI matrix, the inputs/outputs of each area and the schedule process.

- **Definition of governance and S&OP team**
  - Define a team to coordinate the S&OP (not necessarily dedicated) and do the governance of the process and a team of representatives from each area to participate actively in the process (not dedicated). Finally, define the executive body that will take care of the action plans at the Executive Forum.
- **Planning of the 1st round of S&OP**
  - Kick-off meeting
    - Reinforce the objectives and the importance of the process.
    - Reinforce the delivery dates of each information necessary for the process (inputs and outputs of each step).
    - Confirm those responsibilities for each delivery.
    - Align the baseline for comparison and the performance indicators.
  - Align and schedule the checkpoint and validation forums dates with the S&OP executive body.
- **Execution of the 1st round of S&OP**
  - Evaluate the indicators already aligned in the kick-off.
  - Suggest new indicators, if necessary. It is important to conceptualize well the new indicators with everyone involved so that everyone understands what and how it is being measured.
  - Register the action plan suggested by the areas.
  - Register the validated action plan in the executive forum and the next steps.
  - Define when will be the next round.
- **After 1st round of S&OP**
  - Define the indicators to measure the S&OP process and its efficiency in the face of established challenges.
  - Reserve the results of the round to compare with the company's actual result and the result of the next round. Understanding evolutions is fundamental to this process. The more aligned with reality, the more credibility the S&OP process will have.
  - Reinforce the importance of decisions taken in the executive forum and their execution. It is important to create a monitoring process of the execution of the plans that were aligned in the executive forum, in addition to including a chapter on that in the next round (Action plan status - previous round).
  - Map what worked and what didn't in the round and define the improvements/adjustments for the next round. It is important to understand along with the areas which improvements are needed to optimize the process.
  - Encourage the participation of the information technology area in the process to, initially, support the extraction of bases and automation of the most operational processes. As the process grows more robust, the partnership with the sector of information technology will be crucial for the acquisition of software to assist the process.

- Encourage the incorporation of S&OP as an aim for managers. An efficient way to guarantee the commitment of the areas in the process is to associate it with the executives' goal.
- Stimulate the development of cross goals. This guarantees not only the involvement but the commitment to the company and not just to the area.
- Engage the company's leadership in the process. The leadership, once committed to the process, will positively influence its executives to commit as well. For this, it is vital to take to the executive forum only relevant information and indicators that aggregate for a tactical and strategic decision making.

6 Conclusion

This article addresses a research-practice gap regarding the need to aid companies in the hard task of implementing successfully S&OP in their operations. From a theoretical point of view, it sheds lights on new perspectives to the literature on barriers and enablers in S&OP implementations. From a practical point of view, the study provides industries and practitioners with a practical guide that to support on how the process can be implemented clearly and objectively.

With the analysis of the theoretical point of view, the research presents the lessons learnt from the case study, being one of the few longitudinal studies on the S&OP literature. Through it, it is possible to comprehend how the evolution process took place during the implementation. Among them, the study highlights four main pillars for implementation: people, process, performance measures and information technology. In the people pillar, the study addresses the need for support from top management and executive sponsorship; multifunctional collaboration, highlighting the importance of the involvement and engagement of the different areas of the company; and the need to form a team, even if not dedicated, corroborating several authors of the literature (e.g., [1–3, 11, 17–19, 25–27]). In the process pillar, the importance of the process structure that ensures interactions for building a consensus and the engagement of those involved is addressed [2], corroborating [22], who states that the efficiency of S&OP is related to how the activities that are part of the process are organized and managed. The performance measures pillar addresses that the definitions of monitoring metrics must be related to the business (volume of changes in demand and planning), the production model (production peaks and valleys, % of installed capacity utilization and overflow need to the market) and should bring reflections on the financial impact, corroborating [11, 22] on the importance of the effectiveness look of the S&OP process. This pillar also addresses the importance of the alignment of the goals between areas as a competitive differential as highlighted in [2]. Finally, the information technology pillar highlights that, despite not having a system that allows the automation of the information flow and the process as a whole, this did not prevent the implementation of the S&OP and the results already obtained, corroborating some studies that claim that in the initial stages of S&OP, spreadsheets and basic tools are already sufficient to implement the process [11, 18, 28].

From a practical point of view, it was possible to develop a Practical Guide for implementing S&OP in companies considering seven main steps that need to be worked on before, during and after the implementation of the process. The first steps address the need to understand the process, identify how it can be applied in the company, and understand and align which benefits the company expects to achieve with this application. The following steps address the importance of detailing the process in light of the company's production model, establishing the governance of the process, the teams involved and the roles and responsibilities of each. In the last stages, the guide addresses the key points of the 1st round, exploring the planning, execution and post-round, already linking the planning of future rounds. The Practical Guide emphasizes, throughout all the steps, the importance of alignment with the executives involved in the process. This is essential for engagement and commitment to the final result of the round. This guide is summarized in Fig. 1 and aims to be an instrument to guide industry practitioners in future S&OP implementations.

Although this study concretely brings the empirical approach to the application of S&OP with relevant points about the process, it was based only on the observations made in only one company, which is a limitation of the study. Thus, for a deeper understanding of the theme, future researches are needed, expanding the observations to other companies, seeking more generalization of the findings.

References

1. Seeling, M.X.; Kreuter, T.; Scavarda, L.F.; Thomé, A.M.T.; Hellingrath, B. Global sales and operations planning: A multinational manufacturing company perspective. PLoS One, v. 16, p. e0257572 (2021).
2. Wagner, S.M.; Ullrich, K.K.; Transchel, S. The game plan for aligning the organization. Business Horizons 57(2), 189–201 (2014).
3. Goh, S.H.; Eldridge, S. New product introduction and supplier integration in sales and operations planning. International Journal of Physical Distribution & Logistics Management 45 (9/10), 861–886 (2015).
4. Oliva, R.; Watson, N. Cross-functional alignment in supply chain planning: a case study of sales and operations planning. Journal of Operations Management 29(5), 434–448 (2011).
5. Goh, S.H.; Eldridge, S. Sales and Operations Planning: The effect of coordination mechanisms on supply chain performance. International Journal of Production Economics 214, 80–94 (2019).
6. Noroozi, S.; Wikner, J. Sales and operations planning in the process industry: a literature review. International Journal of Production Economics 188, 139–155 (2017).
7. Pereira, D. F.; Oliveira, J. F.; Carravilha, M. A. Tactical sales and operations planning: A holistic framework and a literature review of decision-making models. International Journal of Production Economics, 228, 107695 (2020).
8. Seeling, M.X., Kreuter, T.; Scavarda, L.F.; Thomé, A.M.T., Hellingrath. The role of finance in the Sales and Operations Planning process: a multiple case. Business Process Management Journal (in press). https://doi.org/10.1108/BPMJ-07-2021-0447 (2021).
9. Kristensen, J.; Jonsson, P. Context-based sales and operations planning (S&OP) research: A literature review and future agenda. International Journal of Physical Distribution & Logistics Management 48(1), 19–46 (2018).

10. Shapiro, B.P. Can marketing and manufacturing coexist?. Harvard Business Review, September/October, 104–114 (1977).
11. Grimson, J.A.; Pyke, D.F. Sales and operations planning: an exploratory study and framework. The International Journal of Logistics Management 18(3), 322–346 (2007).
12. Qi, J.; Ellinger, A.E. A conceptual framework of organizational orientation antecedents of sales and operations planning. In Creating Marketing Magic and Innovative Future Marketing Trends, edited by M. Stieler, p. 1319–29. Cham, CA: Springer (2017).
13. Jonsson, P.; J. Holmström. Future of Supply Chain Planning: Closing the Gaps between Practice and Promise. International Journal of Physical Distribution & Logistics Management 46 (1) 62–81 (2016).
14. Swaim, J.A.; Maloni, M.; Bower, P.; Mello, J. Antecedents to effective sales and operations planning. Industrial Management & Data Systems 116(6), 1279–1294 (2016).
15. Lapide L. Sales and operations planning Part III: a diagnostic model. Journal of Business Forecast 24(1) 13–15 (2005).
16. Kreuter, T.; Kalla, C.; Scavarda, L.F.; Thomé, A.M.T.; Hellingrath, B., Developing and implementing contextualised S&OP designs – an enterprise architecture management approach, International Journal of Physical Distribution and Logistics Management 51(6), 634–655 (2021).
17. Tuomikangas, N.; Kaipia, R. A coordination framework for sales and operations planning (S&OP): synthesis from the literature. International Journal of Production Economics 154, 243–262 (2014).
18. Pedroso, C.B.; Silva, A.L.; Tate, W.L. Sales and Operations Planning (S&OP): Insights from a multi-case study of Brazilian Organizations. International Journal of Production Economics, 182, 213–229 (2016).
19. Danese, P.; Molinaro, M.; Romano, P. Managing evolutionary paths in sales and operations planning: key dimensions and sequences of implementation. International Journal of Production Research 56(5), 2036–2053 (2017).
20. Romão, L. O. P.; Scavarda, L. F.; Seeling, M.X. Sales and Operations Planning Case study on the Engineering-To-Order production model in the entertainment industry. Brazilian Journal of Operations & Production Management, 18, e20211037 (2021).
21. Kreuter, T.: Scavarda, L. F.: Thomé, A. M. T. et al. Empirical and theoretical perspectives in sales and operations planning. Rev Manag Sci 16, 319 –354 (2022). https://doi.org/10.1007/s11846-021-00455-y
22. Hulthén, H.; Näslund, D.; Norrman, A. Framework for measuring performance of the sales and operations planning process. International Journal of Physical Distribution & Logistics Management 46(9), 809–835 (2016).
23. Vereecke, A.; Vanderheyden, K.; Baecke, P.; Steendam, T. V. Mind the gap – Assessing maturity of demand planning, a cornerstone of S&OP. International Journal of Operations & Production Management 38(8), 1618–1639 (2018).
24. Voss, C.; Tsikriktsis, N.; Frohlich, M. Case Research in Operations Management. International Journal of Operations and Production Management 22(2), 195–219 (2002).
25. Ivert, L.; Jonsson, P. When should advanced planning and scheduling systems be used in sales and operations planning? International Journal of Operations & Production Management 34(10), 1338–62 (2014).
26. Ivert, L. K.; Dukovska-Popovska, I.; Kaipia, R.; Fredriksson, A.; Dreyer, H. C.; Johansson, M. I.; Chabada, L.; Damgaard, C. M.; Tuomikangas, N. Sales and operations planning: responding to the needs of industrial food producers. Production Planning and Control 26(4), 280–295 (2015).
27. Taşkin, Z.C.; Ağralı, S.; Ünal, A. T.; Belada, V.; Gökten-Yılmaz, F. Mathematical Programming-Based Sales and Operations Planning at Vestel Electronics. Interfaces 45(4), 325–40 (2015).
28. Wallace, T. F.; Stahl, R. A. Sales & operations planning: The how-to handbook. Cincinnati, OH: T. F. Wallace & Company 3rd ed. (2008).

Better Efficiency on Non-performing Loans Debt Recovery and Portfolio Valuation Using Machine Learning Techniques

Jose Tupayachi and Luciano Silva

Abstract The following research is based on a portfolio of non-performing loans (NPLs), which was previously acquired and managed by a collection agency, the company under study is one of the owners of the portfolio. The study compares the efficiency and performance of several machine learning algorithms to develop and implement a forecasting tool to estimate the recovery rate of NPL portfolios. These models help to enhance and support the debt collection operation, allowing to forecast the number of debtors that will be recovered in the lifetime of the portfolio, as well as to efficiently manage resources (recovery task force) by reducing costs and expenses. The application aims to support the valuation process at the time of portfolio purchase. The study shows that the application using a binary ranking approach based on the XGBoost model outperforms other techniques, offering good results. It is also evident that product type was one of the most influential variables among the different models. The model using this algorithm could serve as a decision support tool, precisely in the operation of purchasing a portfolio of unprofitable debts, as it allows the quantification of the client's debt to be recovered by identifying the group of potential debtors with the highest probability of compliance, which would result in a faster and more efficient debt collection process.

Key words Non-performing loans · Machine learning · XGBoost

J. Tupayachi (✉)
Pontifical Catholic University of Peru, Lima, Peru

The University of Tennessee, Knoxville, TN, USA
e-mail: jtupayachi@pucp.edu.pe; jtupayac@vols.utk.edu

L. Silva
Pontifical Catholic University of Peru, Lima, Peru
e-mail: lsilva@pucp.pe

J. Vargas Florez et al. (eds.), *Production and Operations Management*, Springer Proceedings in Mathematics & Statistics 391,
https://doi.org/10.1007/978-3-031-06862-1_3

1 Introduction

1.1 Context and Overview

According to McKinsey non-performing debt recovery is defined as a form of business that generates income. Peru enjoys a well-positioned and credit-towards regulators grating credit penetration which avoids narrowing of the field of application as seen in Eduardo Lizarzaburu and Jesús del Río [15] study. In the social sphere, the so-called unprofitable debts affect both the financial institution (collection agency and the bank or investor) as well as the debtor. On one side, the base of the revenue-yielding activities of these companies depends on the generation and collection of the amounts owed. To maintain its profitability, the business must collect what is owed in the shortest possible time, but inadequate customer forecasting and filtering techniques can affect the business's operations.

After a defined period, financial institutions reschedule defaulted payments. Still, some borrowers do not meet their payment obligations, but entities such as the European Bank [10] encourage banks to dispose of NPLs after three years of management. Many banks or financial institutions encounter obstacles [2] to performing such tasks which include materializing costs in their equilibrium schemes [9]. As Bellotti and Brigo [5] point out, to decrease their defacement, lowering the losses and financial stability concerns, regulators suggest banks cluster their NPLs and sell them to specialized investors [6], called debt collection agencies.

Similarly, debt collection agencies aim to maximize their profits by offering the lowest price at the time of purchase. However, banks are not the only sellers of portfolios, collection agencies are also part of this space and partially worked portfolios are bought by other collection agencies. This research addresses and bases its results on a portfolio previously worked by a collection agency. It should be noted that the non- performing debt portfolio understudy has only one group of debts.

The process of purchasing and evaluating the portfolio is shown in Fig. 1. The diagram shows the NPL portfolio purchase operation as the starting point of the research. The developed solution supports the purchasing decision based on historical information and the methodology addresses how the study is conducted to meet the objectives and proposals. Therefore, a more accurate data-backed forecast of the future revenues of the operation can be obtained. It is important to emphasize that the recovery rate is analyzed from the position of the buyer of the non-performing debt portfolio. That is, from the point of view of the collection agency (Fig. 2).

1.2 Justification

Assessing the correct portfolio pricing is a poorly performed task as the current literature in the field reveals. The disposal of NPLs is hampered by the large bid-ask

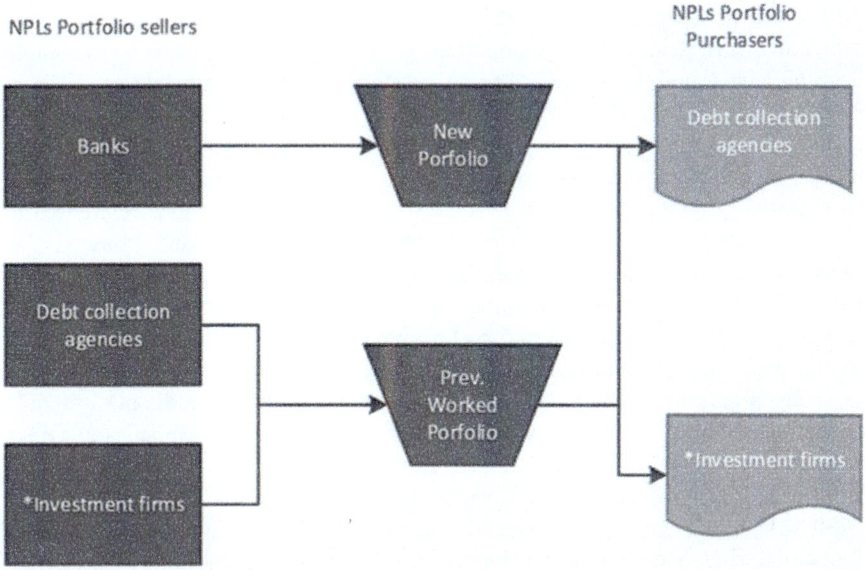

Fig. 1 Portfolio buy and sell operation

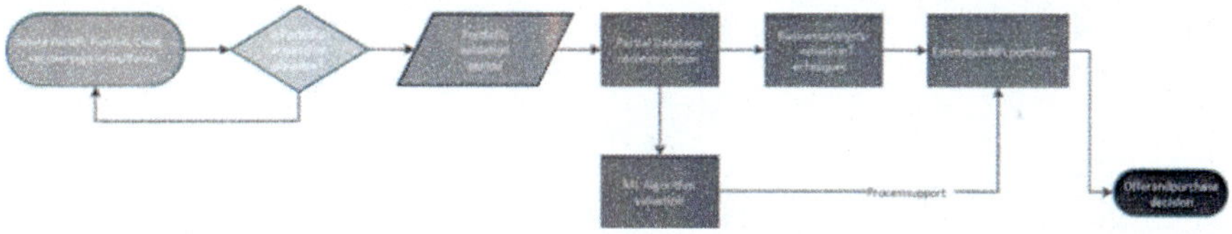

Fig. 2 Portfolio business model purchase operation diagram, showing the interaction of the solution presented for the operation

spreads characterizing their market, determined by discrepancies in data availability between banks and investors, and by poor valuation methodologies as indicated by Ye and Bellotti [4]. From this point, the following inquires emerged: Are these empirical algorithms powerful enough to consider the different interactions that occur between variables? Can they scan the different records and learn from historical data? On top of that, as researched by McKinsey [1] Latin American banks and debt collection agencies are behind on holistic digital transformation, which creates an opportunity for the implementation of new technologies to improve the decision support process. Data related to NPLs are stored by banks in tape backups, which may contain large amounts of information. These are transferred to the appraisers, who have to process the information to make forecasts and estimate their performance but given the complexity and quantity of the data, how efficient are traditional methods of evaluating datasets? According to the ESRB [22] companies still rely on poor valuation methodologies to forecast the recovery rate, therefore, the implementation of a tool to support the management and valuation of a portfolio is an essential task.

The use of machine learning algorithms in source information addressing this problem is still scarce. As Ye and Bellotti [4] explain more recently, machine learning techniques also started to be successfully applied to this field of research. However, most of the existing studies focus on corporate bonds or loans. Models focusing on retail credit products such as mortgages and credit cards still largely need to be investigated. Factors such as [5] the added privacy protection norms and confidentiality policies promoted by banks - did not allow previous references to identify many potential predictors of recovery rates for retail loans. Moreover, as Loterman [3] points out nowadays scholarly resources focusing on loan recovery, have a few outdated machine learning methodologies applied to this problem. The lack of literature in the field of non-performing loans debt recovery shows few and mostly unsuccessful investigations on NPLs. Additionally, the apogee of Machine Learning algorithms in the field of forecasting financial revenues, not yet debt collections, allows companies to base their decisions on such techniques as Gilles Loupe [14] states. While Garrigues [16] confirms an even higher trend noting that retail credits went from a delinquency rate of 3.41% in March to 5.79% in November 2020 [21], retail credits were directly affected by the economic crisis. Correct estimation of the recovery rate of the non-performing debt portfolio is an important component in the debt portfolio purchase transaction, which represents the performance and profit of the collection agency. The use of Machine Learning has not yet been fully developed in this field, leading to new research initiatives and the development of models to improve decision making.

1.3 Purpose and Tools

The purpose of this article is to show the development of a machine-learning algorithm to estimate the recovery rate of unprofitable portfolios [17]. As required by the business, the algorithm must offer a good ability to identify false positives, since these translate into an overestimation of the portfolio in the purchase operation. In other words, more value is invested in the purchase than can be recovered after handling by the collection company. The development compares different machine learning algorithms through the approaches proposed in the study, contributing results to the application area.

A specific field of machine learning supervised learning, is used, in which two approaches are utilized. The regression approach of the target variable (value to be predicted) is continuous and the classification approach in which the target variable is of discrete type. Nine supervised learning models [11] are used to forecast the recovery rates of the mentioned portfolios: Random Forest [13], XGBoost [18], Logistic Regression [20], and Unbalanced Random Forest [19]. For each model, the two aforementioned approaches are evaluated.

1.4 Methodology

A methodology with different approaches is applied to find the algorithm that best fits the business rules. The methodology used supports the way the problem is questioned and allows the conversion of the objectives into applicable approaches used in the experimentation. The classification capabilities of the models are assessed through the customer group approach using the scheme shown in Fig. 7. When an approach does not meet the business needs, a new step in the methodological cascade is taken. Customer prioritization enables the recovery of customer debt using the collection force (dedicated customer contact companies) efficiently and rapidly. The approach is realized through the methodological scheme, which gives a clear picture of what the research can achieve (Fig. 3).

The methodological cascade shows each of the steps followed to produce a model with the ability to differentiate false positives and to predict target variables. The algorithm is based on supervised models, as seen in items six, seven, and eight. The constructed methodology starts from the general requirement of the improvement of the purchasing operation (item one). Then, the reconstruction of the tape backup is performed. The first approach (item three) aims at forecasting the recovery rate. If a second approach (item four) is not possible, it uses multi-class classifiers to group the recovery amount. If the latter fails, it evaluates the binary classification that is expected to recover better scores than the other approaches analyzed. With these premises, the following experimental hypotheses are proposed.

1.5 Hypothesis

Based on the proposed methodology, it is necessary to transform the operational requirement into the desired outcome, i.e. to establish the independent variables and the target variable. The target variable is the result obtained after processing a supervised learning model [8], with dichotomous values.

The four approaches explained in the methodology are used as the basis for the construction of the datasets. Machine learning algorithms allow the formulation of models that can be understood as hypotheses to be tested. It should be noted that the use of classifier algorithms is aimed at discrete variables, while regression algorithms are aimed at continuous variables. In particular, binary classificatory models are used, where the alternative hypothesis assumes that the model can correctly discern between two categories: yes or no (or zeros and ones). The estimation of the amount of debt to be recovered will be covered by regression and classification algorithms; with the latter of the categorical type (more than two categories) and the binary type which, as noted in the previous chapter, prevails in this research.

It is emphasized that the algorithms presented are adapted to the need of each hypothesis. The use of models such as XGBoost and Random Forest together with a binary classification methodology give, through experimentation, superior results compared to the binary classificatory models produced by ImbalanceRFC and

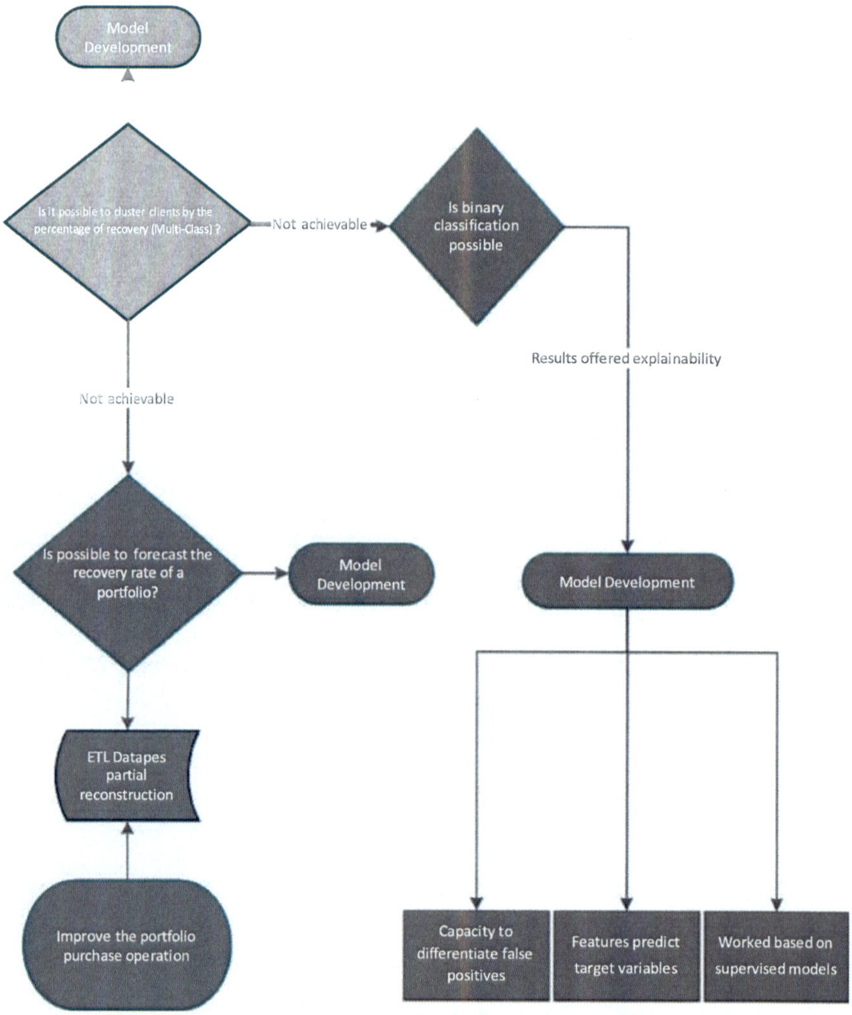

Fig. 3 Methodological cascade. It shows how the problem is approached from the general requirement "Improve the purchase transaction" to "The binary classification of the debtor"

Logarithmic regression. Meanwhile, the regression-based approach for the same models does not explain the target variable, which is usually validated with the coefficient of determination or R^2, a metric used for continuous variables.

2 Data

The characteristics that explain the target variable, according to each of the hypotheses put forward, are extracted from the information provided by the seller of the portfolio in the tape backup. This process is carried out during the portfolio evaluation period (limited time given to buyers to evaluate the tape backup to be purchased). During this period, the partial reconstruction of the portfolio is carried out, which corresponds to the second link in the methodological cascade presented in Fig. 2. These qualities make it possible to explain, after experimentation, the target variable.

The model's input variables include: (a) whether the type of product purchased by the borrower is commercial; (b) whether the credit was purchased by a company and; (c) whether the client is found to have active status with the national tax agency.

It should be noted that the above-mentioned variables belong to the set of categorical variables obtained from the dataset "clients". The aforementioned variables are identified and analyzed in the exploratory evaluation process to finally transform them into data that explain the target variable, after the "partial" reconstruction of the portfolio. Employing variable engineering, the datasets "customers", "payment" and "contact management" are worked on. From these, the 24 input variables for the developed model are obtained. The first dataset contains the inherent information of the debtor and the debt. The last two datasets allow the assessment of the quality of the debt and the opportunity to contact the customer again.

The three datasets arise from the "partial" reconstruction of the portfolio. It is called the "reconstruction process" since the information contained in the tape backups corresponds to flat files with a length of 1000 records each. Given this, it is required that the specialist area gathers and dumps the information in a company repository. It is called "partial" due to the limited time available to make the purchase decision and the non-inclusion of other sources of information that could contain data to complement the information provided.

2.1 *Dataset Clients*

It provides 61 predictive features of personal information containing socio-demographic variables, reported credit rate, and judicial status. This dataset contains 69,683 debtor records from major financial institutions in Peru. Overall, 96.99% of them are retail customers and from those 71.62% represent credit card debts. Each record corresponds to a loan and the identification number of each debtor is used as the primary key in the dataset.

2.2 *Dataset Contact Management*

This dataset contains the communications made by the recovery working group (third-party companies). It contains nine predictor variables that show the historical capacity to contact the debtor, through indicators such as (a) the number of calls; (b) contacts through digital channels, and; (c) visits made by the recovery task force, to collect the debt. It also includes information on customer response, grouped by category. Finally, the quality of the contact is rated by linear coding following business criteria. The dataset contains information on 53,367 customers which is matched with information on each customer using a similar treatment of variables to that used in the previous dataset. The information, which contains both positive and negative responses, captured through contact, allows for an assessment of the willingness of customers to meet their payment obligation. This dataset contains information on debt collection in the time range from January 2015 to October 2019.

2.3 Dataset Payments

The dataset "payments" consists of a dataset showing the recoveries made by the previous owners of the non-performing debt portfolio. This information joins the payments recorded after the purchase transaction occurred dating back to August 2018.

There is a total of 69,335 payment records in the "Payments" dataset. It should be noted that for customers without any payments, a value of zero is assigned. In the portfolio, worked by the previous collection agency, 16,717 customers had not fulfilled their payment obligations. After the purchase transaction, only 3,437 had made at least one payment. This value is part of the target variable which, divided with the capital owed at the time of purchase, forms the target variable. The debt collection period recorded in this dataset is made up of payments made from November 2011 to the first day of August 2020 (Fig. 4).

2.4 Variable Selection and Mapping

2.4.1 Variable Interaction

Figure 5 shows that the variable related to the days that have passed from the date of purchase of the portfolio to the last payment made by the client presents a strong

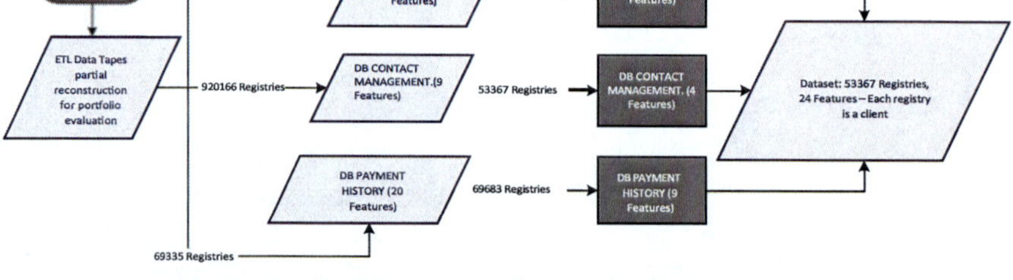

Fig. 4 Joining and transformation tree – partially reconstructed data tapes

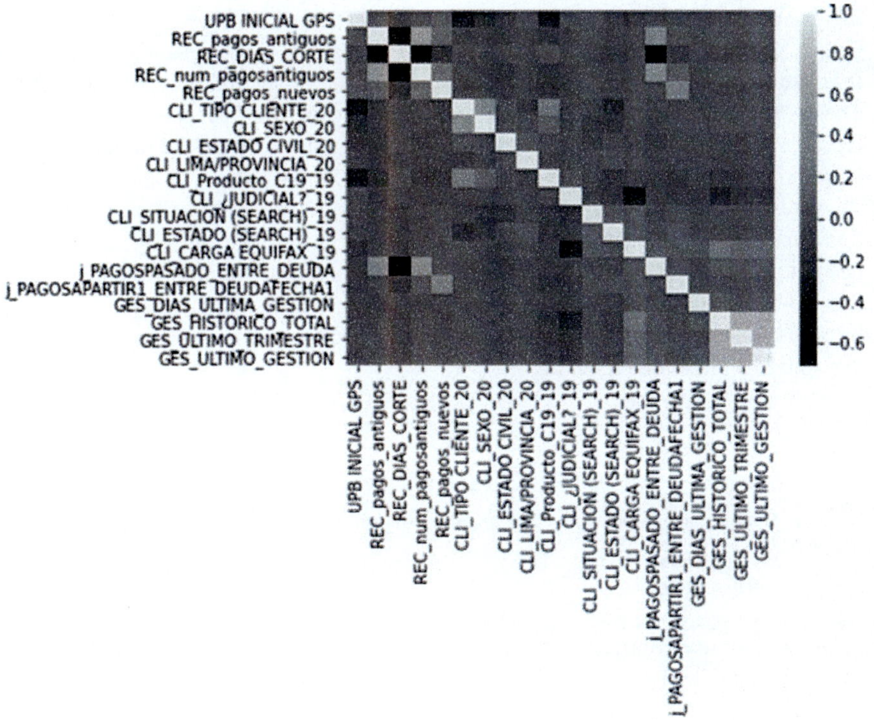

Fig. 5 Variable correlation heat map showing the interactions between variables

inverse correlation with the number of payments made up to the date of purchase. This characteristic shows the clients who have made at least one payment on at least one obligation up to the cut-off date, which is August 9th, 2018. This variable represents the date of the last payment issued by the debtor. It should be remembered that the proportion of customers who have made at least one payment represents only 31.32% of the total portfolio. In addition, it should be noted that this variable has a direct correlation with the variable representing the past payments made concerning the total capital of the debt, the variable is calculated according to the script in (1).

$$\begin{aligned}&\texttt{df['j\_PAGOSPASADO\_ENTRE\_DEUDA']}\\&=\texttt{df['REC\_pagos\_antiguos']/df['UPB INICIAL COVINOC']}\end{aligned} \tag{1}$$

$$\begin{aligned}&\texttt{df['j\_PAGOSAPARTIR1\_ENTRE\_DEUDAFECHA1']}\\&=\texttt{df['REC\_pagos\_nuevos']/df['UPBINICIALGPS']}\end{aligned} \tag{2}$$

In order to model the forecasting algorithm, the variable that measures the number of payments made, up to the date of purchase is removed from the dataset given its correlation with the variables shown in Fig. 5.

2.4.2 Variable Mapping

It is validated that each record in the dataset relates to one and only one customer. The dataset used as input data for the model consists of 53,367 records. This dataset represents the input for the algorithms tested in Table 1 where the target variable calculated in the script (2) is included.

As explained in Sect. 2.4, the cutoff date represents the portfolio purchase transaction and diverges from the equations in scripts (1) and (2). A function needs to be applied to map the classes into the target continuous variable, which corresponds to the output of the script (2). The target variable mapping allows each of the hypotheses set out in Table 1 to be established by transforming the continuous variable to a dichotomous variable.

Table 1 Hypothesis grouped by its corresponding algorithm

Met	Hypothesis description	XBG Class	Imb RFC	Log Reg	RFC	RFC	RFR Lasso	Linear Reg	XGB Reg
Binary	No payment will be received|Does the client pay at least a quantity	x	x	x	x				
3 Multiclass	Does the client will pay more than 50 %|Does the client will pay less than 50 %|No payment will be received	x			x				
4 Multiclass	Does the client will pay more than one third|does the client will pay between one third up to two thirds| Does the client will pay more than two third|No payment will be received	x			x				
Reg 1	What is the recovery rate of those who have paid in the past				x	x	x	x	x
Reg 2	What are the recovery rate of the whole dataset those never payed				x				x

3 Algorithm Selection, Results, and Deployment

3.1 Structure

The following development shown in the flowchart in Fig. 6 shows the different six stages of model development after "partial" portfolio reconstruction: Data processing (also known as extract, transform and load, is a data integration process that combines data from multiple data sources into a single, consistent data store that is loaded into a data warehouse or other target system) allows to reconstruct the data tapes given by the seller.

The data flow shows the modeling process behind the execution of each of the approaches shown in the methodological cascade (Fig. 7).

The data flow shows the modeling process behind the execution. The first link consists of the extraction of the information and loading from the partially reconstructed tape backups. Through the engineering of variables and the transformation of the data, the variables that can describe the target variable are extracted, as explained in Sect. 2. The Implementation Consists of the Iteration of Processes That Follow a Certain methodology until obtaining the model that provides the best

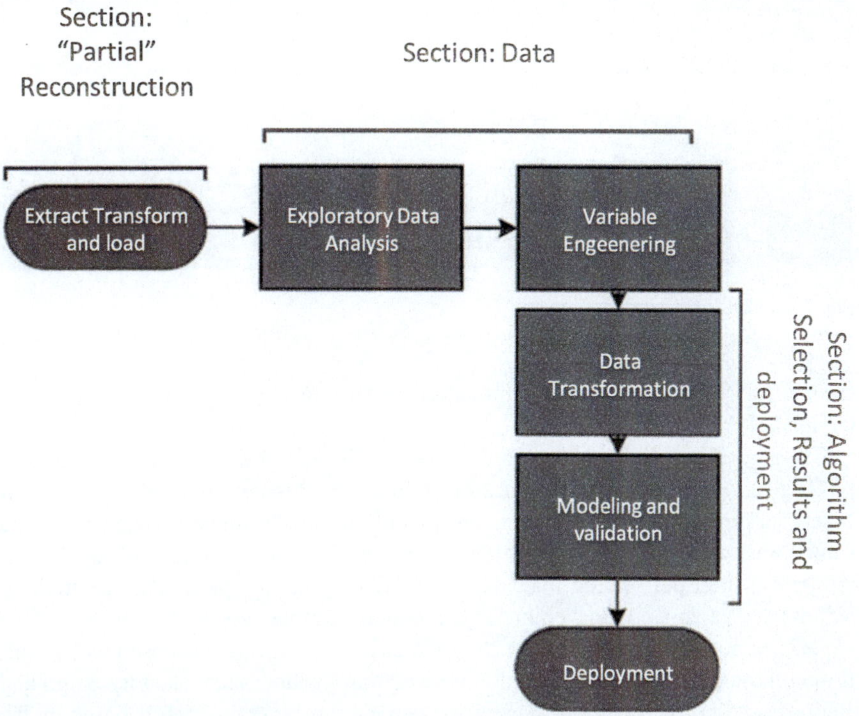

Fig. 6 Data processing scheme

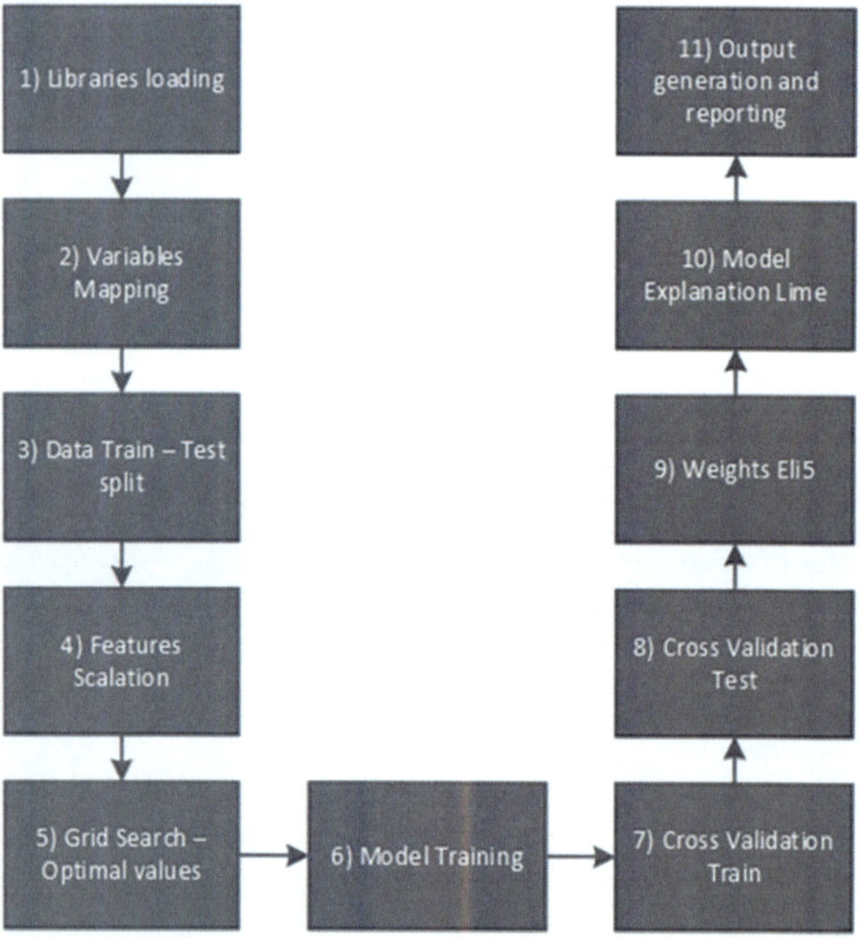

Fig. 7 Modeling scheme

metric, such as the F1- Score; this, after discarding the use of regression-based models.

The algorithms are shown in Table 1 are used inspired by the research carried out by Brigo and Bellotti [5] and the results offered, found that rule-based algorithms of ensemble type, especially random forests and the newly added boosted trees and Cubist, displayed the best forecasting performances. Due to the lack of research in the field of non-performing loans using classifiers, the approach taken by the two authors serves as a starting point. This research becomes the starting point for the experimentation and validation of the worked models, which is subsequently confirmed through the results. It should be noted that the approach taken by Brigo and Bellotti is based on the use of regressors, while the present research is based on the use of classifiers, as the approach taken by the two authors failed to achieve the minimum threshold required by the methodology, therefore the classification approach provides new results for the field under study.

The scheme presented above shows the working route to obtain the results of the experiment. The first link illustrates the loading of the necessary libraries and packages (Python programming language). The mapping of variables allows masking the target variable, obtained from the preprocessing. The "data train-test split" allows the dataset to be split in a proportion of 70% (training) and 30% (testing). After the split, the information is scaled, where the minimum and maximum scaler is used given the different scales in the units presented by the variables. The "grid search CV" package is used to obtain the optimal configuration of hyperparameters of the base model. The information is loaded to the elaborated model and the training is carried out to obtain the results. These mainly consist of obtaining the confusion matrix from the cross-validation process as well as the F1-Score and R^2 performance metrics. The ability to explain the calculations made by the model is sought, together with the ranking of the most relevant variables. Finally, the output file of the information is generated.

3.2 Modeling Algorithms

The methodologies in Table 1 are addressed by categorizing them according to the method used and the hypothesis put forward for each of the approaches. The best-performing algorithms are then evaluated.

3.3 Metrics

The following metrics are used to compare performance between models (Table 2):

3.4 Data Pre-processing and Development

The development follows the scheme shown in Table 1.

Table 2 Used metrics to compare performance between models

Algorithm	Metrics
Regression	R^2
Binary classification	F1 Score (Macro)
Multiclass classification	F1 Score (Macro)

3.4.1 Binary Classification

Base Model Experimentation

This classification methodology offers the best results obtained from the four algorithms used: XGBoost, Random Forest, Logistic regression, and Imbalanced Learning. There is a clear recognition of false positives, which can cause an overestimation of the portfolio, as those defaulters would be counted as future payers. This results in good discrimination in the confusion matrix, as the business rule requires that the number of false positives, which are related to type 1 error, should be minimized. It is also worth mentioning that by expanding a group of customers (cluster) there is an improvement in the performance of the model.

It is evident from Fig. 8 that both tree models give high relevance to the variable that explains the capital owed at the time of the portfolio purchase and the binary variable that explains whether the product is a credit card or not; however, the model with the best F1-score is based on the XGBoost algorithm (Fig. 9).

The algorithm that bests filters out false positives and is selected by the company is binary XGBoost due to the low overestimation that ensures the forecast. True positives make up the group of defaulting customers with the highest probability of recovering the principal owed in a shorter time. Therefore, the recovery task force can be targeted to this group of customers to benefit the company's performance. Unlike

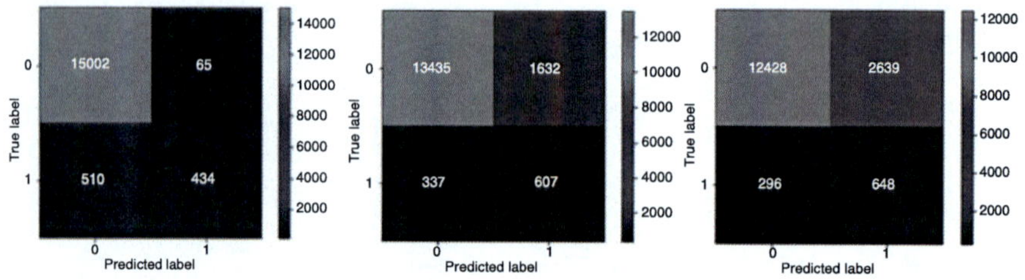

Fig. 8 Binary XGBoost, RFC, ImbalancedRFC Confusion Matrix

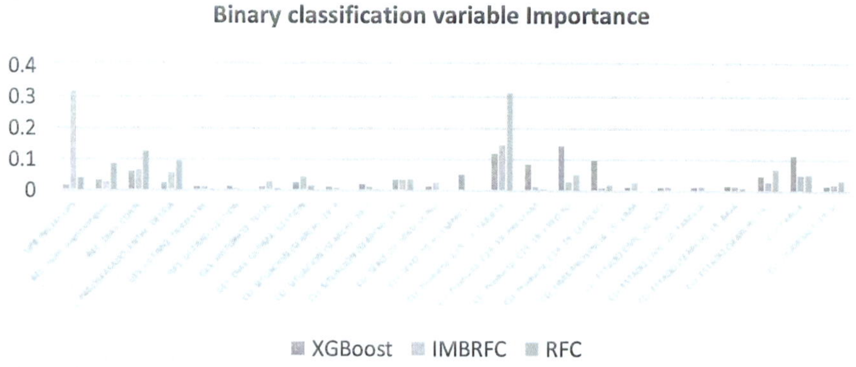

Fig. 9 Variable importance on binary classification algorithm

the XGBoost model, the Random Forest and Imbalanced Learning algorithms show higher type one error, but this generates an increase in positive true values. While this model may appear to have sifted out a better number of customers who meet at least a percentage of their debts, the increase in error type one may decrease and obliterate the profitability of the firm, since a higher bid price may be offered at the time of portfolio purchase; that is, a larger amount of money may have been paid, but never recovered. On the other hand, in the Logistic regression model, which was treated as the base model for the described methodology, the type one error is similar to the two models mentioned above. The unbalanced RFC model in Fig. 8 shows a true identification of 648 but has 2,639 records corresponding to type one error. Therefore, these records are misclassified being this number higher compared to the other binary classification algorithms to which the same analysis is applied (Fig. 10).

The graph shows the performance metrics obtained after the experimentation of the third approach (binary classification) of the methodology. The need for a macro F1- score metric of more than 65% is highlighted as an indispensable requirement. Three algorithms were capable of what was requested, the XGBoost model the one that gave the best performance in the indicated metric among the three algorithms that meet the criterion (Fig. 11).

The ROC AUC curve shows a performance measurement for the classification problems at various threshold settings, the ROC AUC curve achieves a value of 0.842. A model with perfect skill is depicted as a point at (1,1). A skillful model is represented by a curve that bows towards (1,1) above the flat line of no skill. The other two approaches within the classification algorithms used are the three clustered and four clustered, both representing part of the methodology proposed. These approaches offered less performance compared to the binary classification approach.

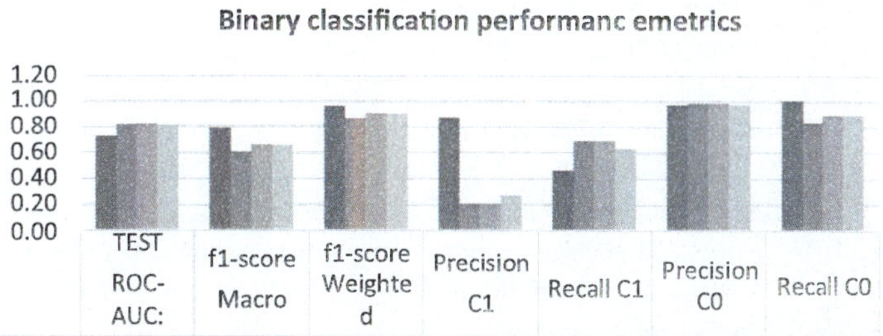

	TEST ROC-AUC:	f1-score Macro	f1-score Weighted	Precision C1	Recall C1	Precision C0	Recall C0
XGBoost	0.73	0.79	0.96	0.87	0.46	0.97	1.00
IMBRFC	0.82	0.60	0.86	0.20	0.69	0.98	0.83
RFC	0.82	0.66	0.90	0.20	0.69	0.98	0.89
LR	0.81	0.66	0.90	0.27	0.63	0.97	0.89

Fig. 10 Performance metrics binary classification

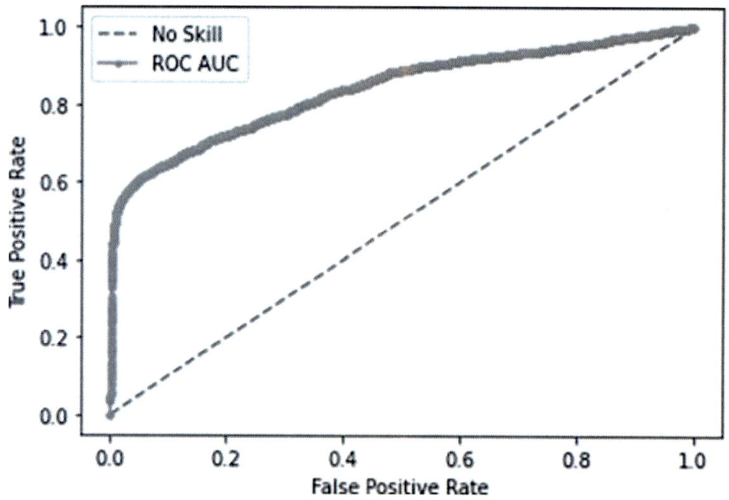

Fig. 11 ROC AUC curve XGBoost binary classification

Further Experimentation (Fine Tunning)

To improve the performance of the presented model and allow a greater generalization using the binary classification approach, the following hyperparameter testing is proposed for the XGBoost model. It should be noted that the hyperparameters presented are specific to each model, library, and programming language in which they are executed. We start with a base reference model to compare 3 Grid Search Method methods (a method that exhaustively searches, from a pre-established list of hyperparameters, to obtain the best performance). The first one, Coordinate Descent, optimizes one parameter at a time, and in the case of having several hyperparameters, there could be several local minima. This method is not optimal as there could be cases where a combined modification of two or more hyperparameters would produce a better score. The second method, Randomized Search, selects hyperparameter values from the search space randomly; and finally, the third method. The technique that calculates a posterior distribution on the objective function based on the data and then selects good test points against this distribution. For the experimentation of the whole data set, it was separated into two data sets (test and training) in a ratio of 3 to 7. The training set is again divided in a ratio of 3 to 7 to obtain a second level of train and test data sets (Fig. 12).

Selected hyperparameters are iterated following a discrete list of values. The graphic shows how a specific parameter (gamma - Minimum loss reduction required to make a further partition on a leaf node of the tree. The larger gamma is, the more conservative the algorithm will be) varies according to its value change. After the experimentation process, the following values of the area under the curve of the AUC curve are obtained. This allows obtaining the discernment capacity of the model. It is important to clarify that the curve shows the performance through the

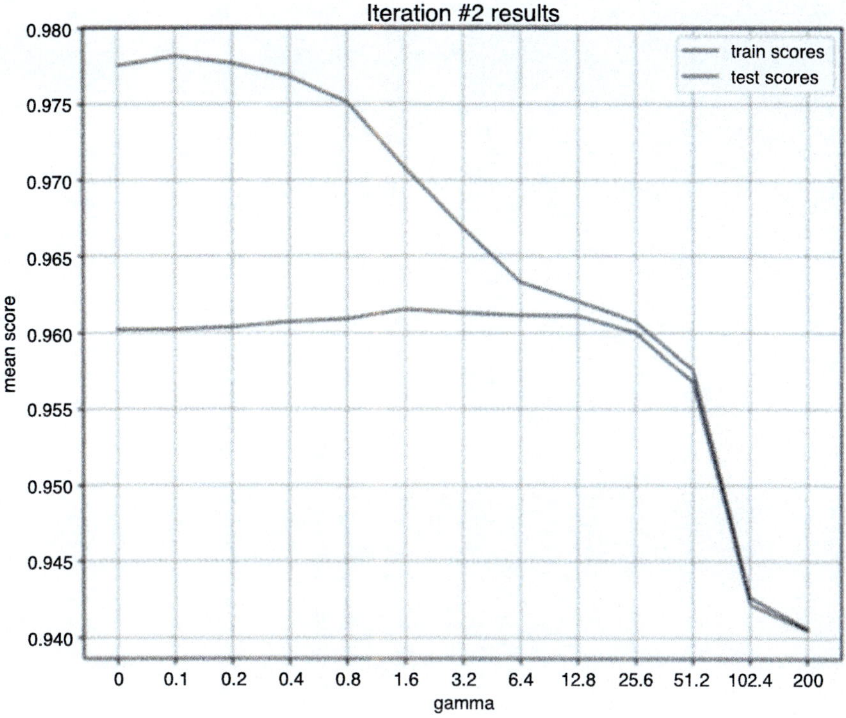

Fig. 12 Iteration parameter

Table 3 AUC fine tunning

	AUC Test	AUC (Never seen)
Benchmark	0.8224	0.8160
Coordinate	0.8277	0.8225
Randomized	0.8424	0.8344
Bayes	0.8408	0.8324

different methods used in the Grid Search CV to find the hyperparameters that give the best performance of the model (Table 3).

It is observed that the tuning of the hyperparameters using the Randomized Search and Bayesian Search methods results in a lower value of type 1 errors, as well as an increase in the AUC of approximately 2%.

3.4.2 Hypothesis Testing

The comparative table with the performance metrics for the models is shown in Table 4 is presented. Regression models do not appear because they do not offer significant results.

Table 4 Results and performance metrics

Classification					Regression				
Appr.	XGB Class	Imb RFC	Log Reg	RFC	Appr.	RFR	Lasso	Linear Reg	XGB Reg
Binary	0.79*	0.6*	0.66*	0.66*	Reg 1	0.3**	0.26**	0.21**	0.28**
3 Multiclass	0.59*			0.51*	Reg 2	0.3**			0.4**
4 Multiclass	0.48*			0.34*					

(*) The values obtained correspond to the F1-score (Macro) metric. The values are comparable to each other
(**) The values obtained correspond to the correlation metric R^2. The values are comparable with each other
(Appr.) Denotes the selected approach

4 Conclusions

It is evident that the preprocessing of the information requires the use of a window of time in which two moments are denoted "pre and post-purchase", this allows the construction of the target variable.

The methodological cascade starts from the specific hypothesis of predicting the recovery rate. After experimentation, it is concluded that the results do not meet the need of the business. To address a new casuistry, the use of the continuous target variable is ended and it is proposed three new approaches that have as their mission to respond to dichotomous target variables (groupings by percentages), culminating as fourth experimentation the approach of binary segmentation.

The approach based on classification models of dichotomous variables allows obtaining better performance metrics. Furthermore, within the classifier models, the binary classification approach is more favorable than the other approaches compared. Within the binary classification approach, the model based on "Boosted Trees" [12] (XGBoost Classifier) had the best performance among the other algorithms (0.79 in the F1-Score macro metric) for the prediction of the binary target variable.

The XGBoost-based binary classifier model obtained the best false positive differentiation rate (Type one error). A value of 0.41% (65/16011) corresponding to type one error was obtained. Thus, it could be concluded that the overvaluation of the portfolio, a requirement requested by the business, is avoided. The ROC AUC shows a value of 0.842 reveals the degree of discerning offered by the model.

The implementation presented allowed the classification of customers who would pay at least some amount of their debts so that the resources of the recovery force could be directed to that group. With this, it is expected that the expense of the debt collection operation will be minimized.

5 Recommendations

To improve the performance of the model, new variables and data transformations can be incorporated so that the model can have a greater capacity to explain the target variable, thus obtaining a better predictive capacity.

The inclusion of variables in the regression models containing data on willingness to pay and ability to pay can allow for better consistency in the model. That is, implementing, for Random Forest-based models, variables targeting characteristics that better detail customers who do not make any debt payments. For XGBoost-based models [7], the focus should be on features that detail customers who pay on time during portfolio operation.

In addition, manual refinement of hyper-parameters can improve predictability along with experimentation with new models that could improve predictability and the ability to predict the debt recovery rate. Plus, by using ensemble techniques, better performance could be obtained from two or three models. The inclusion of different data sources can reduce the bias of the model input data.

References

1. McKinsey & Company. (2019, July). Lessons from leaders in Latin America's retail banking market. https://www.mckinsey.com/~/media/mckinsey/industries/financial%20services/our%20insights/lessons%20from%20the%20leaders%20in%20latin%20americas%20retail%20banking%20market/lessons-from-leaders-in-latin-americas-retail-banking-market.pdf
2. European Systemic Risk Board. (2019, March). Annual Report 2018. https://doi.org/10.2849/042348
3. Loterman, G., Brown, I., Martens, D., Mues, C., & Baesens, B. (2012). Benchmarking regression algorithms for loss given default modeling. International Journal of Forecasting, 161–170. https://doi.org/10.1016/j.ijforecast.2011.01.006
4. Ye, H., & Bellotti, A. (2019). Modelling Recovery Rates for Non-Performing Loans. Risks. https://doi.org/10.3390/risks7010019
5. Bellotti, A., Brigo, D., Gambetti, P., & Vrins, F. (2019). Forecasting recovery rates on non-performing loans with machine learning. *Credit Scoring and Credit Control XVI*. https://doi.org/10.3390/risks7010019
6. Deloitte Hungary. (2019). What's beyond the peak? CEE loan markets still offer new opportunities. https://www2.deloitte.com/content/dam/Deloitte/ce/Documents/about-deloitte/non-performing-bank-loans-npl-study-2019.pdf
7. Friedman, J. (2001). Greedy Function Approximation: A Gradient Boosting Machine. https://doi.org/10.1214/aos/1013203451
8. Shalizi, C. (2008). Statistics 36–350: Data Mining. Carnegie Mellon University. https://www.stat.cmu.edu/~cshalizi/350/2008/
9. ESRB. (2019). The impact of uncertainty on activity in the euro area. European Union: ESRB. https://doi.org/10.2849/224570 (pdf)
10. European Central Bank. (2018). *Guidance to banks on non-performing loans*. European Central Bank. https://doi.org/10.2861/96204
11. Hastie, T., Tibshirani, R., & Friedman, J. (2017). *The Elements of Statistical Learning*. Springer. https://doi.org/10.1007/978-0-387-84858-7
12. Coadou, Y. (2013). Boosted Decision Trees and Applications. EPJ Web of Conferences, 55, 02004. https://doi.org/10.1051/epjconf/20135502004
13. Statistics Department University of California Berkeley, & Breiman, L. (2001, January). RANDOM FORESTS. https://www.stat.berkeley.edu/~breiman/randomforest2001.pdf
14. Gilles, L. (2014). Understanding Random Forests: From Theory to Practice. *Department of Electrical Engineering & Computer Science*. University of Liège. https://doi.org/10.13140/2.1.1570.5928
15. Lizarzaburu, E., & del Brío, J. (2016). Evolución del sistema financiero peruano y su reputación bajo el índice Merco. Período: 2010–2014. *Suma de Negocios*, (págs. 94–112). Lima. https://doi.org/10.1016/j.sumneg.2016.06.001
16. Garrigues. (2020). Transacciones con carteras de deuda (NPLs) y activos tóxicos (REOs) LatAm & Iberia – NPLs Task Force (4T 2020). https://www.garrigues.com/sites/default/files/documents/transacciones_con_carteras_de_deuda_npls_y_activos_toxicos_reos_situacion_a_noviembre_de_2020.pdf
17. Sarker, I. (2021). Machine Learning: Algorithms, Real-World Applications and Research Directions. *SN Computer Science*. https://doi.org/10.1007/s42979-021-00592-x
18. Chen, T., & Guestrin, C. (s.f.). XGBoost: A Scalable Tree Boosting System. https://doi.org/10.1145/2939672.2939785
19. Lemaitre, G., Nogueira, F., & Aridas, C. (2017). Imbalanced-learn: A Python Toolbox to Tackle the Curse of Imbalanced Datasets in Machine Learning. Journal of Machine Learning Research, 18, 1–5. https://www.jmlr.org/papers/volume18/16-365/16-365.pdf

20. Shen, Aihua & Tong, Rencheng & Deng, Yaochen. (2007). Application of Classification Models on Credit Card Fraud Detection. International Conference on Service Systems and Service Management. 1 - 4. https://doi.org/10.1109/ICSSSM.2007.4280163.
21. Garrigues. (2021). Transacciones con carteras de deuda (NPLs) y activos tóxicos (REOs) LatAm & Iberia – NPLs Task Force (3T 2021). https://www.garrigues.com/sites/default/files/documents/transacciones_con_carteras_de_deuda_npls_y_activos_toxicos_reos_situacion_a_octubre_de_2021.pdf
22. European Systemic Risk Board, Suárez, J., & Sánchez Serrano, A. (Eds.). (2018). Reports of the Advisory Scientific Committee (N. 7). https://doi.org/10.2489/617721

Defense Offsets as a Public Policy: A Bibliometric Review in Brazilian Publications

Ricardo Matheus and Rodrigo Antônio Silveira dos Santos

Abstract Brazil uses defense offsets as an important strategy to transfer technology to its industrial basis and reach autonomy on strategic sectors. Although this practice has a consistent legal basis, Brazilian publications about defense offsets are few and dispersed. This paper then realized a bibliometric review about offsets in Brazil and discovered that the number of papers about the theme is far below than the expected.

Keywords Defense offset · Offset practices · Technology transfer

1 Introduction

The SIPRI institute released its survey of military expenditures by country and verified that, in 2020, US$ 1960 billion were spent worldwide (excluding Cuba, Djibouti, Eritrea, North Korea, Somalia, Syria, Turkmenistan, Uzbekistan and Yugoslavia). South America had a total expenditure of US$ 50.7 billion and Brazil had an expenditure of US$ 25.1 billion, that is, 49.5% of the total spent by South America and 1.28% of the total spent worldwide [1]. Even though the expenses generated by offset policies cannot be precise due to their nature, it is estimated that between 1993 and 2013 the US had 54 companies participating in 955 offset contracts, amounting to US$ 99.8 billion [2]. Besides, it is known that Brazil started several high-level technological projects of the Armed Forces that involve offset practices, such as the Integrated Border Monitoring System (SISFRON), the acquisition of the Gripen aircraft and the nuclear submarine, all part of an initiative to promote the defense sector [3].

Based on the Brazilian and global scenarios and in line with the recent publication of a new policy for Technological, Industrial and Commercial Compensations for the Defense Sector (PComTIC Defesa – Normative Ordinance No. 61/GM-MD, of October 22, 2018), it is expected that Brazil's scientific production about offset

R. Matheus (✉) · R. A. Silveira dos Santos
Universidade da Força Aérea (UNIFA), Rio de Janeiro, Brazil
e-mail: ricardorpm@fab.mil.br; silveirarass@fab.mil.br

J. Vargas Florez et al. (eds.), *Production and Operations Management*, Springer Proceedings in Mathematics & Statistics 391,
https://doi.org/10.1007/978-3-031-06862-1_4

policies will follow this reality, not only to validate such projects but also to expand the applicability of these practices. With this in mind, this article presents a survey of Brazilian scientific production on the subject since 2019 up until now.

With the main objective of mapping recent academic production on offset practices, this research contemplates all articles published in the last two years in Brazilian journals with a Qualis A1 concept in the area of "Political Science and International Relations", since it is the area responsible for the majority of scientific discussions about the defense sector. Eleven journals with this specific profile were found based on the Sucupira platform from the 2013–2016 quadrennium, the most recent available data on the CAPES Foundation's website [4].

As for the specific objectives, this article also analyzes the different concepts of offset policies and the Brazilian scenario regarding such policies, both being applied in the methodology used to research the academic production on the subject. The methodology used was the bibliometric review, which will be detailed on Sect. 3.

2 Literature Review

2.1 Offset Definitions

The International Chamber of Commerce (ICC) together with the European Club for Countertrade & Offset (ECCO) inform in their guide to international offset contracts that there is no uniform definition on the subject, but there are two definitions more commonly accepted: "non-standard contracts which require that a form of economic activity is transferred from the seller to the government of the purchasing country as a condition for the sale of goods and/or services on government procurement markets" or "any condition or undertaking that encourages local development or improves the Party's balance-of-payments accounts, such as the use of domestic content, the licensing of technology, investment, counter-trade and similar action or requirement" [5]. As searches for specific terms will be carried out to identify the adherence of the researched articles to the offset theme, it is important to select them from different classifications and definitions of offset from the chosen authors who will make up the Tables 1, 2 and 3 from Sect. 3.3.

Vieira and Alves (2018) define offset, or compensation agreements, as "commercial practices that aim to reduce the negative economic impacts on the trade balance of the contracting State in the face of large acquisitions of goods or services involving foreign suppliers", with compensatory contracts being the legal instrument that makes such acquisitions possible. The authors also define technological compensation as a tool that "aim to promote the technological development of the acquiring States in the sectors that the offset objects refer to" [6].

Ribeira and Júnior (2019) define offsets as "forms of compensation in which the exporting company grants the importing government production-related concessions", thus developing the country's economic and industrial base through access to new markets with the technology transfer derived from these compensatory

Table 1 Selected search terms in Portuguese

Authors	Chosen terms	Terms in the original language	Search terms
Vieira and Alves (2018) [6]	1. Compensatory agreement 2. Offset 3. Business practices 4. Compensatory contract 5. Technological compensation	1. Acordo de compensação 2. Offset 3. Práticas comerciais 4. Contrato de compensação 5. Compensação tecnológica	1. Acord* compensaç* 2. Offset* 3. Prática* comercia* 4. Contrato* compensaç* 5. Compensaç* tecnológic*
Ribeira and Júnior (2019) [7]	***1. Offset*** ***2. Compensatory agreement*** ***3. Technology transfer*** ***4. International contract*** ***5. International acquisition***	***1. Offset*** ***2. Acordo de compensação*** ***3. Transferência de tecnologia*** ***4. Contrato internacional*** ***5. Aquisição internacional***	***1. Offset**** ***2. Acord* compensaç**** ***3. Transferênc* tecnol**** ***4. Contrat* internaciona**** ***5. Aquisiç* internaciona****
Brustolin, Oliveira and Senna (2016) [2]	1. Offset 2. Compensatory practice 3. Technological development 4. Compensatory agreement 5. Compensatory transactions	1. Offset 2. Prática compensatória 3. Desenvolvimento tecnológico 4. Acordo de compensação 5. Transações de compensação	1. Offset* 2. Prática* compensa* 3. Desenvolv* tecnol* 4. Acord* compensaç* 5. Transaç* compensaç*

agreements. The authors claim that offset can be used on both military and non-military contracts, with several cases of compensation agreements reported in the literature on the subject for civil international contracts and that the demand for such offsets in international acquisitions has several different motivations, depending on the objective that the State is trying to achieve [7].

Brustolin, Oliveira and Senna (2016) adopt the definition of offset contained in the Decree No. 7546, of August 2, 2011, as "any compensatory practice established as a condition for strengthening the production of goods, the technological development or the provision of services, with the intention to generate benefits of an industrial, technological or commercial nature". The authors also define the compensatory agreement as a "legal instrument that formalizes the commitment and obligations of the foreign supplier to offset the imports carried out" and that it forms a set of clauses that contains the compensatory transaction [2].

Khan (2010) defines offset as "industrial compensation practices that are required as a condition of purchase", representing an important part of international trade. The author emphasizes that the "offset agreement is implemented by one or more offset

Table 2 Selected search terms in Spanish

Authors	Chosen terms	Terms in the original language	Search terms
Vargas Vergnaud (2005) [12]	1. Offset 2. Co-production 3. Subcontracting 4. Technology transfer 5. Acquisitions of goods	1. Offset 2. Coproducián 3. Subcontratación 4. Transferencia de tecnología 5. Aquisiciones de bienes	1. Offset* 2. Coproduc* 3. Subcontratac* 4. Transferenc* tecnol* 5. Aquisic* bien*
Rodríguez-Gutiérrez et al. (2017) [11]	***1. Offset*** ***2. Compensatory agreement*** ***3. Technological innovation*** ***4. Technology transfer*** ***5. International collaboration***	***1. Offset*** ***2. Acuerdo de compensación*** ***3. Innovación tecnológica*** ***4. Transferencia tecnológica*** ***5. Colaboración internacional***	***1. Offset**** ***2. Acord* compensac**** ***3. Innovaci* tecnol**** ***4. Transferenc* tecnol**** ***5. Colaborac* internaciona****
Zagal (2007) [13]	1. Offset 2. Industrial compensation 3. Direct offset 4. Indirect offset 5. Technological requirements	1. Offset 2. Compensacíon industriale 3. Offset directo 4. Offset indirecto 5. Requerimientos tecnológicos	1. Offset* 2. Compensac* indústria* 3. Offset* direct* 4. Offset* indirect* 5. Requerimien* tecnol*

transactions with a credit value claimed against the agreement" and that "marketers should be aware of offset policies and practices of their foreign customers and governments to better prepare for the competitive bidding process" [8].

Markowski and Hall (2014) define defense offset as "a range of industrial compensation arrangements required by a foreign government as a condition of purchase of US defense articles and services", with these compensatory agreements being demanded by importing governments as a condition for purchasing goods or services from exporting countries. The authors state that thc most common forms of offset transaction types are: countertrade, subcontracting and technology transfer [9].

Petersen (2011) defines offset as "any type of non-monetary compensation that a procuring government requires as exporting firm to provide a condition of the sale", also referring to offset as countertrade or industrial participation. The author states that more than 80 countries require offsets when looking for goods and that these mandatory conditions can take various forms, such as "co-production or licensed production arrangements, foreign subcontracting, technology transfers, foreign investment or purchases, training" [10].

Rodríguez-Gutiérrez et al. (2017) define offset as "commercial agreements demanded by a buyer and accepted by a seller, which oblige the seller to perform actions that offset the cash flow required by the sales contract" and that these compensatory agreements are used by countless nations to introduce technological

Table 3 Selected search terms in English

Authors	Chosen terms	Terms in the original language	Search terms
Khan (2010) [8]	1. Offset 2. Offset agreement 3. Offset policy 4. International trade 5. Industrial compensation	1. Offset 2. Offset agreement 3. Offset policy 4. International trade 5. Industrial compensation	1. Offset* 2. Offset* agreement* 3. Offset* polic* 4. Internation* trade* 5. Industr* compensat*
Markowski and Hall (2014) [9]	***1. Offset 2. Industrial compensation 3. Compensatory agreement 4. Countertrade 5. Technology transfer***	***1. Offset 2. Industrial compensation 3. Compensatory agreement 4. Countertrade 5. Technology transfer***	***1. Offset* 2. Industr* compensat* 3. Compensat* agreement* 4. Countertrade* 5. Techn* transf****
Peterson (2011) [10]	1. Offset 2. Countertrade 3. Industrial participation 4. Foreign subcontracting 5. Technology transfer	1. Offset 2. Countertrade 3. Industrial participation 4. Foreign subcontracting 5. Technology transfer	1. Offset* 2. Countertrade* 3. Industr* participat* 4. Foreign* subcontract* 5. Techn* transf*

innovations, thus strengthening the high-tech industries based on the technological transfers that are received in these agreements. On a historical level, the authors cite that the use of offset agreements mainly took place in the last 50 years, after World War II, and that it comes from a European arms policy of international collaboration between allied countries to pool resources, to obtain economies of scale and to reduce risks when obtaining new weapons systems [11].

Vargas Vergnaud (2005) defines offset as commercial and industrial benefits granted to foreign governments to induce or condition the purchase of military goods and services, such as co-production, subcontracting, technology transfer and acquisition of goods. The author details these types: co-production, a frequently used practice in which an international company uses another company located in the buyer country to assemble, build or produce the items necessary for the production of the acquired goods; subcontracting, that occurs when the exporting country acquires goods and services from a company in the buyer country, but which are not used directly in the production of the acquired goods, as occurs in co-production; transfer of technology, another very frequent modality, in which the companies that sell the goods and services to be acquired already have knowledge of the technologies involved in the production and transfer them to the buyer country; and purchases of goods, in which the selling countries purchase goods and services from the buyer country without direct association with the production of the acquired goods [12].

Zagal (2007) defines and typifies the offsets found in industrial compensation projects as two types: direct offset, in which the buyer country's capacities are used to produce some or several parts of the acquired goods, or indirect offset, in which the buyer country receives benefits from the exporter related to other economic sectors. The author highlights that the choice of offset policies can offer the buyer country the possibility to develop strategic areas, to address the deficiencies of existing national markets, to promote regional development and to develop new technological requirements [13].

These concepts will be used in the bibliometric review without worrying which one is more or less adequate, since such concept is subjective and varies according to the application of the policies. The choices were made based on which set of terms is the broadest, so that it can facilitate the identification of similar articles on the topic during the search.

2.2 *Offsets in Brazil*

Since 2005, Brazil has been adopting changes in the guidelines related to its defense sector in order to revitalize its defense industrial base and the national production of high-tech goods, with this growth heavily depending on offset policies and the transfer of international technology to Brazilian companies, with an estimated expenditure close to US$ 190 billion in investments for the renewal of its armed forces [14]. Several programs were adopted as critical for this revitalization of the defense sector, including: the submarine development program (PROSUB); the Brazilian nuclear submarine program (PNM); the Guarani program for armed vehicles; the Integrated Border Monitoring System (SISFRON), a national border surveillance program; the FX-2 program, to replace tits fighter aircrafts fleet; and the KC 390, for cargo planes. All of these with contracts that involve high costs and transfer of cutting-edge technology, thus proving that it has a clear concern with the renewal of the national industry in order to make it more competitive in the international market and more independent from foreign services and goods [14].

In order to allow these changes in guidelines, it was necessary to review and change the Brazilian legislation so that it brings stability and predictability to these agreements, while allowing their validation from public agents and the society. Brustolin, Oliveira and Senna (2016) summarized a history of some of the most relevant laws [2] and this paper complemented the list with more recent ones:

- Decree No. 6703, of December 18, 2008, which approved the National Defense Strategy, promoting partnerships with other countries with the intention of developing the national technological capacity;
- Law No. 12349, of December 15, 2010, which allows the inclusion of measures in bidding notices for commercial, industrial and technological compensations;

- Decree No. 7546, of August 2, 2011, which defined measures for Commercial, Industrial or Technological Compensation (offset);
- Law No. 12598, of March 22, 2012, which established special rules for purchases, contracts and development of defense products and systems;
- Decree No. 7970, of March 28, 2013, which regulated Law No. 12598, of March 22, 2012, and established additional rules for purchases, contracts and development of defense products and systems;
- Constitutional Amendment No. 85, of February 26, 2015, which expands the treatment given to science, technology and innovation activities;
- Law No. 13243 of January 11, 2016, which encourages scientific development, research, scientific and technological training and innovation, while also aim at international cooperation for technology innovation and transfer;
- Decree No. 9283, of February 7, 2018, which regulates Law No. 13243, of January 11, 2016, and outlines the incentive to technological innovations, including international cooperation projects;
- Normative Ordinance No. 61/GM-MD, of October 22, 2018, which establishes a defense policy for technological, industrial and commercial Compensations (PComTIC Defesa);
- Resolution No. 245, of August 1, 2019, which institutes the innovation policy for the National Nuclear Energy Commission (CNEN) and outlines international cooperation agreements within the scope of the commission; and
- Normative Instruction No. 1, of November 6, 2020, which establishes the innovation management system and outlines the concepts, rules and procedures for the management of the processes that guide transfers of technology.

Although the intention is not to exhaust all the Brazilian legislation about the subject, these norms serve as an example of the importance given to the subject by government entities and by private and scientific institutions, since most legislation follow public consultation and debate. Due to the high costs involved in these contracts and the public policies and legislation on the subject, these projects need to be monitored by society to ensure that the industrial renewal occurs, hence the need for an academic debate about offset policies to ensure reliability in decision-making process and when assessing results.

3 Methodology

For the bibliometric review, five terms detailed in Sect. 3.3 were searched among all articles. The idea behind these terms is for them to be as broad as possible while still being about offset so that it can minimize the probability of an article that cites the topic not being found in the search, hence the necessity that these terms are used routinely when it comes to offset. To cover the specifics of each original language of the articles, five terms were chosen based on three authors with articles about offset in their respective language, not taking into account which of these productions has the most up-to-date content but instead which one has the most varied verbiage used to describe offset practices.

Based on the researched journals, it was necessary to select five different terms for three languages: Portuguese, Spanish and English. When there was more than one idiomatic version available for consultation, priority was given to Portuguese and then to Spanish, with articles in English being consulted only when this was the only option available. This prioritization was given so that the results would reflect more faithfully the Brazilian production in its native language, Portuguese. It should also be noted that only scientific articles were used in the search, excluding any other type of production such as forum, thesis, essay, review, editorial etc.

3.1 Choice of Journals

Based on the search carried out on the Sucupira Platform selecting the evaluation area of "Political Science and International Relations", the classification event of "Classifications of Journals in Quadrennium 2013–2016" and the classification of "A1", 11 Brazilian journals were found:

1. Cadernos de Pesquisa (Fundação Carlos Chagas Online) – ISSN 1980-5314 [15];
2. DADOS – Revista de Ciências Sociais – ISSN 1678-4588 [16];
3. Opinião Pública – ISSN 1807-0191 [17];
4. RAE. Revista de Administração de Empresas – ISSN 0034-7590 [18];
5. Revista Brasileira de Estudos de População – ISSN 0102-3098 [19];
6. Revista Brasileira de Política Internacional – ISSN 1983-3121 [20];
7. Revista de Administração Pública – ISSN 1982-3134 [21];
8. Revista de Economia Política (Online) – ISSN 1809-4538 [22];
9. Revista de Sociologia e Política – ISSN 0104-4478 [23];
10. Saúde e Sociedade (Online) – ISSN 1984-0470 [24]; and
11. Cadernos de Saúde Pública (Online) – ISSN 1678-4464 [25].

The Qualis classification of A1 in the area of "Political Science and International Relations" refers to journals that have: indexing in the SCIma-go/Scopus database; top quartile position (p75); publications with original articles exclusively; 30% of its articles with international collaboration or authorship of researchers with priority institutional affiliation abroad; editorial line, thematic vocation and frequency of publications reported on the Sucupira Platform; Editorial Board formed by leading international authors; SCImago Journal Rank (SJR) indicators to measure citations; double blind peer review system; and at least 85% of articles by authors not linked to the institution that edits the journal [4]. This choice was made based on the classification of journals that would have the greatest academic and social impact on the studied topic.

As explained in Sect. 2.1, the area that most applies to the topic is "Political Science and International Relations", as it is a matter that involves government guidelines, international contracts and public policies for the absorption of acquired knowledge and technology through offset. This is not an exhaustive criterion, as other areas may contain articles about offset policies, but it was the criterion chosen since it is the most objective without widening too much the search, which would be detrimental to the analysis that was carried out.

3.2 *Choice of Search Period*

In order to analyze the most contemporary publications, the promulgation of Normative Ordinance No. 61/GM-MD, of October 22, 2018, which establishes a defense policy for technological, industrial and commercial Compensations (PComTIC Defesa), served as a time reference for the evaluation. The purpose of the article is not to assess whether the normative ordinance caused any kind of impact on academic publications, but simply to serve as a measuring point for public debate about the subject, since the normative ordinance is specifically aimed at technology transfer contracts, as discussed in Sect. 2.2.

The enactment of a specific policy about offset from the Brazilian Ministry of Defense presupposes both a popular consultation with interested parties as well as an analysis of scientific and academic debates on the subject. Therefore, its publication date of October 22, 2018 serves as a parameter to assess the current zeitgeist around the subject and, because of that, articles from the beginning of 2019 up to this date were considered, as each journal has a different periodicity.

3.3 *Choice of Search Terms*

With the definitions reported in Sect. 2.2, it was possible to select from each author five terms that would more broadly define the concept of offset and choose a group of these to carry out the search. As there were articles in Portuguese, Spanish and English in the journals, three authors were selected for each of these languages and a set of specific terms for each one, thus choosing the possible terms that would better identify an article that cites offset, even if superficially.

Tables 1, 2 and 3 were organized in such a way as to separate authors by language, choosing the terms from their definitions that would most probably represent the theme in a search, as well as these terms written in their original language and the search expressions that were used. The authors chosen were Ribeira and Júnior (2019) [7] for the search in Portuguese, Rodríguez-Gutiérrez et al. (2017) [11] for the search in Spanish and Markowski and Hall (2014) [9] for the search in English, with these choices made based on the fact that these were the authors who presented the most diverse related terms in their respective languages so that it could increase the likelihood of the search returning an article that cites offset policy.

Once the specific terms for each language were chosen, all articles published according to the parameters detailed previously were searched for such terms, with the exception of texts related to the bibliographical references, the journal's information and the authors' credentials. Upon finding at least one term, the articles were analyzed to detail the level of adherence to the researched topic, as some of the terms may have been used in another context that differs from the searched theme, and were classified as:

- *Non-Adherent* – At least one of the terms is cited in the text, but none in the context of offset practices;
- *Low Adherence* –One of the terms found in the text refers to offset practices, but it's only mentioned superficially;
- *Average Adherence* – One of the terms found in the text refers to offset practices and the theme is somewhat developed but it's not the main theme of the article; and
- *High Adherence* – The main theme of the article is about offset practices.

4 Analysis and Discussion of Results

Based on the entire search carried out in the 11 Brazilian journals with the terms previously chosen, the results are presented in Table 4.

Among all journals, the search was performed on a total of 1486 articles, with only 58 articles (3.9%) returning with at least one of the terms found. Out of these 58 articles, only 27 articles (46.5%) were considered adherent, that is, at least one of the terms found in the text of the article referred to offset practices. Out of the 11 journals, 6 had at least one article adhering to the topic and the journal REVISTA BRASILEIRA DE POLITICA INTERNACIONAL (ISSN 1983-3121) had the highest number of articles adherents, with 10 articles in total. Regarding the adherence levels defined in Sect. 3.3, from 27 adherent articles, 18 articles (66.7%) had low adherence, 6 articles (22.2%) had average adherence and 3 articles (11.1%) had high adherence, but the ones with high adherence were from the same journal: CADERNOS DE HEALTH PUBLIC (ONLINE) (ISSN 1678-4464).

It is important to emphasize that, although the articles with high adherence are about technology transfer, they are focused on the health area and are in no way related to the defense sector. With that in mind and the fact that 5 journals (45.5%) did not have any quotation of the search terms, these results reflect poorly on the debate about defense offset policies. This can have a correlation with the choice of searching only journals with Qualis A1 classification, which reduces the possibility of obtaining adherent results since none of the journals that met the parameters is more exclusively about the defense sector.

Considering the results in Table 5, "technology transfer" was cited 62 times, "offset" was cited 3 times and "technological innovation" was cited 3 times in the 27 adherent articles, while all other terms weren't mentioned even once. This may be

Table 4 Search performed in the 11 journals

ISSN	Publication	Year	Total articles	Articles with search term found	Adherent articles
1980-5314	CADERNOS DE PESQUISA (FUNDAÇÃO CARLOS CHAGAS. ONLINE)	2019	53	0	–
		2020	50	0	–
		2021	26	2	0
1678-4588	DADOS – REVISTA DE CIÊNCIAS SOCIAIS	2019	28	0	–
		2020	28	0	–
		2021	21	0	–
1807-0191	OPINIÃO PÚBLICA	2019	24	0	–
		2020	25	0	–
		2021	10	0	–
0034-7590	RAE. REVISTA DE ADMINISTRAÇÃO DE EMPRESAS	2019	20	2	1
		2020	13	1	0
		2021	14	0	–
0102-3098	REVISTA BRASILEIRA DE ESTUDOS DE POPULAÇÃO	2019	24	1	0
		2020	17	0	–
		2021	15	0	–
1983-3121	REVISTA BRASILEIRA DE POLÍTICA INTERNACIONAL	2019	23	10	6
		2020	25	3	2
		2021	9	4	2
1982-3134	REVISTA DE ADMINISTRAÇÃO PÚBLICA	2019	51	2	1
		2020	84	1	0
		2021	40	0	–
1809-4538	REVISTA DE ECONOMIA POLÍTICA (ONLINE)	2019	38	6	2
		2020	43	8	5
		2021	32	3	2
0104-4478	REVISTA DE SOCIOLOGIA E POLÍTICA	2019	26	0	–
		2020	31	0	–
		2021	0	0	–
1984-0470	SAÚDE E SOCIEDADE (ONLINE)	2019	95	0	–
		2020	99	3	2
		2021	61	0	–
1678-4464	CADERNOS DE SAÚDE PÚBLICA (ONLINE)	2019	166	2	0
		2020	183	7	2
		2021	112	3	2

related to a poor choice of terms, but it could also be a reflection on the content of the articles, which mostly did not address the subject of offset policy from the perspective of the defense sector. There was also an excess of citations of the term "offset" for articles in English, but this was expected since it is a commonly used word and is only directly related to the searched theme within a specific context, signaling in hindsight that choosing a compound term with the word could have been a better choice for searching articles in English.

Table 5 Adhering articles according to search terms

Authors	Chosen terms	Search terms found	Terms found in adherent articles
Ribeira and Júnior (2019) [7]	1. Offset	7	0
	2. Compensatory agreement	0	0
	3. Technology transfer	36	33
	4. International contract	0	0
	5. International acquisition	0	0
Rodríguez-Gutiérrez et al. (2017) [11]	1. Offset	0	0
	2. Compensatory agreement	0	0
	3. Technological innovation	6	3
	4. Technology transfer	2	1
	5. International collaboration	3	0
Markowski and Hall (2014) [9]	1. Offset	71	3
	2. Industrial compensation	0	0
	3. Compensatory agreement	0	0
	4. Countertrade	0	0
	5. Technology transfer	33	28

Fig. 1 WordCloud of the keywords from the adherent articles

Since several of the search terms used did not obtain any results, both in non-adherent and adherent articles, a survey of all keywords used in the adherent articles was carried out, regardless of their language or level of adherence. All keywords chosen by the authors were gathered on a WordCloud in Fig. 1, in which the most cited keywords are: development, health, policy, innovation, Brazil, technology, industry, China, partnerships, foreign, international and south-south. Although some of these terms are in line with the intent behind offset policies, they are either too broad or not specific enough to ensure that an article is in line with the theme. This implies that, even with the use of these keywords as search terms, the results would be too vague to state that it is an article about the theme, reinforcing the possibility raised that the subject was rarely addressed in the chosen journals, since even the adhering articles do not focus exclusively on the offset policy.

This disassociation between the results found searching the articles and the expectancy of debates about defense offsets does not necessarily imply that there was no scientific research on the subject during this period, given the limited number of journals searched, but it may indicate that there is an interest from others areas in transfer of technology and offset. Based on this, it is possible to open a larger conversation about these offset laws and policies for other sectors of society other than only the defense sector, even contributing to the strengthening of such measures.

5 Conclusion and Suggestions for Future Research

The results obtained from the searches demonstrate that the issue of offset policies has not been treated as a priority in recent studies on political science and international relations, even though Brazilian legislation and government projects demonstrate the importance of debating the subject as an option to contribute for the growth of the national industry. Even though the chosen journals were only the ones with Qualis A1 qualification, this theme is important enough to be dealt with diligence by some of these journals, since it involves diplomacy, technological advances, promotion of industrial production, training of national professionals, public policies, high value public contracts, among others. The fact that the articles with high adherence to the subject were from the health area reinforces this need for a broader conversation on the subject involving more sectors besides defense, even to strengthen the existing practices.

There is a need to apply this research also for other journals to obtain a real temporal picture of how the theme is being treated throughout the Brazilian scientific scene. One possibility is also to limit only to periodicals that deal specifically with the defense sector, but this would inhibit a larger debate about extending the subject to other possible areas of study. It could also increase the search time frame, but it may not add much to the debate on the theme, since they are always in contracts with very contemporary implications and they lose a little urgency over time, given the high contractual expenses and the management varying a lot according to the government of the moment. There is also the possibility of comparing the findings with international journals in the same area of evaluation to assess how the debate is being conducted outside Brazil and understand if they are diverging or converging, as well as why.

Based on the findings of this article, there is a need to increase a national scientific production about offset practices, regarding both Brazilian and foreign realities, something that can bring several benefits to the national industry and to diplomatic relations. A larger collection of debates on the subject would both help the monitoring of current projects as well as a mapping of areas that would benefit from these offset policies, which can help public agents in making decisions about the use of this practice.

References

1. SIPRI Military Expenditure Database Homepage, https://sipri.org/databases/milex, last accessed 2021/08/20.
2. BRUSTOLIN, V., OLIVEIRA, C., SENNA, C.: Análise das práticas de OFFSET nos contratos de defesa no Brasil. Revista da Escola de Guerra Naval 22(1), 169–196 (2016).
3. GOUVEA, R.: Brazil's new defense paradigm. Defense and Security Analysis 31(2), 137–151 (2015).
4. Sucupira Plataform Homepage, https://sucupira.capes.gov.br/sucupira/public/index.xhtml, last accessed 2021/08/20.
5. ICC-ECCO Guide to International Offset Contracts Homepage, https://iccwbo.org/publication/icc-ecco-guide-international-offset-contracts-2019/, last accessed 2021/08/20.
6. VIEIRA, A., ÁLVARES, J.: Acordos de compensação tecnológica (OFFSET). Revista da SEF 1, 20–29 (2018).
7. RIBEIRO, C., JÚNIOR, E.: Política de Offset em Compras Governamentais: uma análise exploratória. Instituto de Pesquisa Econômica Aplicada 2473, 1–34 (2019).
8. KHAN, A.: Market trends and analysis of defense offsets. DISAM Journal 32(1), 138–154 (2010).
9. MARKOWSKI, S., HALL, P.: Mandated defence offsets: Can they ever deliver?. Defense and Security Analysis 30(2), 148–162 (2014).
10. PETERSEN, C.: Defense and commercial trade offsets: Impacts on the U.S. Industrial base raise economic and national security concerns. Journal of Economic Issues 45(2), 485–492 (2011).
11. RODRÍGUEZ-GUTIÉRREZ, I., AMAR-SEPÚLVEDA, P., MIRANDA-REDONDO, R.: Fomento de la innovación tecnológica en Colombia: un análisis desde la experiencia internacional de los Offsets del sector defensa. Revista ESPACIOS 38(51), 3–14 (2017).
12. VARGAS VERGNAUD, M.: Una mirada económica a los acuerdos de offsets en el sector defensa y seguridad en Colombia. Planeación & Desarrollo 35(2), 389–416 (2005).
13. ZAGAL, C.: Impacto de las compensaciones industriales sobre las capacidades de las Fuerzas Armadas de Chile. Revista Enfoques: Ciencia Política y Administración Pública 5(7), 199–225 (2007).
14. GOUVEA, R.: Brazil's defense industry: Challenges and opportunities. Comparative Strategy 37(4), 346–359 (2018).
15. Cadernos de Pesquisa Homepage, http://publicacoes.fcc.org.br/index.php/cp/issue/archive, last accessed 2021/08/20.
16. DADOS – Revista de Ciências Sociais Homepage, http://dados.iesp.uerj.br/edicoes/, last accessed 2021/08/20.
17. Opinião Pública Homepage, https://www.cesop.unicamp.br/por/opiniao_publica/arquivo, last accessed 2021/08/20.

18. RAE. Revista de Administração de Empresas Homepage, https://rae.fgv.br/rae/edicoes-anteriores, last accessed 2021/08/20.
19. Revista Brasileira de Estudos de População Homepage, https://rebep.org.br/revista/issue/archive, last accessed 2021/08/20.
20. Revista Brasileira de Política Internacional Homepage, https://www.scielo.br/j/rbpi/grid, last accessed 2021/08/20.
21. Revista de Administração Pública Homepage, http://bibliotecadigital.fgv.br/ojs/index.php/rap/issue/archive, last accessed 2021/08/20.
22. Revista de Economia Política Homepage, https://centrodeeconomiapolitica.org.br/repojs/index.php/journal/issue/archive, last accessed 2021/08/20.
23. Revista de Sociologia e Política Homepage, https://revistas.ufpr.br/rsp/issue/archive, last accessed 2021/08/20.
24. Saúde e Sociedade Homepage, https://www.scielo.br/j/sausoc/grid, last accessed 2021/08/20.
25. Cadernos de Saúde Pública Homepage, https://www.scielo.br/j/csp/grid, last accessed 2021/08/20.

A Proposal for Collaborative Research Projects Involving Academy and a Brazilian Navy Science and Technology Institution

Leonardo Antonio Monteiro Pessôa, Rodrigo Abrunhosa Collazo, Fernando Muradas, and Helder Gomes Costa

Abstract This study presents a proposal of a framework to guide collaborative research projects between a Brazilian Navy science and technology institution and university labs to effectively contribute to the Brazilian Navy's science and technology strategy. This approach aims to overcome bureaucratic constraints and to adapt the existing management controls to the characteristics of research projects, where it is not possible to define precise goals beforehand. Therefore, the framework also contributes to technology and research management. Moreover, the proposed framework defines a symbiotic relationship between research and ongoing development projects, in order to present feasibility tests for further developments and to identify both methodological gaps and possible new applications. The paper presents two projects following the guidelines of the framework as practical results.

Keywords Brazilian Navy · Science and technology · Research project management

L. A. Monteiro Pessôa (✉)
Centro de Análises de Sistemas Navais, Rio de Janeiro, RJ, Brazil

Universidade Federal Fluminense, Niterói, RJ, Brazil
e-mail: lampessoa@id.uff.br

R. Abrunhosa Collazo
Comando Naval de Operações Especiais, Rio de Janeiro, RJ, Brazil

F. Muradas
Centro de Análises de Sistemas Navais, Rio de Janeiro, RJ, Brazil

H. Gomes Costa
Universidade Federal Fluminense, Niterói, RJ, Brazil

J. Vargas Florez et al. (eds.), *Production and Operations Management*, Springer Proceedings in Mathematics & Statistics 391,
https://doi.org/10.1007/978-3-031-06862-1_5

1 Introduction

CASNAV (Center for Naval Systems Analysis) is a Brazilian Navy science and technology institution created in 1975 [1] with the aim of providing technical support for the Brazilian Navy regarding operational research, cryptology and system engineering development.

Despite the significant number of in-house researchers, recently CASNAV started developing collaborative research projects with some universities, including this innovative approach as one of the main activities of its strategic plan.

> CASNAV, to carry out its activities, is divided into two major segments: Research and Production. The Research segment develops scientific knowledge for methodological and modeling improvement. These elements constitute the basis for the construction of solutions contracted with CASNAV, prototyped and produced by the Production segment [2].

Nevertheless, to overcome the disparities between research and development, there was a need to develop a proposal of a framework to adapt the management procedures to the collaborative research projects, which is the focus of this paper.

The following section presents the theoretical basis and the background of CASNAV, the development projects, and why a separate structure must be established to stimulate and manage research projects there. Section 3 presents the intended framework with its features and processes, while Sect. 4 presents some examples of the results as well as the actual collaborative projects produced so far. Finally, Sect. 5 presents some issues to be discussed and the concluding remarks.

2 Material and Methods

2.1 Collaborative Research Projects

It is widely mentioned in academia that university–industry collaboration helps the development of competence and knowledge in both universities and industries and for the development of innovations [3].

Among the advantages of collaborative research projects are the possibility to "advance technologically, at lower cost and with less inherent risk" and "better access a greater breadth and depth of knowledge and technologies than would normally be possible through internal development" [4].

Although the search for knowledge transfer between universities and companies has increased, there are many examples of failure [3].

Usually, universities and industries face difficulties to understand each other's goals, cultures or constraints [3].

Furthermore, regarding project management, collaborative research projects face many challenges to succeed, since they have some intrinsic characteristics such as high uncertainty and risks, individually oriented project personnel, physically distant

Table 1 Paradoxes on collaborative projects

Topic	Thesis	Antithesis
(1) Uncertainty	"require freedom and flexibility to generate innovative results"	"needs strict management in order to avoid failure (...) while creativity needs firm structures in order to be transformed into widely usable project outcomes."
(2) Heterogeneity	"fosters integration of the research perceptions, ideas and views that are needed in order to solve problems comprehensively."	"heterogeneity (...) leads to problems with respect to inter-cultural, inter-organisational (...) management."
(3) Management	"management of the project, vision and integration of results require the commitment and involvement of all project parties"	"managers' limited authority because of the autonomy of partners and governance structures."

Adapted from Brocke and Lippe [5]

and heterogeneous project partners, and high expectation of creativity and innovativeness [5].

In order to deal with heterogeneity and collective responsibilities, it is necessary to evaluate partners' commitment to the intended research line [4].

Building trust is also very important in these type of projects for many reasons, such as enabling cooperation, promoting adaptive organizational forms, reducing the cost of transactions, and encouraging the formation of ad-hoc groups, among others [6], as summarized below:

> Trust is a psychological state comprising the intention to accept vulnerability based upon positive expectations of the intentions or behavior of others

Regarding applicability in the military field, the challenge is heightened by the need of an experimental idea of military planning and operations.

Due to their nature, collaborative research projects demand a reasonable measure of performance, and adequate goals, because if they are controlled by development projects' standard management logic, these projects tend to fail.

> The model needs to reflect the importance of achieving mutual benefit, i.e. ensuring that an appropriate balance between academic objectives and industrial priorities is achieved, with particular care being taken in defining the role of student researchers [4].

Brocke and Lippe [5] defined some paradoxes regarding collaborative research projects due to their characteristics, as presented in Table 1.

2.2 *Domain Knowledge and Methodological Knowledge*

We start by presenting the knowledge dichotomy necessary to explain how research for CASNAV is peculiar. There are two essential sets of knowledge for CASNAV to successfully carry out scientific development: domain and academic.

Domain knowledge comprises information related to the application universe. CASNAV's research is naturally focused on the military area, which requires that the majority of its in-house research people have considerable experience in this area, guaranteeing reasonably specialized knowledge.

Regarding academic knowledge, CASNAV, as a science and technology institute and service provider, needs to scientifically prepare its staff to carry out its activities. Therefore, the training of personnel for scientific development at CASNAV requires a peculiar management competence that differs from other areas of the Brazilian Navy [7, 8].

The Brazilian Navy's scientific development model, unlike those of the Army and Air Force, is supported by federal universities [7].

Specifically at CASNAV, most of its own researchers received their training either at Rio de Janeiro Federal University (UFRJ) or Fluminense Federal University (UFF) [1, 8]. Such a model offers diversity and the possibility of proposing courses in programs with different vocations [8].

Without the support of domain knowledge (or without a co-advisor who can provide it), the risk of developing academic research misaligned with the core objectives of the institution is higher, increasing the chance of failure.

Meeting the needs of the Brazilian Navy is the main purpose of CASNAV. Thus, the methodological and modeling knowledge acquired during research aims to contribute to the application of CASNAV development projects.

However, development projects have well-defined objectives and products established by contract with a client, while research, as discussed in the previous section, cannot be limited by such a restriction, since what is under study is something whose feasibility has not yet been proven. As a matter of fact, discovering the impossibility of a given hypothesis is a valid result for a research, because it avoids expenditure in search of fruitless development.

Therefore, it is necessary to separate production projects from research projects, since they have different objectives and products. In addition, they must be supported by different execution and control processes suited to their peculiarities.

3 Framework

3.1 Proposed Framework

The proposed research process involves the framing of academic and domain aspects. Consequently, it must be conducted by a member of the technical staff designated to interact with the researchers. It is mandatory that this member be experienced enough to easily understand and explain the domain.

As in any research process, it is essential to have an overview of the interface between methodological and intended domains in order to:

- Verify similar developments previously studied in the literature;
- Identify which research gaps can be explored; and
- Identify application possibilities.

The verification of similar developments can be done through comprehensive literature reviews in order to avoid rework, saving futile scientific effort.

Research gaps represent unprecedented developments, which may require greater scientific effort, possibly addressed by a doctoral thesis.

The identification of application possibilities establishes a more restricted domain and method, with the need of a specific feasibility test. These application possibilities are embodied in studies, usually leading to an article or a university degree conclusion paper, possibly leading to a prototype.

The importance of this work consists of contributing to two main objectives: validating the possibility of conducting any research and reinforcing CASNAV's reputation as a scientific institution that works to find innovative ideas and validate those ideas with scientific rigor.

Based on validation, it is possible for a development project related to the domain to provide a prototype built based on academic validation to the client, that is, a development proposal with previously defined feasibility, reducing the risk. Furthermore, it could be the starting point of a new contract, at the discretion of the development team and the client.

Preferably, these works should be carried out by students, to broaden the research. Ideally, these students should become part of the development team on a win-win basis: CASNAV will have a professional with prior knowledge in research and the student will have a job to improve his/her knowledge.

This approach allows a small guidance team to reach more goals. The condition to reach these goals is for CASNAV to provide a constant number of students.

With regard to the management of CASNAV, it is important for its scientific research representatives to be able to establish a link between research projects and possible development projects that could benefit from scientific results. In addition, it is important for these guidelines to be connected in a coherent manner to CASNAV's areas of interest, as described in the Brazilian Navy Science and Technology Strategy.

Concerning the contractual form, in order to provide better control, collaborative research projects are established in contracts where CASNAV performs both roles: developer and client.

Preferably, projects dedicated to each methodological area are established, identifying the corresponding partners in academia. Since there are incoming students from CASNAV in each postgraduate program, it is possible to allocate them to the development of projects with partners in the same research area.

The participation of students in these projects meets a threefold objective: using research for the benefit of the Brazilian Navy; increasing CASNAV's scientific production; and contributing to a better evaluation of the Partner Program Graduate Courses (recently, in some academic areas in Brazil, there has been a significant change, increasing the importance of students' academic production).

The structure of goals and objectives of research projects, however, is different from development projects. In the first moment, the construction of the research panorama is embodied in a technical report or in a scientific article, constituting the only tangible objective that can be anticipated.

The identification of gaps and possibilities of use in a research project, on the other hand, is not predictable. Therefore, it is desirable in the early phases of these projects for them to be described in terms of the goals of articles and scientific works, allowing the use of a metric for the evaluation of merit (acceptance and publication) and progress (originality), in addition contributing to the academic production of CASNAV. These goals are also used to organize projects according to human and material resources available.

In terms of temporal aspects, due to the fluidity of the objectives mentioned above, scientific projects have an extended duration (2–3 years), in order to reduce unnecessary bureaucratic burdens. The resources used must be dedicated to the goals of scientific publication, if possible, extended to the professors and students of the partner institutions involved.

Consequently, it is possible, with reduced resources, to exercise guidance in the partner institutions aligned with CASNAV's institutional objectives.

These projects, especially in areas with less staff renewal at CASNAV, should be consolidated in an expeditious manner with renowned partner institutions. The consolidation of more solid institutional projects allows part of CASNAV's scientific and domain knowledge also to be preserved in partner institutions.

From the interface between domains of knowledge, actors emerge to assume responsibility for making the connection between domain and methodology.

From the domain perspective, there must be interface between the development project client and CASNAV, as the manager of the production project, defined in a contract, with well-established goals and deadlines. On the other hand, from the methodological perspective, the student him/herself works as the direct interface, as he/she is immersed in the methodological knowledge, with the corresponding scientific advisor, even without having a comprehensive overview of CASNAV projects or the methodological developments achieved by CASNAV.

There is also the need of another interface between these actors and CASNAV, for the purpose of technical guidance. This task should be performed by the research project coordinator at CASNAV. This technical guidance demands confidence and adequate scientific reputation.

Since the process is heterogeneous in nature, trust and reputation are important aspects. The academic reputation comes individually from scientific production. Therefore, consistent scientific production, including participation at events and publication of works in journals of recognized quality, is the only way to obtain individual scientific reputation for both researchers and institutions.

On the other hand, trust is important and favors the building of a strong reputation through several institutional interactions with favorable results with time, so that positive expectations can be satisfied. Hence, it is important for these relationships to be thought of as long-term bridges, in which there are positive aspects for both sides involved.

CASNAV will increase its doctoral programs (PhD and DSc) to provide coorientation so as to comply with recent changes in the evaluation of postgraduate courses by CAPES (Office to Improve Higher Education Personnel, of the Ministry of Education), especially those related to Engineering III, where new parameters for co-orientation were introduced, based not only in publications, but also on citations (for example, factor h). Thus, it reinforces the need for qualified scientific production of ICT researchers.

3.2 *Processes*

Figure 1 depicts the processes to achieve the stated objectives. The figure only represents the processes logic and is not strictly adherent to BPM notation.

The research project must be initiated by the **Domain and Method Identification**, performed by the CASNAV research coordinator, who will identify a portion of the domain and specific methodological knowledge in which he/she has experience for co-orientation, ideally being related to development projects at CASNAV. Thus, the organizational objective is aligned with scientific development, in a feasible way.

Based on the chosen method, potential **partner selection** is possible, and partnership may be initiated with academic advisors (from universities) according to the vocation of the undergraduate program. The research coordinator formally presents the **research project**.

At the same time, the process of **selecting volunteers** for postgraduate courses is being carried out by CASNAV, involving both the internal selection and the selection of the postgraduate program managers.

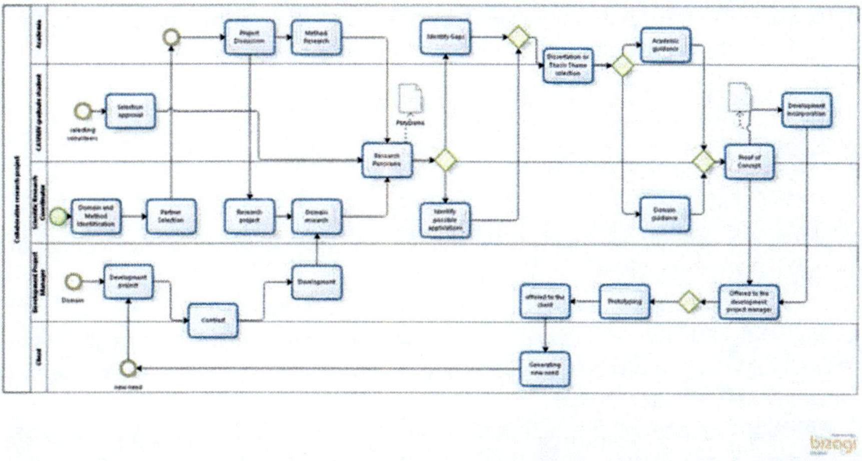

Fig. 1 Framework

Combined with information from the project manager, the research coordinator establishes a **research domain** that will define the application context and general orientation for the objectives of scientific development, while the academic advisor, based on the **research method**, presents the methodological context of the project.

Based on these elements, it is possible for the academic researcher, CASNAV's research coordinator and the student to prepare a description of the **research panorama**, consisting of a joint bibliographic survey of method and domain already present in the literature.

This element should be the first deliverable of research projects and may exist in the form of an article or technical report.

With this panorama in mind, it will be possible to **identify gaps** in the literature, which call for more robust methodological developments, and therefore are candidate topics for doctoral theses.

It is also possible to **identify applications** more geared to the practical needs of development projects, which can be addressed by a master's dissertation.

With the accomplishment of these two processes, it is then possible to **select the thesis theme** with the academic guidance of the university advisor and technical guidance of the research coordinator, as prescribed.

The work can generate a **proof of concept** that exemplifies the use of the method in a particular context, proving or disproving its feasibility. Therefore, this proof of concept is generated in accordance with the tangible product of the research project, which can be included in a scientific article or in the thesis. A viable proof of concept is the step enable **offering the result of the research to the development project manager**.

The development project manager will then have a scientific product developed with reduced risk that, according to his assessment, can go through **prototyping** to be offered to the client. If approved, it will **generate a new development project**, supporting by a contract.

In addition, in this case the development project manager will have the student who participated in the process as a member of the process.

4 Practical Contributions

As practical contributions, there are two ongoing projects following the guidelines of the proposed framework.

The first one is a project with researchers from Pernambuco Federal University (UFPE) and UFF researchers, devoted to the integration of military multicriteria methodology. The second one is devoted to optimized use in the submarine warfare domain, with UFRJ researchers.

Although these projects are in their initial stages, with no dedicated students so far, partly due to COVID issues, they have already produced some relevant results, with coordination officers also performing the students' role. First, there were two submission proposals to obtain funding from agencies (Prodefesa 2018 and Procad Defesa 2019), and papers devoted to use of OR methods in the military domain [9–18].

Moreover, mention should be made of other projects, which despite not being conducted under this framework, are related to enhancing the relationship of CASNAV with the scientific community, in order to identify and build other collaborative projects: CASNAV Special Session on the SBPO 2018; CASNAV Special Session on the SBPO 2019; and CASNAV Special Session on the SBPO 2020.

Those special sessions, devoted to defense and sea power, provided an opportunity to discuss development projects related to the domain of interest of CASNAV, and to meet possible partners for new collaborative projects. This effort will be continued.

CASNAV has just started partnership with the Federal Institute of Minas Gerais (IFMG), which has the potential to become another project using this framework, devoted to ship identification, and has already produced an application [19], and a registered program [20]. Finally, there is another collaboration program, the R package [21, 22], based on [23].

5 Discussion and Conclusion

This proposal is consistent with the necessary technical-normative-legal framework and enables scientific development in line with the CASNAV domain development.

First, regarding the uncertainty aspect [5], the construction of the research landscape is embodied in a technical report or a scientific article, constituting the only objective envisioned. The identification of gaps and possibilities of use, however, is not predictable. Thus, these phases of the research project should be described in terms of the goals of articles and scientific works. These goals address the management control problem, as they make it possible to use a metric for evaluation of merit (acceptance and publication) and progress (originality), in addition to contributing to the academic production of CASNAV.

On the other hand, these projects must be staggered according to the financial and human resources available. Financial resources used in these projects should focus on the production of articles and scientific works with joint benefits to CASNAV and its partners, so that the partnership is perceived as sustainable, and confidence is maintained. Factors such as trust and commitment have been continuously mentioned in the literature as important for successful collaboration [4].

Regarding human resources, the number of students available (approved by both the Brazilian Navy and the partner institution) will also determine the scope of the research project and the respective anticipated results.

Another aspect related to trust that is crucial for the success of this type of project is to start with modest objectives, aiming to meet established commitments, in order to build and maintain solid partnerships with academia [4].

For the execution of collaborative projects, scientific reputation is also an indispensable aspect. Future works can be devoted to study actions for that purpose, including the prioritization of authors' participation at scientific events as well as motivating the publication of collaborative research projects under open access conditions, increasing the possibility of citations.

The resources used in the project must be geared towards the goals of scientific publication, if possible, extended to the professors and students of the partner institutions involved. Thus, with reduced resources, it is possible to establish, in partner institutions, guidance in line with the institutional objectives of CASNAV and the Brazilian Navy.

Finally, it is important to build an organizational structure capable of crystallizing part of CASNAV's organizational knowledge in academia, and the proposed framework is an adequate instrument for this purpose, with practical results.

Acknowledgements The authors acknowledge the CASNAV support.

References

1. Pessôa, L.A.M., da Silva Teixeira, L., Guedes, M.J.M., Martins, E.R., de Souza, A.J.N.: Pesquisa operacional na marinha do brasil: O Casnav, seu passado, presente e perspectivas. (2016)
2. CASNAV: PEO (2016)
3. Wallin, J., Isaksson, O., Larsson, A., Elfström, B.O.: Bridging the gap between university and industry: Three mechanisms for innovation efficiency. International Journal of Innovation and Technology Management 11(01), 1440005 (2014)
4. Barnes, T., Pashby, I., Gibbons, A.: Effective University – Industry Interaction:. European Management Journal 20(3), 272–285 (Jun 2002)
5. Brocke, J.v., Lippe, S.: Managing collaborative research projects: A synthesis of project management literature and directives for future research. International Journal of Project Management 33(5), 1022–1039 (Jul 2015)
6. Rousseau, D.M., Sitkin, S.B., Burt, R.S., Camerer, C.: Not so different after all across-discipline view of trust. Academy of Management Review 23, 393–404 (1998)
7. Pessôa, L.A.M., Collazo, R.A.: Metodologia para identificação de competências necessárias em pesquisa científica na marinha do brasil (2018)
8. Pessôa, L.A.M.: Adaptação Metodológica para GPC EM Pesquisa Científica na MB. Monografia, Escola de Guerra Naval (2016)
9. Pessôa, L.A.M., Silva, R.C., Ferreira, R.J.P., Costa, H.G.: Proposta de aprimoramento do modelo CPC, baseado em FITradeoff durante um jogo de guerra. In: SBPO (2020)
10. Pessôa, L.A.M., Costa, H.G., Mota, C.M.d.M.: Multicritério e Programação Linear Inteira em auxílio à decisão de projetos de ciência e tecnologia na Marinha do Brasil. In: SBPO (2020)
11. Pessôa, L., Ferreira, R.J.P., Lage, C.P.M., de Almeida, A.T.: Understanding aDecision Policy: Using Fitradeoff as a Mirror for a Decision Maker. In: The 25th International Conference on Multiple Criteria Decision Making. Istambul (2019)
12. Pessôa, L.A.M., de Arruda, E.F., Bahiense, L.: Panorama of the use of Operational Research in the submarine war environment. Revista Pesquisa Naval 31, 36–42 (2019)
13. Silva, R.C., Pessôa, L.A.M., Ferreira, R.J.P., Costa, H.G., Almeida, A.T.d.: Uma aplicação do método fitradeoff na comparação de poderes combatentes de unidades de superfície. In: Simpósio de Pesquisa Operacional e Logística da Marinha. pp. 2905–2920 (2019)

14. Pessôa, L.A.M., Arruda, E.F., Bahiense, L.: Um panorama do uso de pesquisa operacional no ambiente de guerra submarino. In: Simpósio de Pesquisa Operacional e Logística da Marinha - Publicação Online. vol. 3, pp. 2865–2876 (Nov 2019)
15. Costa, H.G., Roboredo, M.G., Pessôa, L.A.M.: Metodologias Multicritério no campo militar: um panorama do domínio tático/operacional. In: SBPO (2019)
16. Guedes, M.J.M., Pessôa, L.A.M., Collazo, R.A.: Computer aid for laying submarine artifacts based on set covering. Revista Pesquisa Naval 30, 39–45 (2018)
17. Pessôa, L.A.M., Collazo, R.A.: Methodology for Scientific research competence identification in the Brazilian Navy. Revista pesquisa Naval 30, 46–52 (2018)
18. Botelho, T.A.T., Pessôa, L.A.M., Ferreira, R.J.P., de Almeida, A.T.: Aplicação do método multicritério FITradeoff para escolha de obuseiro para batalhão de artilharia de Fuzileiros Navais. In: XLIX Simpósio Brasileiro de Pesquisa Operacional (2017)
19. dos Santos, F.H.W., Pessôa, L.A.M., Oliveira, B.A.S., Vieira, L.M., Cardoso, G.M.M.: Rastreamento de embarcações em imagens satelitais utilizando metodologia multicritério para a priorização em tarefas de busca e salvamento. Brazilian Journal of Development 6(5), 28245–28257 (2020)
20. dos Santos, F.H.W., Vieira, L.M., Oliveira, B.A.S., Pessôa, L.A.M.: Satellite Imagery Ship Detection (2020)
21. Taranti, P.G., Pessôa, L.A.M., Cosenza, C.A.N.: coppeCosenzaR (2017)
22. Taranti, P.G., Cosenza, C.A.N., Pessôa, L.A.M., Collazo, R.A.: coppeCosenzaR: A Hierarchical Decision Model. Software X 17, 100899 (2022)
23. Cosenza, C.A.N., Doria, F.A., Pessôa, L.A.M.: Hierarchy Models for the Organization of Economic Spaces. Procedia Computer Science 55, 82–91 (2015)

Part II
Production Process Innovation and New Technologies

Assessing the Attractiveness of Onshore Wind and Solar Photovoltaic Sources in Brazil

Marcelo Casagrande and Erick Meira

Abstract This paper analyzes the economic competitiveness of alternative sources of renewable energies in Brazil, notably onshore wind and solar photovoltaic. The LCOE (Levelized Cost of Energy) metric is used as criterion for comparison with other electricity generating sources present in the Brazilian energy matrix. Different time horizons (short, medium and long term) are considered. In doing so, the study aims to understand the current situation of competitiveness and investment attractiveness of wind and solar energies, and identify possible avenues for the development of these sources over time. The findings in both the medium and long term corroborate the hypothesis of cost competitiveness gains with technological advances and economies of scale in the sector.

Keywords Renewable energies · LCOE · Energy planning

1 Introduction

Mankind has predominantly used fossil fuels to generate energy. The main arguments for their dissemination in the last century were the advantages over renewable sources, such as greater efficiency and storability. However, fossil fuels emit greenhouse gases (GHGs) when they generate energy, making them accountable for global warming. Extreme weather conditions, rising ocean levels and higher atmospheric temperatures are some of the harmful consequences that humanity must face due to global warming [1].

M. Casagrande
Federal University of Rio de Janeiro, Rio de Janeiro, Brazil

E. Meira (✉)
Pontifical Catholic University of Rio de Janeiro, Rio de Janeiro, Brazil

Brazilian Agency for Research and Innovation, Rio de Janeiro, Brazil
e-mail: research@erickmeira.com

J. Vargas Florez et al. (eds.), *Production and Operations Management*, Springer Proceedings in Mathematics & Statistics 391,
https://doi.org/10.1007/978-3-031-06862-1_6

In this context, energy transition, which comprises the transition from a carbon-dependent economy to an increasingly decarbonized, sustainable economy, is key. A consensus has already been reached at the international level, such as the guidelines towards environmental preservation policies [2]. Under the Paris Agreement of 2015, for instance, to achieve the goal of controlling the increase in global temperature by 1.5 °C or, at most, 2 °C, one of the most important actions is the reduction of GHGs emissions. In this context, incentive mechanisms for cleaner energy matrices is a pressing priority for policy makers. Brazil, as a signatory, should anchor its policies on guidelines and targets that support the agreement, one of them being the investment in renewable energies.

Renewable energies can be defined as energy sources that do not exhaust. Examples include wind, solar, biomasses, geothermal and hydraulic sources. The latter has been vastly explored in Brazil since the nineteenth century, with the first hydroelectric power plant being built in 1889 [3]. Figure 1 illustrates the importance of hydraulic energy sources in Brazil: in 2019, their share accounted for 60.5% of the energy in the Brazilian energy matrix, followed by wind sources, with 9% [4]. Being a fully established form of generation, with substantial investments made since the nineteenth century and accounting for the largest share in terms of installed capacity for decades, hydroelectric plants are not considered here as an alternative renewable energy source with potential for strong cost reduction in the long term, as opposed to wind and solar plants. Today, solar energy represents 1.5% of the total installed capacity [4].

In terms of GHG emissions, the Brazilian energy matrix is considerably competitive when compared to energy matrices from other global economies. For instance, in Brazil, the share of fossil fuels in electricity generation is estimated at 14.3% [4], while the world average revolves around 64.2% [5]. However, this fact should not

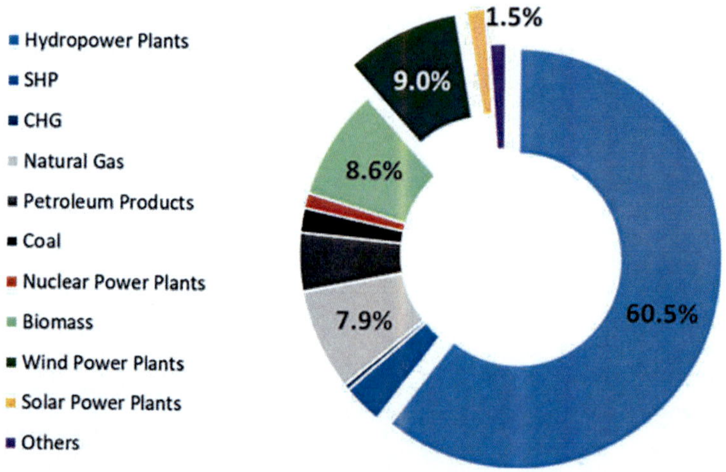

Fig. 1 Installed Capacity in the Brazilian Energy Matrix in 2019. (Source: EPE (Empresa de Pesquisa Energética) [4])

discourage the constant search for more clean energy resources. They can complement hydroelectric plants in ensuring energy security, without requiring more polluting thermoelectric plants to fulfill this function. Particularly noteworthy are wind energy, which presents the highest global Capacity Factors (CFs)[1] in Brazil, and solar photovoltaic energy, which also shows promising potential given the high incidence of direct radiation in certain Brazilian regions, coupled with the possibility of decentralized generation, as in micro and minigeneration projects [6].

In light of the above, this paper aims to assess the competitiveness of wind and solar photovoltaic sources in Brazil, whilst predicting their possible growth in the next decades. For being one of the most relevant factors in investment decisions, particular attention is given to cost competitiveness, comparing the costs associated with wind and solar power with those from other energy sources in Brazil. For this purpose, the LCOE (Levelized Cost of Energy) metric is adopted, which can be applied to any source of energy generation in order to make their costs comparable to each other. It is expected to find values that converge with the hypothesis of technological advances and economy of scale gains for these renewable sources in the long term.

Besides this introduction, the article is divided into four other sections. Section 2 shows how official data is obtained and processed to generate the results. Section 3 presents the results for the LCOE computed in the short-term, i.e., up to 2030. The estimates for the LCOE in the medium (2030–2040) and long-term (2040–2050) are depicted in Sect. 4. Finally, Sect. 5 concludes and suggests directions for future studies.

2 Data and Methods

To investigate the potential increase of onshore wind and solar photovoltaic generation in Brazil in the next decades, official data provided by the Brazilian Energy Research Office (EPE) is considered. A comparison in terms of cost competitiveness with other energy sources is conducted using data from their capital expenditures (CAPEX), their operational costs (OPEX) and the concept of LCOE (Levelized Cost of Energy). Simply put, CAPEX can be understood as the initial investment and purchase of capital goods, OPEX as ongoing maintenance costs and LCOE as the sum of the two parameters (or the total cost of a plant) divided by the total power generation over a plant's lifetime.

The LCOE is a widely accepted and widespread metric in the literature – see, for example, [7–9]. According to IRENA, the LCOE is defined as the investment required to receive a rate of return equal to the discount rate applied over the entire lifetime of an energy plant, without taking into account tax expenditures and

[1] Capacity factor represents the proportion between the real (observed) electricity production of a power plant and its total production capacity.

inflation [10]. Thus, the LCOE can be calculated according to the following equation, taken from [11]:

$$LCOE = \frac{CAPEX + \sum_{t=1}^{T} \frac{OPEX_t}{(1+WACC)^t}}{\sum_{t=1}^{T} \frac{AEP_t}{(1+WACC)^t}} \tag{1}$$

where CAPEX represents the capital expenditure and OPEX represents the operational expenditure. AEP stands for the annual energy production of a plant and WACC is the discount rate adopted (Weighted Average Cost of Capital). The LCOE thus provides an estimate of how much money is required to produce 1 MWh of electricity.

CAPEX and OPEX data for the different energy sources considered were collected from two official studies conducted by EPE: the Ten-Year Energy Expansion Plan 2030 [12] (henceforth, PDE as in its Portuguese acronym), and the National Energy Plan 2050 [13] (henceforth, PNE). The objective of these studies is to supply data for energy policy formulation and present scenario predictions about the future expansion of the Brazilian energy sector [14, 15]. Furthermore, to calculate the AEP for the different sources, we used the official data available at the 2020 Statistical Yearbook of Electricity [4], also prepared by the EPE. Table 1 summarizes the collected data.

Regarding the discount rate, two different scenarios were considered. In the first scenario (hereafter referred to as "Future DI"), aiming to represent the reality of the Brazilian economy as closely as possible, the annual projections for the future interest rates were estimated from the rates of futures contracts for the Interbank Deposit rate (DI, in Portuguese acronym), traded in the Brazilian stock exchange. These rates are depicted in Table 2. In the second scenario (identified as "WACC"), we adopted a fixed discount rate of 10% p.a. as a simplifying measure, in line with recent works [11].

As can be noted, not all years are associated with Future DI contracts ending on January 1st. Thus, to avoid compromising the LCOE calculations, a simple linear interpolation was conducted for the year 2032, that is, we considered for this year's annual rate an arithmetic average between the rate of the previous year and that of the

Table 1 Annual Energy Production (AEP), Installed Capacities (Inst Cap) and Capacity Factors (CF) for different sources in Brazil

Energy source	AEP (MWh)[a]	Inst Cap (MW)[b]	CF (b)
Nuclear	16,128,820	1990	90%
Coal	15,327,230	3228	69%
Natural Gas	60,188,290	13,385	65%
Hydro	375,770,910	102,999	55%
Wind	55,985,620	15,378	47%
Biomass	52,111,160	14,703	33%
Solar	6,650,540	2473	30%

Sources: [a]EPE (Empresa de Pesquisa Energética) [4]. [b]EPE (Empresa de Pesquisa Energética) [13]

Table 2 Selected DI Future Rates

t (years)	Ticker	Due date	Interest rate (% p.a.)
1	DI1F22	01/01/2022	4.975
2	DI1F23	01/01/2023	6.610
3	DI1F24	01/01/2024	7.520
4	DI1F25	01/01/2025	8.010
5	DI1F26	01/01/2026	8.330
6	DI1F27	01/01/2027	8.630
7	DI1F28	01/01/2028	8.810
8	DI1F29	01/01/2029	9.000
9	DI1F30	01/01/2030	9.220
10	DI1F31	01/01/2031	9.240
11	Linear interpolation		9.275
12	DI1F33	01/01/2033	9.310

Source: Infomoney [16] (Accessed on May 27th 2021)

following year. From 2033 onwards, a constant rate of 9.310% p.a. was adopted, the same as observed in the last available year. This was necessary because the service life of the analyzed power plants is at least 20 years, a duration that exceeds the limit of estimates for future DI contracts.

Based on the aforementioned data, the CAPEX, OPEX and LCOEs of the different energy sources considered were computed. When the costs of wind and solar energy are compared to those of other sources, renewable or not, they should theoretically be attractive for investments. On the other hand, one must also take into account the advantages and disadvantages that investing in these sectors can provide, not only in relation to the expansion, security and productivity of electricity generation in Brazil [17], but also in terms of their sustainability, such as impacts on ecosystems, climate and local communities and economies.

We hasten to add that several caveats exist regarding the use of LCOE as a measure of economic attractiveness. First, multiple metrics for this calculation are available in the literature, with different institutions and authors adopting different variables, which limits the comparison between different data sources. Exchange rate variations can also make it difficult to compare LCOEs from different periods, as they directly affect the associated costs [18]. Another criticism comes from the fact that the LCOE disregards any environmental or social externalities, such as waste and pollution arising from energy generation, which can directly impact the decision to prioritize renewable and clean energy sources [19].

3 Short-Term LCOE Results

Table 3 summarizes the CAPEX per installed kW, the OPEX spent per kW in each year and the expected service life of each type of plant selected, according to the corresponding energy source. As previously outlined, all data were obtained from the PDE 2030 and PNE 2050, prepared by EPE. It should be noted, however, that the costs calculations expressed in the PDE 2030 considered the dollar quoted at R$ 4.90. As this relationship does not reflect nowadays the observed exchange rate, an adaptation was made using the median of the market projections expressed in the Focus Bulletin, published on a weekly basis by the Brazilian Central Bank (BCB). The May 24th, 2021 edition was used as reference, arriving at the expectation of an exchange rate priced at R$5.30/US$ at the end of 2021. This explains the difference between CAPEX and OPEX data presented in Table 3 and those depicted in PDE 2030. In addition, the Focus Bulletin provided exchange rates forecasts for the years 2022, 2023, 2024 and 2025 (R$5.30; R$5.30; R$5.08; and R$5.05, respectively) [20]. These were used to calculate the OPEXes for the same years. Due to the lack of data for the following years, the last exchange rate was extended and used as an approximation. This exchange rate variation was imposed on OPEX because this is the variable cost component, applicable to each year of the service life of each power plant. Meanwhile, CAPEX is a fixed component, applicable only in the initial year.

Finally, the last column in Table 3 presents the total investment per kW necessary for the whole service life of a power plant. This is obtained by summing the CAPEX with the resulting multiplication of the OPEX by the years of each power plant service life, as a preliminary comparison. The results indicate that the nuclear source stands today as the most expensive energy source with respect to the capital expenditure necessary to build and operate the energy plant. On the other hand, the least expensive source in this preliminary perspective is the solar photovoltaic.

As can be noted, Table 3 does not take into account the annual energy production to provide a fairer comparison, as considered in the LCOE. Since the CF of the solar photovoltaic energy source is very low (30%), when we consider the comparison using the LCOE, this source loses several positions in terms of economic

Table 3 CAPEX and OPEX of the energy sources considered in the study

Energy source	CAPEX/kW[a]	OPEX/kW/year[b]	Service life (years)[b]	Total
Nuclear	R$ 26,500.00	R$ 530.00	30	R$ 42,400.00
Coal	R$ 10,600.00	R$ 173.06	25	R$ 14,926.53
Hydro	R$ 9663.88	R$ 43.27	30	R$ 10,961.84
Natural Gas (CC)	R$ 4434.69	R$ 173.06	20	R$ 7895.92
Wind	R$ 4867.35	R$ 97.35	20	R$ 6814.29
Biomass (Cane)	R$ 4326.53	R$ 97.35	20	R$ 6273.47
Solar Photovoltaic	R$ 4326.53	R$ 54.08	20	R$ 5408.16

Sources: [a]EPE (Empresa de Pesquisa Energética) [12]. Adapted for an exchange rate of R$ 5.30/US $. [b]EPE (Empresa de Pesquisa Energética) [13]
Notes: CC stands for Combined Cycle

Table 4 Short-term LCOE for different sources in Brazil

	R$/MWh		US$/MWh	
Energy source	Future DI	WACC	Future DI	WACC
Nuclear	382.10	409.64	72.09	72.09
Coal	270.08	292.43	50.96	50.96
Hydro	262.25	280.93	49.48	49.48
Natural Gas (CC)	194.76	208.34	36.75	36.75
Wind	171.45	182.74	32.35	32.35
Biomass (Cane)	159.48	169.78	30.09	30.09
Solar Photovoltaic	144.51	152.83	27.27	27.27

Source: The authors
Notes: CC stands for Combined Cycle

competitiveness, as can be observed in Table 4. The Natural Gas (in the combined cycle generation) is the most competitive energy source from the point of view of the LCOE. This is an advantage for attracting investments in this fuel: besides being efficient in terms of power generation, natural gas is also considered as a less polluting energy source among the fossil fuels [21]. In addition, power plants that operate using natural gas are considered as base energy plants, as they do not rely on intermittent sources like wind and solar. On the other hand, one should note that generating energy through natural gas power plant still contributes to global warming. Therefore, despite being less expensive than other energy sources, natural gas prioritization should be relativized, taking into account the environmental advantages of investing in alternative renewable sources.

For illustration purposes, Fig. 2 shows the short-term LCOEs computed for each energy source, in R$/MWh and considering the 'Future DI' scenario (first column of Table 4).

As can be noted from Table 4 and Fig. 2, onshore wind and solar photovoltaic do not currently rank among the most cost competitive energy sources in the Brazilian market from the viewpoint of the LCOE, at least in the short-term. However, with several technologies still to be improved to generate electricity more efficiently, along with the potential reductions in CAPEX and OPEX, this scenario may change considerably in the next years.

For now, despite these sources not being as economically competitive as some competing alternatives, they cannot be readily labelled as unattractive options. For instance, considering only clean energy sources, wind and solar are only less competitive than biomass (sugarcane-based) power plants. This is a sign that the sources are heading towards more favourable cost competitiveness situations.

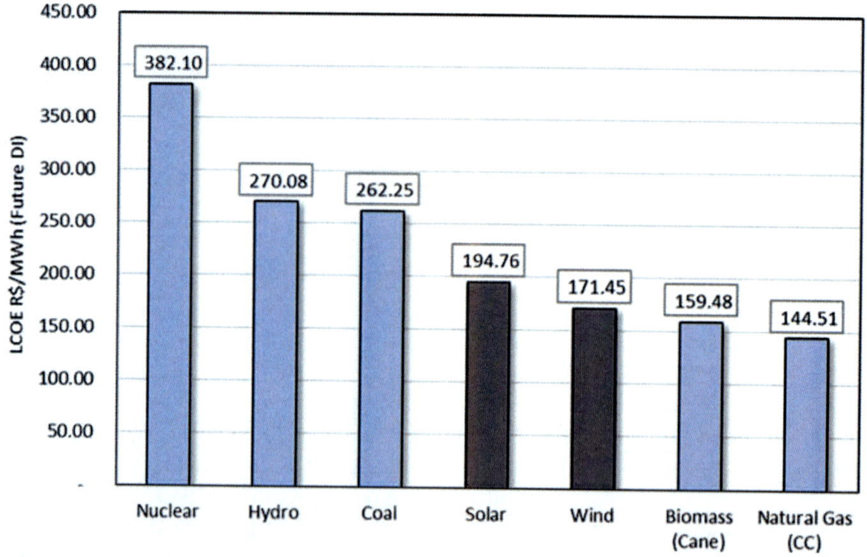

Fig. 2 Short-term LCOEs for the different energy sources considered in Brazil

Table 5 CAPEX and OPEX values for onshore and solar photovoltaic energy sources in the short, medium and long term in Brazil

	CAPEX (R$/kW)		
Energy source	**2020–2030(a)**	**2030–2040(b)**	**2040–2050(b)**
Onshore Wind	4867.35	3853.32	3536.94
Solar Photovoltaic	4326.53	3502.43	2641.24
	OPEX (R$/kW)		
Energy source	**2020–2030(a)**	**2030–2040(b)**	**2040–2050(b)**
Onshore Wind	97.35	97.35	97.35
Solar Photovoltaic	54.08	43.27	32.45

Sources: (a) EPE (Empresa de Pesquisa Energética) [12]. (b) The authors, based on official data from EPE [13]

4 Medium and Long Term LCOEs

To analyze the cost reduction potential in wind and solar generation in Brazil in the next decades, we also computed the LCOEs of these sources for the periods comprising the years from 2030 to 2040 (medium term) and the years from 2040 to 2050, (long-term). The PNE 2050 [13] provides estimates of how much the CAPEX and OPEX costs should decrease in these periods, which was considered in the calculations. Hence, based on data from the PDE 2030 for the 2020–2030 period (short term, as presented in the previous section), and taking into consideration the perspectives of costs reductions suggested in the PNE 2050, the following CAPEX and OPEX values were obtained, as depicted in Table 5:

Table 6 Medium and long term LCOEs for onshore wind and solar photovoltaic in Brazil

	R$/MWh		US$/MWh	
Energy source	2030–2040	2040–2050	2030–2040	2040–2050
Onshore Wind	151.06	140.85	27.47	25.61
Solar Photovoltaic	169.06	127.43	30.74	23.17

Source: The authors

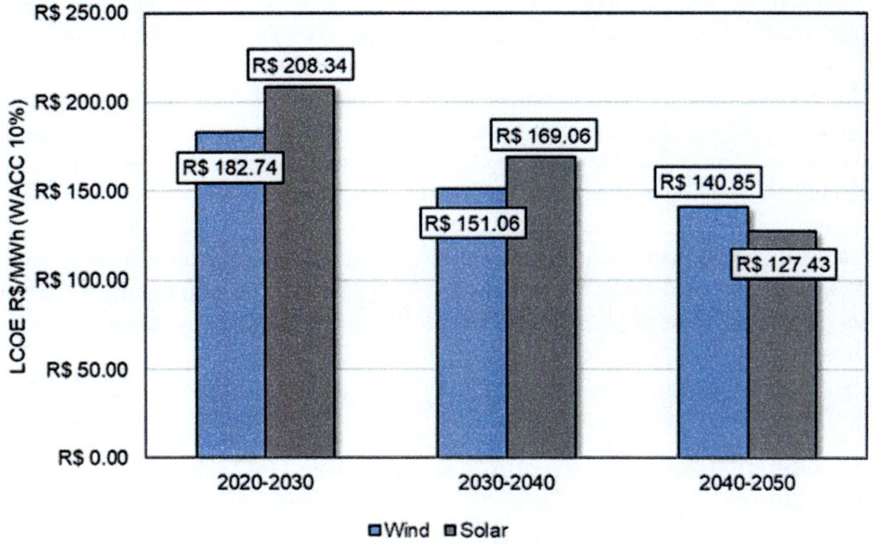

Fig. 3 Short, medium and long-term LCOEs for onshore wind and solar photovoltaic energy sources in Brazil. (Source: The authors)

Among all energy sources herein considered, those with greater prospects of cost reductions according to the 2050 PNE are wind and solar photovoltaic. This perspective is justified because they are renewable sources that have not yet reached their full maturity and expected gains of scale, unlike the established sources in the Brazilian energy matrix. These, notably fossil fuels, hydro, nuclear and biomass, should maintain their cost levels or suffer little significant reductions.

The R$5.05/US$ exchange rate was adopted, according to the longest forecast for this variable available in the Brazilian Central Bank Focus Bulletin of May 24th, 2021. Based on the new data and considering only the WACC discount rate of (10% p.a.) since the Futures DI projections would be insufficient for long term projections, the new LCOEs were computed for two time horizons: the medium term, i.e., the years between 2030 and 2040, and the long term, i.e., the years between 2040 and 2050. The results are depicted in Table 6 and in Fig. 3.

The results indicate that both onshore wind and solar photovoltaic energy sources will gain competitiveness over time. The gains in scale of the national and global industry should contribute to the reduction in CAPEX and OPEX costs. Hence, lower investments will be required to install the same generation capacity. Despite the significant cost reductions, the simplifying hypothesis of Capacity Factors (CFs) remaining unchanged over time was adopted. However, CFs are also likely to improve with technological advances and the better use of wind towers and photovoltaic cells in energy production [8], leading to higher amounts of electricity production. As a consequence, the competitiveness gains for wind and solar power in Brazil may be even higher than herein estimated.

Overall, the results indicate that solar photovoltaic generation will be the least expensive among all sources, renewable or not, in the long term. Onshore wind power, in turn, is likely to remain more competitive in the medium term, i.e., until 2040.

Compared to the other sources, whose levelized costs are likely to remain in the same levels as observed in the short-term, both solar and wind power should become considerably more attractive for long term investments. If electricity generation via biomass and natural gas keep their short-term costs, both above R$ 140.00 according to the 'Futures DI' scenario, they will become less attractive than solar and wind energy in Brazil. In this case, not only the argument of supporting the combat against climate change will favor towards the latter, but also the economic-financial estimates of lower costs and greater competitiveness, since it will be cheaper to invest in these sources than in any other energy resource.

5 Conclusions and Directions for Further Research

The importance of wind and solar power is likely to increase in Brazil in the next decades, following the global trend of energy transition. Even though the country already shows a low-carbon energy matrix, the largest share is attributed to hydraulic sources, which show signs of exhaustion in terms of new installed capacity and are dependent on weather conditions, being subject to climate events such as droughts. With diversification as a goal to be followed in energy expansion policies, renewable alternatives should play a central role in this process. This article aimed to demonstrate, with official data and reasonable estimates, at which pace and how the cost competitiveness of wind and solar sources should evolve in Brazil, thus providing important insights for investors and other stakeholders, who may balance the estimates herein provided in their decisions.

In the short-term, the levelized costs of some traditional energy sources are still lower than those observed for wind and solar power in Brazil. For instance, considering the 'Futures DI' scenario, the LCOEs for onshore wind and solar photovoltaic were estimated at R$171.45/MWh and R$194.76/MWh, respectively. When compared with the least expensive source according to the same calculation, that is, natural gas in combined cycle (R$144.51/MWh), the differences are still considerable, with wind and solar being approximately 19% and 38% more expensive, respectively.

In the long term, the situation is expected to change. Wind and solar power will become considerably more competitive thanks to several technological advances and the gains of scale, as these sources are likely to gain more protagonism in a global scale. From the short to the medium term, solar PV should benefit from a drop of almost 19% in its LCOE and another 25% from the medium to long term. This would result in a decrease of approximately 39% from the short to the long term, a considerable gain that would make solar photovoltaic the most competitive source in the Brazilian energy matrix until 2050. Onshore wind, in turn, is likely to experience a 17% decrease in its levelized costs from the short to the medium term and another 7% decrease from the medium to the long term. Therefore, by 2050, the costs are likely to decrease by approximately 23% considering the current cost levels.

From the results we can infer that alternative renewable sources in Brazil are likely follow the global trend towards considerable gains in competitiveness over the years. With the threat of global warming and the need to invest in environmentally friendly sources, coupled with the exhaustion of traditional hydro sources in Brazil, wind and solar energy are likely to increase considerably in importance, bringing diversification, and ensuring security of supply in the energy matrix. Dependence on hydro resources has been an ongoing issue due to the droughts [22, 23], and this situation is expected to occur again in 2021 [24, 25]. By investing in alternative renewable sources, this situation may change significantly.

In the short term, despite being more competitive than coal-based thermoelectric power, hydro and nuclear power plants, wind and solar energy still need impulses from energy policies to become more attractive, either by credit facilitation for renewable alternatives or by discouraging the production of energy from polluting sources. In the long term, on the other hand, the considerable decrease in expected levelized costs tends to make wind and solar sources naturally more attractive. This should be translated into greater generation capacities and consequently larger gains in representativeness for these sources, with even greater prominence of wind power, which already accounts for the second share in terms installed capacity in Brazil and holds the first position in the Northeast, the region in which wind power is the most competitive. Solar energy, in turn, should also gain increasing importance, thanks to the ongoing trend towards decentralisation of electricity generation and the greater ease to install micro and mini generation plants.

Finally, some improvements can be pointed out for future studies. Firstly, it is worth noting that although studies based on LCOE can be useful in comparing competitiveness among energy sources, investors and policy makers should keep in mind their limitations, balancing other factors in decision making. For instance, geographic, regulatory and public acceptance conditions are also important factors and may be critical in some cases. Another important point, which may stimulate greater interest in investing in clean technologies, is the pricing of environmental impacts. With increasing global environmental concerns, firms and regulators should estimate the 'environmental risk' of investing in carbon intensive technologies, which would result in greater competitiveness gains for renewable sources such as wind and solar power.

References

1. IPCC (Intergovernmental Panel on Climate Change) (2018). Impacts of 1.5°C Global Warming on Natural and Human Systems. In: Global Warming of 1.5°C. Capítulo 3. URL https://www.ipcc.ch/sr15/.
2. Oliveira EMD (2015) Corporate social responsibility and firm performance: A case study from the Brazilian electric sector. Pontifical Catholic University of Rio de Janeiro (PUC-Rio), Rio de Janeiro, Brazil. https://doi.org/10.17771/PUCRio.acad.25647
3. Canal Energia (2020). Primeira hidrelétrica construída no Brasil deixa de operar para o SIN. https://www.canalenergia.com.br/noticias/53131200/primeira-hidreletrica-construida-no-brasil-deixa-de-operar-para-o-sin.
4. EPE (Empresa de Pesquisa Energética) (2020). Anuário Estatístico de Energia Elétrica 2020. URL https://www.epe.gov.br/pt/publicacoes-dados-abertos/publicacoes/anuario-estatistico-de-energia-eletrica.
5. IEA (International Energy Agency) (2018). Global share of electricity generation, 2018. URL https://www.iea.org/data-and-statistics/charts/global-share-of-electricity-generation-2018.
6. Castro N et al. (2018). Perspectiva da Energia Eólica Offshore. URL http://gesel.ie.ufrj.br/app/webroot/files/IFES/BV/castro184.pdf.
7. IRENA (International Renewable Energy Agency). (2019). Renewable Power Generation Costs. URL https://www.irena.org/publications/2020/Jun/Renewable-Power-Costs-in-2019.
8. IEA (International Energy Agency). (2020). Projected Costs of Generating Electricity 2020. URL https://www.iea.org/reports/projected-costs-of-generating-electricity-2020.
9. EIA (U.S. Energy Information Administration). (2021). Levelized Costs of New Generation Resources in the Annual Energy Outlook 2021. URL https://www.eia.gov/outlooks/aeo/pdf/electricity_generation.pdf.
10. IRENA (International Renewable Energy Agency). (2016). Innovation Outlook: Offshore Wind. URL https://irena.org/publications/2016/Oct/Innovation-Outlook-Offshore-Wind.
11. Dos Reis M et al. (2021). Economic analysis for implantation of an offshore wind farm in the Brazilian coast. Sustainable Energy Technologies and Assessments, v. 43, 100955. doi:https://doi.org/10.1016/j.seta.2020.100955.
12. EPE (Empresa de Pesquisa Energética) (2021). Plano Decenal de Energia 2030. URL https://www.epe.gov.br/pt/publicacoes-dados-abertos/publicacoes/plano-decenal-de-expansao-de-energia-2030.
13. EPE (Empresa de Pesquisa Energética) (2020). Plano Nacional de Energia 2050. URL https://www.epe.gov.br/pt/publicacoes-dados-abertos/publicacoes/Plano-Nacional-de-Energia-2050.
14. De Oliveira EM, Cyrino Oliveira FL (2018) Forecasting mid-long term electric energy consumption through bagging ARIMA and exponential smoothing meth-ods. Energy 144:776–788. https://doi.org/10.1016/j.energy.2017.12.049
15. Meira E, Cyrino Oliveira FL, de Menezes LM (2021) Point and interval forecasting of electricity supply via pruned ensembles. Energy 232:121009. https://doi.org/10.1016/j.energy.2021.121009
16. Infomoney (2021). Cotações – Juros Futuros. https://www.infomoney.com.br/ferramentas/juros-futuros-di/. Accessed: 05/27/2021.
17. Oliveira EMD (2020) Getting the most out of the wisdom of the crowds: Improving fore-casting performance through ensemble methods and variable selection techniques. Pontifical Catholic University of Rio de Janeiro (PUC-Rio), Rio de Janeiro, Brazil. https://doi.org/10.17771/PUCRio.acad.48429
18. Johnston B et al. (2020). Levelized cost of energy: A challenge for offshore wind. Renewable Energy. URL: https://doi.org/10.1016/j.renene.2020.06.030.
19. Guimarães L (2019). O Custo Nivelado da Eletricidade e seu Impacto na Transição Energética. FGV Energia. URL https://fgvenergia.fgv.br/opinioes/o-custo-nivelado-da-eletricidade-e-seu-impacto-na-transicao-energetica-0.

20. BCB (Brazil Central Bank) (2021). Consolidated Statistics Series. https://www3.bcb.gov.br/expectativas/publico/consulta/serieestatisticas. Accessed: 27/05/2021.
21. Meira E, Cyrino Oliveira FL, De Menezes LM (2022) Forecasting natural gas consumption using bagging and modified regularization techniques, Energy Economics, 106:105760. https://doi.org/10.1016/j.eneco.2021.105760
22. Dutra R, Szklo A (2006). A Energia Eólica no Brasil: Proinfa e o novo modelo do Setor Elétrico. In: Anais do XI Congresso Brasileiro de Energia-CBE, p. 842–868.
23. Diniz T (2018). Expansão da Indústria de Geração Eólica no Brasil: Uma Análise à Luz da Nova Economia das Instituições. Planejamento e Políticas Públicas, n. 50, p. 233–255.
24. Infomoney (2021). Especialistas falam em risco de apagão energético; veja impactos da crise de energia na Bolsa e na economia. https://www.infomoney.com.br/mercados/especialistas-falam-em-risco-de-apagao-energetico-veja-impactos-da-crise-de-energia-na-bolsa-e-na-economia/. Accessed: 02/06/2021.
25. Valor Econômico (2021). Crise hídrica se agrava e cresce risco de blecautes. https://valor.globo.com/brasil/noticia/2021/05/31/crise-hidrica-se-agrava-e-cresce-risco-de-blecautes.ghtml. Accessed: 02/06/2021.

Forecasting Total Hourly Electricity Consumption in Brazil Through Complex Seasonality Methods

Erick Meira, Fernando Luiz Cyrino Oliveira, and Paula Maçaira

Abstract Accurate electricity demand forecasting, especially in the short run, is critical for the optimal management of power systems. However, high-frequency time series usually contain unique stylized facts, such as multiple seasonal patterns, which imposes an extra challenge in accurate estimation and forecasting. Aiming to contribute to the reliable operation and planning of the Brazilian National Interlinked System (SIN), this work compares the accuracy of several statistical methods when forecasting multiple series of hourly electricity consumption in Brazil up to 7 days (168 h) ahead. Both benchmark methods and complex seasonality models, i.e., methods that consider multiple seasonal patterns of the underlying series, are assessed. We provide robust evidence towards the adherence of two selected complex seasonality models for the Brazilian total electricity consumption in the short term. Results and implications in terms of decision-making are further discussed.

Keywords Forecasting · Complex seasonality · Electricity consumption · Energy planning · Power systems

1 Introduction

Ensuring an adequate energy supply is a pressing national priority in virtually every country. The issue is even more crucial in electricity supply because, unlike other sources, it cannot be stored for large-scale consumption. From an operational point of view, electricity supply must meet the demand at any instant of time. Otherwise,

E. Meira (✉)
Pontifical Catholic University of Rio de Janeiro, Rio de Janeiro, Brazil

Brazilian Agency for Research and Innovation, Rio de Janeiro, Brazil
e-mail: research@erickmeira.com

F. L. Cyrino Oliveira · P. Maçaira
Pontifical Catholic University of Rio de Janeiro, Rio de Janeiro, Brazil
e-mail: cyrino@puc-rio.br; paulamacaira@puc-rio.br

J. Vargas Florez et al. (eds.), *Production and Operations Management*, Springer Proceedings in Mathematics & Statistics 391,
https://doi.org/10.1007/978-3-031-06862-1_7

the consequences of over or underestimation can be costly [1, 2]. For example, when the delivered load is higher than the current demand, the supplier not only wastes resources but may also bear high costs due to the strong regulation of spot energy markets in certain countries. On the other hand, underestimation naturally results in blackouts and rationing, leading to productivity losses and subjecting the supplier to sanctions and penalties.

In the last decades, a considerable number of approaches have been proposed to adequately estimate the dynamics of electricity demand and predict its future behavior [3, 4]. Roughly speaking, the methods can be divided into two broad categories, depending on the forecast horizon: short and mid/long-term. Mid and long-term forecasts are necessary for proper maintenance planning and scheduling. On the other hand, short-term load forecasts are required for the controlling and programming of power systems [5]. In addition, forecasts considering this time horizon are required by transmission companies in cases where a self-dispatch market is in operation.

However, the inherent complexity of short-term electricity consumption data calls into question the effectiveness of traditional methods in providing accurate forecasts, especially when it is necessary to estimate future demand several steps ahead. This is because high-frequency consumption time series usually present stylized facts that distinguish them from other series, such as multiple seasonal patterns. Furthermore, energy consumption time series are typically influenced by external (exogenous) factors, such as the impact of macroeconomic variables and strategic government decisions, which may lead to one or more outliers and/or structural breaks [6]. In this context, the so-called complex seasonality methods, i.e., approaches that allow for multiple seasonal patterns in time series modeling and forecasting, assume a strategic role, aiding in accurate short-term decision-making.

In light of the above, this work assesses the accuracy of different approaches when forecasting total hourly consumption of electricity in Brazil. Accurate forecasts of this particular time series are essential, since they guide strategic decisions for the operation and planning of the Brazilian National Interlinked System (SIN). The experiments are conducted considering a wide forecast window, up to 168 h (a whole week) ahead. Different techniques are assessed, ranging from benchmark methods in the forecasting literature to more sophisticated methodologies, such as those capable of dealing with multiple seasonal patterns in time series. Thus, the aim is to effectively assist in decision-making by different stakeholders, such as power system operators and investors operating in the energy spot and futures markets.

The paper unfolds as follows. Section 2 summarizes the main stylized facts that are present in the electricity consumption time series, highlighting the most appropriate methods for modeling and subsequent forecasting in each case. Section 3 describes the data and how the empirical exercises are conducted. The results are presented in Sect. 4. Finally, Sect. 5 concludes and provides directions for future work.

2 Methods

The choice of the most appropriate methodologies for forecasting a time series is closely related to its stylized facts, that is, its key components [7, 8]. In the following subsections, we provide a brief overview of such elements and argue for selecting specific methods, such as those that consider multiple seasonal profiles, as the most suitable approaches for forecasting the time series under study.

2.1 *Critical Components in Hourly Consumption Time Series*

Time series can be represented by their key components, such as trends, seasonalities, cycles, and remaining components, i.e., the stochastic part of the time series that does not fit the other components. Depending on the phenomenon studied, the time series may have one or a few components.

Generally speaking, a trend exists when there is a long-term increase or decrease in the data. This trend does not need to be linear: in many cases, it is a long-term "change of direction" in the data, going from an increasing trend to a decreasing trend over the time horizon considered. A seasonal pattern, in turn, occurs when a time series is affected by seasonal factors, such as the seasons of the year or the days in a week. Seasonal factors are fixed, i.e., their frequencies are known and do not change over the years. Finally, a cycle occurs when the data exhibits rising and falling patterns that do not have a fixed frequency. These fluctuations are usually due to economic conditions and are often related to a business cycle [9]. In addition, they typically last for at least 2 years. Many practitioners confuse cyclical behavior with seasonal behavior, but they are pretty different. If fluctuations are not of fixed frequency, they will be cyclical; if the frequency is immutable and associated with some calendar aspect, the pattern is seasonal. In general, the average cycle length is longer than the length of a seasonal pattern, and cycle magnitudes tend to be more variable than seasonal pattern magnitudes [10].

Short-term consumption time series, in general, contain multiple seasonal patterns associated with different frequencies. Hourly data, for example, usually show three different types of seasonality: a daily pattern (hourly seasonality), a weekly pattern (daily seasonality), and an annual pattern. The annual pattern can be represented by different behaviors of the series over the quarters (quarterly seasonality), months (monthly seasonality), or weeks (weekly seasonality). The choice of representation for the annual pattern depends on several factors, such as data availability (e.g., it does not make sense to account for quarterly or monthly seasonality if the data is only available for a few weeks or months), the length of the forecast horizon, the easiness of data processing, and the limitations offered by some models.

2.2 Univariate Models for Time Series with Simple Seasonal Patterns

Among univariate forecasting techniques, some approaches stand out for the consistency of their results when forecasting time series of different frequencies and over multiple time horizons. Hence, they are often used as benchmarks in the literature, i.e., reference models for comparison. Well-known examples, which are continuously revisited and improved, are the exponential smoothing models [11–13] and the Seasonal Autoregressive, Integrated, Moving Average (SARIMA) formulations, initially proposed by Box and Jenkins [14].

Exponential smoothing models consist of basically revisiting the forecasts of the time series when novel information is available. In short, the models assign weights to the observations of the time series, and these weights decrease as the observations become "older", that is, less recent. One or more smoothing constants determine the rate of decay. There are several variations of exponential smoothing available in the literature. The works of Gardner Jr. [15, 16] provide a compilation of these variations.

In practice, when forecasting using exponential smoothing models, practitioners tend to opt for using a selection algorithm between state-space forms called ETS [12], an acronym for *ExponenTial Smoothing* or *Error, Trend and Seasonality*. The approach consists of selecting from a set of different available models (called state space formulations), the one that best suits the underlying data. According to the taxonomy presented in [15], 15 different formulations can be obtained by varying trend and seasonality components. These are illustrated in Table 1. Furthermore, the error component can also vary, being either additive or multiplicative. Hence, a total of 30 different formulations can be obtained.

This work selects the most suitable formulation via the ets( ) function of the *forecast* package for the R language [17]. The algorithm fits each ETS formulation to the training set, using the maximum likelihood technique, and selects the one presenting the lowest value of the Akaike information criterion corrected for small sample bias (AICc) [18]. Then, the selected model is used as input in the forecast( ) function from the same package to generate the forecasts throughout the desired forecast horizon/lead time (number of observations ahead to be estimated).

Table 1 Possible variations for the trend and seasonality components of exponential smoothing formulations under the ETS state space framework

	Seasonality		
Trend	None (N)	Additive (A)	Multiplicative (M)
None (N)	N, N	N, A	N, M
Additive (A)	A, N	A, A	A, M
Additive Damped (A_d)	A_d, N	A_d, A	A_d, M
Multiplicative (M)	M, N	M, A	M, M
Multiplicative Damped (M_d)	M_d, N	M_d, A	M_d, M

SARIMA formulations consist of an alternative approach to exponential smoothing. They are similar to exponential smoothing models in the sense that they are adaptive, can model seasonal patterns and trends, and can be automated. On the other hand, SARIMA formulations are based on the autocorrelation structure of the time series, rather than a structural view of its level, trend and seasonal components.

SARIMA formulations can be denoted by SARIMA $(p,d,q) \times (P,D,Q)_S$ and written as follows:

$$\nabla_S^D \nabla^d \phi(B) \Phi(B^S) Z_t = \theta(B) \Theta(B^S) a_t \quad (1)$$

where:

- p, d, and q are non-negative integers referring to, respectively, the order of the autoregressive model, the degree of differencing, and the order of the moving-average model;
- S refers to the number of periods in each season;
- the uppercase P,D,Q refer to the autoregressive, differencing, and moving average terms for the seasonal part of the ARIMA model;
- a_t is the error term;
- B is the backward shift operator (eg. $By_t = y_{t-1}$);
- $\phi(B)$ and $\Phi(B^S)$ are the non-seasonal and seasonal autoregressive polynomials, respectively;
- $\theta(B)$ and $\Theta(B^S)$ are the non-seasonal and seasonal moving-average polynomials, respectively;
- ∇^d and ∇_S^D are the non-seasonal and seasonal differencing operators, respectively.

In this work, the best-fit SARIMA formulation is selected using the auto.arima( ) function, also available in the *forecast* package for R [17]. The algorithm combines unit root tests, minimization of the AICc and Maximum Likelihood Estimation (MLE) to obtain the best-fit ARIMA formulation for the training set sample. As in ETS, forecasts can be later computed for the desired number of steps-ahead using the forecast( ) function.

2.3 *Models for Complex Seasonality Time Series*

The literature concerning electricity consumption forecasting contains a variety of applications with univariate methods [3]. However, for series of high frequencies, such as hourly or sub-hourly series, the presence of multiple seasonal patterns imposes several restrictions to the list of applicable models.

An approach that has become popular for high-frequency time series forecasting is the Double-Seasonal Holt-Winters model proposed by Taylor [19]. This exponential smoothing model was developed to forecast time series with two seasonal patterns: a shorter one, which repeats within a longer one. For instance, with a series of half-hourly data, one would set the first period to equal 48 observations for the daily period and the second period encompassing 336 observations, representing the weekly period. Initially applied to forecasting sub-hourly time series of electricity consumption in England and Wales [19], the model gained notoriety and has been constantly improved over the years. Gould et al. [20] extended the concept to the proposition of a *framework*, in state space forms, which allows the development of several models with additive and multiplicative seasonality. Taylor and Snyder [21] developed an exponential smoothing formulation that enables parts of different days of the week to be treated as identical, resulting in an approach that involves initializing and smoothing fewer terms. Despite the undeniable advances of its variants, the ease of implementation of the original Double-Seasonal Holt-Winters method – which is available in most statistical packages –, coupled with its satisfactory results obtained in different series of electrical energy consumption [22], makes the method a potential candidate for accurate modeling and forecasting the time series herein studied.

An alternative to deal with complex seasonality time series is the extension of SARIMA formulations to deal with more than one seasonal pattern. This approach, known as dynamic harmonic regression with multiple seasonal periods [10], involves the use of Fourier terms to represent multiple seasonal periods and Autoregressive, Moving Average (ARMA) formulations to deal with any remaining serial correlation. The number of Fourier terms for each seasonal period can be optimized by minimizing some information criterion, such as the AICc [18]. Dynamic harmonic regression with multiple seasonal periods also constitute a candidate for modeling and forecasting the time series under study. However, one should note that this method may be infeasible to apply in some cases, given the considerable computational effort (and, consequently, the time) taken during the estimation step.

Finally, another potential candidate is the TBATS approach [23], an acronym for Trigonometric Exponential Smoothing State Space Model with Box-Cox Transformation, ARMA Errors, Trend and Seasonal Components. The approach consists of an estimation framework (and subsequent forecasting, when desirable) that uses a combination of (i) Fourier terms with (ii) an exponential smoothing state space model and (iii) a Box-Cox transformation, in a fully automated way. The TBATS differs from dynamic harmonic regressions for multiple seasonal periods as it allows seasonality to change slowly over time. In contrast, harmonic regression terms force seasonal patterns to repeat periodically without changes. On the other hand, TBATS models are also computationally intensive, which can lead to considerable estimation times, depending on the number of series involved.

3 Data and Empirical Setup

3.1 Forecasting Total Hourly Consumption

The analysis considers hourly observations of total electricity consumption in Brazil, between January 1st, 2020 and December 31st, 2020. The data was retrieved from the Brazilian National Electric System Operator (ONS) database [24]. The series contains multiple seasonal patterns, as illustrated in Fig. 1. Three patterns can be identified: (a) an hourly seasonality, which occurs over 24 observations (hours) in a day; (b) a daily seasonality, which occurs over the 7 days of the week; and (c) the seasonality that occurs over 1 year, which can be represented by a pattern that occurs over the weeks (weekly pattern), as shown in the figure, or over the months (monthly pattern) or the quarters (quarterly pattern).

We split the data into two sets for the experiments: the first, comprising observations between January 1st, 2020 and December 24th, 2020, was defined as the training set and used for model estimation; the second, encompassing the hourly historical data between midnight on December 25th, 2020 and 11:00 pm on December 31st, 2020, was used as test set to assess the accuracy of the involved methods when generating out-of-sample forecasts up to 168 steps-ahead. Given the selected forecast horizon, the methods should ideally be able to deal with two

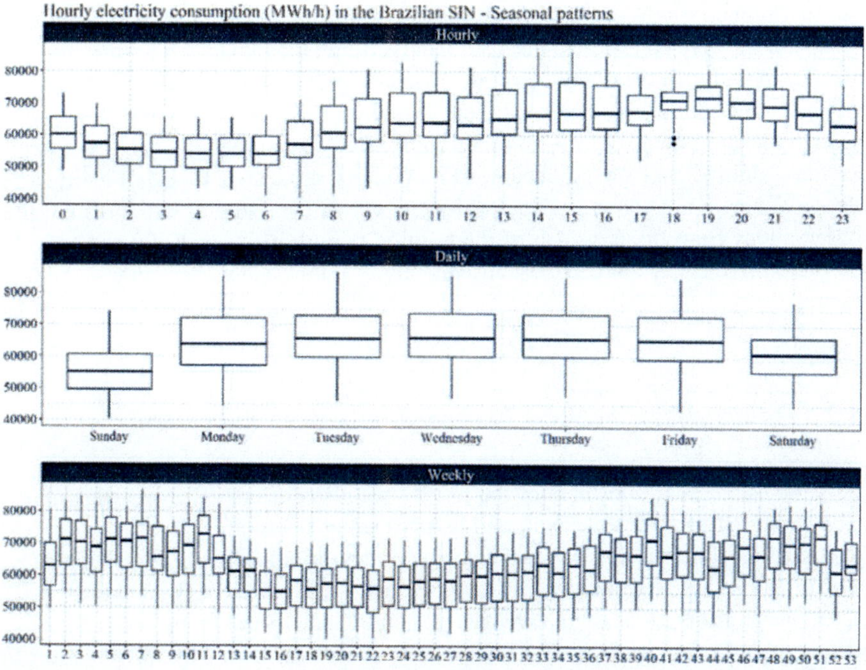

Fig. 1 Seasonal patterns of the total hourly electricity consumption in the Brazilian National Interlinked System (SIN), in MWh/h. (Source: The authors)

Table 2 Error metrics considered for accuracy evaluation

Metric	Formula
Root Mean Squared Error (RMSE), in MWh	$RMSE = \sqrt{\frac{\sum_{t=1}^{h}\left(\widehat{y_t}-y_t\right)^2}{h}}$
Mean Absolute Percentage Error (MAPE), in percentage points (%)	$MAPE = \frac{1}{h}\left(\sum_{t=1}^{h}\left\lvert\frac{\widehat{y_t}-y_t}{y_t}\right\rvert\right) \times 100\%$
Symmetric Mean Absolute Percentage Error (sMAPE), in percentage points (%)	$SMAPE = \frac{2}{h}\left(\sum_{t=1}^{h}\frac{\lvert\widehat{y_t}-y_t\rvert}{\lvert\widehat{y_t}\rvert+\lvert y_t\rvert}\right) \times 100\%$
Mean Absolute Scaled Error (MASE), dimensionless	$MASE = \frac{1}{h}\frac{\sum_{t=1}^{h}\lvert\widehat{y_t}-y_t\rvert}{\frac{1}{n-m}\sum_{t=n-m+1}^{n}\lvert y_t-y_{t-m}\rvert}$

Notes: In the above formulae, y_t and $\widehat{y_t}$ respectively represent the actual (observed) and forecasted values, h is the forecast lead time (number of steps ahead), m is the number of observations in the shortest seasonal period and n is the train set length

seasonal patterns present in the series: the hourly seasonality (over 1 day); and the daily seasonality, over a whole week (including Saturday and Sunday).

Forecasting accuracy evaluation was conducted considering different error metrics, such as the Root Mean Squared Error (RMSE), the Mean Absolute Percentage Error (MAPE), the Symmetric Mean Absolute Percentage Error (sMAPE) and the Mean Absolute Scaled Error (MASE), calculated according to the equations shown in Table 2.

As outlined in Sect. 2.3, three models capable of dealing with multiple seasonalities were considered: the TBATS algorithm; the dynamic harmonic regression with multiple seasonal periods, hereafter referred to as Double-Seasonal ARIMA; and the Double-Seasonal Holt-Winters method. As benchmarks, we considered the traditional ARIMA and ETS approaches, as outlined in Sect. 2.2, as well as the Additive and Multiplicative Holt-Winters models [11, 12]. Finally, the Seasonal Naive method was also considered, which consisted of extrapolating the observations from the last 24 h of the training period, between 12:00 am and 11:00 pm on December 24th, 2020, subsequently repeating them every day in the test period.

3.2 *Robustness Checks – Forecasting Across Subsystems*

To demonstrate the robustness of the results, we also considered forecasting under alternative settings. First, we collected hourly data of electricity consumption across the four subsystems of the SIN, each corresponding to a specific geographic area in Brazil: North (N), Northeast (NE), Southeast and Midwest (SE/CO), and South (S). Then, we considered an alternative test set period encompassing the hours between midnight on December 18th, 2020 and 11:00 pm on December 24th, 2020. Figure 2 illustrates how hourly electricity consumption in Brazil varies across the subsystems. As can be noted, all series share similar seasonal patterns, although their magnitudes

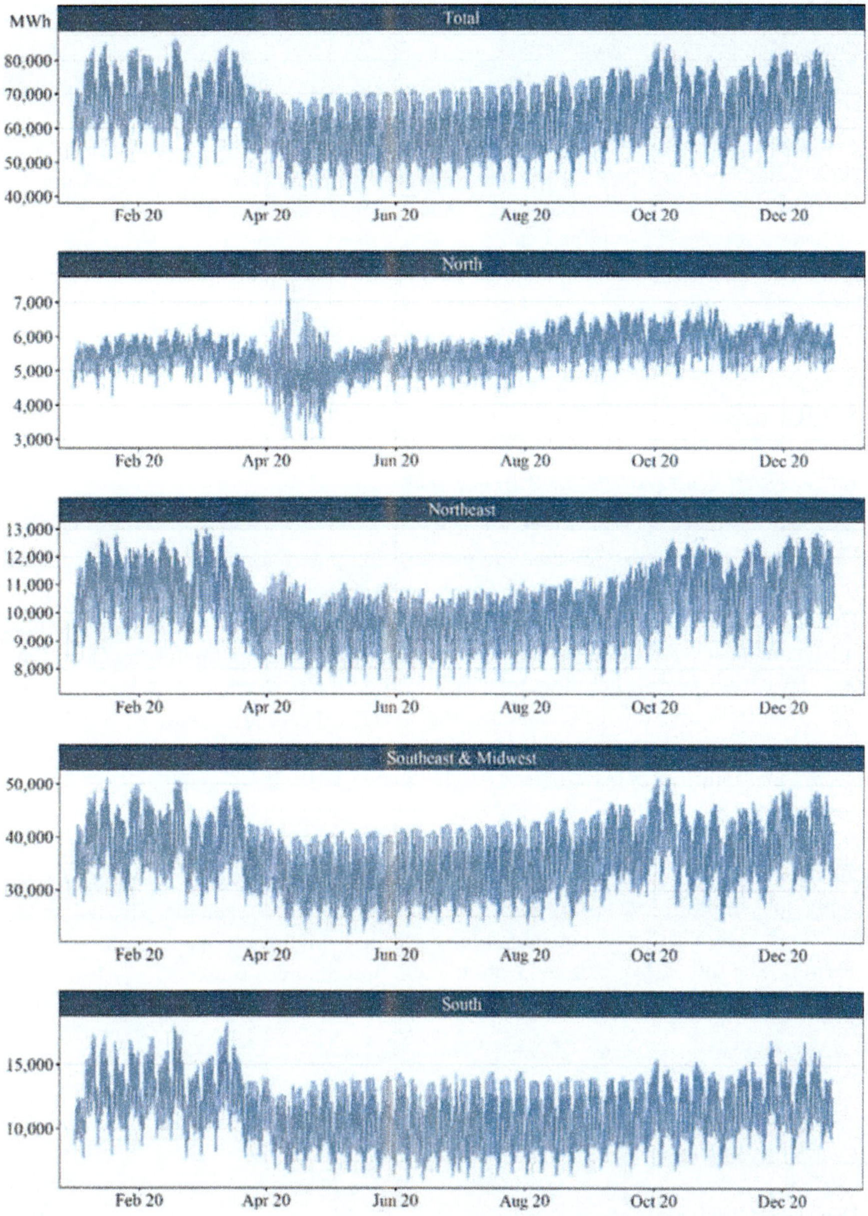

Fig. 2 Hourly electricity consumption in Brazil, in MWh/h – Total (SIN) and across subsystems. (Source: ONS (Brazilian National Grid Operator) [24])

differ considerably. In addition, some series present unique stylized facts: for instance, the volatility in consumption between early April and mid-May 2020 in the Northern subsystem was considerably higher than in the other subsystems. This highlights the challenge of proposing a generalized method that provides reliable forecasts in every case considered.

As four time series were involved in the second experiment – each corresponding to a different subsystem – the results were presented in terms of average of the evaluation metrics across all countries, considering 168 steps ahead forecasts. Forecasting performance was also assessed by considering the distribution (boxplots) of each metric obtained when methods were individually applied to each time series. These plots enable an assessment of which time series are more difficult to forecast, as well as which methods vary their performance considerably across time series.

4 Results

4.1 Total Consumption Forecasts

The results of the error evaluation metrics computed for each approach herein considered when forecasting total hourly consumption are illustrated in Table 3. The best performance is highlighted in **bold** for each metric, while the second-best appears in *italics*. All calculations were performed considering the forecast lead time of 7 days/168 h ahead, i.e., h = 168. To illustrate the performance of each method, Fig. 3 presents the charts containing the actual (observed) values of the hourly electricity consumption series in Brazil over the test set period (hours between December 25th and 31st, 2020) and the forecasted values generated with each method.

The results indicate that the TBATS and the Double-Seasonal ARIMA are considerably more accurate than the other competing approaches for hourly electricity consumption forecasting in Brazil. The difference between these two methods and the benchmarks was considerable, regardless of the observed point of view (metric). The forecasts generated from the two best methods tracked very close, with

Table 3 Forecasting total hourly electricity consumption in Brazil – Error metrics for each method considered

Forecasting method	RMSE (MWh)	MAPE (%)	sMAPE (%)	MASE
Complex seasonality models				
TBATS	*5048.85*	*6.19*	**5.99**	*1.04*
Double-Seasonal ARIMA	**4822.60**	**6.15**	*6.03*	**1.04**
Double-Seasonal Holt-Winters	7584.44	11.08	11.57	1.96
Benchmarks				
ARIMA	6465.45	8.64	8.79	1.55
ETS	8204.41	11.47	12.11	2.06
Additive Holt-Winters	9323.09	12.92	13.84	2.34
Multiplicative Holt-Winters	8499.44	11.38	12.09	2.09
Seasonal Naive	6538.64	8.92	8.76	1.54

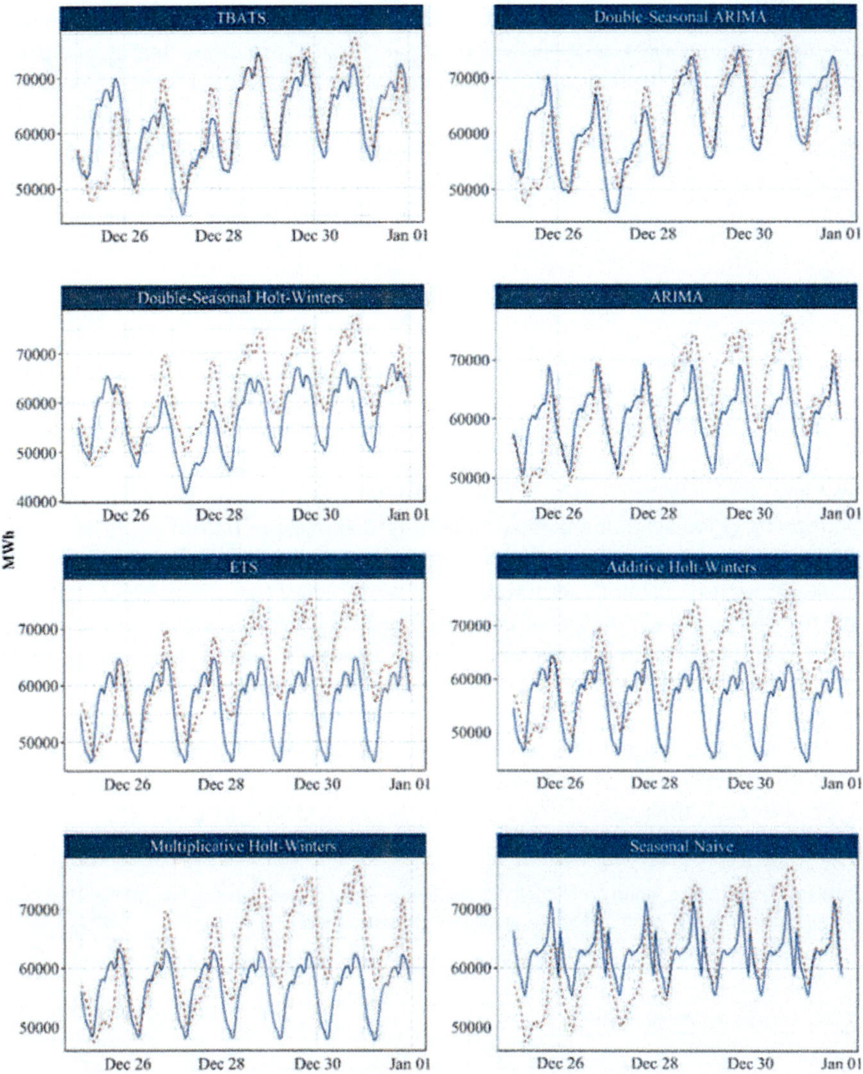

Fig. 3 Actual (observed) values and forecasts of hourly electricity consumption in the Brazilian National Interlinked System (SIN) between December 25th and 31th, 2020 according to each method considered. Forecasts in blue, actual values in red. (Source: The authors)

TBATS being slightly more accurate from the perspective of sMAPE and Double-Seasonal ARIMA providing more accurate results in light of the other metrics.

Figure 3 highlights the greater adherence of TBATS and the Double-Seasonal ARIMA, which delivered forecasts very close to the observed data for the most significant part of the considered forecasting lead time. Among the other methods, the traditional ARIMA formulation also presented competing results, especially in the first 3 days of the test period. In general, while it is true that simple seasonality formulations may deliver satisfactory results, particularly in the beginning of the test period (when the effects of multiple seasonal patterns are not yet present), complex seasonal profiles gradually become more apparent (from day two onwards, in our example), calling for more sophisticated techniques capable of dealing with multiple patterns.

4.2 Consumption Across Subsystems

As outlined in Sect. 3.2, the second empirical experiment provides a more robust view of the forecasting performance of each method considered, as it considers forecasting performance evaluation across all four subsystems in the Brazilian National Interlinked System (SIN). The results are summarized in Table 4, where averages of performance metrics across all four time series are provided. The best performances are again highlighted in **bold** (second-best in *italics*).

The results endorse the superior performance of the TBATS and the Double-Seasonal ARIMA methods for forecasting hourly energy consumption time series in Brazil. Both methods presented considerably lower average values in each error metric considered. The Double Seasonal Holt-Winters ranked third best in most cases, presenting lower values of Average MAPE, sMAPE and MASE than the benchmarks.

Besides the average values computed across subsystems, we also assessed the distribution (boxplots) of each metric obtained when methods were individually applied to each time series. The boxplots for the RMSEs, MAPEs, sMAPEs and

Table 4 Forecasting hourly electricity consumption across the subsystems of the Brazilian National Interlinked System (SIN) – Average of the error metrics

Forecasting method	Avg. RMSE (MWh)	Avg. MAPE (%)	Avg. sMAPE (%)	Avg. MASE
Complex seasonality models				
TBATS	*1502.09*	**5.08**	**5.26**	**0.98**
Double-Seasonal ARIMA	**1438.89**	*5.27*	*5.43*	*0.99*
Double-Seasonal Holt-Winters	1888.96	5.77	6.10	1.25
Benchmarks				
ARIMA	1989.04	6.91	7.45	1.37
ETS	1728.34	5.99	6.35	1.22
Additive Holt-Winters	1998.38	7.09	7.55	1.50
Multiplicative Holt-Winters	1919.22	6.67	7.14	1.34
Seasonal Naive	1871.82	6.44	6.91	1.28

Notes: For robustness checks, the test set period comprises an alternative time window, encompassing the hours between midnight on December 18th, 2020 and 11:00 pm on December 24th, 2020 (See Sect. 3.2 for further details)

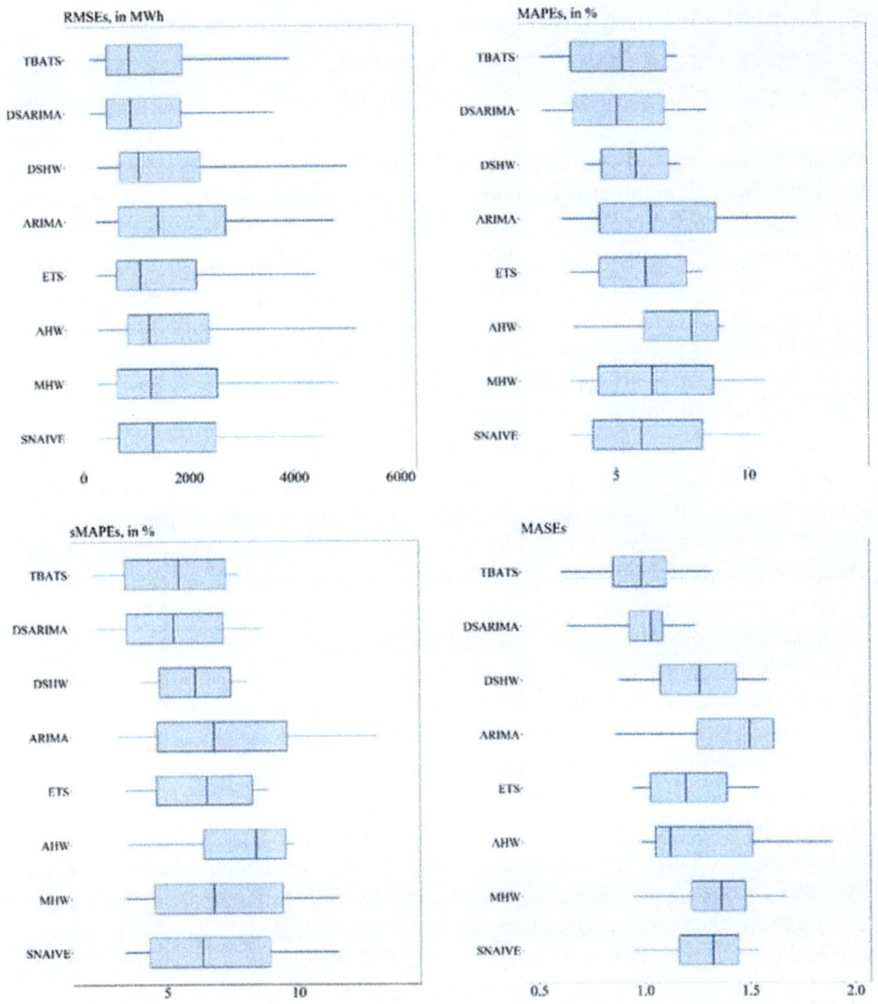

Fig. 4 Boxplots – Distribution of the RMSEs (MWh), MAPEs (%), sMAPEs (%) and MASEs values for each forecasting method considered across the four subsystems. (Source: The authors)

MASEs are depicted in Fig. 4. Overall, the results depicted in the boxplots are consistent with the average values illustrated in Table 4, with complex seasonality methods, particularly TBATS and Double Seasonal ARIMA, generally outperforming benchmarks. Not only did complex seasonality methods presented considerably lower medians, but they were also less sensitive to extreme values.

5 Conclusions and Directions for Further Research

This work assessed the accuracy of several methods when forecasting multiple time series of hourly electricity consumption in Brazil up to 1 week (168 h) ahead. Different techniques were considered, from benchmark methods in the forecasting literature, such as automated exponential smoothing and ARIMA formulations, to more sophisticated approaches capable of dealing with multiple seasonal patterns in a time series. As expected, the methods belonging to the latter group delivered more accurate results than the traditional approaches from the perspective of every forecasting error metric considered, with noteworthy developments for the TBATS algorithm and the Double-Seasonal ARIMA. The results provide essential insights in terms of decision-making for different stakeholders, such as power system operators, policymakers, and investors operating in the energy spot and futures markets.

A promising aspect when estimating future hourly electricity consumption is that a considerable part of its variation depends on univariate stylized facts common to energy demand time series, such as trends and multiple seasonal patterns. In this context, univariate methods capable of dealing with complex seasonality profiles usually show good adherence and performance when forecasting electricity consumption several steps ahead, a generally complex and sensitive task for a large set of forecasting methodologies available in the literature.

Possible extensions of this work include applying alternative complex seasonality methods, such as the Multiple Seasonal-Trend decomposition using *Loess* (MSTL) algorithm [25], as well as combinations of multiple complex seasonality approaches. The use of methodologies that consider exogenous factors when estimating and forecasting time series also constitutes a natural extension of the present work, although this is a more sensitive task, given the need to have reliable forecasts for the independent variables [26]. Finally, future studies can benefit from hierarchical forecasting methods, i.e., approaches that allow for the individual treatment of each consumption class in the Brazilian National Interlinked System (SIN), such as residential, industrial, commercial and others. Such analyzes would contribute to a deeper understanding of electricity demand in Brazil, potentially also improving the quality of final forecasts for the total consumption.

References

1. De Oliveira EM, Cyrino Oliveira FL (2018) Forecasting mid-long term electric energy consumption through bagging ARIMA and exponential smoothing methods. Energy 144:776–788. https://doi.org/10.1016/j.energy.2017.12.049
2. Meira E, Cyrino Oliveira FL, de Menezes LM (2021) Point and interval forecasting of electricity supply via pruned ensembles. Energy 232:121009. https://doi.org/10.1016/j.energy.2021.121009
3. Suganthi L, Samuel AA (2012) Energy models for demand forecasting – a review. Renewable and Sustainable Energy Reviews, 16(2):1223–1240. https://doi.org/10.1016/j.rser.2011.08.014

4. Kuster C, Rezgui Y, Mourshed M (2017) Electrical load forecasting models: A critical systematic review. Sustainable Cities and Society, 35:257–270. https://doi.org/10.1016/j.scs.2017.08.009
5. Weron R (2014) Electricity price forecasting: A review of the state-of-the-art with a look into the future. International Journal of Forecasting, 30(4):1030–1081. https://doi.org/10.1016/j.ijforecast.2014.08.008
6. Meira E, Cyrino Oliveira FL, Jeon J (2021) Treating and pruning: New approaches to forecasting model selection and combination using prediction intervals. International Journal of Forecasting, 37(2):547–568. https://doi.org/10.1016/j.ijforecast.2020.07.005
7. Oliveira EMD (2015) Corporate social responsibility and firm performance: A case study from the Brazilian electric sector. Pontifical Catholic University of Rio de Janeiro (PUC-Rio), Rio de Janeiro, Brazil. https://doi.org/10.17771/PUCRio.acad.25647
8. De Oliveira EM, Cunha FAFS, Palazzi RB, Klotzle MC, Maçaira PM (2020) On the effects of uncertainty measures on sustainability indices: An empirical investigation in a nonlinear framework. International Review of Financial Analysis, 70, 101505. https://doi.org/10.1016/j.irfa.2020.101505
9. De Oliveira EM, Cyrino Oliveira FL, Klotzle MC, Pinto ACF (2021) Dynamic associations between GDP and crude oil prices in Brazil: Structural shifts and nonlinear causality. Emerging Markets Finance and Trade, 57(10):2767–2791. https://doi.org/10.1080/1540496x.2019.1658072
10. Hyndman RJ, Athanasopoulos G (2021) Forecasting: principles and practice. 3rd ed. OTexts, Melbourne, Australia. https://otexts.com/fpp3/
11. Holt CC (1957) Forecasting seasonal and trends by exponentially weighted moving average. O.N.R. Research Memorandum, no. 52. Carnegie Institute of Technology, Pittsburgh (PA), United States. Reprinted in 2004: https://doi.org/10.1016/j.ijforecast.2003.09.015
12. Winters PR (1960) Forecasting sales by exponentially weighted moving averages. Management Science, 6(3):324–342. https://doi.org/10.1287/mnsc.6.3.324
13. Hyndman RJ, Koehler AB, Snyder RD, Grose S (2002) A state space framework for automatic forecasting using exponential smoothing methods. International Journal of Forecasting, 18(3): 439–454. https://doi.org/10.1016/s0169-2070(01)00110-8
14. Box GEP, Jenkins GM (1970) Time series analysis: forecasting and control. Holden Day, San Francisco, United States.
15. Gardner Jr. ES (1985) Exponential smoothing: The state of the art. Journal of Forecasting, 4(1): 1–28. https://doi.org/10.1002/for.3980040103
16. Gardner Jr. ES (2006) Exponential smoothing: The state of the art—part II. International Journal of Forecasting, 22(4):637–666. https://doi.org/10.1016/j.ijforecast.2006.03.005
17. Hyndman R, Athanasopoulos G, Bergmeir C, et al. (2021) forecast: Forecasting Functions for Time Series and Linear Models. Version 8.15. https://cran.r-project.org/package=forecast
18. Sugiura N (1978) Further analysts of the data by Akaike's information criterion and the finite corrections. Communications in Statistics – Theory and Methods, 7(1):13–26. https://doi.org/10.1080/03610927808827599
19. Taylor JW (2003) Short-term electricity demand forecasting using double seasonal exponential smoothing. Journal of the Operational Research Society, 54(8):799–805. https://doi.org/10.1057/palgrave.jors.2601589
20. Gould PG, Koehler AB, Ord JK, Snyder RD, Hyndman RJ, Vahid-Araghi F (2008) Forecasting time series with multiple seasonal patterns. European Journal of Operational Research, 191(1): 207–222. https://doi.org/10.1016/j.ejor.2007.08.024
21. Taylor JW, Snyder RD (2012) Forecasting intraday time series with multiple seasonal cycles using parsimonious seasonal exponential smoothing. Omega, 40(6):748–757. https://doi.org/10.1016/j.omega.2010.03.004
22. Taylor JM, McSharry PE (2007) Short-term load forecasting methods: An evaluation based on european data. IEEE Transactions on Power Systems, 22(4):2213–2219. https://doi.org/10.1109/tpwrs.2007.907583

23. De Livera AM, Hyndman RJ, Snyder RD (2011) Forecasting time series with complex seasonal patterns using exponential smoothing. Journal of the American Statistical Association, 106(496):1513–1527. https://doi.org/10.1198/jasa.2011.tm09771
24. ONS (Brazilian National Grid Operator), Curva de carga horária (MWh/h), http://www.ons.org.br/Paginas/resultados-da-operacao/historico-da-operacao/curva_carga_horaria.aspx, last accessed 2021/06/23
25. Cleveland RB, Cleveland WS, McRae JE, Terpenning I (1990) STL: A Seasonal-Trend Decomposition Procedure Based on Loess. Journal of Official Statistics, 6:3–73. http://bit.ly/stl1990
26. Oliveira EMD (2020) Getting the most out of the wisdom of the crowds: Improving forecasting performance through ensemble methods and variable selection techniques. Pontifical Catholic University of Rio de Janeiro (PUC-Rio), Rio de Janeiro, Brazil. https://doi.org/10.17771/PUCRio.acad.48429

Rural Area Electric Power Distribution Coverage Improvement

Erick Quispe, Jonatán Rojas, Andrea Caballero, Alexia Cáceres, Alessandro Gilardino, Yessenia Morales, and Renzo Benavente

Abstract There is a high correlation between electricity distribution and economic development for poverty reduction (Cabraal R, Barnes D, Agarwal S, Annu Rev Environ Resour 30(1):117–144, 2005). In Peru, some families lack electricity, education, and health services. In this context, this research proposes an eco-friendly electric generator to supply energy to rural areas through sustainable development approaches and life cycle analysis.

Key words Rural area electric power distribution · Recirculated water · An eco-friendly electric generator

1 Introduction

Electric power service stands as one of the basic services that ensure the well-being and economic development of the population because it is essential for street lighting access and electronic devices operation such as radio and television that bring information and entertainment to families. Most of the populations that do not have access to electric power are located in rural areas due to its hard access and the dispersion of the populations with low income. The Peruvian rural electrification general law established the regulatory framework for electric power distribution in those areas. Rural electric power distribution's low rate prompted the government to implement the First Rural Electrification Project (RE1) in 2006, when the rate was 30%, and the second one (RE2) in 2011 [2]. Although according to the National Institute of Statistics and Informatics (INEI), the Peruvian rural electric power distribution rate has reached 83.9% in 2019 [3], it is still below most of the countries in Latin America. Nowadays, bring electric power to the rural areas stands as a challenge for the government due to the hard access areas and the distance among

E. Quispe · J. Rojas (✉) · A. Caballero · A. Cáceres · A. Gilardino · Y. Morales · R. Benavente
Pontificia Universidad Católica del Perú, Lima, PERÚ
e-mail: erick.quispem@pucp.pe; jrojasp@pucp.pe; andrea.caballero@pucp.edu.pe; alexia.caceres@pucp.pe; agilardino@pucp.pe; y.morales@pucp.pe; renzo.benavente@pucp.pe

J. Vargas Florez et al. (eds.), *Production and Operations Management*, Springer Proceedings in Mathematics & Statistics 391,
https://doi.org/10.1007/978-3-031-06862-1_8

them. Given the situation, the Energy and Mining Department has developed the Rural Electric Power Distribution National Plan 2016–2025 that makes top priority the use of renewable energy sources for these areas because connecting them to the electric power main system is almost impossible [4]. Like-wise, there is an international concern for environment preservation; because of this, the United Nations' Sustainable Development Goals established as Objective 7 the renewable energy global access [5].

This research proposes a clustering model that will identify the most critical rural zones of the country according to economic and demographic variables as well as the development of a mathematical model that will improve the electric power distribution rate using an easy to install and low-cost hydraulic generator that will be able to adapt to every type of soil and will be ideal to fulfill the necessities of the rural areas. Finally, it will be evident the reduction of greenhouse gases emissions due to the hydraulic generator which only uses water as its unique source of energy.

2 Literature Review

In this section, the following basic concepts will be described: clustering, operations research, Life Cycle Assessment, and Hydraulic generator.

2.1 Cluster

Clustering is a multivariable tool used to sort multivariable data sets (observations) in groups known as clusters. Even if the observations in each group are close to each other (similar observations); the observations in different groups show different criteria [6]. Clustering also finds the groups that will appear in the analyzed data sets. These groups are subsets separated from the original dataset, that share a characteristic: data that belong to different clusters have a larger difference among them than data that belongs to the same cluster. This means clustering has the objective of finding a certain type of natural structure in the dataset. The means to perform this task, generally consist of a similarity meter. Clustering is not just an important cognitive tool, but a method to reduce big datasets as well [7]. Clustering has been applied to a vast number of research problems where sorting big datasets in groups is a necessity. With that in mind, some algorithms generate groups that reflect data behavior and structure. There are two types of clustering methods: (a) Hierarchic sorting which is a sequential process. It starts by calculating the distance of each observation to the other ones and generates a distance matrix for each of them to create a dendogram. (b) Non-hierarchical methods take into consideration other criteria besides distance, to group the observations in a certain number of groups; an example of this is the k-means method [6]. Clustering algorithms start from the assumption that the cluster structure has a data behavior pattern. Clustering techniques tend to generate sorting criteria for any dataset; however, this can generate groups with invalid criteria when data does not have a behavior pattern and/or the observations are very far from each other [8].

2.2 Operation Research

To solve some of the problems given in this paper, the discipline of operations research will be needed. According to Goodeve, operations research is a scientific method to provide the executive departments with a quantitative basis for decisions regarding operations under their control [9]. Operations research is the application of advanced analytical methods to help make better decisions [10]. There are many techniques to solve the mathematical models and the most prominent is linear programming [11]. A linear program is an optimization problem in which the objective function is linear in the variables and the constraints consist of linear equalities and linear inequalities [12]:

$$max\,imize\ \ c_1x_1 + c_2x_2 + \cdots + c_nx_n$$

Subject to:

$$a_{11}x_1 + a_{12}x_2 + a_{13}x_3 + \cdots + a_{1n}x_n = b_1$$

$$\ldots\ldots$$

$$a_{m1}x_1 + a_{m2}x_2 + a_{m3}x_3 + \cdots + a_{mn}x_n = b_m$$
$$and x_1 \geq 0,\ x_2 \geq 0,\ x_3 \geq 0,\ \ldots,\ x_n \geq 0$$

Where *bi*, *ci* and *aij* are real constants and the variables *xi* are real numbers which should be determined.

2.3 Life Cycle Analysis

Life Cycle Analysis (LCA) started at the same time in the United States and Europe in 1969. Midwest Research Institute (MRI) developed the first LCA in Coca-Cola Company intending to reduce resources usage and in consequence minimizing the amount of emitted air pollution gases [13]. LCA is an environmental management tool that has a scope that limits the analyzed system "from the cradle to the grave", analyzing the process in five steps and finding which of the productive cycle phase generates the most amount of environmental impact and from there finds ways to reduce it [14]. Besides, LCA scores the environmental impact of a product or service. Scoring is based on a particular function and takes into account every life cycle step; on top of that, it helps with the identification of potential environmental improvements during the life cycle of a product. This tool can be used to compare many products, processes, or systems as well as the different steps of the life cycle of a certain product. According to the definitions given by the International Organization for Standardization (ISO) and by the Society of Environmental Toxicology and Chemistry (SETAC), LCA consists of four steps: (1) in this step the objectives and scope of the study, as well as the objectives and scope of the problem are defined. (2) The second step, focuses on inventory analysis and quantifies the amount of air, water, and earth polluting emissions as well as the renewable and non-renewable raw material extraction. (3) The third step, analyzes and scores the environmental impact caused due to the emission found in step 2. (4) Finally, the fourth step also known as interpretation consists of the analysis of the obtained results and the evaluation of certain scenarios with an uncertainty degree. Improvement options can be found through sensitivity studies [15].

2.4 *Hydraulic Generator*

The design and making of an electric power hydraulic generator start from the movement or spin of a turbine caused by water falling from a vessel located at a determinate height. The most common turbines used for hydraulic generators are Francis, Kaplan, and Pelton turbines [16]. Nowadays, electric power generators are the main power source of the hydroelectric central, being the key challenge the maintenance and correct performance for an optimal electric power generation [17]. According to the Hydropower Status Report 2020 of the International Hydropower Association (IHA), approximately 13,000 hydroelectric central had been built in over 150 countries [18].

3 Proposed Model

This research proposes a hydraulic generator which will be located at rural area houses. First, the used dataset for the most critical department selection process will be described (considering many variables such as electric power access, basic services access, academic development, among others); after that, the development of a mathematical model for the optimal recirculate water hydraulic generator installment will start, which will help social development and will reduce the number of environmental pollutants compared to the ones generated by a traditional hydraulic power plant.

3.1 Proposed Hydraulic Generator

The proposed recirculated water energy power generator consists of five important parts: a Pelton turbine, a water pump, a power multiplier system, a velocity multiplier with a flywheel, and a generator. The process begins with the start of the generator which consists of a manual system with a gearbox that speeds up the turbine axis spin velocity. The startup manual system contains four sprockets that have 55, 11, 71, and 50 tooths; it also contains six axes. By rotating the crank at 60 RPM, the system receives 226 RPM which generates electric power, the water pump is connected to the distribution board and the startup manual system is disconnected. The distribution board is then activated, which gives 220 volts to the water pump (1HP, 2500RPM, 3.5 A). The power multiplier produces 20 times the initial power which means that a 1 HP input will have approximately a 20 HP output, this is all shown in Fig. 1.

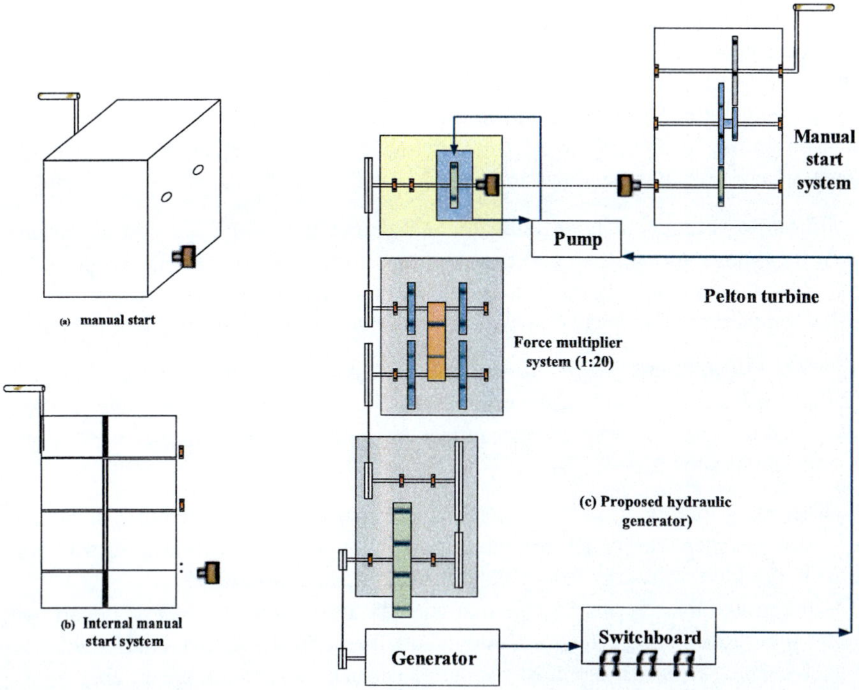

Fig. 1 Proposed generator

3.2 Description of the Dataset Used for Critic Department Selection

The dataset is composed of 36 variables that are related to sociodemographic, economy, population, current electric power access, social development such as mathematics and reading test performance, among others. The variables are numerical and represent each of the 26 departments in which Peru is divided as shown in Table 1.

3.3 Critical Department Selection (PCA and Cluster)

The most critical country department selection process consists of two steps.

In the first step, the tool IBM SPSA Statistics 24 will perform the Principal Components Analysis (PCA) which will reduce the number of variables from 36 variables to 3 components that will represent all of the variables with a high correlation rate.

For the second step, after finding the 3 components for every department of Peru (26 departments). A program for cluster developed in R will be used to detect the critic department.

PCA Simplification To avoid distortions in the results due to the different units and magnitudes in the data as shown in Table 1 it is necessary normalization of the data without affecting the covariance between each variable; the result would be data with binary values (0 and 1). With SPSS, normalized data through the factor reduction analysis brought with it three main components which explain 94.086% of the accumulated variance as shown in Table 2.

The first component is an indicator that shows the people outside of extreme poverty conditions and with good academic performance; the second component is people in extreme poverty conditions considering their electric power usage, and the third component is the availability of the resources for the life quality of people.

Critical Department Selection with the Cluster Predetermined function in R 'hclus' with the 'dist' (distances) parameters and 'ave' method (average method) was used to obtain the clusters that reflect the input data behavior for each department. This is shown in Fig. 2a.

The PCA method results indicated that Lima is one of the most developed departments, and has better results as shown by its electric power distribution, energy production, energy usage and test performance indicators.

This means that cluster 3 is the one with the most average and low department indicators. Considering efficiency among both methods, the results relation indicates that Ayacucho is one of the most critical departments as shown in Fig. 2b.

Table 1 Employed variables for critic department selection

ID	Attribute	Description	ID	Attribute	Description
1	PBI	GDP per cápita (S/)	19	HEP_VAL_AL	Extremely poor houses that use candles for lighting (thousands)
2	POBLAC	Population (thousands)	20	HEP_GEN_AL	Extremely poor houses that use electric power generators for lighting (thousands)
3	HOG	Houses (thousands)	21	HEP_OTR_AL	Extremely poor houses that use other lighting sources(thousands)
4	DENS	Population density (people/km^2)	22	HEP_SIN_AL	Extremely poor houses that don't have lighting (thousands)
5	EXTENSION	Territory extension (km^2)	23	HPO_RPU_AL	Poor houses that use public lighting (thousands)
6	PBZ	Poverty(thousands)	24	HPO_KER_AL	Poor houses that use kerosene for lighting (thousands)
7	PBZ_EXT	Extreme poverty (thousands)	25	HPO_PET_AL	Poor houses that use oil or gas for lighting (thousands)
8	R_LECTURA	Reading performance (thousands)	26	HPO_VAL_AL	Poor houses that use candles for lighting (thousands)
9	R_MATE	Mathematical performance (thousands)	27	HPO_GEN_AL	Poor houses that use electric power generators for lighting (thousands)
10	HCE_GLP_NAT	Houses that cook with electric power, LPG or natural gas (thousands)	28	HPO_OTR_AL	Poor houses that use other lighting sources (thousands)
11	HC_LAV	Houses with washing machine (thousands)	29	HPO_SIN_AL	Poor houses that don't have lighting (thousands)
12	HC_CEL	Houses with at least one cellphone (thousands)	30	HNP_RPU_AL	Non-poor houses that use public lighting (thousands)
13	PEA	Employed labor force (thousands)	31	HNP_KER_AL	Non-poor houses that use kerosene for lighting (thousands)
14	COB_ELECT	Electric power coverage (thousands)	32	HNP_PET_AL	Non-poor houses that use oil or gas for lighting (thousands)
15	PROD_ELECT	Electric power production (Mw.h)	33	HNP_VAL_AL	Non-poor houses that use candles for lighting (thousands)

(continued)

Table 1 (continued)

ID	Attribute	Description	ID	Attribute	Description
16	HEP_RPU_AL	Extremely poor houses that use public lighting (thousands)	34	HNP_GEN_AL	Non-poor houses that use electric power generators for lighting (thousands)
17	HEP_KER_AL	Extremely poor houses that use kerosene for lighting (thousands)	35	HNP_OTR_AL	Non-poor houses that use other lighting sources (thousands)
18	HEP_PET_AL	Extremely poor houses that use oil or gas for lighting (thousands)	36	HNP_SIN_AL	Non-poor houses that don't have lighting (thousands)

3.4 Mathematic Model

In this section, the mathematical model is described. The main variable is QGenij which describes the quantity of generator of type "i" in the village "j". In addition, the variables uj and vj are the slack and surplus, respectively.

$$Min = \sum_{i=1}^{3} \sum_{j=1}^{6925} QGen_{ij} Cost_i + \sum_{j=1}^{6925} M\left(u_j + v_j\right) \quad (1)$$

Subject to:

$$\sum_{i=1}^{3} QGen_{ij} * Power * (1 - f) + u_j - v_j \quad \forall j \in J \\ = Demand_j \quad (2)$$

$$\sum_{i=1}^{3} QGen_{ij} * Power_i \leq Pl_{max} \quad \forall j \in J \quad (3)$$

$$QGen_{ij} \geq 0, \quad \text{integer} \quad \forall i \in I, \quad j \in J \quad (4)$$

$$u_i \geq 0 \quad \forall j \in J \quad (5)$$

$$v_i \geq 0 \quad \forall j \in J \quad (6)$$

Where i is the set of types of generators and j is the set of the 6925 villages in Ayacucho. The objective function shown in (1) represents the total cost of the generators and the goal to achieve the energy demand. Constraint (2) indicates the demand of each village obtained with the generators which should be fulfilled.

Table 2 Explained total variance

Component	Initial autovalues			Load extraction squared sum			Load rotation squared sum		
	Total	% variance	% cumulated	Total	% variance	% cumulated	Total	% variance	% cumulated
1	23.599	65.552	65.552	23.599	65.552	65.552	22.594	62.761	62.761
2	9.058	25.161	90.713	9.058	25.161	90.713	9.958	27.662	90.423
3	1.214	3.373	94.086	1.214	3.373	94.086	1.319	3.663	94.086
4	0.846	2.351	96.437						

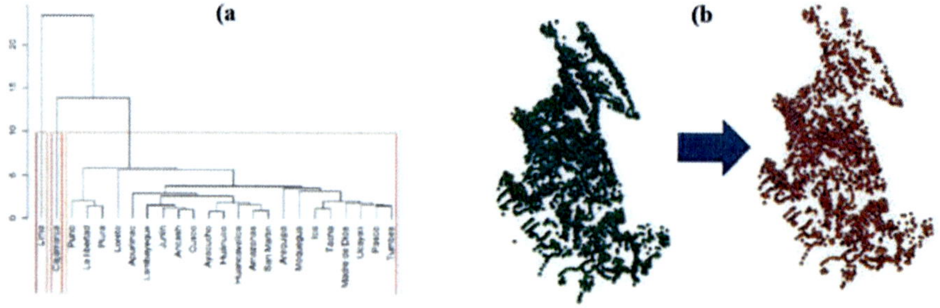

Fig. 2 (**a**) Cluster dendogram. (**b**) Ayacucho – Potential population centers

Constraint (3) imposes the maximum quantity of energy given to a village. Finally, constraints (4), (5), and (6) are the range of the variables.

3.5 Life Cycle Analysis

The traditional way to carry out rural electrification is to transport the electrical energy by cables from the nearest station. The proposal of this research would eliminate the use of this wiring to have only generators. The lifetime of the proposed electric generators is 10 years operating 16 h a day (57,600 h of electric power each).

The electrical energy produced by each type of generator during its lifetime is shown in Table 3, Emitting 0.002 FC, 0.001 FC, and 0.001 FC for each kWh generated as shown in Table 4.

4 Conclusions

The proposed generator can replace the traditional method of rural electrification with the difference that one would be needed for each house and would be independent of a power plant.

The total cost of the proposal is S/. 2 540 240.00, being a cheaper cost than having to put together a wiring network from the nearest electric station. Even cheaper than other options of renewable energy, like a wind energy generator which investment in a project in Cajamarca (similar population as Ayacucho) was more than 26 million dollars [19].

The carbon footprint emitted by the production of one kWh according to the electric mix of Peru is 0.311, with the proposed generators being 99.5% (generator 2 kW), 99.7% (generator 5 kW), and 99.8% (generator 7.5 kW) less pollutant for each kWh produced.

Table 3 Footprint emitted

Material	CF	Units	Generator 2 kW (kg)	Generator 5 kW (kg)	Generator 7.5 kW (kg)
Steel plate	1.76	Kg	40	52	60
Cooper coil	1.86	Kg	8	20	28
Brass	2.46	Kg	2	2	2
Drawn steel h11	2.124	Kg	3.5	3.5	5
Aluminum	1.2	Kg	3.6	3.6	3.6
Cementing steel	1.48	Kg	3.29	3.29	3.29
PVC	3.57	Kg	3	3	3
Steel H	1.76	Kg	22	27	36
Nodular cast iron	1.76	Kg	4.5	4.5	4.5
Cooper coil	1.86	Kg	1.5	1.5	1.5
Flatiron	1.76	Kg	2	2	2
Transporting by boat	0.0040286	tkm	1.92	2.82	3.42
Electric power	0.311	kWh	9.64	9.64	9.64
		Total CF	173.48898	225.7326018	273.7210163
		CF/ kw.h	**0.002**	**0.001**	**0.001**

Table 4 kW H generated

Generator	2 kW	5 kW	7.5 kW
kW.h produced (10 years)	115,200.00	288,000.00	432,000.00

References

1. Cabraal, R., Barnes, D., & Agarwal, S. Productive uses of Energy for Rural Development. Annual Review of Environment and Resources 2005 30:1, 117–144. (2005).
2. The World Bank. Project Performance Assessment Report-Rural Electrification Project Peru. Published on June 30th of 2017. https://ieg.worldbankgroup.org/sites/default/files/Data/reports/ppar-perururalelectrification-09012017.pdf (2017).
3. INEI. Encuesta Demográfica y de Salud Familiar. Accessed February 10th of 2021. https://www.inei.gob.pe/media/MenuRecursivo/publicaciones_digitales/Est/Endes2019/pdf/resumenejecutivo.pdf (2019).
4. Ministerio de Energía y Minas – MINEM. Plan Nacional de Electrificación Rural, PNER 2016-2025. http://extwprlegs1.fao.org/docs/pdf/per153304anx1.pdf (2016).
5. United Nations' Sustainable Development Goals – UNDP. Sustainable Development Goals. Goal 7: affordable and clean energy. www.undp.org/content/dam/undp/library/corporate/brochure/SDGs_Booklet_Web_En.pdf (2015).
6. Alkarkhi, A., & Alqaraghuli, W. Cluster Analysis. Easy Statistics for Food Science with R. Chapter 11: Cluster Analysis. Pages 177–186. Elsevier. https://doi.org/10.1016/B978-0-12-814262-2.00011-X(2019).

7. Wierzchoń, S., & Kłopotek, M. Modern Algorithms of Cluster Analysis. Studies in Big Data. Springer International Publishing AG 2018.Volume 34. https://doi.org/10.1007/978-3-319-69308-8 (2018).
8. Adolfsson, A., Ackerman, M., & Brownstein, N. To Cluster, or Not to Cluster: An Analysis of Clusterability Methods. Pattern Recognition. https://doi.org/10.1016/j.patcog.2018.10.026 (2018).
9. Goodeve, C. Operational Research. Nature No. 4089 – March 13, 1948. Vol 161 (4089) pages 377–384. https://doi.org/10.1038/161377a0 (1948).
10. Institute for Operations Research and the Management Sciences – INFORMS. FAQs About O.R. & Analytics. www.informs.org/Resource-Center/INFORMS-Student-Union/Career-FAQs (2018).
11. Taha, H. A. Operations Research: An Introduction, 10th Edition. Pearson Education. (2017).
12. Dantzig, g. & Thapa, M. Linear Programming. Vol 1 – Introduction. Springer New York. (2015).
13. Romero, B. El Análisis del Ciclo de Vida y la Gestión Ambiental Gestión Ambiental. Boletín IIE, July-September 2003. (2003).
14. Carranza, R. Instrumentos de gestión ambiental. Análisis de ciclo de vida. Pages 37–42. (2014).
15. Jolliet, O., Saadé-Sbeih, M., Shaked, S., Jolliet, A., Crettaz, P. Environmental life Cycle assessment. 2016 by Taylor & Francis Group. (2016).
16. Demirbas, A. Waste Energy for Life Cycle Assessment. Chapter 3: Unconventional Energy Sources (pages 71–118). Springer International Publishing Switzerland 2016. (2016).
17. Xu, B., Chen, D., Li, H., Zhuang, K., Hu, X., Li, J., Skjelbred, H.I., Kong, J., Patelli, E. Priority analysis for risk factors of equipment in a hydraulic turbine generator unit, Journal of Loss Prevention in the Process Industries. (2019).
18. International hydropower association. Hydropower Status Report 2020. Published on May 28th, 2020. https://www.hydropower.org/publications/2020-hydropower-status-report (2020).
19. Cajamarca: Invertirán más de US$ 26 millones en nuevo parque eólico de Huambos. Rumbo minero internacional (2019).

Maintenance Facility Location and Routing Optimization for a Company That Provides Electrical Services

Wilmer Atoche, Renzo Benavente, and Victor Farro

Abstract This research proposes a solution for a company that provides electrical services and outsources maintenance activities in the city. A model was developed to determine the number of outsourced company headquarters according to their location, and another model to set routes that optimize each trip of the maintenance teams.

Key words Facility location · Routing problem · Maintenance activities · Electrical services

1 Introduction

Lima is a city in constant population growth, it is for this reason that the requirements of basic services of water, sewage and electricity are increasingly in demand. This article tries to optimize the attention to the requirements of new medium voltage and low voltage electrification works, whether aerial or underground. The main objective of the research is to determine the adequate number of bases to meet the future demand for crews to electrification works.

Historical data and future demands were obtained from a major electricity company operating in the city of Lima. Currently, the average base-work and base-work travel time is close to 100 min, values that are discounted to the daily workday. The works have a duration of approximately 7 days working an average crew of 5 people, these calculations mean that there are about 60 man-hours not worked due to the daily commute.

W. Atoche · R. Benavente (✉) · V. Farro
Pontificia Universidad Católica del Perú, Lima, Peru
e-mail: watoche@pucp.edu.pe; renzo.benavente@pucp.pe

J. Vargas Florez et al. (eds.), *Production and Operations Management*, Springer Proceedings in Mathematics & Statistics 391,
https://doi.org/10.1007/978-3-031-06862-1_9

2 State of the Art

2.1 Linear Programming

Mathematical programming is a modeling technique used in the decision-making process. When it comes to solving a linear programming problem, the first stage is to identify the possible decisions that can be made; This leads to identifying the decision variables. Normally, the variables are quantitative in nature and the values that optimize the objective function are sought. The second stage involves determining which decisions are admissible; This leads to a set of constraints that are determined with the nature of the problem in mind in mind. In the third stage, the cost-benefit associated with each admissible decision is calculated; this involves determining an objective function that assigns to each possible set of values for the decision variables, a cost-benefit value. The set of all these elements defines the optimization problem. Linear programming, which deals exclusively with a linear objective function, is a part of mathematical programming, and one of the most important areas of applied mathematics [1–3].

2.2 Location Problem

The location problem seeks to obtain the best location for the implementation of a base considering variables such as the number of existing points, the distance between them and the possible location of the base.

In general, there are several strategies to determine the location of a base, such as a center of gravity approach to locate a single warehouse considering the distances between nodes [4]. A variation of this strategy, for cases where you want to locate more than one bucket, is the k-means method. This grouping methodology divides a group of customers into k groups and the store would be in the center of gravity of each group [5].

A warehouse location optimization model requires the following data [6]:

$$Min = \sum_{i=1}^{n} f_i y_i + \sum_{i=1}^{n} \sum_{j=1}^{m} c_{ij} x_{ij} \tag{1}$$

$$\sum_{i=1}^{n} x_{ij} = D_j, \forall j = 1, \ldots, m \tag{2}$$

$$\sum_{j=1}^{m} x_{ij} \leq K_i y_i, \forall i = 1, \ldots, n \tag{3}$$

$$y_i \in \{0; 1\}, \forall i = 1, \ldots, n \quad (4)$$

$$x_{ij} \geq 0, \forall i = 1, \ldots, n \cup j = 1, \ldots, m \quad (5)$$

This general model has the following sets, parameters, and variables:

- n, Number of possible locations and capacities of potential warehouses
- m, number of demand points D_j = demand of point j
- K_i, Capacity of the potential warehouse
- f_i = Fixed cost of the potential warehouse i
- c_{ij}, Cost of moving goods from warehouse i to point j
  Having the following decision variables:
- y_i, 1 if the warehouse is installed, 0 if it is closed
- x_{ij}, Quantity shipped from warehouse i to point j

Regarding the model, the objective function (1) seeks to minimize the fixed cost of building a warehouse and the total cost of distribution to all the demanding points. Constraint (2) defines that demand must be respected, while (3) limits the maximum capacity that each warehouse can supply. Constraints (4) and (5) define the range of existence, "y" as a binary variable and "x" as a nonnegative variable.

2.3 Allocation Problem

The next stage is to determine what tasks each of the bases should perform. In the bibliography we find that engineering services often depend on time windows, which is why a correct assignment of tasks is of utmost importance [7]. The objective of this assignment is to be complemented with the results of the location of bases to be able to evaluate each scenario globally.

3 Methodology and Proposed Model

3.1 Description of the Problem

There is a finite number of bases, from which attention is paid to the different low and medium voltage electrical works at different points of the concession of the electricity supply company. The objective of the model is to find the optimal number of bases that minimize the total transfer time from base-job and job-base for future demand.

As a second objective, once the number of bases has been defined and a demand for the next day is known, it is sought to plan in the best way the assignment of crews to the defined jobs.

3.2 Optimization Model

The integer linear programming model aims to optimize the optimal number of contractor bases to meet the future demand for low voltage and medium voltage works in the different districts of Lima. The parameters and input data required by the program are:

1. The possible reference locations of the outsourced bases (latitude and longitude)
2. The minimum and maximum annual service capacity of each contractor base; this can be set with the fleet size at each base point.
3. The demand projections for the coming years of the low voltage and medium voltage works, transformed into the number of annual trips, located referentially in the different districts of Lima (latitude, longitude)

The objective of the model is the determination of the number of optimal bases, considering the displacement of the possible bases to the different future jobs, considering that the number of bases is an input data. That is, the model will be programmed with different base values and comparing the values of the objective function will take the smallest.

The input parameters are the locations of the possible bases of the outsourcings, the reference locations of the future demands in the different districts of South Lima, and the updated capacities in each base of the contractor. The objective function is to minimize the total travel time of future demands. Figure 1 shows an estimate of the demand in various parts of the city to have a better overview of the problem.

The output of the program is the determination of the number of optimal bases, as well as the future allocation in number of trips to each base. The algebraic model has the following considerations:

- Sets
  - I: Set of location points of the possible N bases.
  - J: Set of points of location of the possible P zones of future demands.
- Parameters
  - MinCapacity: Minimum quantity of annual trips available in base i.
  - MaxCapacity: Maximum number of annual trips available in base i.
  - TotalTime: Total duration of going from base i to point j.
  - K: Number of bases to activate out of N possible available.
- Variables
  - x_{ij}: Decision to assign base i to point j.
  - y_i: Decision to have the database active i.

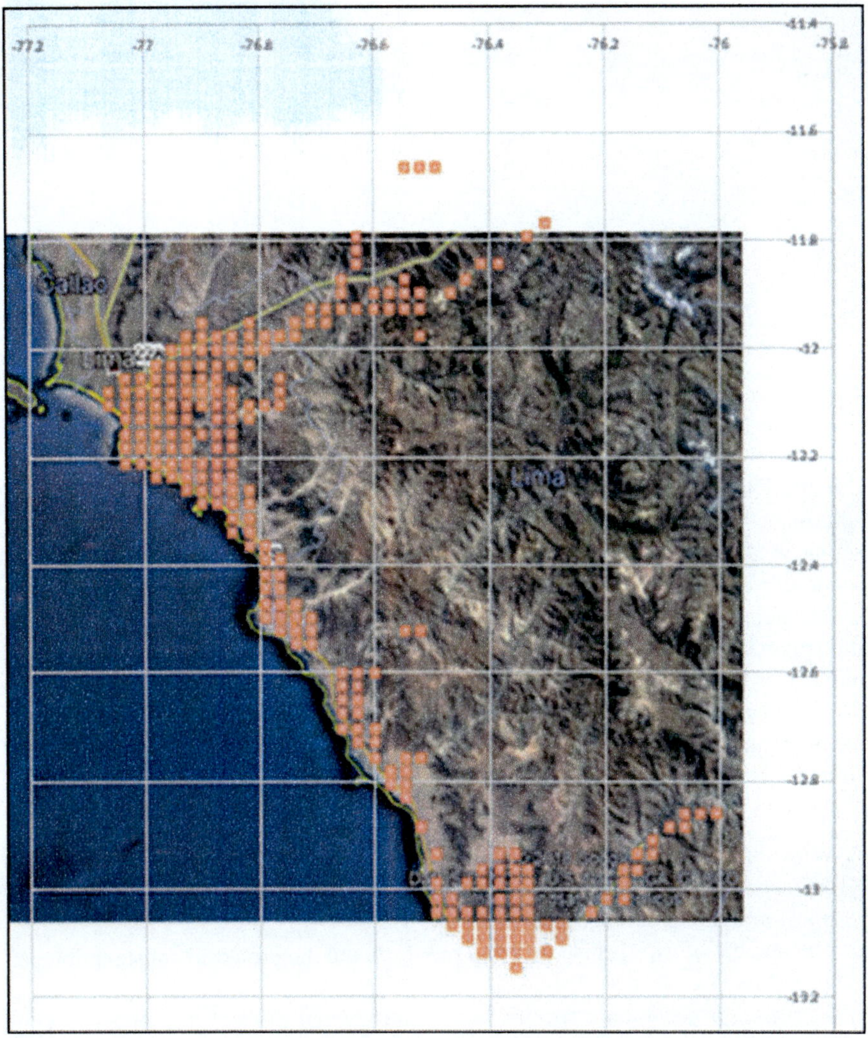

Fig. 1 Geographical location of the demand

$$Min \sum_{i=1}^{I} \sum_{j=1}^{J} x_{ij} TotalTime_{ij} \tag{6}$$

$$\sum_{j=1}^{J} x_{ij} \leq MaxCapacity_i y_i, \forall i \in I \tag{7}$$

$$\sum_{j=1}^{J} x_{ij} \geq MinCapacity_i y_i, \forall i \in I \quad (8)$$

$$\sum_{i=1}^{I} x_{ij} = 1, \forall j \in \mathrm{j} \quad (9)$$

$$\sum_{i=1}^{I} y_i = k \quad (10)$$

$$x_{ij} \in \{0; 1\} \quad (11)$$

$$y_i \in \{0; 1\} \quad (12)$$

For this calculation, 12 existing bases with which we are working in Lima and a proposal of 20 locations for new bases were considered. The idea of this model is to determine how many bases are needed in the next decade to meet the demand. In addition to this, it is also important to determine if we will continue with current suppliers, the contract will be canceled for others, and finally new suppliers will be found that can work in the locations determined by the model.

The objective function is to minimize the total daily commute time. The output of the program is the daily assignment of new construction to base and shows the total number of hours of displacement. The information is obtained from the historical data of a current contractor serving the study company. It is worth mentioning that the minimum capacity of the new bases is equal to the average capacity of the current bases. It is determined based on the quantities by type of works that could be attended in an average year, considering the average duration in days of the same.

1. With the historical information, the demand for 2025, 2030 and 2035 was estimated. This demand is calculated first by number of works and then transformed into number of trips.
2. The number of trips is distributed in proportion to the projected consumption per district and per year.
3. The travel projection of a specific year is taken, and the model is executed varying the number of available bases, the value of "k".
4. Step 4 is repeated for the other projected years.
5. The sensitivity of the variation of the parameter "k" in each of the evaluated years is analyzed to determine which would be the recommended one.
6. Steps 4, 5, and 6 are performed for three different scenarios:
7. Considering only the current 12 bases.
8. Considering the current 12 bases and the 20 proposals (32 in total).

The second part of this analysis is a short-term planning model to assign the jobs to each of the bases. The linear programming model aims to optimize the daily assignment of trips from the outsourced bases to the operations in the jobs, the parameters and input data required by the program are:

1. The locations of the contractors' bases (latitude and longitude)
2. The travel capacity of each contractor base; this can be set with the fleet size at each base point.
3. The daily demand points in the works, located in the different districts of Lima (latitude, longitude)
4. The factor base of standard times per kilometer for each district of South Lima

   Then the demand points are analyzed, following the procedure:

   - If the demand point belongs to a work in progress (it is not the first day), a vehicular unit from the base of the contractor that is attending this work is committed, with its respective resources (crew, machinery, materials). The capacity of the contractor's base will be reduced by one unit.
   - If the demand point is new (it is the first day), the assignment of the base of the contractor that will serve it will be done through optimization using the linear programming model.
   - The variables of the linear programming model is the assignment of the new points to one of the bases, for this the total time of travel from the base to work and from work to base is considered, considering that the vehicle remains on the work all the workday.

The input parameters are:

- Contractor base locations (12 current bases).
- The locations of the new demand points, that is, the locations of the works that are intended to be attended to during the day, in their respective quadrant, previously assigned.
- Up-to-date capabilities at each contractor base.

4 Results

The following results are based on the following scenario: All 32 contractor bases are used (12 existing and 20 new). The results clearly show that the demand can be satisfied with only 7 or 8 bases of the 32, having an average travel time of 34 or 31 min, respectively (Figs. 2 and 3).

It can be seen in Table 1, that for the years 2020 and 2025, the bases to be used must be in Chorrillos, La Victoria, Lurín, Surquillo, Villa el Salvador, Villa María del Triunfo, Cañete, Ate Vitarte.

The result of the second model for a group of days as a sample is shown in Table 2. The transport time is reduced by 43%.

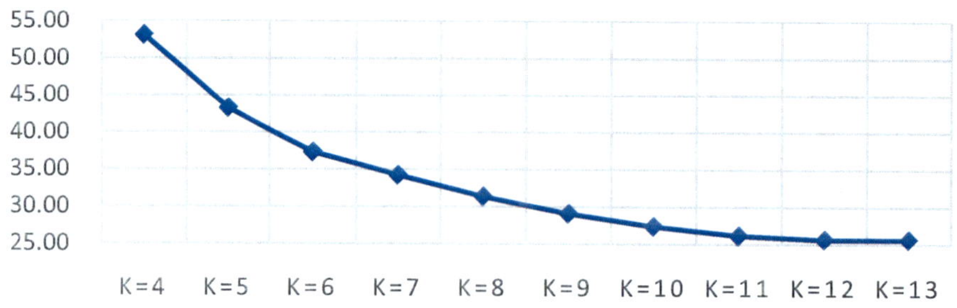

Fig. 2 Number of bases versus minutes per trip, year 2025

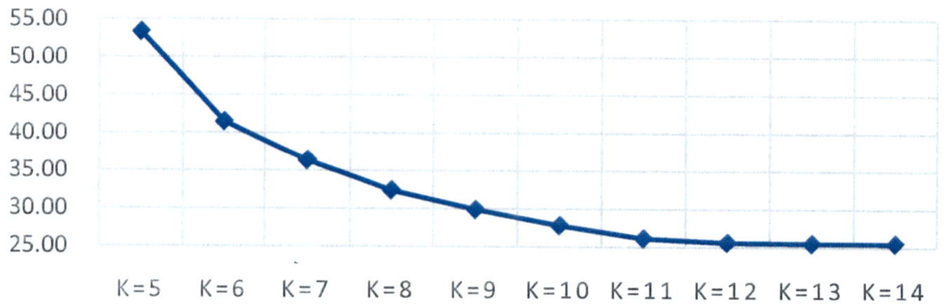

Fig. 3 Number of bases versus minutes per trip, year 2035

Table 1 Demand projection in number of trips

	2025	
District	k = 7	k = 8
Ate Vitarte	9,888	7,709
Chorrillos	5,081	5,081
Surquillo	16,329	10,635
Villa El Salvador	7,143	7,143
Villa Maria del Triunfo	14,098	14,098
Cañete	2,499	2,499
Pucusana	2,627	2,627
La Victoria	0	7,873
Total time (min)	1,973,816	1,808,656
Journey time (min)	34	31

Table 2 Improvement with the allocation model

	Jobs	Stopped	Total time	Average time
Current situation	23	20	1,461	64
Current simulation	43	0	3,712	86
Proposal model	43	0	2,117	49

5 Conclusions

The purpose of this study was to have a scientifically based plan to improve the transfer of personnel from the base to the construction sites. The measurements are carried out under normal working conditions, within the concessionaire's bases and the transfers base-work, work-base and work-work.

The average preparation times observed on the basis are 38 min and converted into standard time would reach 46 min, this is one of the important reductions in working time, which affects the performance of work execution.

The transfer times from base to work is 51 min and the return to work to base is 53 min, this reduces the working day by approximately 104 min, affecting performance.

The demand was projected until 2035, having different future demands in the districts, this served as input data to calculate through the base location model, which, with a limited number of bases, can meet all future demand. This model is used to estimate the daily work-base and base-work travel time, the results being more favorable than the current operating values.

References

1. Winston, Wayne L. (1994). Operations research: applications and algorithms. Belmont, CA: Duxbury Press, 1994.
2. Taha, Hamdy A. (2012). Investigación de operaciones. México D.F.: Pearson Educación, 2012.
3. Hillier, F. S., & Lieberman, G. J. (2010). Introducción a la investigación de operaciones. México, D.F: McGraw-Hill.
4. IACOB, S. V. (2014). Distribution center optimum localization and the gravitational model. Journal of Applied Quantitative Methods, 62.
5. Zhao, Y. (2012). R and data mining: Examples and case studies. Academic Press.
6. Chopra, S., & Meindl, P. (2013) Administración de la cadena de suministro. Quinta edición. México.
7. Johns, S. (1995). Heuristics to schedule service engineers within time windows. Journal of the Operational Research Society, 46(3), 339–346.

Condition-Based Maintenance Program on Lithium-Ion Batteries Using Artificial Intelligence for Aeronautical Operations Management

Fernando Garay, William Huaman, Wilmer Atoche, and Elmar Franco

Abstract On 2013, all Boeing 787 were grounded due to events of deflagration in lithium-batteries installed in these aircraft, it generated subsequently changes in the flight itinerary, dissatisfaction in customers and expenses in maintenance costs in many companies around the world, losing about $22,000 per hour. For this reason, condition-based maintenance program was performed using State of Health and Remaining Useful Life indicator. A new technique Machine Learning was used for solves regression problems in non-parametric data, called Gaussian Processes, this emerging algorithm of Artificial Intelligence generates predictive models based on previous knowledge, giving a probability distribution that follows the current state, allowing interpret the reliability of the component in different cycles of useful life. The paper used the dataset from the NASA repository, due to it has the same internal composition and is tested run to failure. Kernel mixed Matern1.5 + Matern2.5 got good results versus other mixtures during the different test, mapping the real behavior of the battery. The health status diagnostic was quantitatively evaluated and it got results of 98.34% and 1.13% in R^2 and in RMSE respectively, likewise the model served to forecast the remaining useful life of the battery, predicting 64 cycles with a minimum error of 1.53% in reference to the real data. Finally, it helped development a condition-based predictive maintenance program that generated a return on investment (ROI) of 173% and a profit of $331,360 during the first year.

Key words Artificial intelligence · Aircraft maintenance · Lithium-ion battery · Gaussian process

---

F. Garay (✉) · W. Huaman · E. Franco
Faculty of Mechanical Engineering, National University Engineering, Lima, Peru
e-mail: Fernando.garay.v@uni.pe; william.huaman.u@uni.pe; efrancog@uni.edu.pe

W. Atoche
Faculty of Mechanical Engineering, National University Engineering, Lima, Peru

Engineering Department, Pontifical Catholic University of Peru, Lima, Peru
e-mail: watoche@pucp.edu.pe

J. Vargas Florez et al. (eds.), *Production and Operations Management*, Springer Proceedings in Mathematics & Statistics 391,
https://doi.org/10.1007/978-3-031-06862-1_10

1 Introduction

The use of batteries in aircrafts of transport civil requires rigorous maintenance tasks. It is due to the batteries are considered the last source power used in electrical failure when the main sources (engines) aren't available [1].

Nowadays, the management aeronautical has maintenance tasks based on repair periodical and scheduled (planned) or in some cases under condition, it include the management of the battery and the life useful operation. Maintenance predictive helps to identify anomalous behavior from history data of failure and turns it into valuable information for proactive maintenance, avoiding downtime or operational accidents. Thus, is necessary to have technical knowledge of the equipment. It can be achieved by installing sensors to record and monitor the target variables, so the alerts turn on when some values are close to threshold set [2]. By other hand, Airlines monitor health status of aircrafts fleet, provides the opportunity to identify previous failures and perform proactive maintenance to reduce service maintenance unscheduled [3]. Condition-Based Maintenance (CBM) is a strategy maintenance and give information about the condition of the equipment, it provide maintenance tasks effective. The main purpose of CBM is prevent functional failures or prevent a significant decrease in performance asset [4]. Recently, Condition-Based Maintenance strategies have been proposed to reduce the number of maintenance tasks and preserve safety operational [5]. Currently in the era of the pandemic, the automation based in artificial intelligence and big data, helps airlines to be agile and focused on the safety passenger while keeping costs under control. The Covid-19 increase the need to get solution based Artificial Intelligence (AI) in aeronautical operations, as soon as possible. Nowadays, need to be creative and adopt advanced analytics solutions to maintain profitability, while the industry recovers. Carrying out the maintenance schedule earlier based on the failures predicted by AI model will increase safety and help to better plan the maintenance task [6].

2 Aeronautical Maintenance

Currently, Aircraft maintenance usually is divided into two main types: corrective maintenance, also called reactive, it includes all action maintenance (technical or management) that do to return to service the aircraft or their components after of an emergence (downtime). It is performed after a failure or after of an aleatory failure own nature operational, it can't be planned, for this reason, it is also called unscheduled maintenance. On the other hand, preventive maintenance is a way to extend the operation of aircraft, planning and scheduling tasks, the main goal of preventive maintenance is to avoid unscheduled downtime or failures that would result in corrective maintenance [7]. Aircraft maintenance is crucial for safe and efficient aircraft operations and therefore airlines spend almost 9.5% of their operating costs on maintenance, while striving for profitable maintenance. The safety remains a priority for aircraft operators [8]. The costs agreed to maintenance can provide significantly costs in an airline. Historically, the maintenance costs are calculated in a mean of 10% and 15% of overall costs in an Airlines [9]. Maintenance costs are a part important into total operation costs in most industrial sectors, but the real impact is often underestimated. The "Iceberg Model" highlights the impact of hidden (indirect) maintenance costs on the business, which is much higher than the direct costs associated with traditional maintenance [10].

3 Degradation Parameters

The degradation parameters that affect aging of lithium-ion batteries and were used for the development of this research: State of Health (SOH) and Remaining Useful Life (RUL). The SOH and RUL of lithium-ion batteries are two critical factors that typically are predicted using the capacity [11]. Currently these indicators are widely used in different literatures regarding to prognostic and health management on batteries.

3.1 State of Health

The SOH describes the condition of the battery during its operational life. When it stars its operation, the battery is in 100% or otherwise, the SOH is equal to 1. After a cycle, the SOH value is reduced according to the operating profile in which the discharge cycle occurs [1]. SOH is defined as the level of internal degradation of the battery during a period, it represents its internal features, such as the reduction of the useful capacity and the increase of the internal resistance [12]. SOH is related to power or energy fading, as internal resistance increases with age and the available power of the battery fades. Battery health is critical to the reliability and safety of the energy storage system. The prediction of the state of the battery has an important role, this ensures the good operation and decreases the maintenance costs [13]. The resulting health status provides an indication of the condition of the battery during its operation.

3.2 Remaining Useful Life

The RUL is considered as the time that pass from the moment initial of operation to the end of its useful life [14]. Other concept talks about the time remaining or the number of charge and discharge cycles before it reaches SOH at 0% [15]. It is important to understand that SOH is a degradation parameter which provides information to predict RUL, it is a parameter to determine the end of battery life.

Therefore, for the evaluation of energy storage assets, depreciation, warranty, insurance and maintenance, the calculation of the prediction of the remaining useful life at the design stage and during operation is crucial to determine the value of the asset. The future of operation [16]. The remaining useful life is essential for condition-based maintenance and health management, it is usually random and unknown, for this reason it must be estimated from available information sources such as information obtained on health status [17]. The estimated of RUL calculated from the management of the battery health give information about the cycles remaining that the batteries can operate until a reach the threshold of health, considering the failure component over these limits [1].

4 Gaussian Process Regression

Gaussian process regression (GPR) has an approach in the learn supervised. It is used in complex regression problems and in dimensions with few data that show discontinue linear in their characteristics. The main idea is to consider random functions, using multiple random variables to obtain a joint normal distribution of dimensions [18]. Compared to other algorithms like: Neural Network (NN) and Support Vector Machine (SVM), GPR has the advantage of being composed of hyperparameters that can be tuned, it is also simple to implement and has no performance loss. The results [19].

Recently, researchers have developed the GPR method, since it uses a prediction model under the Bayesian framework [20]. Gaussian processes (GP) are very powerful tools provided by artificial intelligence. These allow predictions to be made about the data, incorporating previous knowledge [21]. The Gaussian process is constituted as a probability distribution over random functions, so that the subset of variables follows a joint Gaussian distribution, the GP function f (x) is defined in the Eq. (1), where it is fully specified by the mean function m (x) and the covariance function k (x, x′), both of which are shown in the Eqs. (2) and (3) respectively:

$$f(x) \sim GP(m(x), k(x, x')) \tag{1}$$

$$m(x) = E[f(x)] \tag{2}$$

$$k(x, x') = E[(f(x) - m(x))(f(x') - m(x'))] \tag{3}$$

The GRP are used on random variables that can be scalars or vectors. It is a stochastic process used in the properties of functions and it is based on Bayesian inferences. Therefore, this process has a finite random variable, which has a joint Gaussian distribution, and the GP characters are completely determined by the mean function and the covariance function [21].

4.1 Covariance Function

The covariance function, which is also called Kernel. It describes the shape of the distribution and determines the characteristics of the function that we want to predict, we generate a matrix evaluating the Kernel. The Kernel function is a key in the GPR, it is capable of discover useful characteristics, such as smoothness, periodicity and non-stationary in the data through previous assumptions. For example, the quadratic exponential kernel is appropriate for uniform structures [22]. Over the last decade there has been many projects in Machine Learning using kernel, probably the best-known example is the algorithm SVM, also during this period there has been a lot of researches regarding the application of Gaussian process models to machine learning tasks [21]. The kernel has the function of capture the correlation between the inputs and the dominant factor in the prediction of a Gaussian process, commonly the kernel has capacity of approximate different processes [23].

Radial Basis Function (RBF) It is a stationary kernel. It is usually called squared exponential kernel. It has a size scale parameter $l > 0$, it can be a scalar parameter as a isotropic kernel variant or it can have the same number of dimensions as the inputs as anisotropic kernel [24], it is detailed in Eq. (4).

$$k(x_i, x_j) = \exp\left(-\frac{d(x_i, x_j)^2}{2l^2}\right) \tag{4}$$

Where: d (·,·) is the Euclidean distance, x_i, x_j are values correlatives and l are the parameter scale.

Matern It is a kernel stationary, it is a generalization of the kind RBF Kernel. It has a characteristic special, it adjusts the smoothness of the final resulting function [24], and other hand, it is structured by a size scale parameter $l > 0$, it can be a scalar (isotropic kernel variant). The Matern has similar features to RBF and these have also as parameter the Euclidean distance, the mathematical structure is detailed in Eq. (5).

$$k(x_i, x_j) = \frac{1}{\Gamma(\nu)2^{\nu-1}}\left(\frac{\sqrt{2\nu}}{l}d(x_i, x_j)\right)^{\nu} K_{\nu}\left(\frac{\sqrt{2\nu}}{l}d(x_i, x_j)\right) \tag{5}$$

Where: d (·,·) is the Euclidean distance, x_i, x_j are values correlatives, l is the parameter scale, by other hand, the values such as $K\nu$ (·) and Γ (·) are a modified of bessel and gamma function, Eq. (6) details the equation.

$$(x_i, x_j) = \exp\left(-\frac{1}{l}d(x_i, x_j)\right) \ v = \frac{1}{2} \tag{6}$$

In particular cases ν can be 3/2 and 5/2, both detailed in Eqs. (7) and (8) respectively.

$$k(x_i, x_j) = \left(1 + \frac{\sqrt{3}}{l}d(x_i, x_j)\right) \exp\left(-\frac{\sqrt{3}}{l}d(x_i, x_j)\right) \tag{7}$$

$$k(x_i, x_j) = \left(1 + \frac{\sqrt{5}}{l}d(x_i, x_j) + \frac{\sqrt{5}}{3l}d(x_i, x_j)^2\right) \exp\left(-\frac{\sqrt{5}}{l}d(x_i, x_j)\right) \tag{8}$$

These are commonly chosen for functions more complex, they aren't differentiable (as the RBF Kernel) but can be adjusted, the tow commonly used are $\nu = 3/2$ and $\nu = 5/2$. These flexibilities are used for adjust the smoothness of the function result, through of the parameter "ν", it allows to adapt its properties with the true functional relationship predict.

Rotational Quadratic The RQ kernel is showed as a particular mixture of RBF Kernel with many characteristics in the length scales. It is confirmed by a size scale parameter $l > 0$ and a scale mixture parameter $\alpha > 0$, they are an isotropic variant, where they have a compatible scalar [24]. The kernel is shown in Eq. (9) by:

$$k(x_i, x_j) = \left(1 + \frac{d(x_i, x_j)^2}{2\alpha l^2}\right)^{-\alpha} \tag{9}$$

Exponential Sin Quadratic (ESQ) It allows to model periodic functions. It is structured by a length scale parameter and a periodicity parameter greater than zero (l and p). On the other hand, this kernel has an isotropic variant, which has a compatible scale that allows calculations to be carried out trigonometrically [24]. The function is shown in the next Eq. (10):

$$k(x_i, x_j) = \exp\left(-\frac{2sen^2(\pi d(x_i, x_j)/p)}{l^2}\right) \tag{10}$$

5 Model Development

The model development followed in this work is presented in Fig. 1, first the data was download and preprocessed, later this information was plotted to observe behavior through time and selection the indicators of health keys, such as capacity, temperature and voltage. The stage GPR model was generated combining kernel,

Fig. 1 Model development in stages used to perform the methodology of research based in algorithm of artificial intelligent

finally the best kernel mixture was evaluated with different indicators, for SOH used root mean square error (RMSE) and coefficient of determination (R^2). RUL was evaluated by mean of Error% and Forecast RUL.

5.1 *Dataset Ion-Lithium Batteries*

The Lithium Ion Batteries Dataset (B0005) from the Repository of the National Aeronautics and Space Administration (NASA) was used through 3 different operating profiles (charge, discharge and impedance) to room temperature. This dataset was useful due to that their composition and technology are similar to batteries used in civil and cargo transport aircraft. The data was extracted from the repository NASA dataset.

5.2 *Health Indicator Selection*

Voltage Curve The behavior of the voltage curve during its useful life cycle decreases as the cycles increase, starting from the initial voltage and the life time until it decreases, reaching wear in a period of 3500 s to 2500 s and the voltage for its part drops from 4.2 to 4.0 volts, respectively. Figure 2 shows trend voltage curve.

Temperature Curve On the other hand, the temperature curve during its operation cycle has an opposite behavior to the voltage curve, it increases according to time, from an initial temperature of 24 to 42 °C. On the other hand, the lifetime decreases from 3400 to 2400 s and the final temperature increases from 38 to 42 °C, this behavior indicates that the curve trend is irregular and cannot be analyzed. Figure 3 details behavior temperature curve in different cycles.

Capacity Curve The capacity of lithium ion battery is obtained by measurement of the internal resistance that reflect specific capacity. Due to the impedance of the battery increases with the loss of the capacity. The condition of the impedance is continuous and decrease through life cycle, this remains linear and constant Therefore, the capacity trend has a negative slope and continues, this curve has complex characteristics to monitor and estimate the life of lithium-ion batteries using its capacity to determine a trend in its health status. Figure 4 shows the features observed in the capactity curve.

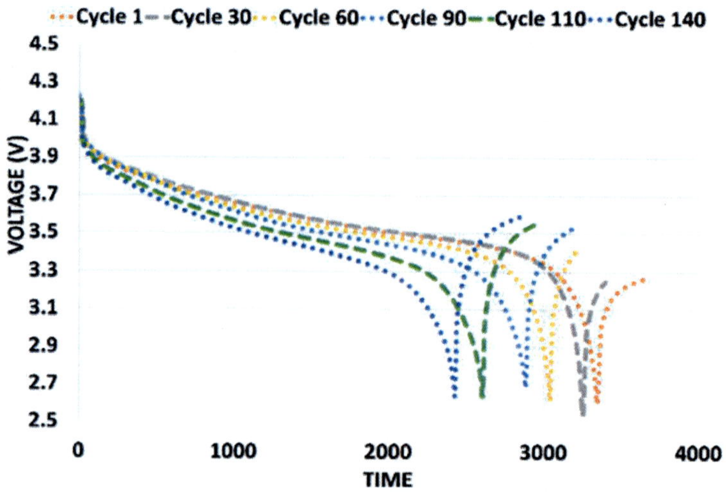

Fig. 2 Voltage curve show the behavior of the voltage in different discharge cycles, it has an irregular behavior and does not maintain an adequate trend; therefore, it was not appropriate to analyze

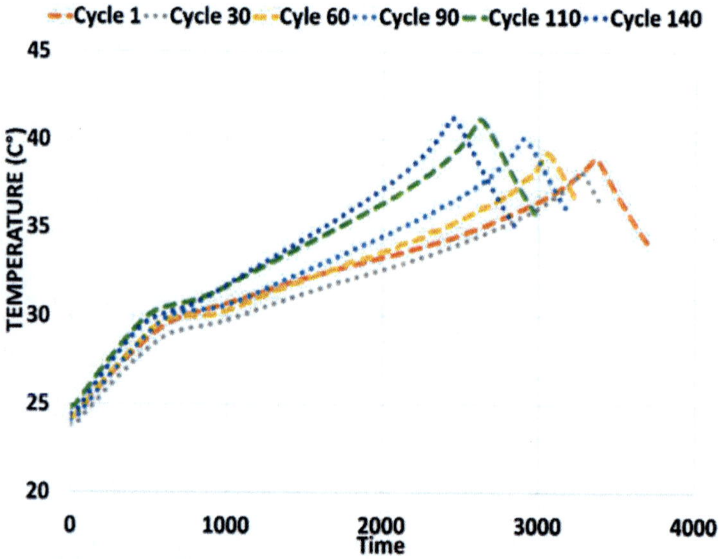

Fig. 3 Temperature curve show the trends of the temperature at different cycles of useful life, the behavior is appositive to the previous curve, making it in the same way not suitable for development of research

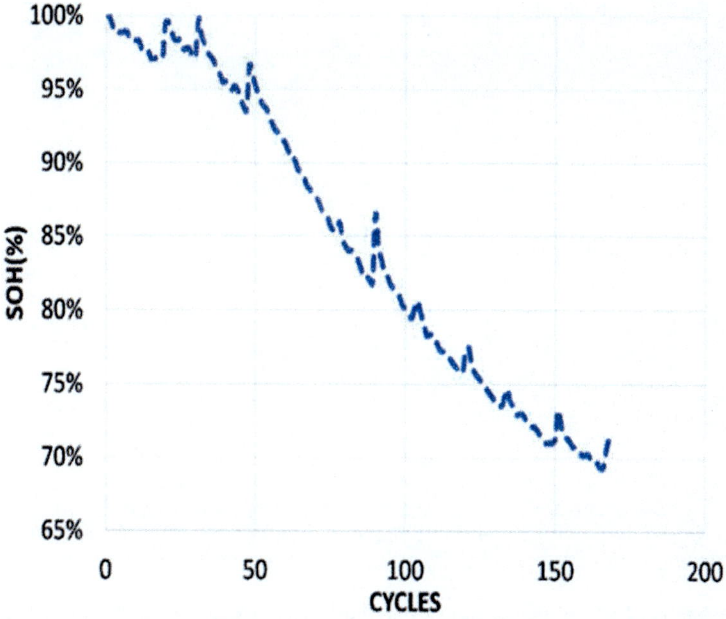

Fig. 4 Capacity curve is optimal to forecast the cycles during useful life, many literatures it uses for condition monitoring and currently is development through of algorithms of artificial intelligent supervised and unsupervised

5.3 *Kernel Mixtures*

In this chapter various combinations were developed; different kernels were tested by setting the optimizer and alpha hyperparameters to 100 and 0.0001 respectively. On the other hand, GPR requires prior data to build insights about the covariance function and the mathematical function. It is usual for covariance functions to contain hyperparameters, these define their specific properties and optimize the results. This is why determining the correct values for each hyperparameter is also a challenge in the elaboration of the predictive model [25].

The metrics quantitative RMSE and R^2 were for the SOH diagnosis and the Error % in the RUL, the cycles are also analyzed and forecast, using the indicator Forecast Cycles. The kernels used for the research were combined and tested at different cycle sizes (60, 68, 82, 90, 108, 109, 110 and 130 cycles) during the test stage to obtain the optimal machine learning model. Table 1 shows the kernel combination used in the tests.

In this research, the nominal capacity of lithium ion batteries is 2 Ah and the failure threshold is set at 1.5 Ah (75%). It can be extended to 1.38 Ah or 70%. When the batteries reach the end of useful life, the capacity of the battery is degraded to

Table 1 Kernel combination

Item	Kernel	Optimizer	Alpha
1	Matern1.5 + Matern2.5 + ESS	100	0.0001
2	Matern2.5 + Matern2.5 + RBF	100	0.0001
3	Matern1.5 + Matern2.5	100	0.0001
4	Matern1.5 + RQ + RBF	100	0.0001
5	Matern1.5 + Matern2.5 + RBF	100	0.0001

Table 2 SOH and RUL results

		Diagnostic SOH		Prognostic RUL	
Test size	Kernel	RMSE test	R^2	Error%	Forecast cycles
60	Matern1.5 + Matern2.5	2.04%	60.27%	11.76%	15
68	Matern 1.5 + Matern 2.5	2.02%	76.19%	12%	22
82	Matern 1.5 + Matern 2.5	4.19%	13.13%	30.76%	27
90	Matern 1.5 + Matern 2.5	2.86%	79.16%	19.15%	38
108	Matern 1.5 + Matern 2.5	1.35%	97.56%	1.53%	64
109	Matern1.5 + Matern 2.5	1.13%	98.34%	−6.06%	70
110	Matern 1.5 + Matern 2.5	1.78%	96.01%	−17.91%	79

approximately 70% of the nominal capacity [26]. As a result, the data helped to generate alarm and alert parameters which improve maintenance by condition with the best optimal model.

6 Results and Discussion

6.1 SOH Diagnostic and RUL Prognostic

Comparing the results through of mixture kernel, the Matern1.5 + Matern2.5 got the best scores versus other mixtures, obtaining 1.13% and 98.34% of RMSE and R^2 respectively for a size of 109 cycles. The results of Error% (difference between real and predicted) of 1.53% and the forecast of 64 cycles (Forecast Cycles) for a size of 108 cycles. Dataset was divided into 35% train and 65% test, getting an algorithm robust, reliable and capable of forecast with short train size. Table 2 shows the results taken with the Matern1.5 + Matern2.5 kernel tests. In addition, the Fig. 5 show the graph of algorithm proposed in python.

6.2 Condition-Based Maintenance Program

The result of the analysis of the monitor and health status of the Lithium ion batteries proposed to generate a Condition-Based Maintenance Program in the Avionics workshop in according to the operation of the 787–8 fleet. The maintenance tasks

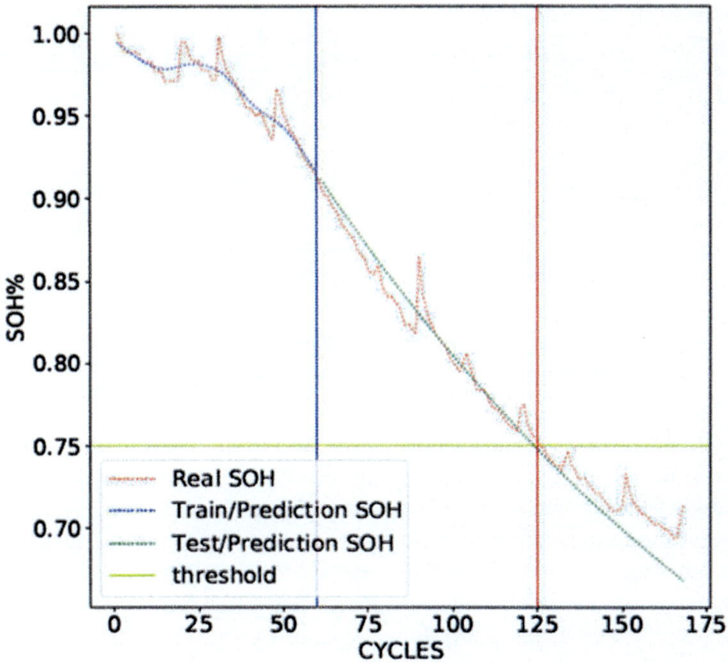

Fig. 5 The model shows the curve of the SOH, providing the forecast of the state of health according to the configuration. On the other hand, this same figure shows the prediction curve of the RUL (difference between the red and blue vertical lines), the red line was established according to the 75% failure threshold. These two curves determined the complexity of the algorithm

by condition focus in the adequate inspection and timely. It does not require stop the aircraft operation for long time and tasks can be perform on service, when the aircraft is in operation during the day or night, carrying out the flight itinerary. On the other hand, is necessary adjust the flight schedule so that the aircraft has an operation on ground, during the services required before to flight (in transit). Three services check were developed through of the data monitoring: periodic check, regular check and Overhaul. Table 3 show the periodical maintenance task for batteries lithium-ion designed.

After that, work orders and tasks were defined for the maintenance program, the followed step was calculated the savings that the implementation generated, as well as the operational logistics that will be added to the development of the program. To calculate these savings, we have to focus on estimating costs that are not yet executed, so used the previous knowledge of the P-F Curve. The project calculation obtained a profit of $331,360.00 per year, financial indicator ROI of 173% and return capital period of 1 year. Table 4 shows in detail the calculation of the savings associated with a failure in the aircrafts B787 fleet, for this analysis the direct costs that can generate atypical maintenance (Cost F) due to these failures are considered in contrast to maintenance based in condition (Cost P).

Table 3 Work orders – MBC

Item	Work order	Task	Interval
1	Periodical check	Visual inspection	Weekly
2	Periodical check	Light cleaning	Weekly
3	Periodical check	Insulation check	Weekly
4	Periodical check	Polarization test	Weekly
5	Regular check	Vent valve cleaning	Biannual
6	Regular check	Electrolyte level adjust	Biannual
7	Regular check	Capacity check	Biannual
8	Overhaul	Disassembly/assembly	Annual
9	Overhaul	Replacement component	Annual
10	Overhaul	Component inspection	Annual

Nevertheless, the algorithm lost efficiency in unsupervised learn and no lineal regression models, on the other hand, is necessary adjust periodically the alert and alarm levels in according to fleet's operation. Currently some investigations show major result using indicator health mixture (i.e. temperature and voltage), Principal Component Analysis and Grey Analysis method for extract main characteristics. At least, this paper will development in the future Gaussian Mixture model for clustering failures in batteries, through historical data of Boeing 787–8 fleet.

7 Conclusions

This paper showed Gaussian Process model as tool for development predictive maintenance programs in lithium-ion batteries for the aeronautical operation management, through efficient tasks, such as: regular, periodical and overhaul check. The best kernel mixture was Matern1.5 + Matern2.5, by other hand, 109 and 108 cycles are obtained for SOH and RUL respectively. Also, the model showed that needs short training data to development efficient prognostic. Finally, return of investment, return period capital and profit calculated for this Condition-based Maintenance Program, assures that project is viable and it is possible to be implemented in others critical components installed in the aircraft (i.e. tires, actuators, valves and bearing).

Table 4 Savings calculation through P-F curve analysis

		Cost P			Cost F			Fleet B787
Cost operational	Cost/h	Qty.	Qty service/year	Total cost	Qty.	Qty service/year	Total cost	Save cost
Downtime Op.	$22,000	2	120	$5,280,000	6	48	$6,336,000	$1,056,000
Mntto work	$25	6	120	$1800	8	48	$9600	−$8400
Extra hours/Mntto	$35	0	120	–	2	48	$3360	$3360
Spare parts			120	$24,000		48	$96,000	−$144,000
Other services			120	$48,000		48	$96,000	−$384,000
Save cost total				$6,018,000.00			$6,540,960.00	$522,960.00

References

1. Moreira, J., Nascimiento, C., & Ramos, L.: Health monitoring and remaining useful life estimation of lithium-ion aeronautical batteries. IEEE Aerospace Conference, (2012).
2. Dangut, M. D., Skaf, Z., & Jennions, I. K.: An integrated machine Learning model for aircraft components rare failure prognostics with log-based dataset. ISA Transactions, (2020).
3. Torhorst, S., Hozel, N., & Gollnick, V.: Identification and evaluation of the potentials of prognostics and health management in future civil aircraft. European conference of the prognostics and health management society, (2014).
4. Teixeira, H. N., Lopes, I., & Braga, A. C.: Condition-based maintenance implementation: a literature review. Procedia Manufacturing, 51, 228–235, (2020).
5. Alaswad, S., & Xiang, Y.: A review on condition-based maintenance optimization models for stochastically deteriorating system. Reliability Engineering & System Safety, 157, 54–63, (2017).
6. Shukla, B., Fan, I., & Jennions, I.: Opportunities for Explainable Artificial Intelligence in Aerospace Predictive Maintenance. 5th European Conference of the Prognostic and Health Management Society, (2020).
7. Sheng, J., & Prescott, D.: A coloured Petri net framework for modelling aircraft fleet maintenance. Reliability Engineering & System Safety, 189, 67–88, (2019).
8. Lee, J., & Mitici, M.: An integrated assessment of safety and efficiency of aircraft maintenance strategies using agent-based modelling and stochastic Petri nets. Reliability Engineering & System Safety, 202, 107052, (2020).
9. Verhagen, W. J. C., & De Boer, L. W. M.: Predictive maintenance for aircraft components using proportional hazard models. Journal of Industrial Information Integration, 12, 23–30, (2020).
10. Wienker, M., Henderson, K., & Volkerts, J.: The Computerized Maintenance Management System an Essential Tool for World Class Maintenance. Procedia Engineering, 138, 413–420, (2016).
11. Jia, J., Liang, J., Shi, Y., Wen, J., Pang, X., & Zeng, J.: SOH and RUL Prediction of Lithium-Ion Batteries Based on Gaussian Process Regression with Indirect Health Indicators. Energies, 13(2), 375, (2020).
12. Wang, J., Liu, P., Hicks-Garner, J., Sherman, E., Soukiazian, S., Verbrugge, M., Tataria, H., Musser, J., & Finamore, P.: Cycle-life model for graphite-LiFePO4 cells. Journal of Power Sources, 196(8), 3942–3948, (2011).
13. Li, X., Yuan, C., Li, X., & Wang, Z.: State of health estimation for Li-Ion battery using incremental capacity analysis and Gaussian process regression. Energy, 190, 116467, (2020).
14. Li, X. Shu, J. Shen, R. Xiao, W. Yan, Z. Chen.: An on-board remaining useful life estimation algorithm for lithium-ion batteries of electric vehicles, Energies 10 (5), (2017).
15. Nejad, S.: Adaptive Techniques for Estimation and Online Monitoring of Battery Energy Storage Devices (Degree of Doctor of Philosophy). The University of Sheffield, England, The United Kingdom, (2017).
16. Richardson, R. R., Osborne, M. A., & Howey, D. A.: Battery health prediction under generalized conditions using a Gaussian process transition model. Journal of Energy Storage, 23, 320–328, (2019).
17. Cui, L. R., Loh, H. T., & Xie, M.: Sequential inspection strategy for multiple systems under availability requirement. European Journal of Operational Research, 155(1), 170–177, (2004).
18. Chen, T., Morris, J., & Martin, E.: Gaussian process regression for multivariate spectroscopic calibration. Chemometrics and Intelligent Laboratory Systems, 87(1), 59–71, (2007).
19. Yang, D., Zhang, X., Pan, R., Wang, Y., & Chen, Z.: A novel Gaussian process regression model for state-of-health estimation of lithium-ion battery using charging curve. Journal of Power Sources, 3S4, 387–395, (2018).
20. He, Y. J., Shen, J. N., Shen, J. F., & Ma, Z. F.: State of health estimation of lithium-ion batteries: A multiscale Gaussian process regression modeling approach. AIChE Journal, 61(5), 1589–1600, (2015).

21. Rasmussen, C., & Williams, C.: Gaussian Processes for Machine Learning. Massachusetts Institute of Technology, Cambridge, The United States of America, (2006).
22. Xiao, F., Li, C., Fan, Y., Yang, G., & Tang, X.: State of charge estimation for lithium-ion battery based on Gaussian process regression with deep recurrent kernel. International Journal of Electrical Power & Energy Systems, 124, 106369, (2021).
23. Kristjanson, D.: Automatic Model Construction with Gaussian Processes (Degree of Doctor of Philosophy). University of Cambridge, England, The United Kingdom, (2014).
24. Garay, F., Huaman, W., & Vargas-Machuca, J.: State of health diagnostic and remain useful life prognostic for lithium ion-battery by combining multi-kernel in Gaussian Process Regression. 2021 IEEE XXVIII International Conference on Electronics, Electrical Engineering and Computing (INTERCON) (2021).
25. Saha, S., Saha, B., Saxena, A., & Goebel, K.: Distributed prognostic health management with gaussian process regression. IEEE Aerospace Conference 2010, (2010).
26. Yu, J.: State of health prediction of lithium-ion batteries: Multiscale logic regression and Gaussian process regression ensemble. Reliability Engineering & System Safety, 174, 82–95, (2018).

The Impact of Electricity Consumption During the COVID-19 Pandemic

Mariana Rodrigues Carvalho Muniz Santos, Camila Pereira de Rezende, Paula Maçaira, and Erick Meira

Abstract The COVID-19 pandemic impacted several sectors of the economy, including the Brazilian electricity sector. Motivated by the considerable changes in energy consumption behavior in Brazil during the pandemic, the present study conducts an application of several statistical models to quantify the impact of restrictive measures imposed by the government on energy consumption throughout the country. In this way, monthly forecasts of electricity consumption over the year 2020 were generated by considering two scenarios: the actual pandemic scenario and a hypothetical, counterfactual scenario in which the pandemic did not take place. The study used electricity consumption data in Brazil extracted from the Brazilian energy research company (EPE). Forecasts in the counterfactual scenario were generated using exponential smoothing methods and hierarchical reconciliation. The differences in consumption between the two designs were then computed and compared against two indexes published by the Oxford COVID-19 Government Response Tracker: the stringency index and the risk of opening the economy. The correlation results suggest that restrictive measures affected consumption in some regions and classes more than others. In particular, the Northern and Northeastern regions were more likely to be more affected by the adoption of restriction measures, along with commercial consumption. Implications are further discussed.

Key words Brazilian electric sector · COVID-19 pandemic · Energy consumption · Time series · Statistical models · Exponential smoothing · Hierarchical and grouped time series · Pearson correlation

M. Rodrigues Carvalho Muniz Santos · C. Pereira de Rezende (✉) · P. Maçaira
Pontifical Catholic University of Rio de Janeiro, Rio de Janeiro, Brazil
e-mail: paulamacaira@puc-rio.br

E. Meira
Pontifical Catholic University of Rio de Janeiro, Rio de Janeiro, Brazil

Brazilian Agency for Research and Innovation, Rio de Janeiro, Brazil
e-mail: research@erickmeira.com

J. Vargas Florez et al. (eds.), *Production and Operations Management*, Springer Proceedings in Mathematics & Statistics 391,
https://doi.org/10.1007/978-3-031-06862-1_11

1 Introduction

In December 2019, in Wuhan City, China, a highly contagious virus, SARS-CoV-2, popularly known as COVID-19, was discovered. The pandemic generated by the virus has reverberated in a global recession for various segments of the economy [1]. Macroeconomic perspectives were reversed in a short period, given the impossibility of predicting the duration and severity of the crisis.

One of the first measures taken by the World Health Organization (WHO) adopted by the governments of most countries in the world was promoting mass social isolation to prevent the spread of the virus, and the consequences of these decisions soon appeared. According to a publication in the Brazilian Official Gazette of September 14, 2020, several sectors were affected, including the electricity sector.

In Brazil, as highlighted by the National Electric System Operator (ONS), since the beginning of the adoption of restrictive measures by the states and municipalities, a rapid power reduction took place, coupled with notable changes in consumption behavior [2]. In this context, residential use increased considerably, while commercial and industrial use plummeted, causing maximum energy demand in the night period [2].

In a survey conducted in conjunction with the Energy Research Company (EPE) and the Chamber of Electric Energy Commercialization (CCEE), the ONS highlighted a 0.9% decline in energy consumption for the year 2020 (ONS, 2020). However, it points out that the annual planning before isolation predicted an increase of 4.2% [2], as shown in Table 1.

The new consumption scenario, revised after considering the impacts of the COVID-19 pandemic, contradicts the increasing trend in energy consumption projected over the years. With the heating up of the economy, the energy demand had been increasing, and the observed trend was upward, as can be seen in Fig. 1.

Table 1 Impact of COVID 19 on energy consumption

Projection	**2019**	**2020**	**2021**	**2022**	**2023**	**2024**
Annual Planning	67.975	70.825	73.453	76.204	79.013	81.931
Variation %	**2,1%**	**4,2%**	**3,7%**	**3,7%**	**3,7%**	**3,7%**
Real	**2019**	**2020**	**2021**	**2022**	**2023**	**2024**
1st Quadrant 2020	67.835	67.249	70.057	72.745	75.385	78.112
Variation %	**1,9%**	**-0,9%**	**4,2%**	**3,8%**	**3,6%**	**3,6%**
[Real] - [Projection]	**-140**	**-3.576**	**-3.396**	**-3.459**	**-3.628**	**-3.819**

Source: Own Authorship

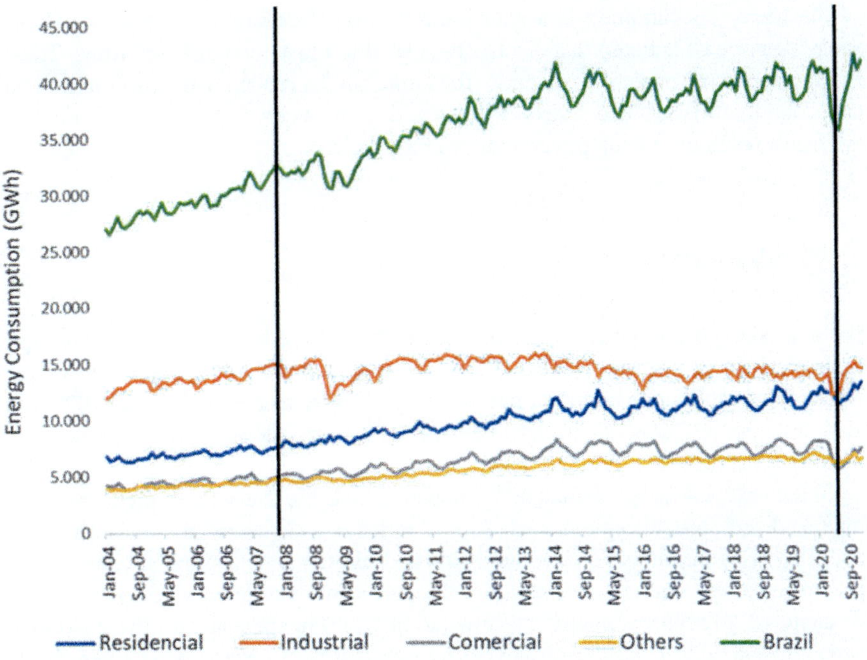

Fig. 1 Energy consumption behavior. (Source: Own Authorship)

Energy demand time series, in general, present a trend and seasonality. However, the graphic showed it is possible to notice two sharp drops in consumption, one in 2008/2009 and one in 2020. The two vertical lines can quickly identify these.

The first fall was caused by the subprime financial crisis that began in the United States and spread worldwide. The second sudden drop in energy consumption relates to the new coronavirus pandemic. This decline in consumption can be justified by the first restriction measures imposed in the country in mid-March.

This paper used the historical values of electricity consumption in Brazil to generate monthly forecasts over the year 2020 in a pandemic-free scenario. The resulting values were then compared with the actual, observed consumption for the same year.

In short, base forecasts were first generated using exponential smoothing models. These models were independently applied to the time series representing electricity consumption across five geographic regions of Brazil (North, Northeast, South, Southeast, and Midwest) and four different classes (residential, industrial, commercial, and others). Forecast reconciliation was then implemented through the Minimum Trace method, an optimal reconciliation approach to time series that follow a hierarchical structure. The differences between the reconciled forecasts and the actual, observed values of electricity consumption were computed. Finally, a correlation analysis considering these differences and the observations for two social isolation indices was conducted.

Electricity consumption is a valuable indicator of economic fluctuations [3, 4] since it can be considered that electricity is used in most economic activities. Thus, the purpose of the paper is to diagnose the impact of the reduction of consumption on the economy of Brazil and assess the relationships between energy consumption and the measures taken by the government to face the crisis.

2 Methodology

For analysis and interpretation of the data, statistical methods were used. Through the assistance of software R [5], it was possible to perform the prediction of the behavior of energy consumption in a pandemic-free scenario for the year 2020 and compare it with what was performed for the same year. The steps of this process are organized according to Fig. 2.

In the first stage, the 25-time series depicted in Table 2 were analyzed, and their main stylized facts were identified. A time series is a sequence of values observed overtime at equal intervals [6]. The analysis of these series applies in cases where past observations contain information about persistent or systematic patterns that can be captured through parametric representations [7]. For the analysis, the 2020 data were considered test set, as this will be the year that the forecasts will be generated.

Exponential smoothing methods are one of the several approaches to generate time series forecasts. These methods make predictions using a weighted average of past observations, with recent observations receiving higher weights than the older ones [8].

Before selecting an exponential smoothing formulation, previous analyses regarding the type of seasonality and trend of the series are suggested [6].

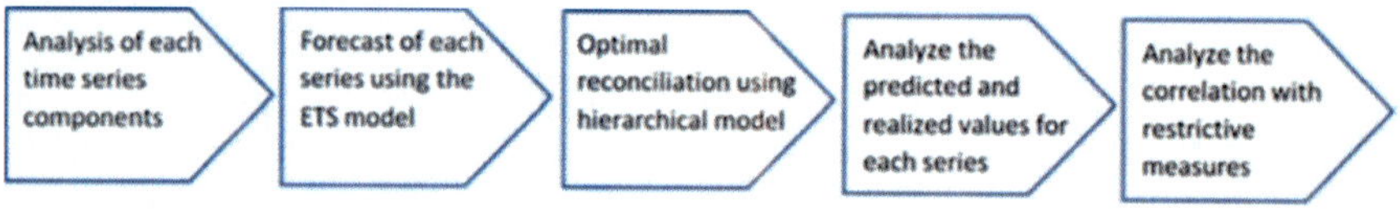

Fig. 2 Data analysis steps. (Source: Own Authorship)

Table 2 Time series of energy consumption

Time series of energy consumption				
North (total)	North (industrial)	North (commercial)	North (residential)	North (others)
Northeast (total)	Northeast (industrial)	Northeast (commercial)	Northeast (residential)	Northeast (others)
South (total)	South (industrial)	South (commercial)	South (residential)	South (others)
Southeast (total)	Southeast (industrial)	Southeast (commercial)	Southeast (residential)	Southeast (others)
Midwest (total)	Midwest (industrial)	Midwest (commercial)	Midwest (residential)	Midwest (others)

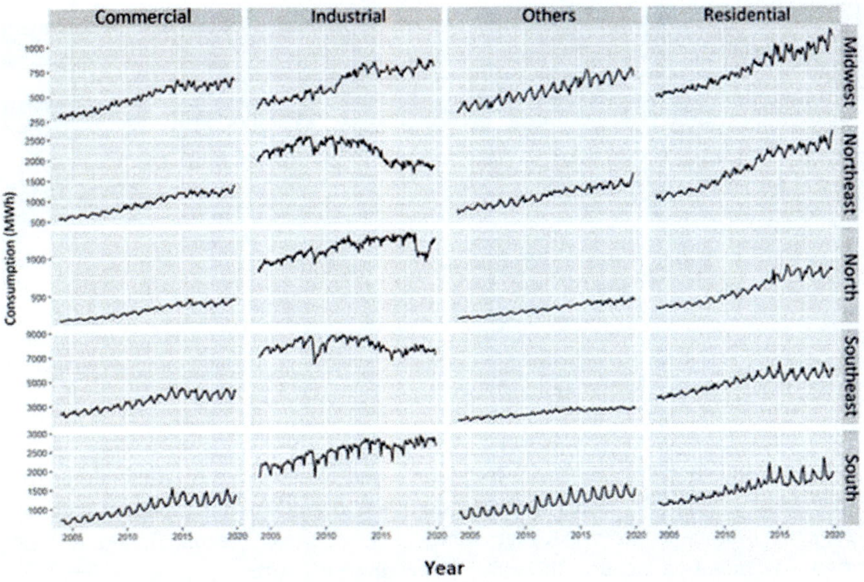

Fig. 3 Brazilian energy consumption by class and region. (Source: Own Authorship)

In order to analyze the time series of electricity consumption, their charts were generated, as shown in Fig. 3. Each time series behaves differently. For example, the time series related to the Northeast region and the industrial class do not present trends and show additive seasonality. On the other hand, the time series representing Northeast residential consumption presents an additive trend and multiplicative seasonality. By correctly identifying the components, it is possible to select the best-fit model in each case.

The state-space model representation of exponential smoothing models is called Error-Trend-Seasonal (ETS) [9] and is present, in software R [5], in the ETS function from package fable [11]. This function returns the best ETS model according to the behavior of each series. The models are then used to generate monthly forecasts for the year 2020. In the next step, the predictions were optimally reconciled using the Minimum Trace (MinT) method [9].

Time series can often be disaggregated naturally into a hierarchical structure, like the Brazilian electricity sector (Fig. 4); total consumption is at the highest level, the regions are in the middle level, and the consumption classes are at the bottom. The most disaggregated time series usually have high volatility (therefore, more challenging to forecast). In contrast, the most aggregated time series is generally smooth and less noisy (consequently easier to forecast) [10]. There are two approaches to deal with time series that have a hierarchical structure. The first concerns using traditional methods that use forecasts from a single hierarchical level and then aggregate or disaggregate to obtain forecasts at the other remaining levels. The second comprises optimal reconciliation approaches, like Minimum Trace (MinT), that minimizes the total forecast variance of the set of forecasts [9].

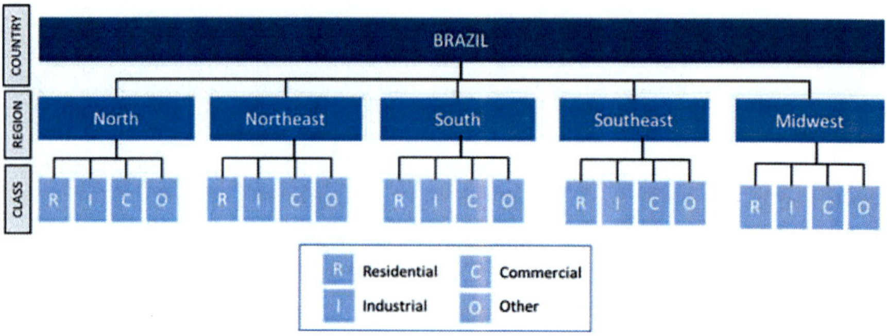

Fig. 4 Hierarchical representation of the Brazilian electricity consumption across regions and classes. (Source: Own Authorship)

In order to increase the forecast accuracy and to reconcile the time series, the MinT was applied by using the function *min_trace()* in conjunction with the *reconcile()* function from package fable [11]. However, reconciliation did not bring significant changes in the model, attesting to a sufficient level of accuracy. The reconciled forecasts are then used for comparison with the 2020 observed values.

In the final step, the estimated impact of the pandemic in the Brazilian electric sector is analyzed by relating these findings with the restrictive measures imposed by the government to suppress the dissemination of the new coronavirus. The Oxford COVID-19 Government Response Tracker (OxCGRT) database [12] collects systematic information on which governments have taken which measures, and when, is used to analyze patterns between common policy responses governments have taken and the impacts on the electricity sector.

3 Result Analysis

As previously presented, this section shows the results of applying hierarchical time series forecasting approaches combined with government restrictions data imposed in Brazilian regions to slow down the COVID-19 disease.

3.1 Data Analysis

The first data analysis is related to the historical electricity consumption values during 2020 compared to the counterfactual scenario. Then, the possible impacts of restriction measures in Brazilian regions were observed to analyze if such measures could affect consumer behavior.

3.2 *Energy Consumption*

The percentage difference values between the estimated and realized values for the classes, when compared, show that until May, the difference tends to increase. From this month ahead, it gradually decreases, making the realized values closer to the predicted, as shown in Fig. 5.

While the residential sector had slight variations and exceeded by 8.40% the October projection, the commercial sector was the most impacted by the pandemic. It demanded less energy than expected for the whole year of 2020.

The analysis of the percentage difference by geographical regions can be seen in Fig. 6. The Southeast region was the one that had its consumption most impacted, taking longer to recover when compared to other regions. On the other hand, the Midwest region had its energy consumption less affected, with a difference of approximately 8% of the designed consumption for June.

3.3 *Oxford Government Restriction Index*

The Blavatinik School of Government and the University of Oxford developed the "COVID-19 Government Response Tracker" project. It can be described as a tool that tracks and compares policy responses rigorously and consistently worldwide.

According to the Blavatinik School of Government [12], the Oxford index was created to assist governors, mayors, lawmakers, and their advisers by providing helpful information when facing decisions about the pandemic.

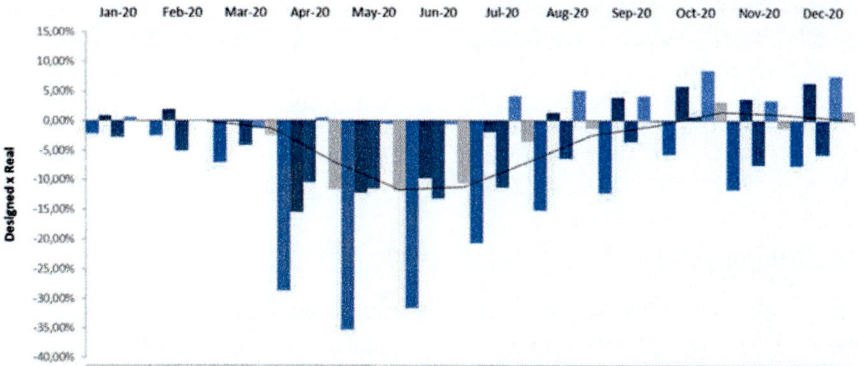

	Jan-20	Feb-20	Mar-20	Apr-20	May-20	Jun-20	Jul-20	Aug-20	Sep-20	Oct-20	Nov-20	Dec-20
Commercial	-2,21%	-2,60%	-7,10%	-28,65%	-35,23%	-31,54%	-20,65%	-15,15%	-12,23%	-5,74%	-11,66%	-7,51%
Industrial	0,78%	1,92%	-0,20%	-15,49%	-12,19%	-9,74%	-1,95%	1,35%	3,86%	5,74%	3,66%	6,27%
Outros	-2,79%	-5,09%	-4,14%	-10,48%	-11,55%	-13,21%	-11,26%	-6,38%	-3,62%	0,55%	-7,39%	-5,69%
Residential	0,61%	-0,11%	-1,42%	0,49%	-0,66%	-0,73%	4,09%	4,99%	4,07%	8,40%	3,44%	7,44%
Total	-0,22%	-0,13%	-2,58%	-11,60%	-11,87%	-10,56%	-3,57%	-1,36%	-0,32%	3,14%	-1,24%	1,67%

Fig. 5 Monthly variations in consumption by class. (Source: Own Authorship)

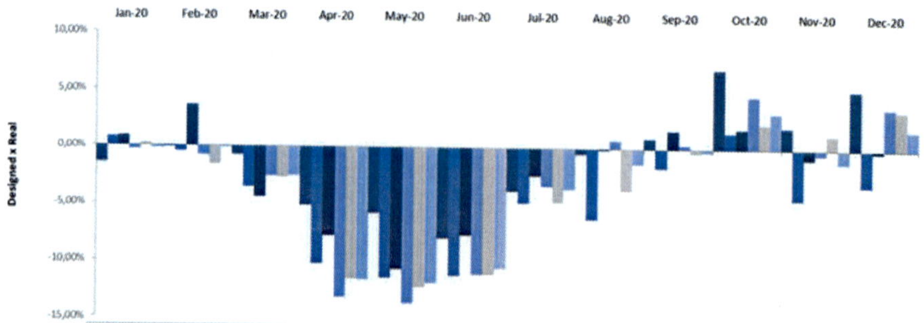

	Jan-20	Feb-20	Mar-20	Apr-20	May-20	Jun-20	Jul-20	Aug-20	Sep-20	Oct-20	Nov-20	Dec-20
Midwest	-1,51%	-0,13%	-0,77%	-5,12%	-5,77%	-7,94%	-3,82%	-0,53%	0,91%	6,95%	1,92%	5,19%
North East	0,78%	-0,51%	-3,56%	-10,26%	-11,46%	-11,24%	-4,84%	-6,20%	-1,72%	1,41%	-4,42%	-3,22%
North	0,86%	3,58%	-4,43%	-7,77%	-10,71%	-7,68%	-2,47%	-0,15%	1,58%	1,77%	-0,88%	-0,22%
Southeast	-0,34%	-0,82%	-2,63%	-13,19%	-13,65%	-11,12%	-3,37%	0,70%	0,34%	4,61%	-0,49%	3,62%
South	0,22%	-1,62%	-2,72%	-11,56%	-12,27%	-11,11%	-4,77%	-3,69%	-0,43%	2,19%	1,25%	3,43%
Total	-0,22%	-0,13%	-2,58%	-11,60%	-11,87%	-10,56%	-3,57%	-1,36%	-0,32%	3,14%	-1,24%	1,67%

Fig. 6 Variation in consumption by geographical region. (Source: Own Authorship)

All the information collected by the Oxford COVID-19 Government Response Tracker [12] is public. Using such information, the university produced 18 government response indicators and provided a measurement system capable of understanding how these responses have evolved. The project traces government policies and interventions through a standard set of indicators and creates a set of indices to measure the effectiveness of each policy and intervention.

The 18 indicators are divided as follows:

- **Eight** general government policy restriction indicators (C1–C8): containment and closure policies, such as school closures and movement restrictions;
- **Four** economic policy indicators (E1–E4): support for citizens' income or the provision of external assistance;
- **Five** health system policy indicators (H1–H5): such as the COVID-19 testing regimen or emergency health investments;
- An indicator of different policies (M1), a "free response" indicator, which records other information of interest.

An analysis was performed using the data from the Oxford government indexes to explain such variations.

(a) **Stringency Index**

The **stringency index** represents composite measures based on twenty response indicators, including school closures, workplace closures, and travel bans, scaled to a value from 0 to 100 (with 100 being the most restrictive). Figure 7 shows how these responses indicators behave in Brazil with the adoption of restrictive measures by the government.

The behavior of the stringency index throughout 2020 remained the same from March to July. However, from that moment on, it is possible to observe the flexibilization of these same measures until the beginning of December 2020.

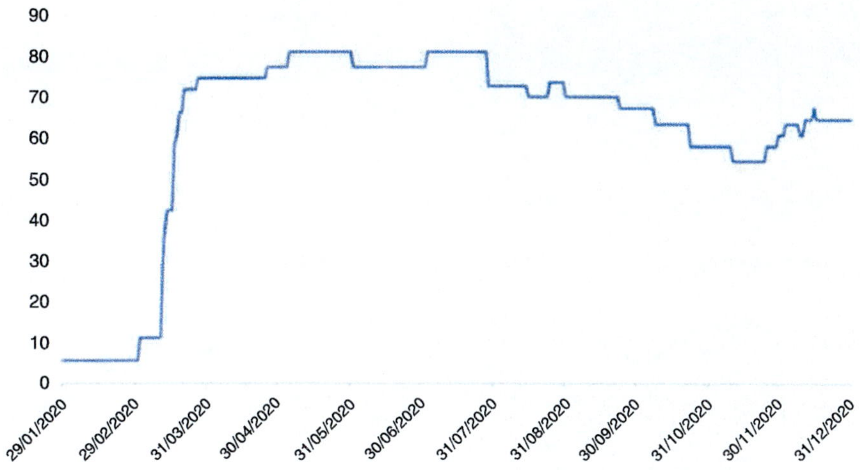

Fig. 7 Stringency Index for the year 2020. (Source: Own Authorship)

The data analysis for the formation of the index, in absolute numbers, measures the restriction imposed by governments is calculated considering the components observed in Fig. 8.

In Fig. 9, it is possible to reiterate a previous analysis, observing the impact of these restriction measures obtained in the map. The most significant impact on restrictions is between March and July, represented by the warmest colors on the map (darker blues). Likewise, lighter blues can be observed after this period, representing a drop in the imposition of restrictions by different governments.

(b) **Opening Risk Index**

Another relevant factor used for this study was the **Opening Risk Index** (RoOI), also derived from Brazilian subnational data. This data is based on recommendations published by the World Health Organisation (WHO). Its goal is to assist the government in making decisions at the right time for the opening of the economy. This measure varies between 0 and 1 (1 being the level when the greater risk of opening the economy is maximum). The evolution of this index over the year 2020 can be seen in Fig. 10. It is important to emphasize that the data in this analysis come from a short period, only until November 2020, which will influence the generations of subsequent graphs.

The calculation of the RoOI involves the use of four sub-indices under the recommendation of the World Health Organisation (WHO) document: controlled cases, test and screening, imported cases, and community cases. The calculation process is thoroughly available in the Oxford database [11]. According to the graphic, the risks of opening the economy were the highest between July and September, suggesting higher rates for the proliferation of the disease. Therefore, the recommendations were most restrictive.

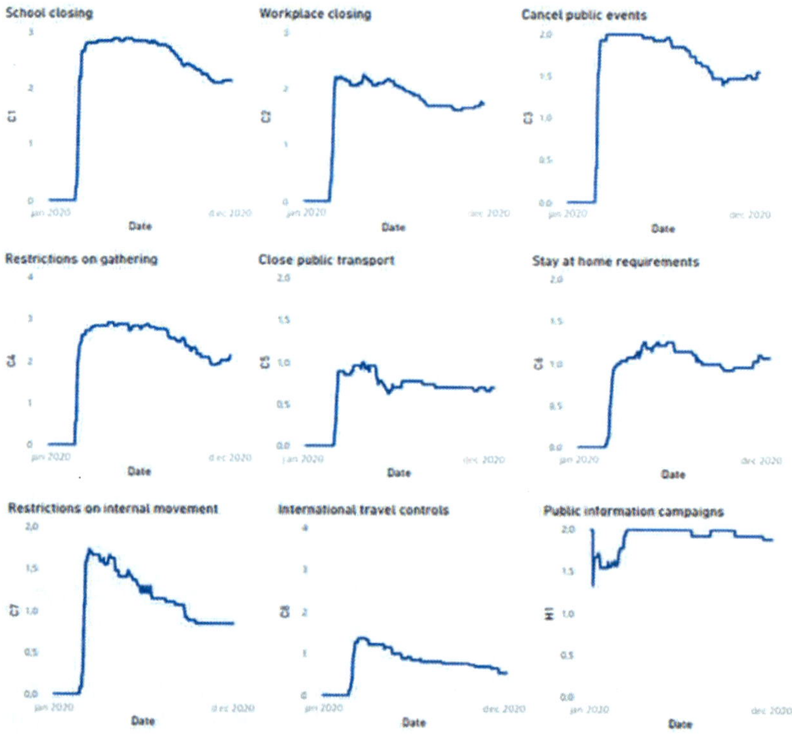

Fig. 8 Constraint Index components. (Source: Own Authorship)

3.4 *Interaction Between Bases*

For the analysis purposes, the correlation between three variables was calculated: the risk of the opening index, the stringency index, and the percentage difference between the values of the designed series for the scenario without pandemic and the realized values, which will be called "Difference % Consumption".

Table 3 shows the three variables. We used a time series that considers national values, divided into 10 months of the year – except for January, before the beginning of the pandemic in Brazil, and December, for which data for the Opening Risk Index were not available.

From the data provided in Table 3, the linear correlation method can be applied to study the behavior of two quantitative variables. Using Excel's "Data Analysis" tool, it was possible to extract a Pearson's Correlation matrix, represented in Table 4. The result of this matrix can vary between -1 and 1, so that:

- $r = 1$: There is a perfect and positive correlation between the two variables;
- $r = 0$: There is no correlation between the variables;
- $r < 0$: A negative correlation between the two variables is observed, that is, they are inversely proportional, and when one increases, the other decreases;

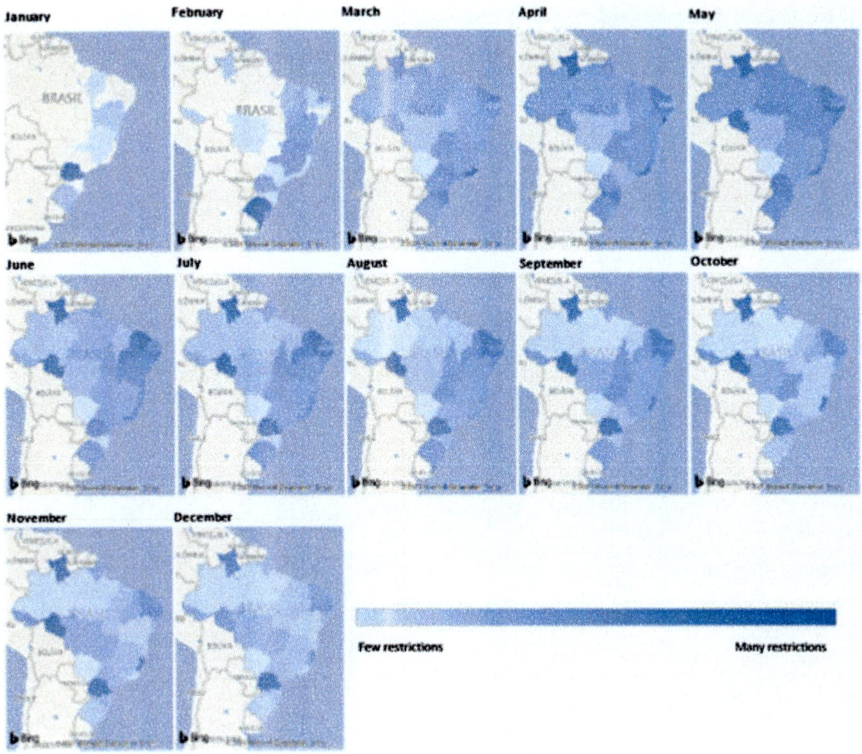

Fig. 9 Restriction index by region. (Source: Own Authorship)

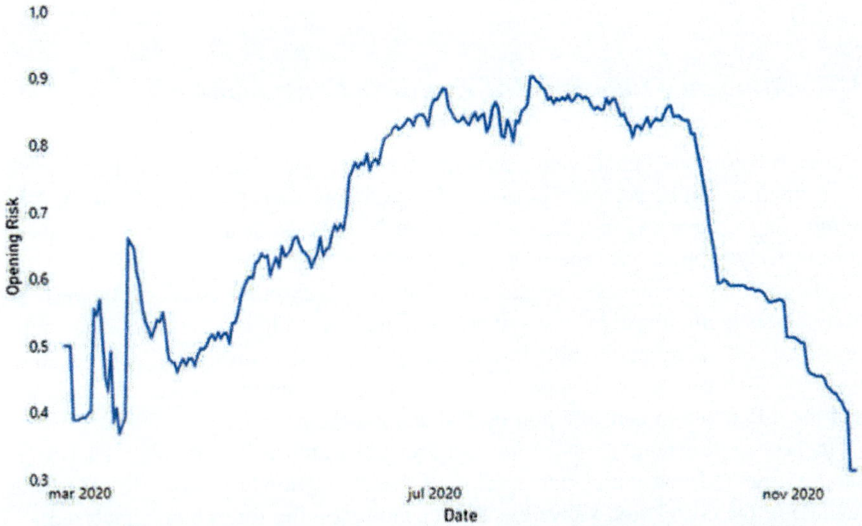

Fig. 10 Opening Risk Index for the year 2020. (Source: Own Authorship)

Table 3 Final data generated from the three variables

Month	Índice de Risco de Abertura	Índice de Rigidez	Diferença % Consumo
February	0,50	11,11	0,00
March	0,81	84,39	−0,03
April	0,75	85,72	−0,12
May	0,91	87,88	−0,12
June	0,99	87,39	−0,11
July	1,00	85,94	−0,04
August	0,99	76,66	−0,01
September	0,97	74,46	0,00
October	0,86	75,96	0,03
November	0,59	67,45	−0,01

Source: Own Authorship

Table 4 Pearson correlation matrix

	Opening Risk Index	Stringengy Index	Difference % Consumption
Opening Risk Index	1,00	0,74	−0,21
Stringengy Index	0,74	1,00	−0,45
Difference % Consumption	−0,21	−0,45	1,00

Source: Own Authorship

- $r > 0$: A positive correlation between the two variables exists; that is, they are directly proportional, and when one grows, the other also grows.

3.5 *Stringency Index* vs. *Difference % Consumption*

The correlation coefficient value between the Stringency Index and Difference % Consumption can be seen in Table 4 and has an approximate value of −0.45. The number is negative and less than 0.5 in absolute value, indicating that the variables are inversely proportional and have a small relationship.

The negative value was expected since the stringency index considers the restriction measures imposed by the government and, the higher it shows, the more restricted the economy is, and less electricity is being consumed. The negative value means that the association is the opposite between the growth of the index and the difference in consumption (actual and expected).

In Table 5, this number and values for each class and region are more detailed. It can be observed that the negative value of the correlations remains. However, the module of the correlation varies greatly depending on the series being analyzed.

Regarding class correlation, two extremes can be observed. There is a high correlation between the commercial consumption variable, which can be considered

Table 5 Correlation between Stringency Index and Difference % Consumption

		Stringengy Index
Difference % Consumption by region	Midwest	−0,33
	North East	−0,53
	North	−0,65
	Southeast	−0,38
	South	−0,42
		Stringengy Index
Difference % Consumption by class	Commercial	−0,61
	Industrial	−0,42
	Residential	0,07
	Outros	−0,34

Source: Own Authorship

the most affected due to the restriction measures, which means that the imposed measures had a significant impact on the electricity consumption of this class. It is also possible that the residential class does not seem to have been affected by the Stringency Index since there is no correlation between the variables.

The North and Northeast regions show a significant impact on electricity consumption. On the other hand, the Southeast and Midwest regions do not present a strong correlation between the Stringency Index and the Difference % Consumption.

3.6 *Opening Risk Index* vs. *Difference % Consumption*

The correlation coefficient value between the Opening Risk Index and the Difference % Consumption shown in Table 4 is −0.21. The negative sign of the value indicates that the variables are inversely related, and the modulus less than 0.5 represents a slight relationship between the variables. Therefore, analyzing the series, it is not possible to infer that the risk of opening index affects electricity consumption behavior.

In this sense, opening the analysis beyond the total consumption series is also important. Table 6 shows that all correlation values are negative, representing two variables that grow in opposite directions. As expected, this index has little or no relationship with the variation in electricity consumption in Brazil, which happens because all results add to an absolute value smaller than 0.5, with the commercial class being the series with the highest index.

Table 6 Correlation between Opening Risk Index and Difference % Consumption

		Opening Risk Index
Difference % Consumption by region	Midwest	−0,29
	North East	−0,31
	North	−0,31
	Southeast	−0,14
	South	−0,31
		Opening Risk Index
Difference % Consumption by class	Commercial	−0,46
	Industrial	−0,17
	Residential	0,23
	Outros	−0,22

Source: Own Authorship

3.7 *Opening Risk Index* vs. *Stringency Index*

The correlation value between the risk of openness and the stringency index, according to Table 4, is 0.74. The correlation found between these two data is positive. It has a value bigger than 0.5 in modulus, indicating that the variables are directly proportional and related. The increase in one variable considerably influences the increase in the other, as shown in Fig. 11. The charts depicted in the figure provide the monthly stringency and opening risk indices per Brazilian state. It should be noted that, in February, the only data available were for São Paulo, the first state to implement restrictive measures.

The Brazilian states decided to take more restrictive measures throughout the whole national territory from March onwards, particularly after the second half of the month. The charts highlight the increase in restriction and risk of opening in all states.

Until July, all states were under the effects of very restrictive measures, with values between 80 and 100 and on the border of maximum risk of opening. These numbers improved from July onwards, as can be seen in the charts in Fig. 11.

4 Conclusions and Future Directions

Throughout this work, techniques were implemented to analyze the impact of the pandemic caused by COVID-19 on the behavior of electricity consumption throughout 2020 and the relation between consumption and government proposals to suppress the virus dissemination.

The ETS method was applied to quantify the government's restrictive measures on electricity consumption behavior by comparing a pandemic-free scenario with the observed time series in 2020.

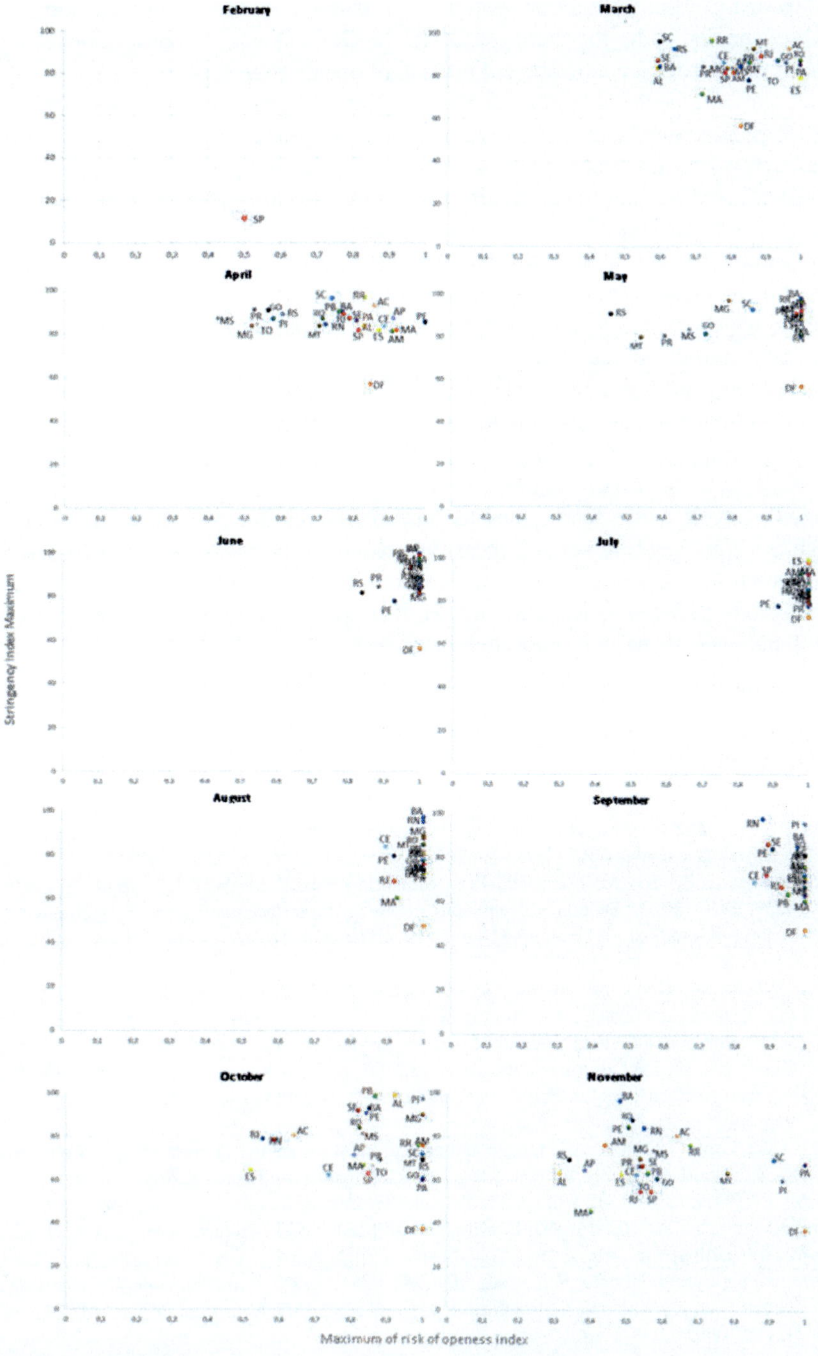

Fig. 11 The dispersion between indices. (Source: Own Authorship)

To analyze the impacts of restrictive measures imposed by governments, two indexes published by the Oxford COVID-19 Government Response Tracker were essential: the stringency index and the risk of opening the economy.

A more detailed study of the relation between stringency index and the difference between estimated and realized electricity consumption showed that restriction measures could affect consumption in some regions and classes more than others. The North and Northeast regions are more likely to be more affected by the adoption of restriction measures, as well as the commercial class.

Therefore, it was possible to analyze that the restrictive measures imposed by the states could change electricity consumption behavior, especially in the commercial sector, which presented a 35% deficit in consumption in May, compared to the estimated non-pandemic scenario.

However, restrictive measures are necessary since it is strongly correlated with the risk of opening the economy to the population, leading to a possible increase in the number of cases of infection by SARS-Cov-2019. The higher the risk, the more necessary it is the use of such measures.

Different strategies can be used for a counterfactual scenario for future research, such as using other time series forecasting models. It is also possible to analyze other indicators that may affect electricity consumption behavior, such as variations in temperature and electricity rationing. Furthermore, it is possible to check if some days and times were more impacted than others.

References

1. CRAVEN, M.; LIU, L.; MYSORE, M.; WILSON, M. COVID-19: Implications for business. McKinsey & Company, Vol. 57, June 2021.
2. Coronavirus disease 2019 (COVID-19) – Situation Report – 51. WORLD HEALTH ORGANIZATION, 2020. Disponível em: https://www.who.int/docs/default-source/coronaviruse/situation-reports/20200311-sitrep-51-covid-19.pdf?sfvrsn=1ba62e57_10. Acesso em: 20 de Abril de 2021.
3. DESTEK, M.; SINHA, A. Renewable, non-renewable energy consumption, economic growth, trade openness and ecological footprint: Evidence from organisation for economic Co-operation and development countries. Journal of Cleaner Production, Vol. 242, 2020.
4. Oliveira EM (2015) Corporate social responsibility and firm performance: A case study from the Brazilian electric sector. Pontifical Catholic University of Rio de Janeiro (PUC-Rio), Rio de Janeiro.
5. R Core Team (2021). R: A language and environment for statistical computing. R Foundation for Statistical Computing, Vienna, Austria. URL https://www.R-project.org/.
6. MORETTIN, P.; TOLOI, C. Previsão de series temporais. São Paulo: Atual, 1987.
7. PINDYCK, R.; RUBENFIELD, D. Econometric Models and Economic Forecasts. New York: McGrawHill, 1991.
8. Meira, E., Cyrino Oliveira, F. L., & de Menezes, L. M. (2021). Point and interval forecasting of electricity supply via pruned ensembles. Energy, 232, 121009. https://doi.org/10.1016/j.energy.2021.121009

9. HYNDMAN, R.; ATHANASOPOULOS, G. Forecasting: principles and practice. Australia: Otexts, 2021. Access: May 21, 2021.
10. Fliedner, G (2001). Hierarchical forecasting: issues and use guidelines. Industrial Management and Data Systems 101(1), 5–12
11. O'Hara-Wild, M.; Hyndman, R.; WANG, E. **Forecasting Models for Tidy Time Series**. Disponível em: https://cran.r-project.org/web/packages/fable/index.html. Acesso em: May 17, 2021.
12. THOMAS, H. et al. A global panel database of pandemic policies (Oxford COVID-19 Government Response Tracker). Nature Human Behavior, Vol. 5, pp. 529–538, 2021.

Offering Wind Farms: Types of Service and Their Characteristics

Gustavo C. Pedrinho, Paulo A. Cauchick-Miguel, and Suzana R. Moro

Abstract Services infusion has expanded in typical product-centered manufacturing companies, in a process named as "servitization". In this context, this study aims to identify which services compose the portfolio of a business unit of power-generation sector (wind farms), as well as to analyze the service types (levels) for such products, based on the literature. Case-based research was adopted as methodological approach at exploratory level. Through six interviews with employees and access to institutional documents, this work identified activities that are related to the servitization process. Besides the product (wind turbine), the business model considers design and operation services, operation and maintenance, spare parts, monitoring and performance systems, as well as operational consulting. After identifying such, categories that represent service levels from the literature were used: (i) basic, (ii) intermediate, and (iii) advanced services aiming to fit the investigated business unit's services into those categories. Results show that the category 'spare parts services' was classified as a basic level service. The 'monitoring and performance system services' was categorized as an intermediate service, while the 'operational consulting services' falls into the advanced service level. The continuity of this study foresees its further development considering how technology can support the creation of new service offerings in wind farms.

Key words Servitization · Service strategy · Service implementation · Wind farms

1 Introduction

The participation of services has expanded in typical manufacturing companies, especially those located in developing countries [1]. This trend occurs in various industries that are driven by customer demands perceived by companies [2]. Rendering services provide attractive profit margins, adding value to the production of

G. C. Pedrinho (✉) · P. A. Cauchick-Miguel · S. R. Moro
Present Address: Universidade Federal de Santa Catarina, Florianópolis, SC, Brazil
e-mail: paulo.cauchick@ufsc.br

J. Vargas Florez et al. (eds.), *Production and Operations Management*, Springer Proceedings in Mathematics & Statistics 391,
https://doi.org/10.1007/978-3-031-06862-1_12

manufactured goods and allowing businesses to differentiate themselves from competitors without reducing operating costs to remain competitive [3]. Thus, many manufacturing companies have added services to their product portfolio [4].

The literature on services has emphasized offering value to consumers as a key differentiation factor in the strategy to increase revenue and competitiveness [3]. For manufacturing sectors, it is not only a matter of offering it, but the whole company needs to change the focus of its organization [4]. This movement of offering services and solutions along with products is known as "servitization". Baines et al. [5] define the term as the innovation of organizational capabilities and processes in a business model from "selling products" to "selling products and services that add value in integrated use". The services offered to add value in integrated use are classified into three levels [6]: (i) basic; (ii) intermediate; and (iii) advanced. Also, according to the cited authors, these levels allow business models to offer an increasing variety of services. Some companies with greater maturity in service management are already evolving to advanced services, given that servitization tends to generate revenue growth and increased profitability [7].

In this context, the present study aims to identify which services compose a company's portfolio in a power-generation segment and evaluate how such a portfolio fits into the classification of service levels associated with the offerings identified in the literature [6]. To achieve this objective, a single case study was conducted in a business unit that includes several services – such as maintenance, spare parts, and a monitoring system, among others – the supply of wind turbines. The present study also aims to provide continuity to a previous pilot study [8].

The remainder of the paper is structured as follows. Section 2 outlines a brief overview of the servitization concept and its classification of service levels in servitization. This is followed by Sect. 3 that highlights the research methods adopted to achieve the results showed in Sect. 4. Finally, Sect. 5 draws some concluding remarks and research implications of this study.

2 Servitization Concept and Its Types of Services

As previously mentioned, servitization is the process in which manufacturing companies offer services associated with their products [6]. From a strategic change, typically manufacturing organizations begin to incorporate services to obtain competitive advantages [9], creating value by incorporating services to products [2], besides bringing changes to the industrial bias, which was once associated with mass consumption for the fulfilment of personalized needs on demand [10]. The cited authors add that the literature often conceptualizes servitization as the result of a change process, not as the change process itself. Nevertheless, other studies (e.g., Rabetino et al. [11]) consider servitization to be the result of a change. To design and deliver services on demand, manufacturing companies need to transform their organization [2]. The role of organizational culture is essential for successful transformation [12]. A target of great attention from the academia over the past decade, servitization is then related to the movement in which companies extend their offerings through integrated packages of products, services, support, self-service, and knowledge to add value to their core businesses and thus rapidly improve their core competencies [5].

2.1 *Classification of Service Levels in Servitization*

Baines and Lightfoot [6] characterize customer demand for services as: (i) customers performing their own activity operations; (ii) manufacturers providing some services; and (iii) manufacturers engaging in an operational activity through services. Even when customers perform some typical service activities, there are likely to be several "basic services" required from the manufacturer, such as product deliveries, spare parts supply, and warranties. Customers may also require manufacturers to provide "intermediate services" such as a technical helpdesk, training, and/or maintenance. To meet growing customer demands, manufacturers also find it necessary to develop "advanced service" offerings, which can enable deeper customer relationships and address more complex requirements [13].

To sustain a competitive advantage throughout a product's life cycle, manufacturing companies can develop competencies to provide advanced services [5]. Figure 1 shows a classification of service levels (basic, intermediate, and advanced services), with some examples in each category.

As Fig. 1 illustrates, basic services are related to product sales and parts replacement. Intermediate services have a bit more structure, in the sense that they allow an even greater variety of services to be offered, consisting of maintenance, operation support, product condition monitoring, periodic overhauls, or repairs in general. Finally, advanced services are at the "top of the service list", as they may even

Fig. 1 Classification of service levels according to servitization criteria. (Constructed based on Ref. [6])

(3) Advanced	**(3.1)** Support agreement with customer **(3.2)** Rental agreement **(3.3)** Risk and revenue sharing **(3.4)** Revenue per use
(2) Intermediate	**(2.1)** Help Desk **(2.2)** Field Service **(2.3)** Monitoring condition **(2.4)** Overhaul services **(2.5)** Repair services
(1) Basic	**(1.1)** Product **(1.2)** Spare parts

consider risk and revenue sharing between the service provider and the customer, as well as various contractual arrangements, such as revenue per use, among others.

There are motivations for servitization that can be exploited forms of service offerings [6]: (i) customer service to improve relationship quality (demand-based motivations); (ii) product-related services (product support services); (iii) services that support business needs (customer support service). Many companies begin their activities with manufacturing and repairing their own products to ensure that customers receive support after their purchase. In addition to warranty services, companies have identified other opportunities in service offerings [5]. Next, the research methods are presented.

3 Research Methods and Procedures

The study began with a bibliographic search and organization, aiming to select literature on the subject from peer-reviewed journals. The focus of the search was on articles related to servitization and/or the transition of typical manufacturing companies to include services in their products with evidence of or influence on the organization's business model. Table 1 summarizes the main elements of the literature search.

After the literature review, and in order to understand how (and if) the services at wind farms offered by the analyzed business unit were aligned with servitization-related characteristics, this study considered a qualitative data analysis. According to Creswell [14], the focus on this type of analysis considers the adoption of an intentional sample based on selection criteria, data collection in the field, and the interpretation and analysis of the data collected. This study considered an object and a unit of analysis, described next.

Table 1 Summary of the literature search

Descriptor	Contents
Data base	Web of Science, Scopus, and ScienceDirect
Period	From 2009 to 2019
Language	English
Keywords	"*service*" AND "servit*"
Scope of keywords	Service Ecosystem, Service Maturation, Services Strategies, Service Implementation, Service Infusion, Service Innovation, Service Transition, Advanced Services, Product-Service System, Servitization
Locations of term search	Titles, abstracts and keywords
Publication type	Article and review

3.1 Object and Unit of Analysis

The object of analysis is a large manufacturing company with global operations in a dozen economic sectors: agriculture, infrastructure and construction, textiles, oil and gas and petrochemicals, maritime and offshore, metallurgy, food and beverages, mining and cement, power generation, pulp and paper, sugar and ethanol, and water and sanitation. The "power-generation" sector is active with biofuel power, transformers, solar power, and wind power. In this segment several products are manufactured and marketed, such as: wind turbines, turbo generators, turbines (hydraulic and steam), among others. The wind-turbine business unit for wind farms falls within this segment and is the analysis unit in terms of research. The company started manufacturing and marketing wind turbines in 2013.

Concerning the product, one-third of the wind turbine is developed internally (represented by the turbine and electrical transmission), and two-thirds of the product is developed by subcontracted partner companies (blades/propellers and support base). The services that are contractually sold with the wind turbines are related to operation and maintenance, call center, spare parts, monitoring system administration, and advanced operational consulting.

3.2 Data Collection

To build the data collection instrument, an institutional material of the power-generation segment was considered, which discusses the commercialization process and administration of wind farms, in addition to data on the management of wind turbines and customer relations. Based on the instrument, six different interviews were carried out with nine employees of the business unit, each one lasting one hour and thirty minutes. Based on the interviews and document analysis (institutional material) the data were analyzed, as described next.

3.3 Data Analysis

After collecting data from the interviews and internal institutional data, the process of data triangulation began by following Creswell's [14] recommendations for data analysis: (i) collection (literature review data, and interview and institutional data); (ii) consolidation; and (iii) construction of the relationship among the data.

4 Findings

The object of analysis had already offered the manufacturing of products such as solar inverters, central inverters, energy storage in batteries, turbo generators, hydro generators, hydraulic turbines, steam turbines and transformers. When starting the design of the strategy and planning for the sale of wind turbines, the company identified that their competitors were also responsible for installing, operating the wind farms and assuring the generation and distribution of electricity, in addition to selling the wind turbines. Gurtu [4] points out that manufacturing companies have added service offerings to their product portfolio to obtain competitive advantages. In this particular case, the business unit had to adapt and start providing services, both in the assembly and operation of a wind farm, as competitors already offered such services.

The company service portfolio is presented next, followed by a categorization of service level agreement from the literature with the service portfolio.

Service Portfolio The structure of the service portfolio is based on each customer's needs because, according to Sayar and Er [15], customers' involvement in understanding their needs should be considered. The purpose is to increase the value of the service experience by grouping services into two categories: (i) products and services and (ii) maintenance and continuity. Table 2 presents the service offerings as each of these categories is divided. Figure 2 illustrates a employee monitoring a wind turbine, an 'intermediate service' category (see Fig. 1).

One of the company's concerns that emerged in the interviews is related to how long the contract is established between the customer and the company. This issue is addressed by Rymaszewska et al. [16], who state that when planning the execution of services, a clear contractual period there must be considered, in this case, to be offered for each category. For example, for the "products and services" category, the contractual period may vary from 2 years for service provision in the acquisition of a wind farm, with the possibility of renewal and extension for another 8 years. For the "maintenance and continuity" category, the contract can be for 8 years with the possibility of renewal at the end of it. The company can also absorb all the operation

Table 2 Service offerings that compose each category

Categories	Services
Products and services	Design and assembly
	Maintenance and operation
	Parts replacement for wind turbines
	Control and performance center
	Advanced operational consulting
Maintenance and continuity	Maintenance and operation
	Parts replacement for wind generators
	Control and performance center
	Advanced operational consulting

Fig. 2 Example of a wind turbine monitoring in the control and performance centre. (Photography from company internal report, 2020)

and support continuity for wind farms manufactured and marketed by other competing companies. Besides adding the "maintenance and operation" services, and "spare parts" services, it is possible to add the "control and performance center" services, as well as "advanced operational consulting" services.

Service Level Agreement In both service categories in the portfolio (Table 2), a service level agreement can be established between the provider and the customer. This type of service level agreement was in the company within the three categories of services accordingly to Baines and Lightfoot [6] (basic, intermediate, and advanced service). In both intermediate and advanced services, a service level agreement must be established between the customer and the supplier, such as call center services, maintenance, or usage-based services (i.e., revenue per use). This agreement penalizes the company if it does not reach 97% availability in energy generation at the wind farm. The availability is calculated based on the failure rate that a wind turbine can present during the year; however, according to the collected data, this percentage may vary from contract to contract, since wind farms have some variables that can influence equipment availability.

The size of the wind farm, the number of wind turbines, the amount of contracted power generation, possible wind turbines from competing companies that were already operating at the wind farm, and unexpected wind turbine failures were mentioned throughout the interviews and are considered, by the company, to be variables that can influence service level agreements.

Monitoring is discussed by Frank et al. [17], whereby service level agreements based on information technology provide more support for innovations in service delivery. Examples of such monitoring are [18]: the availability of operation, preventive maintenance, and equipment condition monitoring.

To comply with the service level agreements established in the contract, the existence of documentation that present the field teams' standardized and approved internal procedures as well as that include information concerning the wind turbines sold by the company, as well as how service should be provided to the wind farms, was identified in the interviews. All documents are available and freely accessible to the employees in a knowledge database. According to Cui et al. [10], it is essential to have extensive knowledge regarding service and how it is included in the service-provision, scope, and a knowledge base is the site where all the information considered relevant is stored. For the company investigated, this is important content concerning services and service-delivery processes.

5 Concluding Remarks

This study identifies the existence of activities related to the servitization process in the investigated business unit. According to the data collected, it was observed that the maintenance, support and parts replacement services do not stand out from those of competitors, that is, they can be considered standard services for wind farms, since other organizations that sell wind farms also offer similar services. However, among the services, monitoring and performance and advanced operational consulting services can be highlighted. Together, they work with client customization, offering not only a product or service, but an integrated solution aiming to meet customers' purposes. The business unit investigated offers services of diverse nature aiming at expanding its offer, possibly to remain competitive in the wind farm market. As a continuity of this study, this research intends to further develop the analysis performed so far, aiming to evaluate the characteristics of the service offerings already identified. Moreover, how technology can support the creation of new service offerings in wind farms could be also explored in the future.

References

1. Valtakoski, A. Explaining servitization failure and deservitization: A knowledge-based perspective. *Industrial Marketing Management*, 60(1), 138–150 (2017).
2. Baines, T., Bustinga, O.F., Vendrell-Herrero, F. Service implementation in manufacturing: an organisational transformation perspective. *International Journal of Production Economics*, 192, 1–8 (2017).
3. Eggert, A., Ulaga, W., Frow, P., Payne, A. Conceptualizing and communicating value in business markets: From value in exchange to value in use. *Industrial Marketing Management*, 69, 80–90 (2018).

4. Gurtu, A. The strategy of combining products and services: a literature review. *Services Marketing*, 40, 82–106 (2019).
5. Baines, T., Bigdeli, A., Bustinza, O.F., Shi, V.G., Baldwin, J., Ridgway, K. Servitization: revisiting the State-of-the-art and Research Priorities. *International Journal of Operations & Production Management*, 37(2), 256–278 (2017).
6. Baines, T., Lightfoot, H.W.: Made to Serve: How manufacturers can compete through servitization and product service systems. John Wiley & Sons, Chicester, UK (2013).
7. Lee, S., Yoo, S., Kim, D. When is servitization a profitable competitive strategy? International Journal of Production Economics 173, 43–53 (2016).
8. Pedrinho, G.C., Cauchick Miguel, P.A. Oferta de aerogeradores para parques eólicos: um estudo piloto no contexto da servitização. In: Anais do XXXIX Encontro Nacional de Engenharia de Produção – ENEGEP, Santos (2019).
9. Ren, G., Gregory, M. Implementing service strategy in manufacturing companies: the case for a 'whole business' approach. Proceedings of the 16th International Annual EurOMA Conference, vol.1(5), pp. 14–17. Gotemburg (2009).
10. Cui, L., Su, S, I., Feng, Y., Hertz, S. Causal or effectual? Dynamics of decision making logics in servitization. Industrial Marketing Management 82, 15–26 (2019).
11. Rabetino, E., Harmsen, W., Kohtmaki, M., Sihvonen, J. Structuring servitization related research. International Journal of Operations & Production Management 38(2), 350–371 (2018).
12. Bigdeli, A., Bustinza, O,F., Vendrell-Herrero, F., Baines, T. Network positioning and risk perception in servitization: evidence from the UK road transport industry. International Journal of Production Research 56(6), 2169–2183 (2018).
13. Dachs, B., Biege, S. Borowiecki, M., Lay, G., Jäger, A., Schartinger, A. Servitisation in European manufacturing industries: empirical evidence from a large-scale database. The Service Industries Journal 34(1), 5–23 (2014).
14. Creswell, J.W.W. Projeto de Pesquisa: Métodos Qualitativo, Quantitativo e Misto. Bookman, Porto Alegre (2010).
15. Sayar, D., Er, O. The transformative effects of digital technologies on the product design practices of servitizing manufacturers. The Design Journal 22(1), 51–68 (2019).
16. Rymaszewska, A., Helo, P., Gunasekaran, A. IoT powered servitization of manufacturing: an exploratory case study. International Journal of Production Economics 192, 92–105 (2017).
17. Frank, A.G., Dalenogare, L.S., Ayala, N.F. Industry 4.0 technologies: Implementation patterns in manufacturing companies, International Journal of Production Economics 210, 15–26 (2019).
18. Coreynen, W., Matthyssens, P., Van Bockhaven, W. Boosting servitization through digitization: Pathways and dynamic resource configurations for manufacturers. Industrial Marketing Management 60(1), 42–53 (2017).

Coffee Value Chain Cost Logistic Analysis in Chanchamayo Peru

Diana Llanos, Mario Chong, Clara Orellana-Rojas, and Bernardo Puente-Mejia

Abstract Coffee is one of the three most consumed drinks and one of the most traded products globally. In 2019, it represented 9% of the total agricultural exports of Peru, ranking fourth with US$ 637 million. In addition, the coffee sector has been an essential source of income for more than 200 thousand Peruvian families, 34% of which have been in a situation of poverty or extreme poverty, aggravated by the COVID-19 pandemic. Unfortunately, during the last years, the coffee industry in Peru has faced a severe crisis due to low international prices, the low productivity per hectare intensified by the plague in the 2014–2015 period, and the high logistic costs within its value chain. In general terms, the Peruvian logistics performance is very deficient and directly impacts the profitability of the coffee sector; for this reason, this research presents the more relevant components of logistics cost. This research identified the shipping from the farms to the storage facilities and then to the processing plants in Lima, and the post-harvesting treatment generates more than 88.68% of the total cost. Afterwards, we proposed scenarios to minimize the overall logistic costs. As a result, 2.10%, 2.71%, and 3.94% of improvement were obtained, impacting their profitability and the sector's survival. The results show a logistics costs reduction from 25.24% to 15.76%.

Key words Coffee · Supply chain · Logistics cost · Road infrastructure

D. Llanos (✉) · M. Chong
Facultad de Ingeniería, Universidad del Pacifico, Lima, Peru
e-mail: di.llanosl@alum.up.edu.pe; m.chong@up.edu.pe

C. Orellana-Rojas · B. Puente-Mejia
Universidad San Francisco de Quito, USFQ, Colegio de Ciencias e Ingeniería, Departamento de Ingeniería Industrial and Instituto de Innovación en Productividad y Logística CATENA-USFQ, Quito, Ecuador
e-mail: corellanar@usfq.edu.ec; bpuentem@usfq.edu.ec

J. Vargas Florez et al. (eds.), *Production and Operations Management*, Springer Proceedings in Mathematics & Statistics 391,
https://doi.org/10.1007/978-3-031-06862-1_13

1 Introduction

Coffee is one of the principal commodities traded in the world. This commodity, in Peru, is leading in agricultural exports and the ninth in the world with 2.4% of the global coffee production (IndexMundi 2021). The cultivated area in Peru is approximately 425,400 hectares, the smallholder farmers have around three hectares (7.4 acres) (United States Department of Agriculture Foreign Agricultural Service [USDA] 2018), giving jobs to 223 thousand families in 15 regions (PNUD 2017). This fragmentation causes poor credit access, generating different credits form; from coffee stakeholders to informal market, and consequently low future prices, high-interest rates and high transaction costs. For this reason, around 25% are a member of associations or cooperatives to improve negotiation power, postharvest production handling techniques (Díaz and Willems 2017; United States Department of Agriculture Foreign Agricultural Service [USDA] 2018) and the sector competitiveness. The crop and harvest zone is the initial node of an extensive internal distribution network with a final node in the port of Callao, the Constitutional Capital of Peru. Peruvian agro-industrial companies logistics cost is one of the main cost components, from 20% to 50% of the total cost (Caballero et al. 2021).

In recent years, the coffee sector has been facing a severe crisis caused by low prices in the international market, low productivity per hectare and high logistics costs. Furthermore, Peru keeps recovering from the rust infestation of 2014 that damaged nearly half of its coffee plantations (United States Department of Agriculture Foreign Agricultural Service [USDA] 2018). Chanchamayo province in the Junín department is a traditional coffee producer in Peru. This crisis had several repercussions in this province; their market share participation reduced from 31% to 22.7% (Cámara Peruana de Café y Cacao 2017). Another consequence of this crisis is that coffee growing is becoming less profitable, worsening the economic situation of small producers and forcing them to abandon or replace their crops with other products.

Peru has three logistics corridors: the northern Tocache-Zarumilla, the central Satipo-Callao and the corridor Puno-Callao; Callao is the principal exportation port. These corridors have certain conditions and characteristics for cargo transportation in three stages production nodes, collection centres and plants (Caballero et al. 2021). Unfortunately, most coffee growing areas have weak and expensive connectivity for the inadequate road network (PNUD 2017) and poor logistics performance that directly impacts the profitability of this sector. Transportation cost from the plants to Callao is around 17–24%; for highway infrastructure and distribution facilities (Hummels 2007; Caballero et al. 2021). In the same line, Villalva-Cataño et al. (2019) proposed a model considering supply chain management costs associated with the coffee-growing activity.

The present study aims to exhibit the coffee logistics cost components and propose strategies to improve this supply chain. This research emphasizes the critical

Peruvian road network and infrastructure. Considering this component, we could develop competitive logistics activities, an adequate transportation system (Engström 2016), and global production networks with certainty to foreign markets (Hummels 2007).

2 Methodology

The proposed methodology has three stages: identify problems and limitations, understand the supply chain, and propose logistics cost improvement alternatives. The first stage recognizes relevant issues in the coffee industry in Peru: Local and international coffee market, challenges of climate change, production techniques, and main inefficiencies in the supply chain. The second stage identifies the main actors, such as farms and producers, transportation activities, processing plants and exporters, and their primary supply chain activities. This stage has interviewers contributions and studies as national or international reports. Finally, propose alternatives as implementing collection centres and consolidating transportation routes. Figure 1 presents an overview of the methodology.

For this analysis, the information presented in the study "Comprehensive Analysis of Logistics in Peru" of the World Bank and the Ministry of Foreign Trade and Tourism and ten formal interviews with experts between September 2020 and February 2021.

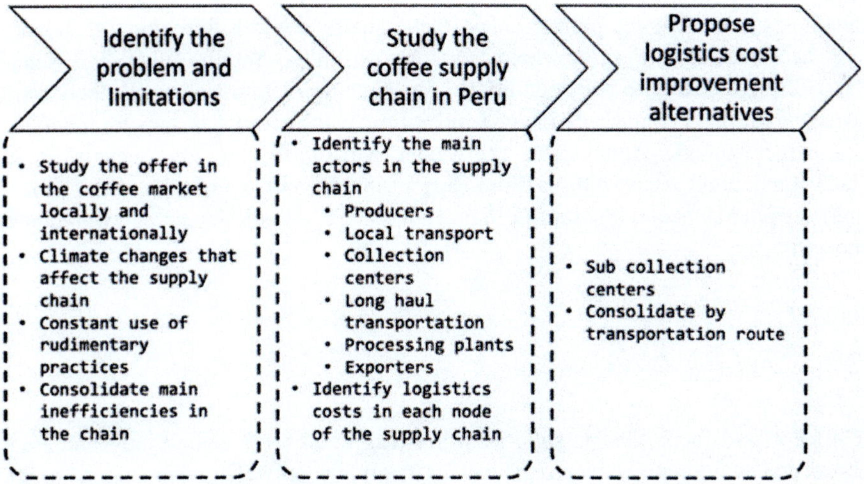

Fig. 1 Methodology proposed for the study

2.1 Coffee Supply Chain in Peru

The first part is production; coffee producers organize as committees, cooperatives, native or peasant communities, contribute with their labour and own or financed capital and apply specific technology and management to manage their productive units: raw material suppliers as seeds, fertilizers, pesticides, tools, bags, blankets, etc. Some entities provide technical assistance for installation, renovation, rehabilitation, and management of coffee plantations, such as NGOs and International Technical Cooperation Agencies. Other entities offer financial aids for raw materials, supplies, labour, certifications, and equipment and machinery (pulpers, dryers). Exporters can also participate in the funding of the harvest (Díaz and Willems 2017).

The harvest has these activities: pulps, ferments, washes, and dries cherry coffee on a small scale. The objective is to obtain high-quality coffee using dryers or solar radiation, considering its humidity levels, from 60% to 12%. Inadequate drying generates ochratoxin A products that produce fungi and toxins (Díaz and Willems 2017). Collectors, intermediaries, producer groups, exporters, and brokers join in the coffee purchase and sale processes. The selection and classification are often manual and complete with a mechanic or electronic machine. The quality index considers bean genetic constitution, environmental factors, drying and storage. Finally, the coffee in bags exportation.

The Peruvian international coffee market represents 95% of its production (Díaz and Willems 2017), considering the Arabic coffee price on the New York Stock Exchange in dollars. The non-exported coffee is commercialized in the local market by national traders and the soluble coffee industry, exporting the coffee to other Latinamerica countries as well-known brands. From farm to ports, the commercialization constraints are inadequate infrastructure to the cultivation areas, low level of technology and training, production cost, and quality controls throughout the chain.

The corridors in Peru are three: Tocache-Zarumilla, Satipo-Callao, and Puno-Callao. Production has poor conditions; for example, Satipo-Callao connects Bajo Anapatí with San Antonio de Pangoa and Satipo, using small-size vans for the roads infrastructure and requirements. The second part, collection, are concentrated in Jaén, Lima and Cusco, with medium-capacity trucks and asphalted roads. The last part is from plants to ports; transportation companies, logistics operators, warehouse operators.

2.2 Logistics Costs

Coffee exports activities are significant in Peru, involve 1.5 million jobs, and are a factor in developing rural areas. The central coffee supply chain corresponds to the Satipo-Callao logistics corridor. It ranges from the production phase, which takes place in the department of Junín, through the collection centres, processing plants until reaching the port of Callao for export.

The first stage for the logistic costs is from the production farms to collection centres. The main customers purchase 96% of the crops to the cooperatives or producer associations during this stage. The total logistics costs per shipment in the production phase amount to S/0.95 per kilo, where 42.60% corresponds to the cost of postharvest treatment, followed by transportation, which represents 36.92%. In addition, the total time from postharvest to deliver in a collection centre is about six days, the time required for postharvest treatment for sale or delivery stands out with 93.40% of the total time invested.

Regarding the information and the results of the interviews published in the "Comprehensive Analysis of Logistics in Peru" coffee growers in the area considering 67% of the surveyed coffee growers usually send their production to cooperative facilities, 25% to a collection centre and only 6% to a processing plant. In addition, 64.90% indicated that they usually use vehicles contracted by the producer for a two to four hours trip. Regarding road infrastructure, 97% has a unique road, and 81% mentioned that the road is a carriageway (World Bank Group & SECO 2016).

From the collection centres to the plants, for example, in 2013: collected 645,192 kg, 98% for export, with a value of S/4 million. About the collection centre: 94% have drying infrastructure, 85% dry storage, and none have refrigerated warehouses. In addition, they have an average capacity of 333,167 ton. As in the production phase, there have been no cases of non-payment by customers. The total time used in this phase, which elapses between the postharvest and the delivery of the merchandise at a collection centre, determined that it takes about 13 days. The main one is related to storage or waiting time until delivery, which takes almost seven days.

Furthermore, it is important to indicate that 1.4% of the product is lost in all the processes in this phase, mainly due to its storage time. Finally, 53.90% of the transports to the collection centres and plants are carried out with vehicles hired by the producers, which usually take between 10 and 12 h for each shipment. In addition, 92% of those surveyed indicated that they use only one route, and it is paved.

The costs will include all those incurred through the coffee value chain until they are transported to the export terminal. It has been determined to consider the sea route as the most important. The average volume of coffee sales during 2013 amounted to 556,006 kg, of which 95% is destined for export. This was valued at approximately S/7.2 million. As a final figure, 45% of the time, safety is the prevailing criterion for choosing a route to the port terminal, 27% is considered the infrastructure of the road and 27%, the shortest distance. In this phase, the roads are 100% paved.

Within the processes mapped through these three phases, during 2018 a total logistics cost of S/1.86 was obtained for each kilo of exportable coffee, which at the exchange rate of that year amounted to USD 0.57, representing 25.24% of the production. total cost. The following figure presents the logistics costs per kilo, disaggregated by each phase updated to 2018.

As identified in Fig. 2, logistics costs are close to representing a third of the total value of the product, including transport, storage and others (they do not include the

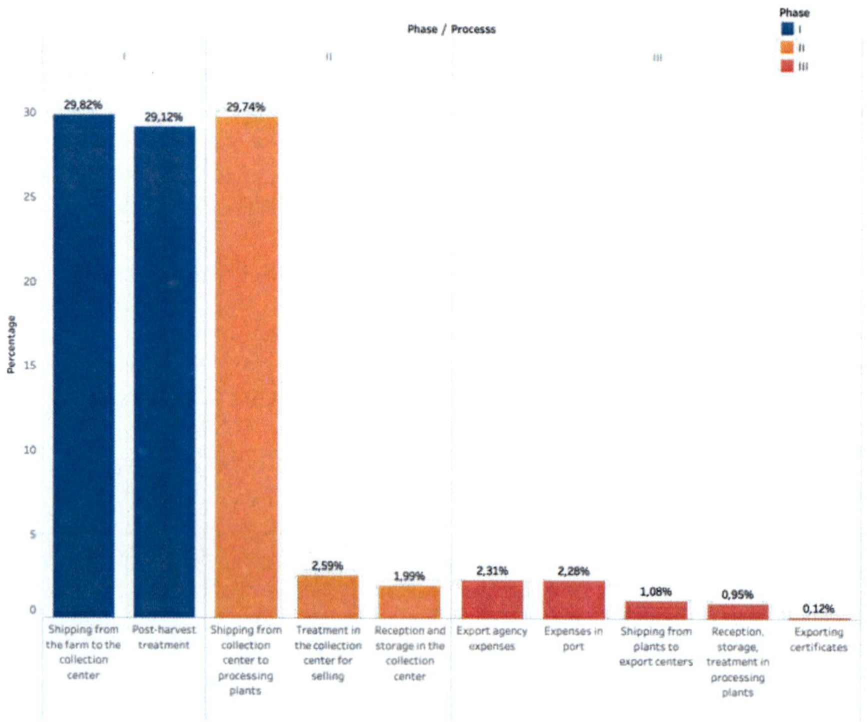

Fig. 2 Logistic costs by activity in the coffee supply chain in Peru

freight to the final destination, only the price posted next to the ship), not including benefits. This participation in logistics costs is worrying if we compare it with the efficiency of other countries. Without going too far, we have Colombia as an example where its cost percentages range between 8% and 14% for the same activities (World Bank Group & SECO 2016). Those mentioned above, together with the low production per hectare, have reduced the competitiveness and profitability of the Peruvian coffee export sector. Therefore, improving logistics infrastructure and reducing costs along the chain are issues that should be among the main pending on the country's political agenda. It was identified that the sending of the farms to the collection centres and from these to the processing plant in Lima and the postharvest treatment of the coffee influence 88.68% in the final logistics costs.

We evaluated some improvement scenarios in these three processes to minimize total logistics costs with the last results. First, the cost of shipping from the farm to the collection centres and cooperatives is related to Phase I and is composed not only of the freight but also of the costs of loading and unloading the merchandise, amounting to S/0.555 of S/1.86, which represents the total logistics cost. Currently, small producers are in areas with difficult access and mainly primitive roads, so cars or trucks that carry between 500 and 1000 kilos are used. During the peak harvest months, such as June, July, August, and September, on average, 12,000 kilos are

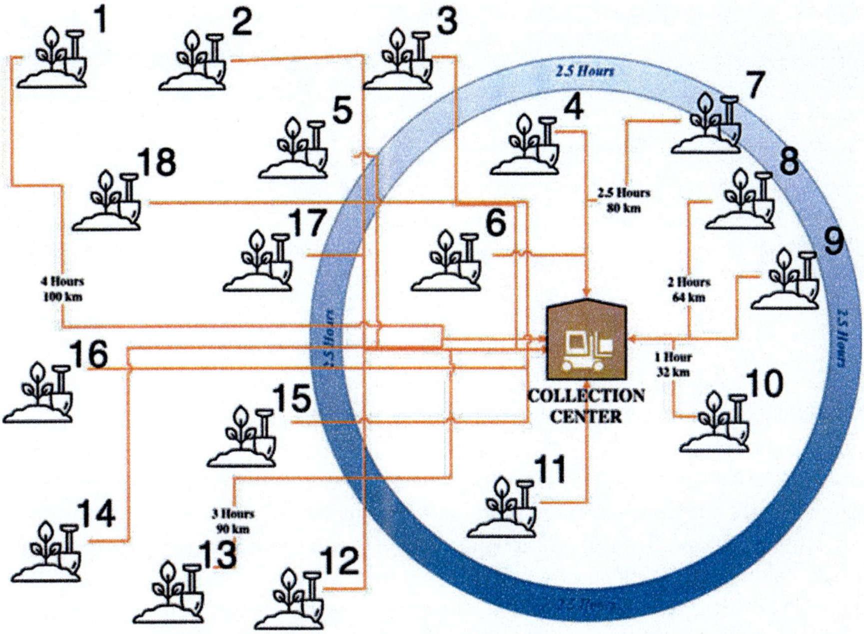

Fig. 3 Spatial distribution from farms to collection center

collected per day, which means 24 shipments of 500 kilos or 12 loads of 1000 kilos each. There are 18 trips from the producers' farms to the largest collection centre or a cooperative with an average load of 750 kilos. Each farmer produces an average of 2700 kilos (considering average productivity of 15 sacks of 60 kilos per hectare and a holding of 1–5 hectares per producer); that is to say, they produce a minimum of 900 kilos or a maximum of 4,500 kilos of cherry coffee. The cost of transportation to the cooperative or collection centre for each trip during 2018 was between S/277.31 and S/554.62 per shipment depending on the number of kilos transferred –considering loading and unloading per point in addition, the average transit time is 1–4 h.

Figure 3 shows a diagram of the current location of the coffee farms and the collection centres or cooperatives. The concentric ring to the cooperative or collection centre is 80 kilometres, based on the information obtained through different interviews with Félix Marín Ludeña, who is a coffee leader from the Chanchamayo area with more than 40 years of experiencia in this field (Marín 2020). Farm 1 represents the furthest point for practical purposes, approximately 100 kilometres to the collection centre or cooperative. Point 10 is one of the closest points, with an average of 32 kilometres. Still, in reality, the average distance between the producers' farms is 5–15 kilometres; in fact, some share boundaries. For the analysis, we will consider that all the land roads outside the concentric ring are gauge and the speed to travel them is ten effective kilometres per hour; in counterpart with the farms located within the arc, which would be connected by better quality roads that

Table 1 Kilometres travelled per day, considering the representativeness of the areas

Zone	Average distance to the collection centre
Zone 1	18.67 km
Zone 2	58.06 km
Average kilometres travelled per day	76.72 km

Table 2 Transportation cost per zone

Zone	Cost/Trip		Cost/Day
Zone 1	S/302.34		S/2116.36
Zone 2	S/488.27		S/5371.02
		Total	**S/7487.39**

would allow traffic at 32 kilometres per hour. To calculate the costs related to this analysis, average conditions will be considered based on the number of kilometres per zone; that is, for Zone 1, an average route of 48 kilometres will be considered. Zone 2 will consider an average route of 95 kilometres.

Around 39% of the farms from which coffee is collected daily are within a radius of 80 kilometres, while the remainder is outside. Next, considering the average kilometres in the area, Table 1 shows the corresponding distances for each zone.

Once the transportation cost per kilometre has been obtained, we can assign the costs making a difference for the conditions of each area. If it is considered that, although the roads from the farms in Zone 2 to the collection points are of primitive and semi-paved roads, the data used for the calculations are average figures for both zones. Therefore, no differentiation will be made regarding the cost of transportation per kilometre. In addition, the loading and unloading costs must be calculated by dispatch and that for this, the average weight of each pick-up has been considered at 750 kilos. Thus, the costs associated with transporting the farm to the intermediate point amount to S/7487.39 per day during the harvest season (Table 2).

Second, the shipping cost from the collection centres and cooperatives to the processing plants. This cost is related to Phase II; they are the expenses incurred when the merchandise is transferred from the collection centers or cooperatives located in the coffee zone to the processing plants or the premises of the exporters in Lima. Its cost is S/ 0.553 for each coffee kg and represents 29.74% of the total logistics cost. The road used for this journey is the Central Highway, a national transversal road connecting the city of Lima with Junín. The average time on this road without traffic is 7 h and 16 min, and it is an entirely paved road. However, the congestion on this highway ends up generating over times of up to 4 and a half hours for this journey, meaning that many trips end up adding 12 h of travel. This causes freight and security expenses (such as protection and insurance) because the vehicle becomes a target for robberies and assaults when moving slower. Around 92% of coffee producers only know and use the Central Highway to transport coffee to Lima. The cost of transferring coffee from the collection centre to the processing plant in Lima has three concepts. The most important is the costs directly related to the transfer, loading at origin, and unloading at destination.

Table 3 Transportation times by route

Routes	Time [hours]	Time [min]
Central Highway	11	660
Vencedores de Sángar or Lima-Tinyahuarco Highway	8.8	485
Road Huaral	8.7	521
Road Oyón	11.55	693

The cost of the kilograms of coffee transferred from one point to another for each minute is determined by the total cost per kilogram divided by the total time in minutes. The basis of the calculation is the travel time on the Central Highway. As the traffic is ten to twelve hours, an average of eleven has been considered for the calculations shown (Table 3).

One container fills up with twelve tons of parchment coffee. The average capacity of a trailer truck that transports merchandise from Junín to Lima is eleven to thirteen tons, which coincides with the capacity of the container. The insurance price per section is USD 3 for a sack. Still, many coffee farmers prefer not to pay it due to the demanding processes they must follow in case the merchandise is stolen and mainly because they do not prevent it from being stolen from the vehicles. In addition, for consistency reasons, throughout this document, sacks of sixty kilos each have been considered. In addition, the considered exchange rate in 2018 was S/3288 for every US dollar.

The cost per kilo per minute was calculated for each of the routes, and the expenses related to transportation were assigned. To calculate the insurance, the cost per kilo was calculated to be S/0.16. Loading and unloading activities were calculated per kilo and per shipment. Finally, as indicated above, the Central Highway is a very unsafe road, where due to the slow traffic of vehicles, it becomes a place used by criminals to steal merchandise from trucks, so many of the coffee growers hire a safeguard or custody to prevent these thefts, amounting to USD 1000 per half container, that is, every six tons.

Finally, we have the postharvest treatment in the farm composed mainly of the coffee bean's pulping, washing, and drying. This cost represents 29.12% of the total logistics costs with S/0.542. These processes can be carried out by farmers, cooperatives, or collection centres, but farmers generally do not have the ideal infrastructure to carry them out.

The postharvest is appropriately composed of the pulping, washing, and drying of the coffee bean. This analysis has also included packaging since the coffee is not delivered loose but in bags. It takes a little more than five days, and it is usually the farmers who do it, often under conditions that are not ideal. This process directly affects the aroma and quality of the final coffee bean, and that is why it is of great importance that it is carried out in places that can guarantee its proper handling.

Pulping is a process that is done with mechanical help. The pulpers can be manual, motorized, and automated; depending on this, they can process from 350 kilos of cherry per hour to 1500 kilos per hour, and some machines can pull up to 4000 kilos simultaneously. The most used, of course, are manual and motorized since their price is much more accessible for farmers. The washing can be done in tanks or some equipment for pulping also, wash the parchment coffee, leaving it without the mucilage. Coffee can be dried in the sun or using mechanical dryers. The first method takes an average of five days, but sometimes in good weather, it can be reduced to just three. The process requires the most labour and control to stir the coffee and thus prevent the coffee from fermenting and acidifying. Using the mechanical dryer can reduce the time from five to as little as one day. These teams can be rudimentary or more technological. The rudimentary ones work with firewood. They must cut and burn trees, which damages the environment and takes shade from the coffee crops, affecting their quality and making them sensitive to the attack of pests. The more technologically advanced options use motors or the force of water or wind to operate, making them eco-friendlier. We seek to assign the costs for each sub-process of the postharvest treatment, considering current practices and times.

3 Scenarios and Results

3.1 Scenario 1

Considering this information, a first improvement alternative has been developed based on a shipping network with an intermediate point to collect the coffee from the farms in Zone 2, which are the most remote and have more difficulties getting their coffee to the main collection centres due to the rough roads. It is essential to mention that this intermediate collection point must have adequate road infrastructure so that a truck with greater capacity can transit between its location and the main collection point. Therefore, it must be located within the perimeter that does have a better quality of roads. In this coffee growing area, vehicles weighing up to 4 tons can travel up to 80 kilometres around the main collection centre without significant inconveniences. Note that in Fig. 3, all the farms in Zone 2 would take their coffee beans in cherry or parchment to the intermediate collection point shown in orange and located on the edge of the circle (Fig. 4).

For the calculations, the following data will be considered, an average distance of 48 kilometers between a farm in Zone 1 to the main collection center. This is obtained by averaging the distances traveled in one and two hours, which have a higher quality of roads. For Zone 2, the same operation was carried out, but considering the kilometers traveled in three and four hours from the main collection point to each farm, achieving an average distance of 95 kilometers for this zone. At this second point, let's not forget that the quality of the access roads is lower. Therefore, the time to travel each kilometer is less. To calculate the distance between the farms in Zone 2 to the intermediate collection point, the distances within their zone were considered, obtaining an average of 15 kilometers. Finally, as data we have, the quality of the access roads is sufficiently accessible up to 80 kilometers around the main collection center, which takes up to two and a half hours to travel.

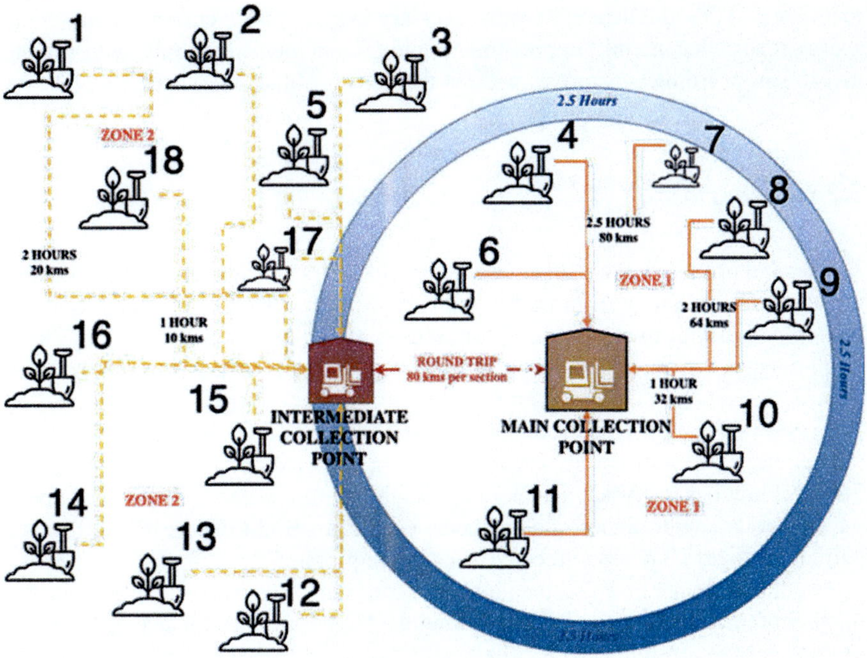

Fig. 4 Alternative 1 with intermediate collection point

Table 4 Transportation costs for scenario 1

Zone	Per trip	Per day
Transportation cost for farms in Zone 1	S/302.34	S/ 2116.36
Transportation cost from Zone 2 to intermediate collection point	S/171.79	S/ 1889.64
Transportation cost from intermediate collection point to main collection point	S/ 1553.37	S/ 1553.37
TOTAL PER DAY		S/ 5559.38

Below are the transportation costs by zone, considering the intermediate collection point and the average number of trips generated in the high harvest season (Table 4):

With this proposed alternative, a daily transportation cost of S/ 5559.38 can be obtained, 25.75% less expensive than the current situation.

But in order to carry out this alternative, it is necessary to generate alliances or agreements with other groups that allow concentric points to the farms that serve to temporarily collect the coffee beans and vehicles that transfer them from the intermediate point to the main collection point. It is important to bear in mind that to implement this alternative, it is not necessary to invest in the current quality of the access roads; that is, this improvement could be implemented only with private investment or from cooperatives without the need for State intervention.

3.2 Scenario 2

The second proposed alternative, unlike the first, does consider an improvement in the access roads, going from trail to partially asphalted at least allowing access to remote points with trucks of four to six tons capacity so that the collection center. The principal can implement a collection system with arrival at each farm that is in Zone 2.

In general terms, for the calculation, in both scenarios it has been considered that the number of kilos of coffee on average that is collected or sent from each farm is 750 kilos, but it is important to make an indication for this second scenario and that is what to count. with higher quality roads, each farm could deliver more than the current average. Thus some fixed costs could be prorated.

For this particular exercise, the number of trips will be calculated, multiplying the number of pick-up points in Zone 2, during the high season, by the average number of kilos per farm, resulting in 8250 kilos. In turn, if we consider the capacity of the vehicles, at least two trips must be made. In real life, the number of trips or vehicle capacity to be used will be determined by the number of kilos that can be collected per day.

If all roads had equal conditions, the average speed of traffic would be 32 kilometers per hour. We will use the exact cost per kilometer obtained previously because the savings will occur over time. In addition, this alternative would be compatible with the improvement of the exportable yield of the grain, since the collection center could guarantee the use of javas for the collection of coffee to avoid wastage (Fig. 5).

The same average distances of the first scenario will be used. An average distance between farms of 10 km, considering that in the community groups that are members of the cooperatives, the distance between the farm's ranges between 5 and 15 kilometers. The costs obtained for the second scenario are presented below (Table 5).

With this second scenario, a daily expense of S/4764.97 is generated, which means a saving of 36.36% over the expenses currently generated.

4 Results and Discussion

Considering the alternatives presented, we introduce the new cost calculated per kilogram in soles for the shipping process from the farm to the collection center / cooperative to evaluate its real impact on the total logistics cost. As seen in Table 6, a saving of up to 2.10% is obtained related to the original logistics cost.

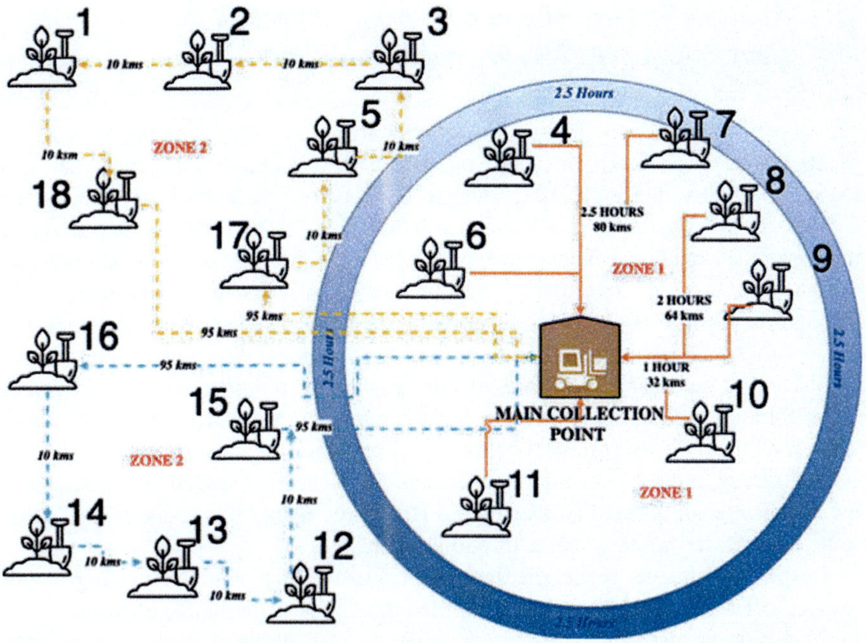

Fig. 5 Alternative 2 with specific routes

Table 5 Transportation costs for scenario 2

Zone	Per trip	Per day
Transportation cost for farms in Zone 1	S/302.34	S/2116.36
Transportation cost between farms Zone 2 – Trip 1	S/152.01	S/912.03
Transportation cost to and from the main collection point – Trip 1	S/488.27	S/488.27
Transportation cost between farms Zone 2 – Trip 2	S/152.01	S/760.03
Transportation cost to and from the main collection point – Trip 2	S/488.27	S/488.27
Total per day		S/4764.97

Table 6 Summary of the analysis of both alternatives

	Cost per kilogram (PEN)		
Processes	Original	Improved 1-A	Improved 1-B
Shipping from the farm to the CC	S/0.555	S/0.412	S/0.353
The new cost per kilogram (USD)		US$ 0.522	US$ 0.504
Difference from the original logistics cost		1.48%	2.10%

4.1 Analysis of Cost Alternatives from Collection Centers/Cooperatives to Processing Plants

This process is related to the transfer of coffee from the coffee growing areas to the plants where the coffee is threshed and classified, concentrated in Lima, Jaén and Cusco. Currently, the most used route to transport coffee from both areas is the Central Highway, presented in the first row of Table 7. The following are alternate routes for the same route. The next step is to calculate the costs of the main route and the alternative routes, considering that the coffee growers that move their cargo through the Central Highway do not usually invest in insurance for their cargo but prefer to pay the custody service to protect their merchandise. On the contrary, it would not be necessary to pay the custody service for the merchandise that would move through the alternative routes, but the cargo insurance would be necessary. Therefore, that information will be included to calculate the total costs per shipment.

In this way, it can be seen that the route of the Vencedores de Sángar highway is the cheapest, even more than the Central Highway, mainly due to the fact that this route has a lower influx of vehicles and is safer.

Despite the changes made, the final result doesn't show significant changes with respect to Table 7. Therefore, it is confirmed that the best alternative continues to be the Vencedores de Sángar Highway. This process offers a saving of 46.64% compared to the current cost of this process.

4.2 Analysis Summary

Like the previous process, it is necessary to analyze the impact that this saving would have on the total cost; therefore, it must be included in the current cost mix in an isolated manner, that is, without incorporating improvements in the cost that corresponds to other processes. Finally, this improvement translates into a saving of 2.71% in the total logistics cost (Table 8).

Table 7 Costing of the Central Highway and alternate routes

Routes	Cost kilogram/ minute	Freight	Insure	Guard	Carry & unload	Total
Central Highway	S/0490	S/ 5.880,82	–	S/ 6.575,40	S/756,53	S/ 13.212,75
Vencedores Highway	S/0360	S/ 4.321,51	S/ 1.972,62	–	S/756,53	S/ 7.050,66
Route through Huaral	S/0387	S/ 4.642,29	S/ 1.972,62	–	S/756,53	S/ 7.371,43
Route through Oyón	S/0515	S/ 6.174,87	S/ 1.972,62	–	S/756,53	S/ 8.904,01

Table 8 Summary of logistics costs per kilo in the current situation and the proposed alternative

Processes	Cost per kilogram	
	Original	Improved 2
Shipping from the CC/cooperative to the PP	S/0.553	S/0.295
The new cost per kilogram		US$ 0.487
Difference from the original logistics cost		2.71%

Table 9 Comparison of current costs and proposed costs

Process	Cost allocation	
	Real	Improvement
Pulping – washing	S/0.032	S/0.011
Drying	S/0.435	S/0.087
Postharvest treatment itself	S/0.467	S/0.098
Packaging	S/0.075	S/0.075
Total process time	S/0.542	S/0.173

4.3 Postharvest Treatment on the Farm

This process is expensive because it is done by hand. Poor infrastructure, poor inputs, high demand for labor, and small-scale production on each farm contribute to the over-cost. The collection centers and cooperatives carry out the process on a large scale, but they don't have the necessary technology to automate the process due to heavy investment and lack of financing.

In this process, the pulping – washing and drying activities will be considered. Using a pulping machine that processes 4000 kilos per hour, it is possible to go from 8.70 h invested in pulping and washing activities to just 3 h; and in the case of the drying activity, the time can be reduced from 120 h (five days) to only 24 h.

With this information, we can reallocate the costs of the process and verify the impact of the reduction of the times in the saving of the total cost (Table 9).

Finally, the projected savings is 68.13% concerning the initial cost of this process. Although the numbers are very encouraging, we mustn't forget that they could only be achieved with a strong capital investment for equipment, infrastructure, and training for their operation and maintenance. Trying to automate these processes with each of the producers would be a project of great magnitude. Still, the solution could be directed towards the empowerment of cooperatives and collection centers, establishing quality parameters for the treatment of coffee in such a way that have better control over costs, reducing them with optimal batches without neglecting the quality they are seeking to achieve and providing loans that allow them to capitalize without being eternally indebted (Table 10).

With this process, an impact on the total logistics cost of 3.94% is achieved, being the process that presents a greater impact on the total cost. Finally, with the proposed improvements, the share of logistics costs was reduced from 25.24% to 15.76%, managing to improve it by 9.47%.

Table 10 Analysis summary

	Cost per kilogram	
Processes	Original	Improvement 3
Postharvest treatment	S/0.542	S/0.173
New cost per kilogram		US$ 0.453
Difference from the original logistics cost		3.94%

5 Conclusions and Recommendations

In 2018, the logistics cost of coffee increased 25.24% of its total value. Three processes represented 88.68% of the total costs, among which are: the shipment from the farms to the collection point with a participation of 29.82%, the shipment from the collection point to the processing plant with 29.74% and, the postharvest treatment of coffee represents 29.12%. The first two are closely related to the area's road infrastructure, and the latter is associated with the lack of technology. These three processes were analyzed to obtain significant improvements within the total logistics cost and to identify which policies should be prioritized to achieve greater efficiencies in the Chanchamayo coffee cluster. Moreover, transport from farms to the collection points is mainly characterized by poor conditions of their routes, where 80% of the roads are not paved. The most remote farms in the mountains generally produce the best coffee due to the privileged conditions of their geography. However, the precariousness of the roads and the low capacity of the vehicles that can travel through forcing them to move coffee in small batches. Two alternatives were presented; the first one assumed that there was not an improvement in the access roads and, the second alternative assumed that there was an improvement. The results showed improvements would be achieved; the first alternative represented a 25.75% saving and the second, a 36.36%, which finally translates into a savings of 1.47% and 2.10% in the total cost, respectively.

On the other hand, the transit between the collection points to the processing plants in Lima is another point in which deficiencies in the road infrastructure generate overall logistics costs. Central Highway improvement has been projected for many years. The former Minister of the Ministry of Transport and Communications (MTC), Eduardo González, announced in 2020 the signing of the Government-to-Government contract for the project to expand this road in March 2021. In addition, he told the completion of the works in the alternate routes of the highways via Oyón and on the Vencedores highway. In the analysis, the Vencedores highway stood out compared to other alternatives by representing a saving of 46.64% to moving the merchandise via the Central highway. Finally, this caused an improvement in total logistics costs of 2.71%. Regarding the postharvest treatment of coffee, the overall cost was determined by the lack of access to technology, drinking water and infrastructure. This process is of vital importance in determining the quality of the coffee bean, which affects the final price of the coffee. The projected savings in this process were 68.13% concerning current costs, positively impacting the final cost, with 3.94%; achieving this improvement significantly impacted the total cost.

Finally, with the proposed improvements, the participation of logistics costs went from representing 25.24–15.76%, reducing them by up to 9.47%.

In addition, improving the access roads that connect the farms with the main collection points, from unpaved roads to at least semi-paved roads, will facilitate the transit of trucks of greater capacity. In conclusion, the expansion of the Central Highway may decrease congestion as well as transit time. Similarly, completing the construction of the Vencedores highway may represent alternative routes to move coffee from Chanchamayo to Lima while improving security.

Another aspect that can benefit is implementing projects to improve the postharvest treatment process and training coffee growers, enhancing their profitability and competitiveness in the foreign market. The support of cooperatives and collection points may improve quality standards, technology, capital and training to reach all its members and promote coffee clusters. For this purpose, the provision of equipment, spaces for production and storage, quality control, training and capital is necessary. The strength of collection points and cooperatives concentrating on postharvest treatment will generate economies of scale that reduce costs and better control quality standards.

References

Programa de las Naciones Unidas para el Desarrollo [PNUD] (2017). Línea de Base del Sector Café en el Perú.

IndexMundi. (2021). Peru green coffee production by year. Peru Green Coffee Production by Year (1000 60 KG BAGS).

United States Department of Agriculture Foreign Agricultural Service [USDA]. (2018). Peru's Coffee production continues recovering.

Caballero, E, Coz Del Castillo, S, Veliz Huamantica, I, Vicente, W, & Galarza, C. (2021). Analysis of internal logistic cost on exports of peruvian coffee in the period 2015–2019. *Acta Logistica, 8*(1), 73–81.

Díaz, C & Willems, M. (2017). Baseline of the Coffee Sector in Peru. State Secretariat for Economic Affairs of the Swiss Confederation (SECO) and Green Commodities Program of the United Nations Development Program (PNUD). Lima: PNUD.

World Bank Group and Secretary of State for Economic Affairs of the Swiss Confederation [SECO]. (2016). Comprehensive analysis of logistics in Peru – Product Coffee. Lima: World Bank Group.

Villalva-Cataño, A., Ramos-Palomino, E., Provost, K., & Casal, E. (2019). A model in agri-food supply chain costing using ABC costing: an empirical research for Peruvian coffee supply chain. In 2019 7th International Engineering, Sciences and Technology Conference (IESTEC) (pp. 1–6). IEEE.

Engström, R. (2016). The roads' role in the freight transport system. Transportation Research Procedia, 14, 1443–1452.

Hummels, David. (2007). Transportation Costs and International Trade in the Second Era of Globalization. Journal of Economic Perspectives, 21 (3): 131–154.

Marín L., F. (2020). *Personal interview by Diana Llanos.* Cooperativa Coopchebi.

Cámara Peruana de Café y Cacao. (2017). Estudio de mercado del café peruano. Programa SECOMPETITIVO de la Cooperativa Suiza – SECO.

Operational Planning Model for Harvesting of Fresh Agricultural Products

Nestor E. Caicedo Solano, Guisselle A. García LLínás, and Jairo R. Montoya-Torres

Abstract In this work, we present an operational planning model for harvesting of fresh agricultural products. This model contains the resources and factors that determine the operational cost of harvesting. These resources are labor, time windows for operations, use of soil, demand of products, types of markets, loss for quality, variety of fruits and productivity of crops, which is founded on the use of lean manufacturing (LM) principles as wastes management tool. This model minimizes the cost of wastes generates by resources used for harvest and allowing for get an operational planning through of mixed linear programming. The results show operational improvements and indicating significant savings can be obtained in cost for harvesting. A study case is show for illustrating the results, savings expressed through the objective function, and a sensibility analysis is described for determining the best levels of production factors for harvesting.

Keywords Operational planning · Harvesting · Lean manufacturing · Wastes · Agricultural products

1 Introduction

1.1 Problem and Background

Given the need for increasing the production and quality of agricultural products for human consumption and industry, as of today the agricultural research has occupied an important place in science. Formal research in process optimization in agriculture began with applications of Operations Research (OR) developed by many

N. E. Caicedo Solano (✉) · G. A. García LLínás
Universidad del Norte, Barranquilla, Colombia
e-mail: nestorc@uninorte.edu.co

J. R. Montoya-Torres
Universidad de La Sabana, Chía, Colombia

J. Vargas Florez et al. (eds.), *Production and Operations Management*, Springer Proceedings in Mathematics & Statistics 391,
https://doi.org/10.1007/978-3-031-06862-1_14

agricultural companies and their administrative areas [1], for example in decision making on crop rotation, selection and location of areas and farms for sowing, combining crops of fruits, vegetables and industrial agricultural products [2].

OR has been important for agricultural planning, considering elements such as water use, soil, meristems volume, type of demand and quantities, life cycles and maturating. Most research has been carried out in order to optimize production in terms of yields and compliance with product quality characteristics, but have avoided the waste generated at all stages of agricultural production. In the agricultural production chain, an important amount of food waste is generated. Since the 1990s there have been a decrease in crop yields. For example, wheat yields grew at an annual average rate of 3.8% between 1961 and 1989, but only 2% annually in the period 1989–1999. The growth rates of rice decreased to less than half, from 2.3% to 1.1% in the same period. In the first decade of the 2000s, increases in world cereal production were minimal. The yield growth will remain the dominant factor of underlying increases in crop production in the future. Overall, it is estimated that approximately 80% of future increases in crop production in developing countries will have to come from intensification with higher yields, increased multiple cropping and period's shorter fallow, coupled with resource and wastes optimization obtained from the operation of agricultural chains [3].

For a mid-size and developing country, such as Colombia, the National Planning Department (NPD) and the Association of Food Banks (ABACO, for its name in Spanish) conducted a study that found that about 9.7 metric tons of food (fruits, vegetables, tubers, cereals, grains, etc.) were wasted or lost each year. This corresponded to about 34% of the total food available in the country. About 22% of the total losses corresponded to the stages of agricultural production, post-harvest and storage, and industrial processing. The remaining (3.4 million metric tons) was lost to waste at distribution, retail and consumption [4].

In this work we present a mathematical model that includes resources used for harvesting. The resources generate wastes in any agricultural production processes, for this reason, our contributions are based on minimizing the costs on wastes looking a sustainable operation and an operational model for planning the agricultural production combining optimization and lean manufacturing principles.

For illustrate this model, we present a case study of Bananas farms in Caribbean area in Colombia. The results show different scenarios for decision making of farmers and planners through the use of experimental design and response surface methodology.

1.2 Literature Review

For this work, we reviewed papers, technical documents, books and other documents that we consider relevant in our research.

The agricultural production as referred to this research focuses the operational logistics plan used for sowing, cultivation, and harvesting. Each of the activities

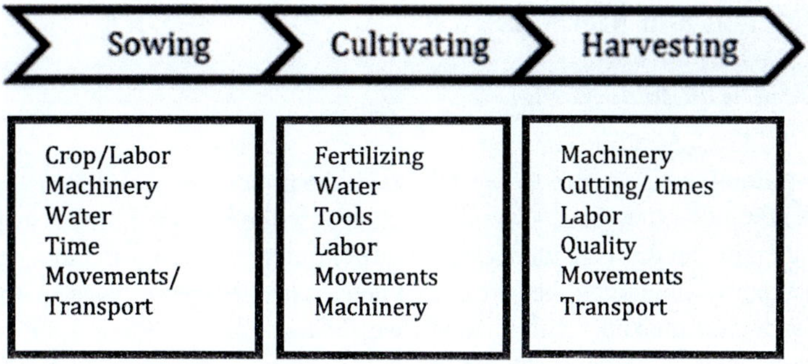

Fig. 1 Resources in agricultural production

depends on resources that must be optimized. At the same time, these resources, unless optimized, can generate waste and consequently, increase costs. All production systems can be described as a set of inputs and relationships, which not only include the food system, but also the food production. Indeed, programming production under certain financial risks [5] or setting the production volumes in line with market behavior [6] require the use of decision-making models for agricultural chains [7–10]. In general, agricultural chains require a structured analysis to find possibilities for optimization and improvement at all stages being the science, data and mathematical models a great option for solving problems about operations, products and agricultural process [11–13].

We also found that wastes are generated in agri-chains, which have special features such as perishing risks, freshness requirements, physical and aesthetic attributes [14, 15], optimal delivery times, external uncertainty variables (i.e. climatic factors and market conditions) [16–18] and crop characteristics [19], all of which explicitly account for the complexity of managing these chains. Decision-making processes under such uncertainty is as such one of the main issues to be resolved [20].

Figure 1 show the resources used for agricultural production in all stages, in special case the harvest process need machinery, labor, movement, transport, times and preserve quality of products.

According to previous highlights, there is an urgent need for efficient and effective decision-making processes in agricultural production. A basic illustration of activities that require critical decision-making in agricultural chains is presented in Fig. 1., which shows an agricultural operations logistics chart illustrating the activities of sowing, cultivating, harvesting and the processes of making decisions at tactical and operational level. In this paper, we focused in harvesting operations in agricultural production systems.

2 Material and Methods

2.1 Mathematical Model

In this section is propose a model that includes the principles of lean manufacturing, understanding that they have several aspects that will be adapted, fitted, reallocated for any crop model that considers agronomic conditions. The crop-planning model is presented below, including the factors that determine the types of waste at this stage of the agricultural chain. In this model, we looking for less cost related to labor, supplies and loss of quality in the harvest. In this model is illustrated the final stage in the agricultural production systems before post harvesting. The following assumptions are made:

1. There are different types of variety to harvest.
2. The lots or hectares to be harvested are known.
3. The costs for harvesting are known.
4. The effect of crop failures is known.
5. Crop yields are known.
6. A maximum waste of waste can be estimated by harvest quality.
7. Penalties for harvesting fruits outside quality specifications are known.
8. Harvest modes are known.

Description Model

Indexes

m: Sets of harvest modes, m = {1: Mode 1, 2: Mode 2}.

l: Set of hectares to be harvested on the farm, $L = L_1$ or L_2, where L_1 corresponds to the hectares to be harvested with mode 1 and L_2 corresponds to the hectares to be harvested with mode 2.

o: Sets of types of labor contracting, O = {1: fixed labor, 2: variable labor}.

f: Type of markets according to product quality, F = {1: Export, 2: National}.

t: Planning horizon of cut order (harvest).

Parameters

Df_t: Demand of fruits to be satisfied for market f in period t

$KgMax_{l,t}$: Maximum fruit production that can be harvested in hectare l in period t

$KgMin_m$: Minimum production of fruits that can be harvested using the mode m.

W_l: Weight of harvested production waste

$Per_{l,t,f}$: Percentage of loss (reduction of production) due to defects in hectare l in the period t defined for the market f.

$Prd_{m,l}$: Productivity of the harvest in mode m in lot l.

CM_o: Labor cost by type of contract o.

CH_o: Cost of hiring an additional unit of labor type o.

CD_o: Cost of dismissal of a fixed labor unit.

$CD_{l,t}$: Cost of production waste harvested on hectare l in period t.

W_2: Penalty value for harvesting fruits out of specifications.
MP: Maximum loss allowed per harvest from hectare l in period t.
W_3: Penalty value for harvesting with poor practices.

Decision Variables

$X_{l,t,m}$: Amount harvested in hectare l, in period t according to mode m.
$WD_{l,t}$: Number of workers available for harvesting of hectare l in period t.
$WCF_{l,t}$: Number of workers hired for the harvesting of hectare l in period t
$WRF_{l,t}$: Number of workers fired at the end of period t in hectare l.
$WCV_{l,t}$: Number of workers hired for harvesting of hectare l in period t in contract type o.

Binary Variables

$$Y_{lt} = \begin{cases} 1, \textit{if harvested in hectare } l \textit{ in period } t, t \in T \\ 0, \textit{otherwise} \end{cases}$$

Model formulation

$$Min\ Z = \sum_{t \in T} \sum_{l \in L} WD_{lt} - WCV_{lt} * (CMO_1 + D_1) * TCS + \sum_{t \in T} D_2 * TDV_t + \sum_{t \in T} C_2 * THV_t + \sum_{l \in L} \sum_{m \in M} \sum_{t \in T} X_{lmt} * CD_{lt}$$

Subject to:

$$\sum_{l \in L} \sum_{m} X_{ltm} + Per_{ltf} \geq D_{ft} \qquad \forall f \in F, t \in T \tag{1}$$

\scale90%

$$\sum_{m \in M} X_{ltm} \leq KgMax_{lt} \qquad \forall l \in L, t \in T \tag{2}$$

$$X_{ltm} \geq KgMin_{lt} \qquad \forall m \in M, l \in L, t \in T \tag{3}$$

$$Prd_{mt} * WD_{lt} \geq X_{ltm} \quad \forall m \in M, l \in L, t \in T \tag{4}$$

$$WD_{lt} = WD_{lt-1} + WCF_{lt} + WCV_{lt} - WCV_{lt-1} - WRF_{lt} \quad \forall l \in L, t \in T \tag{5}$$

$$WD_{lt} \geq 2WCV_{lt} \quad \forall l \in L.t \in T \tag{6}$$

\scale90%

$$Y_{lt} \in \{0, 1\} \quad \forall l \in L, t \in T \tag{7}$$

$$WD_{lt}, WCF_{lt}, WCV_{lt}, WRF_{lt}, \in \mathbb{Z}^{+} \quad \forall l \in L, t \in T \tag{8}$$

The objective function minimizes the total cost of harvesting a farm, taking into account the hectares in which they are divided. The first term expresses the costs per harvest with current modes of cutting and transfer of bunches to the cable track; this term represents the cost of operation and tacitly includes the movements of the harvest crew. The second term includes the cost of hiring and firing of fixed labor, which implicitly reduces the use of labor compared to the operation and which in turn affects the costs of hiring and firing temporary staff, represented in the third and fourth term. The last term determines the cost associated with the loss of production per crop.

Constraints (1) ensure that the harvest does not exceed the market requirements, that is, compliance with demands. Constraints (2) represent the harvest in maximum values, while Constraints (3) are the minimum quantity to harvest. Constraints (4) illustrate the capabilities of labor, Constraints (5) correspond to the flow of labor and Constraints (6) are the maximum number of variable workers that can be hired for the harvest. Constraints (7) and (8) correspond to the type of decision variables.

This work does not intend to physically eliminate mudas, what is sought is to economically quantify losses reflected in the costs of production in order to minimize that cost and thus, to reduce the consumption of resources that can be considered as mudas. Fo this reason, the constraints and objective function include resources and parameters of agricultural production systems.

2.2 *Description Model and Input Data*

This model describes how is possible a planning process for harvesting; which is based on the results of the previous models, according to indicators of production areas, number of lots and potential crop yields. The harvesting processes are generally carried out a few days a week, since the calving intervals of the plants oscillate between 9 and 12 weeks, with conditions that generate variation and dispersion between the harvest dates by the plant. In this model it is considered that the harvest only includes a portion of areas of the total farm, for this reason, only two days of the week are included in which the harvest activity is carried out and the resources that are available, as well as opportunities to reduce waste that has been controlled with crop maintenance activities and that seek to optimize the quality and expected crop volume of the crop. The input data for validation is presented in Table 1. The values of parameters were defined from crop maintenance and sowing models, considering that the values should be maintained through of the models for task.

This model was run in a version of GAMS® 24.4.6 for x86 64bit/MS Windows. The solver used was Lindo, an exact solver for MIP problems that contains GAMS library installed in a computer with an Intel® Core™ i5-5300 CPU 2.3 Ghz Processor, 8 GB of RAM and 500 Gb hard drive. After compilation, the model generated 69 single equations, 6 blocks of variables, 49 single variables, Non zero elements 241 and 32 discrete variables.

3 Results

3.1 *Optimization and Case Study Results*

To illustrate the potential of the model a case is provided of data available of banana farmers in Colombia with parameters described above. Table 2 present results, where period 1 in both modes cover those quantities to satisfy demands. The total harvested in order to Table 2 is 24.614 Kgs in 2 days of harvest. Table 3 shows the labor necessary for harvesting, as well as labor hired in Table 4.

According to Table 3, the necessary quantity of workers for harvesting is 8, considering that each crew for this operation has a formation by 4 workers. Results indicates that workers will not be fired, as well as will not be hire variables workers. The minimum cost obtained by the mathematical model for the agriculture production systems considering these conditions for harvesting is COP$ 61.978.461 (US $ 16.500). The above cost can be obtained if we suppose that all plants produced bunches and generated wastes of fruit by quality problems and lost production.

Table 1 Input data for experimental validation in harvest

	Notation	Meaning	Values
Indexes	*m*	*Harvest Modes*	*1: Elevator Mode* *2: Cradle Mode*
	l	*Lots to be harvested*	*1,2,3,4*
	o	*Type of Labor hired*	*1: Fixed labor* *2: Variable Labor*
	f	*Markets according to quality of products*	*1: Exportation* *2: National*
	t	*Horizon of planning (days for harvesting)*	*1,2*
Parameters	Df_t	*Daily demand of fruits for markets*	*1:8.000* *2: 2.000*
	$KgMax_{l,t}$	*Maximum fruit production that can be harvested in hectare in any period*	*16.500 for all hectares*
	$KgMin_m$	*Minimum production of fruits that can be harvested using the harvest mode.*	*1:3.000* *2: 5.000*
	W_1	*Weight of harvested production waste*	*100*
	$Per_{l,t,f}$	*Percentage of loss (reduction of production) due to defects in any hectare in any period defined for the market.*	*0,65% for all hectares*
	$Prd_{m,l}$	*Productivity of the harvest in any mode in any lot.*	*25.000 kgs*
	CM_o	*Labor cost by type of contract.*	*COP $ 35.000 (US $ 10)*
	CH_o	*Cost of hiring an additional unit of labor type.*	*COP $ 40.000 (US $ 12)*
	CD_o	*Cost of firing of a fixed labor unit.*	*COP $ 130.000 (US $ 40)*
	$CD_{l,t}$	*Cost of production waste harvested on hectare in period.*	*COP $ 2.500 (US $ 0,8)*
	W_2	*Penalty value for harvesting fruits out of specifications.*	*120 Kgs*
	MP	*Maximum waste allowed in harvest.*	*0,23 Kgs/Ha*
	W_3	*Penalty value for harvesting with poor practices in crop traceability.*	*COP $ 36.000*

3.2 Statistical Results and Scenarios

In addition, a full factorial experimental design was also developed with five parameters as factors (A: Daily Demand of fruit for markets, B: Maximum Fruit

Table 2 Fruit (Kgs) for harvesting

	Mode 1		Mode 2	
Lot	Period 1	Period 2	Period 1	Period 2
1	11.922	55	11.922	55
2	55	55	55	55
3	55	55	55	55
4	55	55	55	55

Table 3 Labor required fixed for harvesting

	Period	Period
Lots	1	2
1	1	1
2	1	1
3	1	1
4	1	1

Table 4 Labor hired for harvesting

	Period	Period
Lots	1	2
1	1	1
2	1	1
3	1	1
4	1	1

production harvested, C: D: Percent of waste by quality of fruits in harvesting, E: Hiring cost per worker, 33 runs, and one response variable (Zmin) extracted from mathematical model in each run. These factors were chosen by interest of farmers in cost that can affect the benefits and modify decisions on areas for demands, quality, wastes, productivity and labor.

In this case, a sensibility analysis was possible as method for evaluating scenarios; however, it was required to test changes in specific parameters and cost because some costs are fixed from sowing. In addition, the results can be illustrated in a surface of responses, having a better level of revision on the objective function versus changes in parameters.

To evaluate sensitivity of the model, a response surface methodology (RSM) was built, in this case, RSM predicted the value minimum of the objective function for activities per week, while other parameters is maintained in original values for runs. Data were analyzed in Minitab 19®. In Table 5 summarizes the results of the model. The R-square (95,96%) indicator is obtained, which gives a good adjust and correlation between factors and response variable (Z min).

Table 5 presents statistical evidence that quality of fruits and fruit harvested affect the cost. This result is important to define aspects of planning of resources for minimizing wastes for quality in stage of harvest of the production agricultural systems. For this reason, we present a response optimizer based in a response surface methodology developed in Minitab® 19. The results for five solutions and its

Table 5 Anova for Harvesting

Fuente	Contribution (%)	F	P Value
Model	100.00	1.51E-08	0.0000
Lineal	99.83	4.82E-08	0.0000
Dayli Demand	0.00	0	1.0000
Max Harvest Prod	4.04	908895.29	0.0000
Perc per quality	95.96	2.41E-09	0.0000
Waste fruit cost	0.00	0	1.0000
Hiring cost per person	0.00	7506.49	0.0000
Interactions 2 terms	0.00	4910.14	0.0000
Fruit Vol*Harvest Prod	0.00	0	1.0000
Fruit Vol*Perc per quality	0.00	0	1.0000
Fruit Vol*Waste fruit cost	0.00	0	1.0000
Fruit Vol*Hiring cost per person	0.00	0	1.0000
Harvest Prod*Perc per quality	0.00	48400	0.0000
Harvest Prod*Waste fruit cost	0.00	0	1.0000
Harvest Prod*Hiring cost per person	0.00	301.37	0.0000
Perc per quality*Waste fruit cost	0.00	0	1.0000
Perc per quality*Hiring cost per person	0.00	400	0.0000
Waste fruit cost*Hiring cost per person	0.00	0	1.0000
Curvature	0.17	4015441.43	0.0000
Error	0.00		
Total	100.00		

Table 6 Prediction for multiple responses in harvesting

Solution	Dayli demand	Max harvest prod	Perc per quality	Waste fruit cost	Hiring cost per worker	Z min Ad (US $)	Desirability
1	18000	30000	0.8	60	38000	2.900	0.999
2	15000	30000	0.8	60	38000	2.900	0.999
3	18000	30000	0.8	50	38000	2.900	0.997
4	15000	30000	0.8	50	38000	2.900	0.995
5	18000	30000	0.8	60	42000	2.900	0.994

configurations are in Table 6, which specifies the possible values for optimizing the cost according model and response surface methodology.

The quantities cover the needs for demand with 2 days for harvesting and generates a total cost near to COP $ 10.180.250 (US $ 2.900) even with high percentage out specifications. The proposed model can be applied to different types of crops and farms that produce fresh fruit, being able to apply it to support strategic and tactical decisions and as tool for harvesting. This model defines the possibilities of obtaining a significant reduction of wastes of quality.

4 Conclusions and Discussion

Agricultural production systems are very complex. When implementing Operations Research / Analytics techniques, the literature has traditionally approached each stage of production independently. This research proposed the integration of operations research and lean manufacturing wastes. Objective function and constraints are hence introduced in the proposed mathematical model.

From the literature, it was possible to characterize the factors, the conditions and the resources of harvests that affecting the agricultural production, in order to identify wastes in regards of Lean Manufacturing principles. Due to interactions between resources throughout agri-chains, this model is solved as a Mixed Integer Programming Model. Although the model was conceived to reduce the costs generated by the waste of agricultural production, it also allows into account that the farmer is interested in farming according to profit projections made with harvest and production.

To illustrate the sensitivity of models, we designed experiments with parameters of interest for farmers to know their influence in the total cost. The parameters were chosen according to uncertainty of resources and cost of production, allowing the evaluation of sensitivity with the analysis of variance and Response Surface Methodology for multiple solutions and configuration of levels of factors (parameters) for harvesting. The RSM defines guidelines of volume of production if we review the configuration of values of solutions proposed.

In many farming environments, and especially in Colombia, planning agricultural production is done based on the experience and practical knowledge of managers or farmers. Moreover, data is collected but not properly processed nor analyzed, which results in frequent planning errors that cause higher operational costs in crop maintenance, and risks in fruits quality.

As pointed out through this discussion, the designed and validated model is the first in its kind addressing the planning problem of agricultural production with the minimization of wastes, which are identified based on the application of lean manufacturing principles. The model can offer harvesting strategies for a short time, becoming a great tool for decision-making in real operational and tactical planning settings. Indeed, the proposed methodology for analysis and modeling has been a novel contribution to the planning of agricultural production, covering the gaps evidenced in the literature review and the objectives of this research, especially in reducing waste and production costs. With this reduction, a sustainable agricultural production approach is achieved, since it seeks to optimize the use of soil and water without generating risks in the yields and quality of fruits.

This methodology of modeling and analysis serve as the first approach for further developments in this area. In future researches, the model can also be extended to post-harvest or transport and distributions operations, in the context of lean supply agri-chains. Future extensions also could include sources of uncertainty, as well as multiple products, or combination of products in the same crop, agricultural planning by geographic areas.

References

1. Thornthwasite C.W: Operations Research in Agriculture. Journal of the Operational Research Society 1 (2), 33–38 (1953).
2. French C.: Application of operations research in farm operations and agricultural marketing, Operations Research 6 (5), 766–775 (1958).
3. FAO: Cereal supply and demand brief. http://www.fao.org/worldfoodsituation/csdb/en/, last accessed 2016/12/09.
4. Departamento de Planeación nacional de Colombia (DNP),: Pérdidas y desperdicios de Alimentos en Colombia. Bogotá (2016). https://colaboracion.dnp.gov.co/CDT/Prensa/Publicaciones/P%C3%A9rdida%20y%20desperdicio%20de%20alimentos%20en%20colombia.pdf, last accessed 2019/07/17.
5. Osaki M., Batalha M.O,: Optimization model of agricultural production system in grain farms under risk, in Sorriso, Brazil. Agricultural Systems 127,178–188 (2014).
6. Mason AN, Villalobos JR,: Coordination of perishable crop production using auction mechanisms. Agricultural Systems 138, 18–30 (2015).
7. Weintraub A, Romero C,: Operations research models and the management of agricultural and forestry resources: A review and comparison. Interfaces **36**(5), 446–457 (2006).
8. Cardin M, Alvarez C,: Model for decision-making in agricultural production planning. Computers and Electronics in Agriculture 86, 131–139 (2012).
9. Tselikas N, Keramydas C, Toka A, Aidonis D, Iakovau E,: Agrifood supply chain management: A comprehensive hierarchical decision-making framework and a critical taxonomy. Biosystems Engineering 129, 47–64 (2014).
10. Sopegno A, Busato P, Berruto R, Romanelli T,: A cost prediction model for machine operation in multi-field production systems. Scientia Agricola 73(5), 1–14 (2016).
11. James J, Antle J, Basso B, Boote K, Conant R, Foster I, et al.: Toward a new generation of agricultural systems data, models, and knowledge products: State of agricultural systems science. Agricultural Systems 155, 269–288 (2017).
12. Caicedo N., García G., Montoya-Torres J.: Towards the integration of lean principles and optimization for agricultural production systems: a conceptual review proposition, Journal of the Science of Food and Agriculture 100 (2), 453–464 (2020).
13. Rika A, Tjahjono B,: A framework for managing sustainable palm oil supply chain operations: A case of Indonesia. Production Planning and Control 28(13), 1093–1106 (2017).
14. Lodree A, Uzochukwu E,: Production planning for deteriorating item with stochastic demand and consumer choice. International Journal of Production Economics 116, 219–232 (2008).
15. Amorim P, Gunther H, Almada B,: Multi-objective integrated production and distribution planning of perishable products. International Journal of Production Economics 138, 89–101 (2012).
16. Itoh T, Ishii H, Nanseki T,: A model of crop planning under uncertainty in agricultural management. International Journal of Production Economics 81, 555–558 (2003).
17. Ahumada O, Villalobos JR, Mason N,: Tactical planning of the production and distribution of fresh agricultural products under uncertainty. Agricultural Systems 112, 17–26 (2012).
18. Da Silva AF, Silva FA, Ximenez E, Addressing uncertainty in sugarcane harvest planning through a revised multi-choice goal-programming model. Appl Math Model **39**:5540–5558 (2015).
19. Riveiro JA, Marey MF, Marco JL, Alvarez C: Procedure for the classification and characterization of farms for agricultural production planning: Application in the northwest of Spain. Computers and Electronics in Agriculture 61, 169–178 (2008).
20. Borodin V, Bourtembourg J, Hnaien F, Labadie N, Handling uncertainty in agricultural supply chain management. European Journal of Operational Research 254, 348–359 (2016).

Optimization Model to Consolidate the Hose Load in a Peruvian Agribusiness

Cristhian Giancarlo Aradiel Abad, Diego Ángel Dávila Vilchez, Tania Sthefany Gamboa Rojas, and Gabriela Veliz Ponce

Abstract Transportation represents around 30% of the total cost of a product. Therefore, the distribution of the load and the route must be optimized and chosen correctly, respectively. The present investigation addresses the optimization of the hose load in 40-foot containers by using mathematical optimization.

Key words Optimization · Linear programming · Cubic space

1 First Section

1.1 Introduction

In Peru, the agroexport sector reflects a participation of 14.3% on average with US $ 10 billion for 2021, which reflects a sufficient expectation of opportunity to invest [1]. The present case study is about a Peruvian agroexport company, which has five business groups that offer innovative solutions in various industries. These include sectors such as construction and infrastructure, data communications, chemicals, and more. The main activity of the company is focused on offering technical irrigation products and solutions, monitoring, digital control of open fields, orchards and protected crops [2]. Currently, there are several exporting companies that plan the consolidation of their offices and the distribution of their products, empirically from their main plant to different points of sale internationally [3]. The case study of this research is about a Peruvian exporting company, which has various products (line of drippers, sprinklers, emitters, pipes, connectors and accessories) for the irrigation of crops. In order to satisfy the demanding demand of the markets of different regions, they require having a relevant international logistics, so the flow of containers is the main motivation that mobilizes the supply chain, in that sense the main disadvantage

C. G. Aradiel Abad (✉) · D. Á. Dávila Vilchez · T. S. Gamboa Rojas · G. Veliz Ponce
Pontificia Universidad Católica del Perú, Lima, Peru
e-mail: cristhian.aradiel@pucp.edu.pe; diego.davila@pucp.edu.pe; tania.gamboa@pucp.edu.pe; a20162719@pucp.edu.pe

J. Vargas Florez et al. (eds.), *Production and Operations Management*, Springer Proceedings in Mathematics & Statistics 391,
https://doi.org/10.1007/978-3-031-06862-1_15

of the consolidation of products It requires moving on pallets and containers, which is managed according to the experience of the company's logistics operator. This practice incurs high logistics costs since the total capacity of the containers is not used and there is no palletizing standard, in that sense the space and the allocation of the load is not used optimally according to the type of container 40 High Cube (FCL or LCL). The initial strategy is to optimize the palletizing and consolidation of the load and palletizing. For this reason, the creation of a mathematical model through linear programming is pro-posed, which optimizes the consolidation of the products in the containers and reduces the costs of logistics operations of the maritime shipment to its final destination [4].

1.2 State of Art

Optimization models are used to make decisions that maximize or minimize an un-defined representation from a deduction, which are subject to a set of variables and restrictions. In consequence, the application of linear programming encompasses the resolution of marketing problems, energy reduction, staffing, operations planning, and so on [5]. The selection of a tool, in this case, linear programming, is limited by de-mand, destination and load sizing constraints. Real-world problems require the capabilities of the aforementioned, therefore, it can be concluded that both planning and load consolidation complement each other in order to arrive at the best solution to a real problem [6, 7].

2 Description of Current Situation

2.1 Process Mapping

This activity allowed us to know the process of exporting hoses, accessories, sprinklers, etc. Involving various areas from the request of the client's order, through areas of production, planning, customer service, foreign trade, warehouses and distribution to the dispatch and shipment of documentation to the client. In addition, the flow of the hoses to be exported comes from a forecast planning of the supply of the destination countries in order to minimize the impact of stock breakages. It should be noted that this research and improvement will focus on the activities of the logistics operator (in house) related to the consolidation of palletized cargo in containers (Fig. 1).

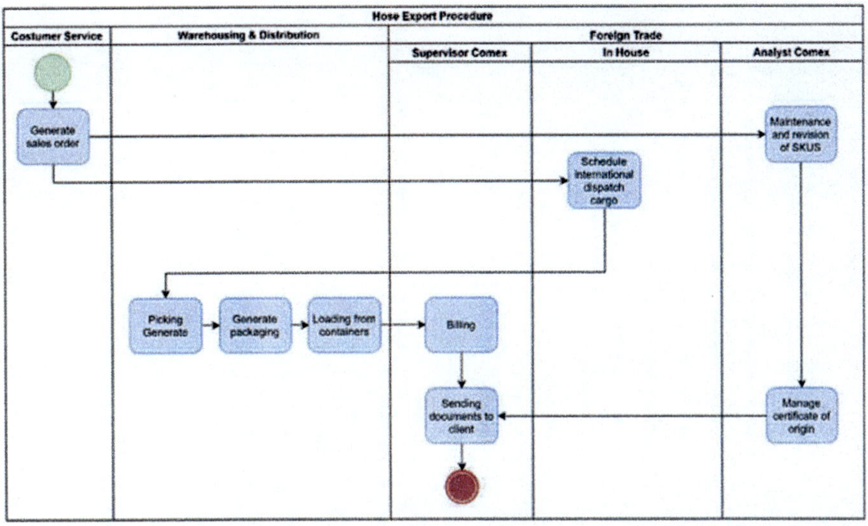

Fig. 1 Hose export process

2.2 *Information Collection and Analysis*

Information was collected with the weights (kg), volume (m^3) and sale of SKUS in the different channels through international demand. For this, the structure of the logistics network was identified with respect to the times related to the shipment of products by destination and by SKU for practical reasons of the agro-industrial company. In summary, the information and an analysis of the demand were consolidated to determine in which SKUs generate greater value for the company. In that sense, it was established that, of 38 products, only 20 types of hoses generate the highest profit according to the Pareto shown (Fig. 2).

2.3 *Construction of the Optimization Model*

In order to determine the set of SKUs and pallets to be assigned to the type of container. It should be noted that the objective of this model is to take advantage of the cubic spacing of the containers through palletized cargo, thus reducing the costs of the flow of containers to be exported. The main decision variable is the forecasted quantity of SKUs to be exported to destinations in Chile, Ecuador, Colombia, Argentina, Mexico, Central America and Brazil, on the other hand the restrictions are linked to customer demand, mode of transport and costs temporary storage, local freight, associated with the container, and use of the port yard, etc. (international logistics costs).

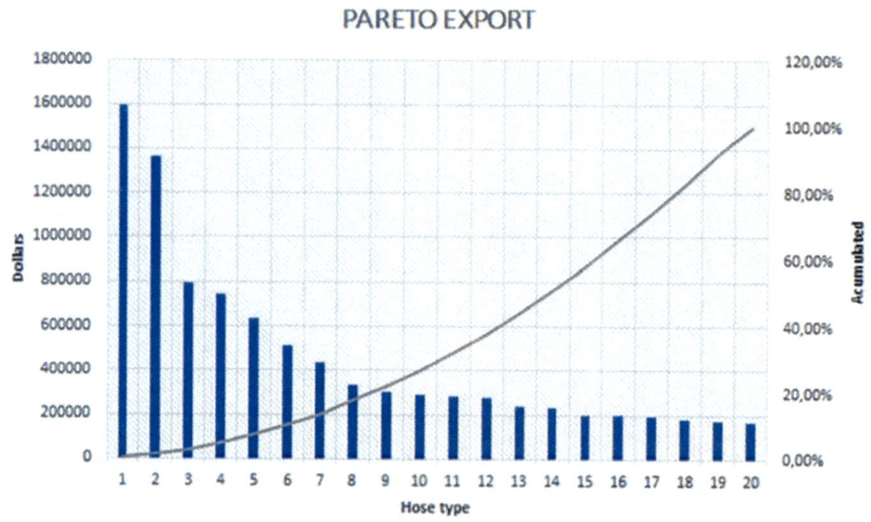

Fig. 2 Pareto of hose sales

Table 1 Variable export costs

Variable costs			
Cargo agency	Temporary deposit	Freight	Associated with the use of the container
Stowage	Gate Out	Half freight	Container opening
	Handling Delivery	Stand by	Container control
	Appointment rescheduling	Transport	Container coverage free
	Delivery service		Container delivery free
	Extraordinary service		Container shipment
	Integrated DT service		Empty delivery
	Comprehensive void delivery service		Container dispatch management
	Use of web platform		Container administration service

2.4 *Analysis of Results and Scenarios*

The results of the current scenario will be presented, that is, of the current empirical situation in which the logistics operator consolidates the load and the results of the proposed mathematical model will be compared. The cost structure related to the export of containers (international logistics in the supply chain) is presented below (Table 1).

This table represents the fixed costs in the export of merchandise and independent of the flow of containers (Table 2).

Table 2 Fixed export costs

Fixed Cost		
Customs agency	Port agency	Administrative transaction
Agency fee	Mechanical assistance to inoperative external trucks	Administrative expense
Dispatch expenses	Unused appointment fee	Administrative expenses
Seal fee	40 High cubic capacity containers	Documentary expo process
	Late delivery of documents	Approval
	Crane handling	
	Mobilization for boe	
	Extra movements in the yard	
	Scac expo	
	Special temporary warehouse service	

Fig. 3 Volume of each type of hose (m^3)

	Volumen (m3)
Manguera tipo 1	0,068
Manguera tipo 2	0,064
Manguera tipo 3	0,067
Manguera tipo 4	0,073
Manguera tipo 5	0,064
Manguera tipo 6	0,066
Manguera tipo 7	0,073
Manguera tipo 8	0,064
Manguera tipo 9	0,071
Manguera tipo 10	0,070
Manguera tipo 11	0,063
Manguera tipo 12	0,070
Manguera tipo 13	0,037
Manguera tipo 14	0,065
Manguera tipo 15	0,070
Manguera tipo 16	0,065
Manguera tipo 17	0,075
Manguera tipo 18	0,066
Manguera tipo 19	0,057
Manguera tipo 20	0,068

In addition, the volume of each product will be taken into account in order to be able to take advantage of the cubic spacing of the containers (Fig. 3):

According to the historical demand of the 20 hose types, it was forecast according to the moving average method with moving trend. In this sense, the forecast table for the monthly demand of the 7 countries to satisfy is presented (Table 3).

Table 3 Predicted demand for each type of hose (units)

	Chile	Ecuador	Colombia	Argentina	Mexico	Centro America	Brasil
Manguera tipo 1	6288	170	2510	0	0	4630	0
Manguera tipo 2	9360	0	192	720	0	1360	0
Manguera tipo 3	0	0	0	0	0	9360	0
Manguera tipo 4	0	0	0	0	0	3360	0
Manguera tipo 5	3180	0	0	0	0	1280	0
Manguera tipo 6	0	1185	720	0	0	0	0
Manguera tipo 7	0	0	600	0	0	2420	0
Manguera tipo 8	0	0	0	0	0	2778	0
Manguera tipo 9	720	0	0	720	0	1575	0
Manguera tipo 10	0	0	0	0	0	2719	0
Manguera tipo 11	2740	0	0	0	0	25	0
Manguera tipo 12	1494	0	0	0	0	0	0
Manguera tipo 13	6100	0	0	0	0	0	0
Manguera tipo 14	3360	0	0	0	0	0	0
Manguera tipo 15	1056	0	0	0	0	580	0
Manguera tipo 16	0	470	0	0	100	0	0
Manguera tipo 17	0	0	640	0	0	1460	0
Manguera tipo 18	0	0	0	0	0	1120	0
Manguera tipo 19	2720	0	0	0	0	0	0
Manguera tipo 20	0	191	0	0	0	100	1440

3 Methodology

Next, the sets, parameters, decision variables, objective function, restrictions to determine the minimum total costs of the exports of the containers to be distributed to the international market will be explained (Table 4).

The optimization model is presented below

Sets

I Set of SKUs

J Set of countries

K Set of container types

Parameters

d_i Demand SKU i

v_i Volume SKU i

$pLCL$ Maximum volume allowed for an LCL container

$pFCL_{min}$ Minimum volume allowed for a FCL container

$pFCL_{max}$ Maximum volume allowed for a FCL container

cag Cargo agency cost

cf Freight cost

Table 4 Diagram of mathematical model

Inputs	Mathematical model	Outputs
Delivery time Types of containers Modes of transport Demand and preferences	Fulfillment of the demand and management of cubic space of FCL and LCL containers	Quantity of containers to be used Type of containers to be used Percentage of closed container sales

cdt	Temporary deposit cost
cac	Costs associated with the use of containers
caa	Customs clearance costs and commissions
cap	Costs and commissions of the port of shipment
cta	Administrative costs

Variables

X_{ijk}	Quantity of SKU i exported to country j in a container type k
W_{jk}	Quantity of type k containers exporting to country j

Description

Objective Function

$$Min\ (caa + cap + cta) + \sum_{j \in J,\ k \in K} \left(W_{jk} * (cac + cf + cdt + cac) \right)$$

The objective function seeks to minimize the total costs of the exports of the containers to be distributed to the international market.

Production demand

$$\sum_{k \in K} X_{ijk} = d_i \qquad \forall i \in I, j \in J$$

The quantity of SKUs to be exported to each country in each type of container must be equal to the monthly production demand per SKU in general.

Maximum capacity limit with LCL containers (k=1)

$$\sum_{i \in I} X_{ij1} * v_i \le pLCL * W_{j1} \qquad \forall j \in J$$

The total volume of products to be shipped in Less Container Load (LCL) containers must be less than or equal to the maximum volume policy of the LCL containers used.

Minimum capacity limit with FCL containers (k = 2)

$$\sum_{i \in I} X_{ij2} * v_i \leq pFCL_{min} * W_{j2} \qquad \forall j \in J$$

The total volume of products to be shipped in Full Container Load (FCL) containers must be greater than or equal to the minimum volume policy of the FCL containers used

Maximum capacity limit with FCL containers (k=2)

$$\sum_{i \in I} X_{ij2} * v_i \leq pFCL_{max} * W_{j2} \qquad \forall j \in J$$

The total volume of products to be shipped in Full Container Load (FCL) containers must be less than or equal to the maximum volume policy of the FCL containers used

4 Results

Minimization of containers to be consolidated to export customer orders (countries). Knowing the demand of the 7 countries and the development of the model, the consolidations must be carried out as follows: The entire order from Chile must be exported in 30 Full Container Load containers, the entire order from Ecuador in 2 Full Container Load containers, the entire order from Colombia in 5 containers Full Container Load, the entire order from Argentina in 2 containers Full Container Load, the entire order from Mexico in 1 container Less Container Load, the entire order from Central America in 30 containers Full Container Load, all the order from Brazil in 2 container Full Container Load. Reduction of costs of international supply of products according to the demand of the Latin American market ($). The main logistics costs that would be reduced are those of cargo agency, freight, temporary storage costs and those associated with the use of containers. In the case of the exporting company, if this linear programming model was implemented, its total logistics costs would be 231,273.21 dollars, much lower than its current costs, which amount to 337,921.11 dollars.

5 Conclusions

The empirical criterion of the logistics operator does not necessarily imply the efficient consolidation of palletized cargo, so the use of containers can be reduced by 31.56%. Additionally, the change from LCL to FCL containers was increased since the quantities and dimensions of each type of hose were essential for the improvement in cubic spacing of 70.26%. This research and improvement determine a relevant decrease in costs as a starting point and planning for the export and negotiation of containers. The model mathematic suggests the integration of sales forecasts in accordance with the Sales and Planning Operation (S & OP) methodology, since the information shared between the areas involved considers the characteristics of international markets.

References

1. Ministry of Agriculture and Irrigation. Available at https://www.proyectosapp.pe/modulos/JER/PlantillaStandard.aspx?ARE=1 & PFL=2 & JER=8168 (2020). Accessed date 29 Aug 2021.
2. COMEXPERU. Principales Preocupaciones de los Agropecuarios. Available at https://www.comexperu.org.pe/articulo/principales-preocupaciones-de-los-agroexportadores (2019) Accessed date 29 Aug 2021.
3. Agro-exporter, precision irrigation changes the economic landscape of agriculture. Available at https://www.agroexportadora.com/es-pe/ (2021) Accessed date 29 Aug 2021.
4. Ulloa. Development and Codification of a Mathematical Model for Optimizing a Vehicle Routing Problem with Multiple Deposits M. Available at http://tangara.uis.edu.co/biblioweb/tesis/2015/156420.pdf. Accessed date 29 Aug 2021
5. C.E.E. Abanto L. Proposal of improvement applied to the scheduling of teller shifts and sizing the number of sales assistants in a department store that increases productivity and customer satisfaction https://repositorioacademico.upc.edu.pe/bitstream/handle/10757/625637/Abanto_lc.pdf?sequence=1 & isAllowed=y. Accessed date 29 Aug 2021
6. F. Juliana. "Review of the state of the art of reverse logistics and adaptation to the technical study for the disposal end of expanded polystyrene" (2016). Available at https://core.ac.uk/download/pdf/71399656.pdf. Accessed date 29 Aug 2021
7. García-Hernández, M. de G., & Garrido, A. (2006). Integración de Planificación y Scheduling. Available at https://www.redalyc.org/pdf/416/41600304.pdf. Accessed date 29 Aug 2021

Blockchain, Innovation to the Value Chain and Improvement in the Management of Peruvian Family Farming

Alan E. Fráquita Maquera

Abstract Blockchain technology has been defined as an ensure and transparency tool for all interested parties. This study aims to identify key points within agro-industrial value chain using blockchain, to add trust and eliminate economic development barriers supported by the literature review.

Key words Agro-industrial · Value chain · Blockchain

1 Introduction

The economic development of any country depends on technological innovation in its different productive sectors. Agribusiness is one of the most relevant, since its productivity will result from the participation of several agents: social, public, investment and innovation [1]. Furthermore, the development of this industry will continue to grow due to the demand sustained by the growth of the world population.

Peru is a region oriented to the international commercialization of its natural resources, which promotes foreign investment and the promotion of public policies to increase the traffic of its operations, presenting a competitive market in the main productive actors.

In this way, the core of the agro-industrial system is made up of its supply chain, which includes all production, manufacturing, and distribution activities up to the final consumer [2]. This chain is represented by nodes of integral participation by farmers and the trade chain, where there is a constant exchange of products and services between the agents that comprise it.

However, the best development of its operations shall be subject to the flow of information, affecting decision making in the correct assignment of resources. In addition, the complexity level for its correct transmission increases due to number of actors, level of demand, environmental, cultural, and sanitary norms.

A. E. Fráquita Maquera (✉)
Pontificia Universidad Católica del Perú, Lima, Perú
e-mail: afraquim@pucp.edu.pe

J. Vargas Florez et al. (eds.), *Production and Operations Management*, Springer Proceedings in Mathematics & Statistics 391,
https://doi.org/10.1007/978-3-031-06862-1_16

Consequently, agricultural development policies fail in joining industrial production with rural production to compete in the international market, in parallel with the continuous globalization of international markets and the need to ensure the transparency of information. This is a vulnerable market because of lack of confidence in the origin, production standards and quality of its production. Resources that are significant in the political and business decision-making process [3].

In response to this issue and the development of digital technologies, it presents a global trend to move to a model of smart agriculture [4]. The purpose of their implementation is to be an improvement in business management, to simplify the flow of documents and to develop a social-sustainable production. As Figorilli [5] says, the introduction of this technology in productive activities brings transparency and efficiency, allowing the development of new levels of communication thanks to a higher level of information in transactions.

In industries where the relevance of transparence is key for the execution of operations within the supply chain, Blockchain is emerging as an alternative for safeguarding the information that constitutes it [6]. This innovation is attractive by its data collection model that collects, links, and recovers all the information distributed in its network, where each node can be coordinated without a unified data center [1].

Although its development by Satoshi Nakamoto was related to the service of carrying through transactions to solve the problem of double spending [7], this technology can be applied within agribusiness value chains for the elimination of asymmetric information among its agents. The joining of information increases agricultural productivity and includes small-scale producers in export chains.

This represents an agricultural development approach in the smallholder value chain that focuses on the development of differentiated, quality products. The integration of small farmers results in an improvement in the social, economic, and cultural well-being of small producer families. As well as establishing optimal performance of value chains and improving Peru's international trade competitiveness [8].

The objective of this research is to analyze the elements along the value chain of Peruvian agriculture production and to find factors that generate development barriers in the trade of its products in both the domestic and international markets. In addition, to propose applications of Blockchain technology as an improvement in the management of the agro-industrial sector and added value for the actors in it.

The article is structured in 4 sections. The first section will introduce the influence of technologies to production chains. The second section will present applications of Blockchain technology to agribusiness. The third section will present new proposals to increase the participation of Peruvian agricultural producers in the value chain. Finally, the conclusions will be presented as the last section.

2 Literature Review

The development of Blockchain technology is considered a revolution in recent years. Although its development and study has been done to a greater extent in the financial sector, improvement points have been identified in other productive sectors such as agribusiness [6]. The opportunities for improvement in operations and product management are presented according to the literature review (Fig. 1).

2.1 *Traceability*

The performance of a supply chain will be dependent on the correct management of the different logistic operations among its agents, which requires an open and constant flow of information. This is necessary to plan, allocate and redirect resources to the activities with the highest priority. However, these chains are inefficient because they are vulnerable to document falsification, which hinders the verification process regarding the origin of the resources to be used in the different production stages. This is due to the complexity of the chain, decentralization, and transparency of information.

Traceability applied by Blockchain makes it possible to identify and verify the status of individual components within a chain as well as the transparency of information among all participants [6]. In addition, within the blockchain

Fig. 1 Tools that improve the small farmer's production capacity

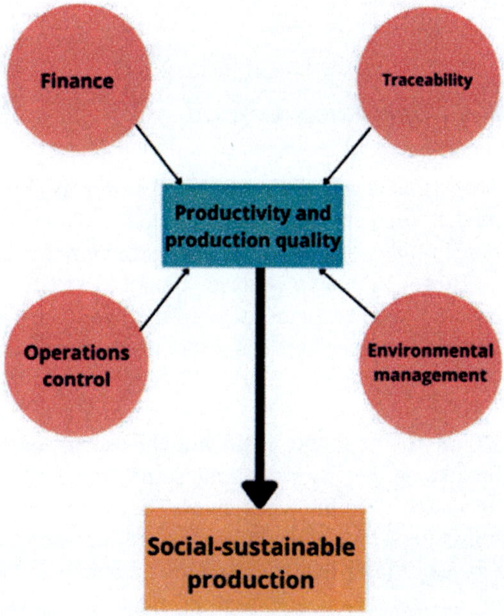

information is obtained data regarding security, pricing, invoicing and distribution of products [9].

The partnership between Walmart and IBM, developed this tool with use of two-dimensional barcodes (QR code) for efficient tracking of products in their disposal. In the same way, Behnke and Janssen [10] adapted the processes within a dairy industry, considering all levels of production with results in improving the traceability of their batches.

In addition, with a focus on developing a relationship within the value chain with consumers. Thanks to the public access of information within Blockchain, the final consumer can ensure the differential of the selected product with the direct consultation in the database regarding the origin as ingredients of their product and execute a fair-trade practice with all participants in the chain.

2.2 *Environmental Management*

The production and industrialization phases of agricultural chains have the greatest environmental impact. It is necessary to evaluate the management of these residues and develop environmental sustainability policies regarding their environmental externalities and contributions to global warming [2]. From the monitoring of data in the blockchain and information technologies (IT) in the different production links, new agricultural production schemes are developed focused on the analysis of conditions, carbon footprint and energy used [11]. This allows a better control of waste with respect to the life cycle of the product and adapting utilization schemes in search of a sustainable agriculture and production model.

2.3 *Operations Control*

A relevant stage in product traceability is the production stage. The production models used in industries focus on avoiding human errors applied to operations based on the flow of information stored in them.

However, these models are ineffective to the alteration of information and do not ensure a total authentication of the operations. The resulting effect of this lack of authenticity generates economic losses and threatens public health with vulnerable food safety [11].

Blockchain opens up value and production chains to all stakeholders, adding security to the supply chain with the use of smart contracts for constant validation of operations. Decision making shifts from IoT technologies (sensors) to the data network within the blockchain. The development of this type of projects increases transparency in the quality control of operations, offering a verification regarding the change of information within the incoming and outgoing products of its productive chain, eliminating costs to the manual review of its operations and information systems for its registration [1]. As a result, product guarantees are assured and verifiable by external and governmental agents.

2.4 *Finance in Agribusiness*

The financial trend of Blockchain technology revolutionized commercial transactions, eliminating third party participants, and adding reliability to the information with its constant verification. The application of Blockchain from the agribusiness sector takes advantage of its main characteristic of having an open information system on its operations, resources, and commercialization of all participants in the supply chain.

The use of this technology improves efficiency and reduces barriers to access to financing for all actors in the production chain, offering new financing models and efficient transmission of transactions [1]. As a result, effective and secure payment platforms are being developed to promote credit in rural sectors, such as Agri Digital and Oliva Coin.

3 Proposed New Value Chain

In Peru, according to Maletta [12] most small farmers belong to the family farming category, from traditional production units to capitalized units. It is these small farms that make food security for consumers possible, with the state being the main responsible for adding improvements to their productivity along with entrepreneurs for their participation in the internationalization of their products [3].

The impact of producers within the value chain has been affected only to the provisioning of inputs within the supply chains. To improve their productivity, Blockchain applications are proposed in 3 improvement factors to sustain an integrative agriculture and export model with its other participating agents

3.1 *Environmental Traceability*

Today, the full certification of a product can be altered by third parties at different stages of the logistics chain. To ensure the quality of these products, a third-party verified quality control of their origin is necessary.

The use of Blockchain in the certification of agricultural products ensures full traceability within the supply chain and is an effective solution to ensure consumer confidence in rural farmers. At the same time, it eliminates product traceability and switches to constant storage in a single public record. This facilitates audits and traceability at different stages of the product life cycle.

The application of Blockchain with environmental traceability builds a social conscience where sustainable and ethical production is valued, guaranteeing support to rural communities that offer quality products.

3.2 Operations Control

From the autonomy and immutability of data Blockchain technology develops a new level of automation towards processes. This with the use of smart contracts that are computer protocols that are executed by validating the requirements of the contract, monitoring products, assets, and their data.

The application of this technology as a validator between the different stages of agro-industrial commercialization adds transparency to the products and allows both parties to be notified about the conditions of the traded good. With its use, the value chain is improved by using secure trade practices through a decentralized verification system, eliminating the cost of double-checking operations (Fig. 2).

3.3 Finance

Today, the difficulty of adapting farmers within the financial system affects the development of new technologies within their production activities, training in production and post-harvest skills. Blockchain, with its constant exchange of information between producers and consumers, allows the development of an open financing model.

Crowd farming adapts the crowdfunding model to agriculture, where it is the consumer who acquires the production of a specific area and can monitor the status of their production and guarantee its origin at the time of receiving it [4].

This alternative seeks for the client to have a closer relationship with the producer and find a greater variety of products.

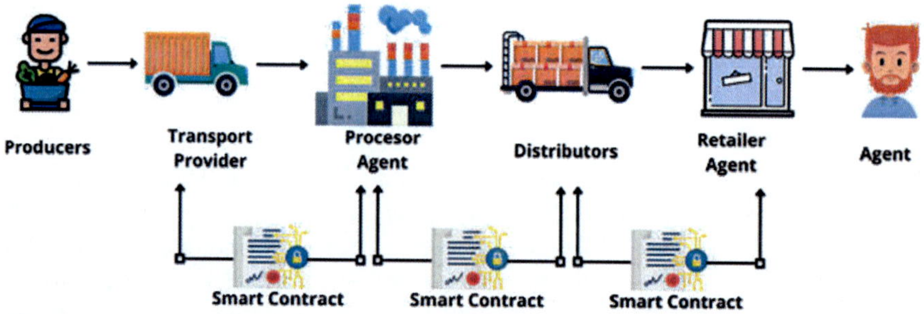

Fig. 2 Agricultural supply chain with the inclusion of smart contracts in the operations of participating agents

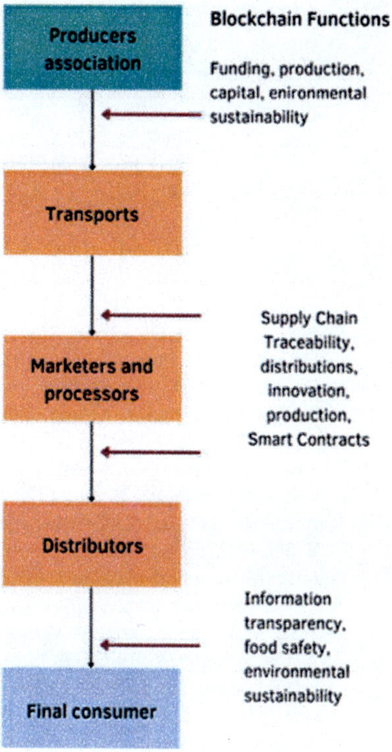

Fig. 3 Blockchain value chain of a producer's agricultural product

Based on an analysis of the different Blockchain tools that prioritize the improvement of agricultural production, a value chain model integrated with Blockchain is proposed to achieve a social-sustainable agro-industrial sector for its participants (Fig. 3).

4 Conclusions

- The global economic development of farmers requires meeting production and environmental sustainability standards. The development of applications through IoT and Blockchain provides new management methods for their products. This allows the farmer to add value from their participation in the life cycle of the product.
- The commercialization of agricultural products requires efficient logistics and quality control in the production chain for all participants. The use of Blockchain allows transparency and storage of information for all participants in the value chain. This makes it an important tool for solving resource allocation problems in the different stages of production.

- The innovation of these new technologies to participation by farmers requires a factor of financing and collective participation of consumers. Blockchain develops a model of trust incentives between consumers and producers, empowering farmers with financing and thus developing more efficient value chains.

References

1. Rocha G da SR, de Oliveira L, Talamini E (2021) Blockchain applications in agribusiness: A systematic review. Future Internet 13:. https://doi.org/10.3390/fi13040095
2. Martinho VJPD (2021) Agri-food contexts in mediterranean regions: Contributions to better resources management. Sustainability (Switzerland) 13:1–17. https://doi.org/10.3390/su13126683
3. Barrientos Felipa P (2019) Estrategia de integración del pequeño agricultor a la cadena de exportaciones. Semestre Económico 22:83–123. https://doi.org/10.22395/seec.v22n51a5
4. Marinchenko TE (2021) Digital Technology in Agricultural Sector. IOP Conference Series: Earth and Environmental Science 666:. https://doi.org/10.1088/1755-1315/666/3/032024
5. Figorilli S, Antonucci F, Costa C, et al (2018) A blockchain implementation prototype for the electronic open source traceability of wood along the whole supply chain. Sensors (Switzerland) 18:1–12. https://doi.org/10.3390/s18093133
6. Xu P, Lee J, Barth JR, Richey RG (2021) Blockchain as supply chain technology: considering transparency and security. International Journal of Physical Distribution and Logistics Management 51:305–324. https://doi.org/10.1108/IJPDLM-08-2019-0234
7. Nakamoto S Bitcoin: A Peer-to-Peer Electronic Cash System
8. Tobin D, Glenna L (2019) Value Chain Development and the Agrarian Question: Actor Perspectives on Native Potato Production in the Highlands of Peru. Rural Sociology 84:541–568. https://doi.org/10.1111/ruso.12251
9. Eashwar S, Chawla P (2021) Evolution of Agritech Business 4.0 – Architecture and Future Research Directions. IOP Conference Series: Earth and Environmental Science 775:0–12. https://doi.org/10.1088/1755-1315/775/1/012011
10. Behnke K, Janssen MFWHA (2020) Boundary conditions for traceability in food supply chains using blockchain technology. International Journal of Information Management 52:101969. https://doi.org/10.1016/j.ijinfomgt.2019.05.025
11. Antonucci F, Figorilli S, Costa C, et al (2019) A review on blockchain applications in the agri-food sector. Journal of the Science of Food and Agriculture 99:6129–6138. https://doi.org/10.1002/jsfa.9912
12. Maletta H (2017) La pequeña agricultura familiar en el Perú – Una tipología microrregionalizada

Proposal to Improve the Consolidated Copper Mineral in a Warehouse, Using Lean Manufacturing Tools

Nelson E. Chambi Quiroz, Jhon Chacón, and Pedro Prada

Abstract This article performs the diagnosis of the mineral consolidation process using the Value Stream Mapping (VSM), improvement opportunities were determined and then using Lean Manufacturing tools, significant improvements were achieved in the process, increasing productivity and improving the work environment. The tools that we have used after the VSM diagnosis are 5'S (Seire, Seiton, Seiso, Seiketsu, Shitzuke) and Total Productive Maintenance (TPM), which give us favorable results of 36.73% reduction in reprocessing work and 33.3% reduction in cycle time in shipment, respectively.

Keywords Copper mineral · Lean manufacturing · Consolidated · Warehouse

1 Introduction

Peru is a mining country that produces silver, copper and zinc worldwide. We have enormous geological potential, the presence of the Andes Mountains throughout the territory, constitutes our main source of mineral resources [1].

Prior to the export of mineral, a process of consolidating the mineral in the warehouse is carried out, which begins with the entry of mineral to the warehouse, then the mineral is unloaded, and then goes through different processes, such as, mixing of mineral to improve the grade, humidification process to reach the desired humidity value, then storage in containers is carried out so that the shipment of the mineral is finally re-alized, in these processes it will be sought to eliminate waste and activities that do not generate value. In Fig. 1 the mineral consolidation process is shown.

Mineral warehouses are receiving more and more for export, so many of them are increasing competitive in their storage processes, in order to promote constant improvement so that companies are increasing profitable and with less waste of

N. E. Chambi Quiroz (✉) · J. Chacón · P. Prada
National University of Engineering, Lima, Perú
e-mail: nchambiq@uni.pe

J. Vargas Florez et al. (eds.), *Production and Operations Management*, Springer Proceedings in Mathematics & Statistics 391,
https://doi.org/10.1007/978-3-031-06862-1_17

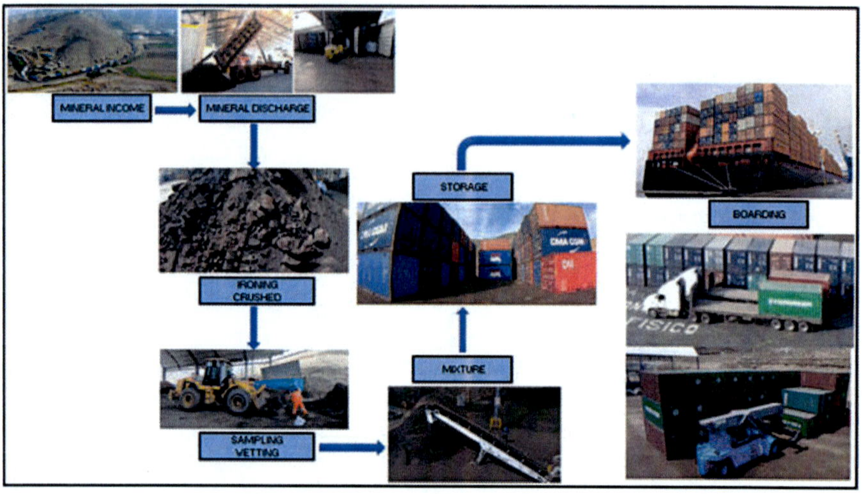

Fig. 1 Mineral consolidation process

hours in your processes, this is necessary for eliminate the causes that generate delays in the copper consolidation process. With the use of lean manufacturing tools we seek to find and eliminate delays in the process; for this we are going to use the VSM to identify the tack times and then we are going to use the 5's and TPM which aim to reduce equipment stoppages and the greater distribution of copper ore.

The rest of the document is organized in 3 sections, where Sect. 1 describes the state of the art for the development of the work, Sect. 2 describes the applied methodology and Sect. 3 describes the results obtained with a future VSM and finally the conclusions are raised where the obtained values are highlighted.

2 State of the Art

2.1 History of Lean Manufacturing

After the Second World War, with the challenge of rebuilding their economy, the Japanese automobile companies, mainly Toyota, assumed the challenge of competing with the American giants of the industry such as Ford, GM Company and Chrysler, in order to achieve this objective Toyota would have to work smarter. The term Lean was used for the first time in the book "The Machine that Changed the World" [2], in which they describe Lean as a development of the Toyota system.

The Toyota Production System (TPS) is considered synonymous with lean manufacturing because its methodology and principles are precursors of Lean Manufacturing, its development is attributed to three people (Sakichi Toyoda, Kiichiro Toyoda and Taichi Ohno).

2.2 Lean Manufacturing

Lean Manufacturing or Lean Manufacturing is a philosophy of management of the operation in the company based on the principle of doing more with less (less stress on the work team, less effort, use less machinery, less work space, fewer resources and deliver the final product in less time), applying the tools of lean manufacturing to our processes allows us to identify waste in each of the process activities, eliminate or reduce them and with this we get closer each time to delivering products to our customers or services exactly requested at the required time. The heart of lean manufacturing focuses on the motivation of the members of the work team, with great flexibility and the ability to continually solve problems that may arise in the process.

The lean manufacturing proposal seeks a way to optimize the production system, trying to eliminate or reduce all the tasks that do not add value within the production process, becoming a productive philosophy that, when considering valuing and controlling waste or waste, focuses its attention on the optimization of the company's resources [3]. Lean Manufacturing is a philosophy that has arisen with the aim of reducing waste throughout the entire process [4]. On the other hand, Lean Manufacturing is a very recognized methodology for its success and effectiveness in the manufacturing world during the last decades, it also focuses on the elimination of worthless activities, with the aim of improving the productivity and efficiency of operations [5]. Lean manufacturing tools are very effective when a correct application study is carried out, correct valid data, the participation of personnel is important to accept the change in the work method which will improve the environment [6]. The following Lean manufacturing tools are shown below, which will be used as an application in the mineral consolidation process.

2.3 Lean Manufacturing Tools

For our case, we will use the Value Map (VSM – Value Stream Mapping).

The value map or VSM is a graphic technique that allows to visualize an entire process, allows to detail and fully understand the flow of both information and materials necessary for a product or service to reach the customer, with this technique the activities are identified that don't add value to the process to later start the activities necessary to eliminate them, VSM is one of the most used techniques to establish improvement plans, being very precise because it focuses the improvements at the point of the process from which the best ones are obtained results. For the application of the value map, the following steps must be followed:

- Establish families
- Draw the current map
- Make the spaghetti diagram
- Draw the future map

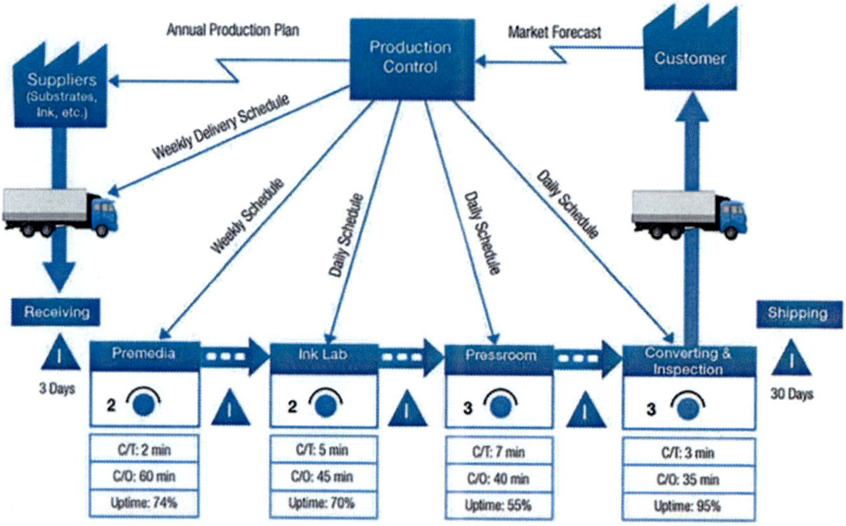

Fig. 2 Value stream mapping

With Value Stream Mapping (VSM) it is possible to detect the causes of delays, rework, process waste and even limitations that affect production, and then apply improvements in each process. In a company dedicated to the electronics industry, Value Stream Mapping (VSM) was applied to the entire production process and it was possible to find several delays [7] (Fig. 2).

2.4 The 5'S

The 5's is a work methodology, originally from Japan, after the Second World War, and is based on the principles of increasing productivity, reducing material consumption and working times. They are called 5'S by its acronym in Japanese and it means:

Seiri (sort)
Seiton (order)
Seiso (clean up)
Seiketsu (standardize)
Shitsuke (self-discipline)

The 5'S propose work behaviors dedicated to having more productive work areas, comfortable, clean and orderly environments, so that the worker performs their activities more efficiently and adopts better work practices [8]. Likewise, the bibliography related to the development of the VMS was reviewed, where it is considered a method applied in industrial practice, due to its robustness of simple analysis, whose purpose is to eliminate waste throughout the value chain but considering delivery time, quality and flexibility [9].

2.5 *Total Productive Maintenance (TPM)*

The equipment used in industries has been increasingly automated and sophisticated, and Japan is now a world leader in the use of industrial robots. This trend towards automation, combined with "just in time" production, stimulated interest in improving maintenance management in the manufacturing and assembly industries. This gave rise to a uniquely Japanese approach called Total Productive Maintenance (TPM), a form of productive maintenance that involves all employees [10]. TPM is a lean tool that involves managers and operators, with the aim of maximizing efficiency by eliminating waste [11].

3 Methodology

The methodology that are used consists of 3 stages:

Stage 1: Diagnosis of the problem, in this stage the problem of the copper mineral consolidation will be evaluated using the VSM.

Stage 2: Proposal for improvement, from the VSM they plan to use some lean manufacturing tools such as, 5's, which consists of eliminating the unnecessary, putting everything in its place, cleaning and inspecting, standardizing and finally continuous improvement is carried out.

Another tool that is used is the TPM, for which it is important to apply the following pillars:

- Identification and elimination of equipment problems: It consists of identifying the main causes and problems, and they must also be eliminated or reduced by making use of techniques such as the 5 reasons, the Root Cause Analysis (ACR).
- Apply autonomous maintenance: With this pillar we are going to ensure that the operator assumes part of the responsibility for maintenance.
- Carry out planned maintenance: It consists of systematically planning and executing maintenance activities which are carried out by qualified technical personnel, in order to obtain the equipment in optimal conditions and avoid unexpected stops.
- Training of maintenance personnel: Consists of training technical personnel on the use and care of the equipment and avoiding losses or damages.

Stage 3: Analysis of Results, in this stage we are going to estimate the results that will be expected after implementing the lean manufacturing tools.

4 Diagnosing the Problem

From the month of August to December of the year 2020, you can observe the high overtime hours of the staff, as well as the non-compliance and excessive use of equipment and to this we add the restriction of operational personnel by the covid-19 protocol.

From Fig. 3 you can see the points where there is high consumption of additional man-hours that is used in each ore discharge process.

In the months of August to December there were delays and postponements in the continuous work, such as cleaning, cutting the big bag, covering the mineral pile, moistening, etc. This has generated dissatisfaction on the part of the client due to the delay in the delivery of the process, which is why there is a need for an increase in staff, an increase in overtime to be able to comply with what is required by the client.

The processes where the delays found in the current VSM are carried out are detailed below.

Mineral Intake Process

In this stage, the units delay their documentation management process, which consists of the following: delivery of the referral guide, validation of the load to receive, confirmation of load, validation of the security belt, weighing of mineral and emission as guide.

It is observed that during the entry the delays are given by the process of receiving guides since they are received by guard station personnel which are delivered to the operations personnel, then this validate and verifies the cargo shipping email, He hands it over to scale personnel so that it can be weighed. In this part, mistakes are

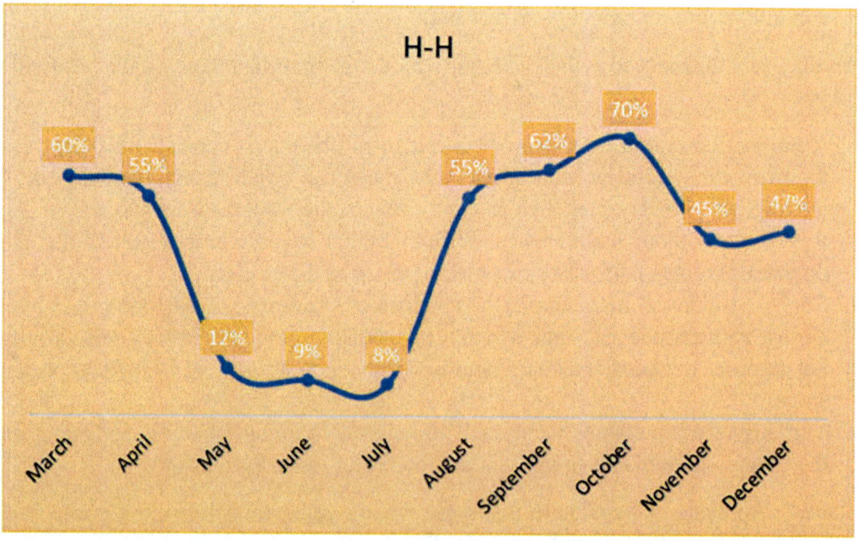

Fig. 3 Man-hours of operating personnel

also made, such as bad weighing, delay in checking the tape, bad location of the encapsulated truck. To improve this process, we are going to propose the study of the purchase of a portable scale as well as the construction of an additional scale.

Wetting Process
In this process, delays are generated by the mineral drying process in the kiln, in order to verify the moisture of the mineral, which must be between 7% and 10%. In this stage, 3 shifts will be implemented, to be able to take mineral samples in the different shifts that are worked.

Boarding Process
In this process, the constant failures of the equipment cause delays, these failures occur due to a matter of bad operation, lack of inspection, on the other hand, the delay is also caused by the lack of transport trucks due to poor planning and distribution of driver personnel. In this process for improvement, TPM will be implemented.

In addition to these delays, we also noticed delays in the ironing process, this is the result of a poor distribution of mineral location, which on more than one occasion didn't have adequate space to carry out the ironing and to achieve this space had to be made available or charged had to be moved to another location. In this process, the 5'S will be implemented, in order to improve the order and distribution of mineral in the slabs where the ironing is carried out. In Fig. 4 you can see the percentages of non-compliance of the mineral consolidation process.

This methodology will allow to plan production and establish limits for each individual process that goes from the mineral input stage to the shipment process (Fig. 5).

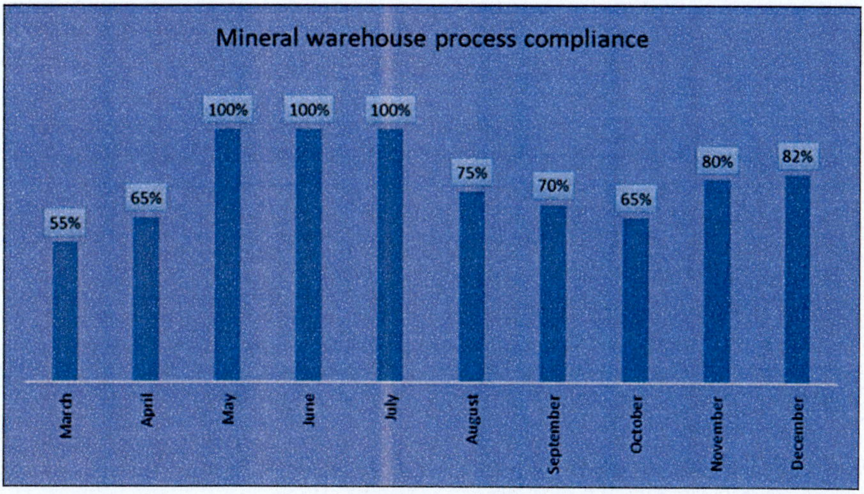

Fig. 4 Compliance control of continuous mineral storage processes

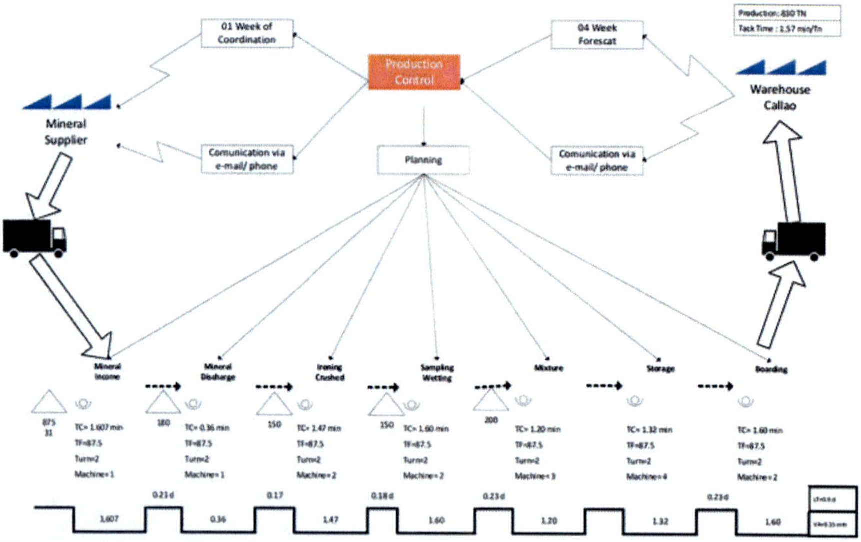

Fig. 5 VSM of the copper ore consolidation process

The problem of the company under study is the delay in dispatching it's sales orders on time, which is due to unproductive times. For this, engineering tools are selected based on research articles with similar problems. Thus, the application of 5's, plant distribution, work study and application of TPM is proposed.

5 Improvement Proposal

Ante la problemática que se presenta en el consolidado de mineral cobre sobre las demoras en los demás procesos, el aumento de horas extras del personal para tratar de cumplir con el proceso de consolidación, estos van afectando las ganancias de la empresa generando sobrecostos durante el proceso de consolidado de mineral cobre. Para ello vamos a desarrollar la metodología de fabricación ajustada el cual nos va a permitir reducir los problemas mencionados en el proceso de consolidación de mineral cobre.

For the mineral entry stage, the study of the implementation of the purchase of a portable axle scale was carried out, as well as the construction of an additional scale, with this we will achieve a greater flow of units at the time of entry (Figs. 6 and 7).

In Table 1 we can see the times of each activity that is carried out in the mineral entry process, from which it can be seen that the weighing process represents more than 30% of the total time, so a way to reduce this time is with the implementation of an additional scale, which will reduce the weighing time from 15 to 7.5 min which represents 31.1% of the total time.

Fig. 6 Portable scale

Fig. 7 Fixed scale platform type

Table 1 Comparative table of reduction of mineral weighing times

Sequence of activities	Time (Minutes)	Percentage (%)	Time (Minutes)	Percentage (%)
1 – Guide reception	17	37.78	17	54.84
2 – Validate Load	5	11.11	2.5	8.06
3 – Confirm Upload	5	11.11	2.5	8.06
4 – Weighing of Mineral	15	33.33	7.5	24.19
5 – Guide Issue	3	6.67	1.5	4.84
	45		31	

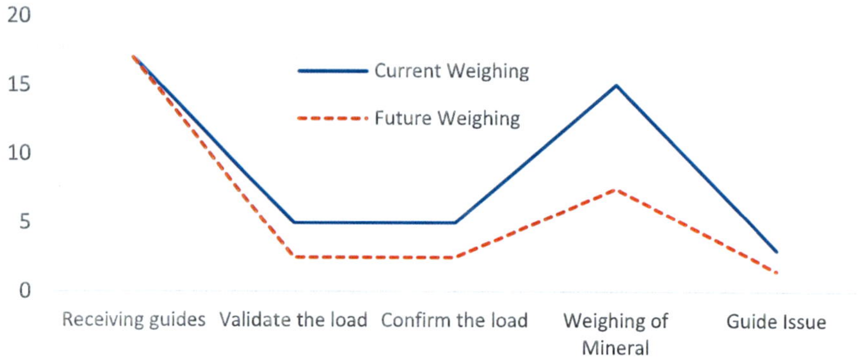

Fig. 8 Comparison of current and future weighing times

Fig. 9 Implementation of additional weighing system

Figure 8 shows a comparative estimate of the total time of the current weighing and the future weighing with the implementation of a scale. It can be noted that the cycle time of the process will be reduced by 0.5 min/Ton, and the entry time of each unit will be 31 min.

With the implementation of another scale, 02 units will be weighed in parallel, for this it will have to require 01 additional scale operator, 01 weighing system, 01 scale display and 01 additional printer, with In this regard, for this process the following stages of the implementation process will be required (Fig. 9).

5.1 *Preparation of the 5'S Work Plan*

Based on the current VSM diagnosis, we are going to use lean manufacturing tools like 5'S and TPM.

For the ironing and crushing process, the 5'S is implemented, not having the mineral pile ordered and located causes delays in the ironing processes, and in seasons of high demand for this process, it happens that due to the same requirement unloads the material in different parts of the slab. The implementation of the 5'S in the storage slab will be carried out in stages and will have the support of the operations management, likewise this implementation will last 01 month (Table 2).

In the sort stage, the mineral is classified by type, be it copper, zinc or lead, another classification factor is by way of storage, either in bulk or in a bigbag, with

Table 2 5'S implementation schedule

Sequence of activities	Week 1	Week 2	Week 3	Week 4
1 – Sort	X			
2 – Order		X		
3 – Cleaning		X	X	
4 – Standardize			X	
5 – Discipline				X

Table 3 Mineral storage slab cleaning schedule

Personal	Week 1	Week 2	Week 3	Week 4	Week 5
1 – Patio Assistant	X				
2 – Straddle Carrier		X			
3 – Girdle Helper			X		
4 – Belt Operator				X	
5 – Charger Operator					X

respect to the residual mineral that is generated by cleaning is relocated on another warehouse slab.

In the ordering stage, according to the mineral classification, it is ordered by priority of shipment and storage, likewise a label is placed which will be easier to locate when carrying out the crushing or crushing process. In the absence of a storage label, an acrylic slate is placed where the type of mineral, arrival date, humidity percentage and customer are detailed, with this it is possible to facilitate their location to the staff and avoid delays.

In the cleaning stage, it focuses on cleaning activities (Table 3) of the mineral storage slab, it is observed that there are disused bigbag bags which are removed on a daily basis.

In the standardization stage, a mineral location format is implemented which is delivered to the supervisor, manager and yard assistant, in order to locate the mineral and verify compliance with correct storage.

In the discipline stage, the correct performance of the mineral storage is managed and supervised by the operating personnel, as well as unannounced inspections to validate what has been implemented.

With the implementation of the 5'S, it was possible to notice improvements in time savings, especially at the time of the location of the type of mineral, the reduction occurred in a 20% time saving in the copper mineral consolidation process, achieving an increase in The fulfillment of the works is reduced by 10% and the cycle time is reduced by 0.54 min and with respect to the wetting process, it decreases by 0.4 min since the delivery of mineral increases for it, in the same way it occurs with the other processes, therefore the reprocessing works and the reduction of overtime hours were eliminated by 36.73%.

5.2 *TPM Application*

En el proceso de embarque se observan paradas de equipos, esto ocurre debido a una deficiencia con el mantenimiento que se realiza al equipo, para reducir las para-das y mejorara la disponibilidad del equipo se planteó como propuesta de mejora aplicar el TPM, por lo que se va considerar las siguientes etapas (Fig. 10).

The implementation time of the TPM will be 03 months, in the fourth month the improvements should already be noticed, see Table 4. With the application of the TPM the following benefits will be achieved:

- Increased availability of loading and transportation equipment.
- Operator training so that they can participate in routine maintenance and inspections.
- Fleet renewal.
- Quality maintenance, improvement of supplies and spare parts.

5.2.1 Initial Diagnosis Before Applying TPM

At this stage we must capture everything that is currently in the place, such as what type of maintenance is being applied, if the tasks are well distributed, frequency of inspections, reports of malfunction, a root cause analysis is carried out, has a stock control of critical spare parts and if any software is used for maintenance. Likewise, it was found that preventive maintenance tasks were not well distributed according to the frequency of work. There is no analysis of equipment failures, as well as proactive actions so that the equipment stop occurs again, nor has the impact of economic loss generated by the stop or mechanical failure been quantified, that is, there is no precedent cost for breakdowns and production.

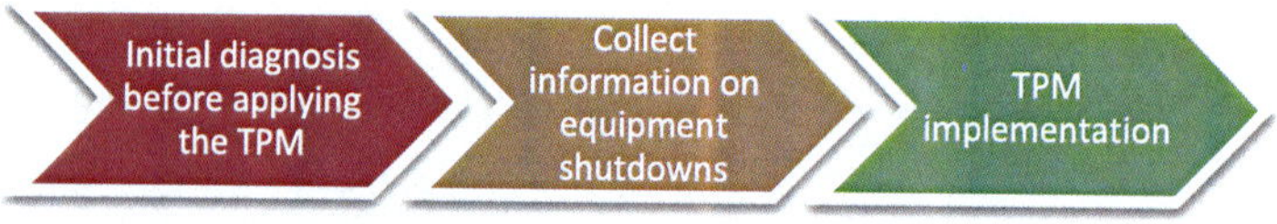

Fig. 10 Stages of TPM implementation

Table 4 TPM stages implementation schedule

Stages of implementation	Month 1	Month 2	Month 3	Month 4	Month 5	Month 6
1 – Initial diagnostic	X	X				
2 – Gather information		X	X			
3 – TPM implementation				X	X	X

5.2.2 Collect Information on Equipment Shutdowns

In this stage we are going to collect all the stops of the equipment in the boarding process, we must take note of the reasons why the equipment stops occurred and select the stops according to the classification that is detailed:

- Stops due to bad operation.
- Stoppages due to lack of preventive maintenance.
- Stoppages for useful life.

These faults must be consolidated in a table to hove them mapped.

5.2.3 TPM Implementation

At this stage we will implement the established pillars:

5.2.4 Identify and Eliminate Problems

In this stage we are going to define the preventive actions that are going to be used in order to obtain improvements and act proactively to reduce equipment stoppages, for this we propose the following steps:

- Analyze the causes of failures, ACR, 5 Why, etc.
- Determine the losses with the greatest impact.

At this stage, it was found that the maintenance programs were not up-to-date, routine inspections had not been implemented, so preventive maintenance tasks had to be updated, and preventive maintenance and inspection books were implemented. The ACR was applied for the failures that were reported by the operators and the cause was determined and improvements were proposed.

5.2.5 Apply Autonomous Maintenance

Steps to follow to carry out the correct application of autonomous maintenance are proposed.

- Carry out inspections and cleaning of the equipment, with this we will prevent and detect leaks and therefore avoid unexpected stops.
- Standardize cleaning and inspection frequencies, work frequencies are created, daily, weekly, for the inspection activities we must rely on the manufacturer's manual.
- Perform maintenance activities, it is detailed what maintenance activities the operators can carry out so that they can carry out the maintenance. As well as improve communication between operators and maintenance personnel.

5.2.6 Perform Planned Maintenance

Implement planned maintenance, make improvements if there is planned maintenance, enforce equipment shutdowns when maintenance frequency is met.

5.2.7 Personnel Training

Maintenance personnel are trained so that they can perform activities with good quality, operators are also trained so that they can identify or report anomalies and that technical personnel can act quickly.

Once the TPM is implemented, it is estimated that in the following months the causes that generate the shutdowns will reduce by 25% after 3 months of implementing the TPM, (Table 5).

In the same way, with the implementation, it will be possible to increase the availability of the transport fleet and therefore decrease the cycle time by 0.53 min/TN, which represents a 33.33% reduction for me.

6 Evaluation of Results

With the implementation of the 5's, it was possible to reduce the time when locating the mineral to be consolidated, likewise during the implementation stage, the opinion of the operating personnel was taken, which was considered in the implementation. Unexpected biweekly inspections are currently being carried out on the storage slabs to ensure that the methodology is successfully maintained.

The future VSM is shown below, where it can be seen that the total cycle time manages to decrease by 27% of the initial time (Fig. 11).

Throughout the process of implementing the TPM, the cooperation of operating personnel was achieved, as well as the autonomy and technical capacity of the operators was developed, in addition to this, it was possible to improve communication between technical personnel and equipment operator, achieving awareness of responsibility for operators in the correct operation and care of the equipment. In the process part, the cycle time is reduced by 0.53 min (Table 6).

Table 5 Equipment downtime

Cause of failure	March	April	May	June	July
1 – Stop by Bad Operation	28	25	20	13	10
2 – Lack of Preventive	48	45	40	23	16
3 – Shutdown for Useful Life	25	22	20	16	13
Total	101	92	80	52	39

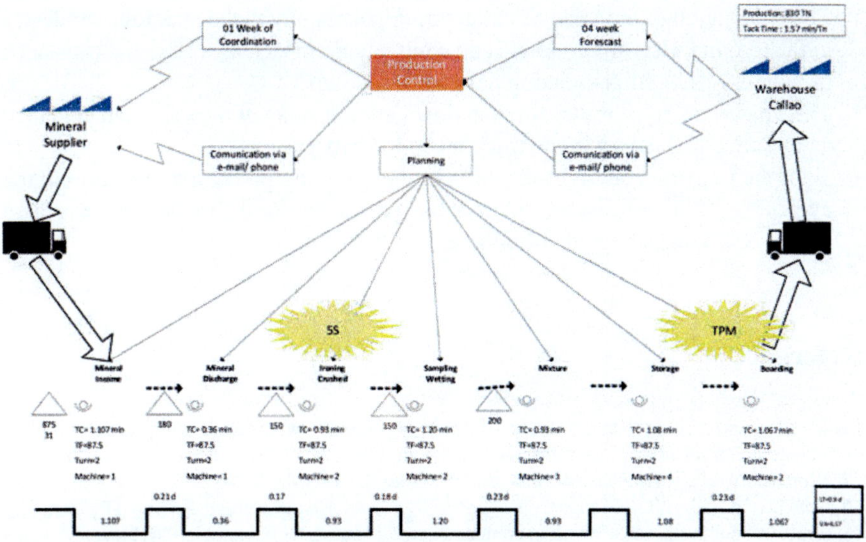

Fig. 11 VSM Future of the copper ore consolidation process

Table 6 Comparative table of expected results

Process	Before (Min/Tn)	Later (Min/Tn)	Diff (Min/Tn)	%
1 – Mineral Income	1.607	1.107	0.50	31.11
2 – Ironing and shredding	1.47	0.93	0.54	36.73
3 – Moistening	1.60	1.20	0.40	25.0
6 – Shipment	1.60	1.067	0.533	33.31

7 Conclusions

- The main objective of this work was to improve the mineral consolidation process with the implementation of lean manufacturing tools, once the lean tool is implemented, an improvement in productivity is obtained, equipment stops are reduced, they reduce reprocessing and this reduces costs, making higher profits are obtained.
- The use of VSM allowed us to identify delays in the process as well as find dead times, with this it will be possible to propose improvements in processes where there are delays.
- There were several benefits after the implementation of the tools. The application of 5'S generated a cleaner and more orderly work environment that resulted in the reduction of unnecessary forklift movements.
- The improvement in the distribution of the warehouse reduced the transport times of freight loads between each zone and the study of the work also allowed to reduce the time of the processes, increasing the productivity of the dispatches.

- The application of the Lean Manufacturing philosophy to the operating processes of the terminals has an economic and commercial impact, as well as the resistance of workers when implementing new work routines.
- The improvement proposals broken down at each stage of the value chain helped us optimize four hours in the truck reception process.
- With the implementation of 03 shifts in the sampling process, a reduction in the cycle time of the process will52 be achieved, so that for the mixing process there would be a greater volume of mineral.

References

1. Ministry of Energy and Mines Homepage, http://www.minem.gob.pe/_detalle.php?idSector=1&idTitular=159&idMenu=sub149&idCateg=159, last accessed 2009/07/03.
2. Womack, Roos, Jones: The Machine that Changed the World (1990).
3. Carrillo, M., Alvis, C., Mendoza, Y., Cohen, H.: Lean Manufacturing: 5 s and TPM, Quality improvement tools. Metalworking company case in Cartagena, Colombia (2019).
4. Neves P., Silva JG., Ferreira LP., Gouveia RM, Pimentel C.: Implementing Lean Tools in the Manufacturing Process of Trimmings Products (2018). Procedia Manuf 2018;17:696–704. doi: https://doi.org/10.1016/j.promfg.2018.10.119.
5. Anosike A., Alafropatis K., Garza JA., Kumar A., Luthra S., Rocha-Lona L. : Lean manufacturing an internet of things – A synergetic or antagonist relationship?. Computers in Industry 129 (2021) 103464. https://doi.org/10.1016/j.compind.2021.103464
6. Martin, N., Dér, A., Herrmann, C., Thiede S.: Assessment of Smart Manufacturing Solutions Based on Extended Value Stream Mapping (2020). Procedia CIRP 93 (2020) 371–376. https://doi.org/10.1016/j.procir.2020.04.019
7. Venkat B., Prathap P., Sivaraman P., Yogesh S., Madhu S.: Implementation of lean manufacturing in electronics industry. Materials Today: Proceedings. https://doi.org/10.1016/j.matpr.2020.02.718
8. Alvarez, M., Paucar P., Róger P., Alberto M.: Development and implementation of the continuous improvement methodology in a metalworking mype to improve productivity (2021). Link to Item http://hdl.handle.net/10757/337910
9. Palange A., Dhatrak P. : Lean manufacturing a vital tool to enhance productivity in manufacturing. Materials Today: Proceedings 46 (2021) 729–736. https://doi.org/10.1016/j.matpr.2020.12.193
10. Suzuki, Tokutaro: TPM in Process Industries (1995).
11. Diaz Aracely: Application of the TPM Philosophy in the Machinery and Equipment of Empresa Minera del Norte S.A. de C.V. Manzanillo Unit (2014).

Application of Sustainable Livelihoods Approach in the Tea Filter Production

Lady D. Infante-Acosta and Jonatán E. Rojas-Polo

Abstract The demand for Peruvian native plant infusions has increased given the trend towards the consumption of nutraceutical products that strengthen the immune system against COVID19. Therefore, this research focuses on improving the stages of the value chain of tea filter production in rural communities using the sustainable livelihoods approach.

Keywords Sustainable value chain · Livelihood approach · Sustainable infusion production · Sustainable development

1 Introduction

The present research project has taken into consideration two situations that are currently occurring in Peru, which are detailed below.

On the one hand, the trend of consuming healthy foods and beverages in Peru has been increasing in recent years, and Peruvian consumers are becoming more conscious and careful about what they eat [1]. These habits have been strongly reinforced by the COVID-19 events, as people have been forced to consume foods and beverages that strengthen their immune system and contain properties that treat possible symptoms of the coronavirus [2]. This situation, analyzed from a beverage consumption perspective, suggests that there has been an increase in the demand for tea since Peruvians consider it a natural, healthy option that helps prevent and improve basic health problems [3]. As an example, Fig. 1 shows the increase in bottled tea consumption, according to data reported by Euromonitor International.

On the other hand, according to reports from the National Institute of Statistics and Informatics (INEI, for its acronym in Spanish), monetary poverty in Peru was 30.1% of the population in 2020, affecting 47.5% of the population in rural areas and 26.0% of the population in urban areas [4]. An analysis of participation by economic

L. D. Infante-Acosta (✉) · J. E. Rojas-Polo
Pontificia Universidad Católica del Perú, Lima, Perú
e-mail: lady.infantea@pucp.edu.pe; jrojasp@pucp.pe

J. Vargas Florez et al. (eds.), *Production and Operations Management*, Springer Proceedings in Mathematics & Statistics 391,
https://doi.org/10.1007/978-3-031-06862-1_18

Fig. 1 Consumption in millions of liters of RTD Tea in Peru. (Adapted from "RTD in Peru – Analysis" by Euromonitor International)

activity among the population with scarce economic resources shows that 55.9% of them are related to agriculture, fishing and mining, which indicates that farmers are part of the most affected population [5]. Therefore, promoting the improvement of the value chain is a clear strategy to tackle poverty. It is important to emphasize that this action is in line with the sustainable development objectives promoted by the United Nations [6], such as the following: 1. No poverty, 2. Zero hunger, 11. Sustainable cities and communities, 13. Climate action, 15. Life on land.

Given these two facts, this project seeks to address the production of filtering tea from medicinal plants such as eucalyptus, West Indian lemongrass, peppermint, Peruvian peppertree, and thyme. The objective of the research is aimed at responding to consumer needs and seeks to support the producers of these plants through the Sustainable Livelihoods approach (SL). This strategy will facilitate sustainable development in terms of production and will provide support to the less favored population by strengthening human, natural, financial, physical, and social capital.

2 Literature Review

This section details the main concepts that support the research.

2.1 Medicinal Plants

Plants defined as medicinal are those that contain active principles, which have pharmacological properties. These properties are used to treat a disease or health problem for which the plant has a beneficial action. According to Peru's Ministry of Agriculture (MIDAGRI, for its acronym in Spanish), 890 species come from the Amazon out of a total of 1109 medicinal plants known in Peru [7].

Some relevant attributes of medicinal plants are related to the improvement of the proper functioning of the immune system and the treatment of respiratory viral and bacterial infections. Among the main uses is the treatment of conditions such as asthma, bronchitis, flu, colds, and cough [8]. The following are some of the main plants that have these types of health benefits.

1. Eucalyptus (*Eucalyptus globulus*)
2. West Indian lemongrass (*Cymbopogon citratus*)
3. Peppermint (*Mentha piperita*)
4. Peruvian Peppertree (*Schinus molle*)
5. Thyme (*Thymus vulgaris*)

2.2 Sustainable Livelihoods

A sustainable livelihood is defined as a scheme in which there is good management of capacities and assets to maintain present and future levels of performance and use to cope with stresses and shocks, without affecting its resources. The main objective is to reduce the impact of external factors that could place a population in a situation of vulnerability [9]. About livelihood, some related capitals and assets are distinguished into five major groups, which are detailed below [10].

- **Human capital**. Related to levels of health, nutrition, education, and knowledge.
- **Social capital**. Related to connections that increase the ability to work together.
- **Natural capital**. Related to the natural resources on which they depend.
- **Physical capital**. Related to the necessary infrastructure and equipment.
- **Financial capital**. Related to the financial resources available for strategies.

2.3 Vulnerability Context

The vulnerability context is associated with unpredictable events that can affect livelihoods and thus cause rural households to fall into poverty. Factors may cause immediate effects and others may be more gradual, but in the end, both cases end up being detrimental [11]. Some situations are presented below as examples.

- Epidemic pests and diseases
- Natural disasters
- Civil disputes
- Environmental crisis
- Economic disruptions

2.4 Sustainable Agricultural Value Chain

Agricultural Value Chain The agricultural value chain refers to the set of actors and activities related from the production of agricultural products in the field to their provision for consumption, adding value to the product at each stage. A value chain is not only limited to production but may also include other stages such as processing, packaging, storage, transportation, and distribution [12], as shown in Fig. 2.

Sustainability Sustainability refers to three dimensions on which efforts are focused: environmental (environmental health), social (social equity), and economic (profitability). According to the Food and Agriculture Organization of the United Nations (FAO), there are five fundamental principles of sustainability for agriculture [13]:

Principle 1 – Increase productivity, employment, and value addition in food systems
Principle 2 – Protect and enhance natural resources
Principle 3 – Improve livelihoods and foster inclusive economic growth
Principle 4 – Enhance the resilience of people, communities, and ecosystems
Principle 5 – Adapt governance to new challenges

Fig. 3 shows the interaction of the above-mentioned principles.

2.5 Filter Tea Production Process

The filtering tea production process involves different stages, according to the particularities of each product. The following are the main operations related to this project.

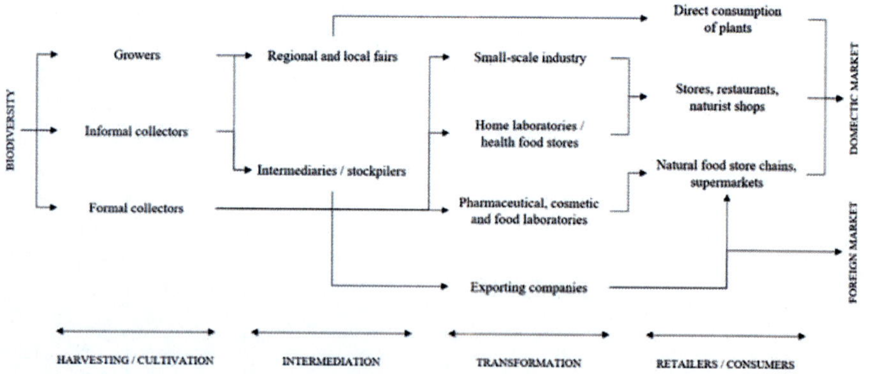

Fig. 2 Value chain of medicinal plants. (Adapted from "Las plantas medicinales en el Perú. Etnobotánica y viabilidad comercial" by M. Puelles, V. Gómez, and J.M. Gabriel y Galán, 2010)

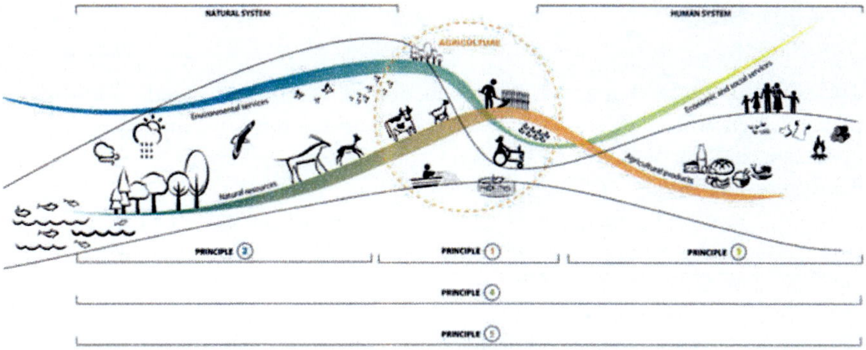

Fig. 3 Application of the vision and the five principles of sustainable agriculture. (Adapted from "Building a common vision for sustainable food and agriculture. Principles and approaches" by Food and Agriculture Organization of the United Nations (FAO), 2014)

- **Reception**. The raw material (medicinal plants) is received.
- **Selection**. The organoleptic characteristics of the raw material are verified.
- **Weighing**. The raw material is weighed and a record is taken to later obtain performance indicators and the percentage of humidity.
- **Washing, disinfection, rinsing, and draining**. The raw material is initially washed with potable water; then, it is disinfected with a sodium hypochlorite solution, whose dosage is estimated between 0.5% to 2.0% and the immersion time ranges from five to ten minutes [14]; then, the plants are rinsed in potable water to eliminate any excess of the disinfectant; and, finally, they are drained to eliminate the excess water.
- **Drying**. The raw material is placed in trays, then placed inside the dryer and the equipment is set to work at 60 °C. The time of permanence will depend on the characteristics of each plant. At this stage, a sample must be taken from each plant to calculate the percentage of humidity.
- **Quality control**. Humidity control is done in order to evaluate and guarantee that the expected conditions are fulfilled during the process.
- **Conditioning**. The edible part of the plant, such as the stem or leaves, is separated and preserved since the properties associated with it can be found in different parts, depending on the type of plant.
- **Grinding**. The plant is ground using a hammer mill with a 1 mm screen for the final filter. This process reduces the volume of the plant and increases the surface area, which ultimately results in the possibility of better extraction of the constituents [15].
- **Packaging**. The ground plant is placed in a tea packing machine to be filled in cotton filter sachets of 1 gr net weight, with cotton thread, label, and outer paper sachet. As the last step, groups of twenty units are formed, so that each of them is placed in boxes that are later closed and sealed.

3 Case Study

The object of study is aimed at the production of filtering tea based on medicinal plants, which is in line with the current trend towards the consumption of healthy beverages and the objective of improving the immune system, mainly focused on combating viral and bacterial respiratory infections.

The scope includes the analysis and improvement of the value chain from plant cultivation to the final product through the application of sustainable livelihood concepts and related sustainability principles. It should be noted that the Andean zone of Peru is where the main medicinal plants with the characteristics mentioned are cultivated, so the scope of the project will be focused on the department of Junín, which is one of the regions that produce a large part of the plants mentioned.

3.1 *Human Capital*

The statistics provided by the Human Development Index prepared by the United Nations Development Program (UNDP) are used to analyze human capital. Through the variables considered, Table 1 shows the areas of health (life expectancy at birth), education (population aged 18 with completed high school education and the average years of schooling of adults over 25), and income (per capita household income).

Table 1 Variables of the Human Development Index

	Population	Human development index	Life expectancy at birth	Population aged 18 with completed high school education	Years of schooling (Population over 25)	Per capita household income
Junín	Inhabitants	IDH	Years	(%)	Years	PEN per month
Year	*VAR_1*	*VAR_2*	*VAR_3*	*VAR_4*	*VAR_5*	*VAR_6*
2003	1,260,773	0.3384	70.10	64.80	8.16	286.00
2007	1,225,474	0.3399	71.80	59.23	8.05	278.11
2010	1,301,844	0.4227	72.09	63.13	8.76	461.60
2011	1,311,584	0.4485	72.41	68.80	8.51	528.30
2012	1,321,407	0.4542	72.64	68.60	8.52	545.54
2015	1,350,783	0.4745	73.10	66.68	8.53	619.43
2017	1,246,038	0.4954	73.18	65.88	8.38	710.32
2018	1,254,639	0.5083	73.41	65.96	8.40	759.83
2019	1,323,404	0.5107	72.94	67.30	8.63	757.26

Adapted from "Índice de Desarrollo Humano – IDH" by Instituto Peruano de Economía, 2020

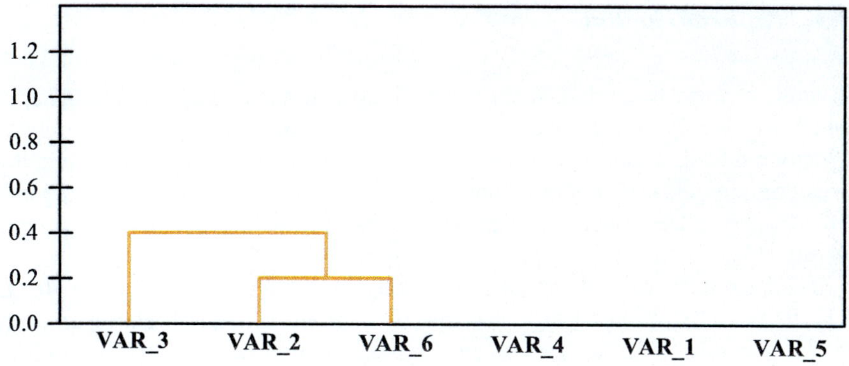

Fig. 4 Multivariate dendrogram based on HDI variables

In order to identify the variable that represents the most relevant relationship with the quality of life of the inhabitants of Junín, the records by year are normalized and Fig. 4 shows a dendrogram in which it is observed that per capita family income is most correlated with the Human Development Index.

Strengthening employability and economic conditions will improve per capita household income, which ultimately aligns with sustainability principles 1 (Increase productivity, employment, and value addition in food systems) and 3 (Improve livelihoods and foster inclusive economic growth).

3.2 Social Capital

In terms of social capital, the role played by farmers' organizations is fundamental in promoting the economic and social objectives of their members. These organizations coordinate with local authorities to obtain credit, inputs, training, and other resources to help them strengthen their efforts [16]. In Peru, 96.0% of the peasant communities are engaged in agricultural activities, according to the results of the I Census of Peasant Communities [17].

To strengthen this capital, MIDAGRI promotes Good Agricultural Practices in farming communities through the Phase II of the Agricultural Health and Agri-Food Safety Development Program. The goal is to break the dependence on the use of pesticides and promote socioeconomic development for sustainable family agriculture [18].

The training provided and economic access facilities will allow farmers to improve their production process, which will guarantee compliance with sustainability principles 1 (Increase productivity, employment, and value addition in food systems) and 2 (Protect and enhance natural resources).

3.3 Natural Capital

In terms of physical capital, Junín has 465,880.4 hectares of agricultural land, of which only 13.3% is irrigated and the remaining 86.7% is classified as rainfed agricultural land, which is considered to be land that uses rainfall as a resource for irrigating the fields. It should be noted that crops are grown during rainy periods from August to December and crops with access to irrigation are grown from April to July [19].

There are about 132.1 thousand agricultural units, of which 70.8% are from 0.1 to 5 hectares; 10.8%, from 5.1 to 10 hectares; 6.4%, from 10.1 to 20 hectares; and only 0.8%, more than 100.1 hectares. In addition to this information, 25.9% use tractors for agricultural work, 1.1% use electric power, and 43.0% do not apply any type of fertilizers or chemical inputs [19].

Promoting the appropriate use of the available land and the climatic conditions of the area will strengthen the production chain, while guaranteeing the protection and enhancement of natural resources, principle 2 of sustainability.

3.4 Physical Capital

In terms of physical capital, one of the main aspects is irrigation canals. According to the information provided by the General Direction of Agrarian Infrastructure and Irrigation in its report, in 2020, 741.60 km of irrigation canals were maintained with an allocated budget of 4′414,004 soles. In addition to this, it is known that for 2019 there were a total of 2876 producers and 11,158 agricultural hectares with irrigation infrastructure installed, recovered, or improved [20].

According to the Ministry of Transportation and Communications, Junín accounts for 7% (11,985 km) of the total national road network. It is specified that 76% corresponds to the neighborhood network, 15% to the national network, and 9% to the departmental network. Likewise, of the total kilometers available in Junín, 89% are not paved, which corresponds to 83% of the neighborhood network, 10% to the departmental network, and 7% to the national network. This situation places Junín below its peers in the highlands of Peru [19].

It is important to promote these physical resource improvements, such as irrigation canals and paved roads, in order to provide facilities for farmers. This is why the alignment to sustainability principle 5 (adapt governance to new challenges) should be strengthened, as these improvements are mainly linked to public sector projects.

3.5 *Financial Capital*

In terms of financial resources, one of the main sources of access to credit is Agrobanco, a state-owned financial institution that provides loans to the agricultural sector. In line with its financial inclusion plan, Junín has received the highest number of disbursements in the last decade with 17.8% and ranks first among the other regions [21].

According to the report on the characterization of the department of Junín prepared by the Central Reserve Bank of Peru (BCRP, for its acronym in Spanish), the number of agricultural producers is 135.8 thousand, 99.3% of whom are individuals [19]. Table 2 shows the report on agricultural financing in terms of the number of products with at least one loan, the total amount allocated through this active product, and the number of loans allocated. Concerning the most recent year, the number of producers with a financial credit represents approximately 19.8%.

Providing facilities to access financial resources will allow farmers to be able to manage unexpected events, such as delays in the supply chain, malfunctioning of their machinery, among others. This will ultimately ensure compliance with principle 4 of sustainability (enhance the resilience of people, communities, and ecosystems).

4 Conclusions

Thanks to this study, the potential points to be strengthened in the farmers of Junín have been identified in order for them to have a better control of their capacities and assets in the event of a vulnerability event that could affect their productive chain.

- In terms of human capital, per capita income is one of the most important factors in ensuring a good quality of life for farmers in the area; in addition, there has been an improvement in factors associated with education, meaning that farmers are potentially more willing to acquire new knowledge or use new technology.

Table 2 Agricultural finance report

	Number of producers with financial loans	Total amount allocated in financial loans	Total amount of financial credits allocated
Year	Persons	PEN	Credits
2016	24,516	–	–
2017	23,989	–	–
2018	31,849	3,492,639.28	675
2019	20,383	2,492,226.50	424
2020	26,869	2,114,209.62	610

Adapted from "Dashboard Temáticos – MIDAGRI. Perfil Departamental" by Ministerio de Desarrollo Agrario y Riego del Perú, 2021

- With respect to social capital, there are campaigns that promote better farming practices associated with sustainable agriculture; however, something that still needs to be emphasized is the creation of more organizations of farmers, which can function as a point of support when improvements are required for their community.
- With regard to natural capital, there is good management of the characteristics of the season, since according to the conditions of the terrain, rainfall is used effectively to irrigate agricultural land; furthermore, one point in favor is that four out of every ten croplands do not use any type of fertilizer or chemical inputs.
- Regarding physical capital, there is monitoring and maintenance of irrigation canals; however, a clear disadvantage compared to other departments is that there is a low percentage of paved road networks.
- Finally, in terms of financial capital, Junín is one of the regions in Peru that receives the highest number of loans for the agricultural sector; however, approximately one out of every five farmers makes use of this resource, for which reason it is important to inform about the benefits of acquiring a loan to improve their farmland.

References

1. Kantar Worldpanel: Hogares peruanos se orientan hacia consumo saludable, https://www.kantarworldpanel.com/pe/Noticias/Hogares-peruanos-se-orientan-hacia-consumo-saludable (2019)
2. elEconomistaAmérica: Cinco nuevas tendencias del consumidor en el sector de alimentos y bebidas, https://www.eleconomistaamerica.pe/empresas-eAm-peru/noticias/10738697/08/20/Cinco-nuevas-tendencias-del-consumidor-en-el-sector-de-alimentos-y-bebidas.html (2020)
3. Euromonitor International: Tea in Peru. Increased demand in quarantine due to affordability and perceived health benefits, https://www.euromonitor.com/tea-in-peru/report (2020)
4. Instituto Nacional de Estadística e Informática (INEI): Pobreza monetaria alcanzó al 30,1% de la población del país durante el año 2020, https://www.inei.gob.pe/prensa/noticias/pobreza-monetaria-alcanzo-al-301-de-la-poblacion-del-pais-durante-el-ano-2020-12875/ (2021)
5. Instituto Nacional de Estadística e Informática (INEI): Evolución de la Pobreza Monetaria 2009–2020 – Informe Técnico, https://www.inei.gob.pe/media/MenuRecursivo/publicaciones_digitales/Est/pobreza2020/Pobreza2020.pdf (2021)
6. United Nations: Sustainable Development Goals, https://www.un.org/sustainabledevelopment/sustainable-development-goals/ (s.d.)
7. Ministerio de Desarrollo Agrario y Riego: Plantas medicinales, https://www.minagri.gob.pe/portal/59-sector-agrario/plantas-medicinales (s.d.)
8. Bussmann, R., Sharon, D.: Plantas medicinales de los Andes y la Amazonía. La flora mágica y medicinal del norte del Perú. https://docs.bvsalud.org/biblioref/2018/10/916684/plantas-medicinales-de-los-andes-y-la-amazonia-la-flora-magica-_Qa3dgqr.pdf (2015)
9. Food and Agriculture Organization of the United Nations: Los medios de Vida Sostenible: Análisis a Nivel Hogar, http://www.fao.org/in-action/herramienta-administracion-tierras/modulo-1/propuesta-metodologica/medios-vida-sostenibles/es/ (s.d.)
10. Food and Agriculture Organization of the United Nations: Glosario, http://www.fao.org/in-action/herramienta-administracion-tierras/glosario/m/es/ (s.d)

11. Food and Agriculture Organization of the United Nations: Módulo 1. Medios de vida, pobreza e instituciones, http://www.fao.org/3/a0273s/a0273s04.htm (s.d)
12. Food and Agriculture Organization of the United Nations: El desarrollo de las cadenas de valor agrícola: ¿amenaza u oportunidad para el empleo femenino?, http://www.fao.org/3/i2008s/i2008s04.pdf (2010)
13. Food and Agriculture Organization of the United Nations: Sustainable Food and Agriculture, http://www.fao.org/sustainability/en/ (s.d.)
14. Carballo, C., Alfaro, T., Palazón, Z., et al.: Desinfección química de plantas medicinales II. *Plantago lanceolata L.*, http://scielo.sld.cu/pdf/pla/v7n3/pla04302.pdf (2002)
15. Colina, L.: Reducción de tamaño de alimentos, http://sgpwe.izt.uam.mx/files/users/uami/mlci/red_tam_solidos_intro.pdf (s.d.)
16. Food and Agriculture Organization of the United Nations: Las organizaciones campesinas en América Latina, http://www.fao.org/3/t3666s/t3666s04.htm (s.d.)
17. Instituto Nacional de Estadística e Informática: Resultados Definitivos de las Comunidades Nativas y Campesinas 2017, http://censo2017.inei.gob.pe/resultados-definitivos-de-las-comunidades-nativas-y-campesinas-2017/ (2017)
18. Servicio Nacional de Sanidad Agraria: MINAGRI: SENASA fomenta las Buenas Prácticas Agrícolas en comunidades campesinas de Huánuco, https://www.senasa.gob.pe/senasacontigo/minagri-senasa-fomenta-las-buenas-practicas-agricolas-en-comunidades-campesinas-de-huanuco/ (2020)
19. Banco Central de Reserva del Perú: Caracterización del departamento de Junín, https://www.bcrp.gob.pe/docs/Sucursales/Huancayo/junin-caracterizacion.pdf (2020)
20. Ministerio de Desarrollo Agrario y Riego: Dashboard Temáticos – MIDAGRI, https://siea.midagri.gob.pe/portal/siea_bi/index.html (s.d.)
21. El Peruano: Acceso al financiamiento en el campo se incrementa, https://elperuano.pe/noticia/119981-acceso-al-financiamiento-en-el-campo-se-incrementa (2021)

The Limit of the Environmental and Productive Performance of Closed-Loop Production: Evaluation in the Wood Pellet Industry in Brazil

Flavio Numata Junior and Helena Navas

Abstract The closed-loop production system can reduce environmental impacts on industries. However, reprocessing has different characteristics from the primary process, which can generate environmental effects superior to the original production. Therefore, this works explores the end of life of industrialization in order to determine the number of remanufacturing cycles that provides the best environmental performance of the closed-loop production system. The research took place in a Brazilian wood pellet industry. Although the volume of closed-loop production is lower than that of primary manufacturing, the CO_2 emissions generated in the pelletizing process are 11% higher than in the original process. The results also demonstrated that the properties of the raw material and the energy flow are factors that interfere in the number of remanufacturing cycles.

Keywords Closed-loop production · Remanufacturing cycles · Energy analysis · CO_2 emissions

1 Introduction

Closed-loop production uses materials recovered after primary manufacturing, hence has links with the concepts of life cycle and eco-efficiency. By-products that are reused have specific properties according to their reprocessing rate. Another positive aspects is the remanufacturing conditions, with processes different from primary industrialization and requiring an additional energy charge to the original process.

F. Numata Junior (✉)
Federation of Industry of Parana State (FIEP), Curitiba, Brazil

The NOVA School of Science and Technology (FCT NOVA) Universidade NOVA de Lisboa, Caparica, Portugal

H. Navas
UNIDEMI, The NOVA School of Science and Technology (FCT NOVA), Universidade NOVA de Lisboa, Caparica, Portugal
e-mail: hvgn@fct.unl.pt

J. Vargas Florez et al. (eds.), *Production and Operations Management*, Springer Proceedings in Mathematics & Statistics 391,
https://doi.org/10.1007/978-3-031-06862-1_19

These factors demonstrate favorable characteristics for production in a closed network, but there are other elements, with low performance, that can cause effects on the environment. In this context, this article aims to calculate the limit amount of closed-loop production cycles to maintain the balance between production and environmental efficiency. The research had an applied phase for the evaluation of the study. The energy sector was chosen because it is the industrial area with the greatest responsibility for environmental effects and shows a trend of high market growth for the coming years. Regarding the approach to manufacturing, a product was defined that uses renewable raw material and closed-loop production. Therefore the experimental work was carried out in a wood pellet industry located in Brazil. The technological level of the plant's processes ensures high quality products and environmental protection, nevertheless the study has added more information for the industry to evolve towards an increasingly sustainable manufacturing.

2 "X" Times Closed-Loop Production Cycles

The Closed-loop production is linked with the concepts of life cycle and eco-efficiency of products and processes. By-products that are reused for manufacturing have specific properties according to their reprocessing rate [16].

Sustainable manufacturing integrates two concepts. The one that involves the operations of transforming material and energy resources into product production, and the concept associated with sustainability, which aims to meet the needs of the present without compromising the needs of future generations [37]. The concept of sustainability is broader when connecting the elements of the Triple Botton Line which, currently, have become requirements in the planning of industrial operations. Nowadays, worldwide, it is verified by the of the 2030 Agenda of the United Nations (UN) on the 17 Sustainable Development Goals (SDG) [5, 21, 36].

The SDG expand and strengthen the recommendations proposed by the Brundtland Commission more than twenty years ago with targets until 2030. Therefore, the manufacturing industry, even though it has a strong socioeconomic relationship, should increase its responsibility for the environment. The historical evolution of manufacturing is highlighted by the use of technology to optimize, control and automate processes to increase quality and production standards. On the other side, industrialization processes demand a high energy load and are responsible for a high level of greenhouse gas (GHG) emissions [9]. Regardless of a country's level of technological development, polluting emissions occur. In the United States of America, about 28% of GHG discharges come from industries [28]. In Brazil this value is even higher, reaching about 35% of the total contribution of emissions [15]. The manufacturing models have evolved towards the context that involves sustainability as shown in Fig. 1:

This transformation of manufacturing demonstrates the application of control over materials, energy, water and by-products generated, from extraction, use and reuse, applying ways to manage the life cycle of its products, to minimize

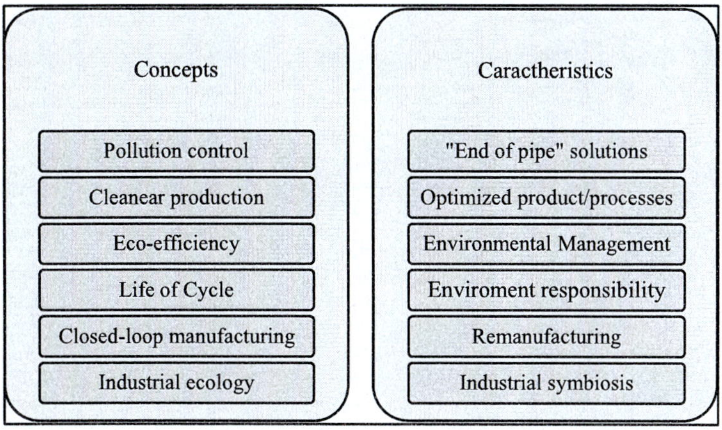

Fig. 1 Conceptual evolution of manufacturing

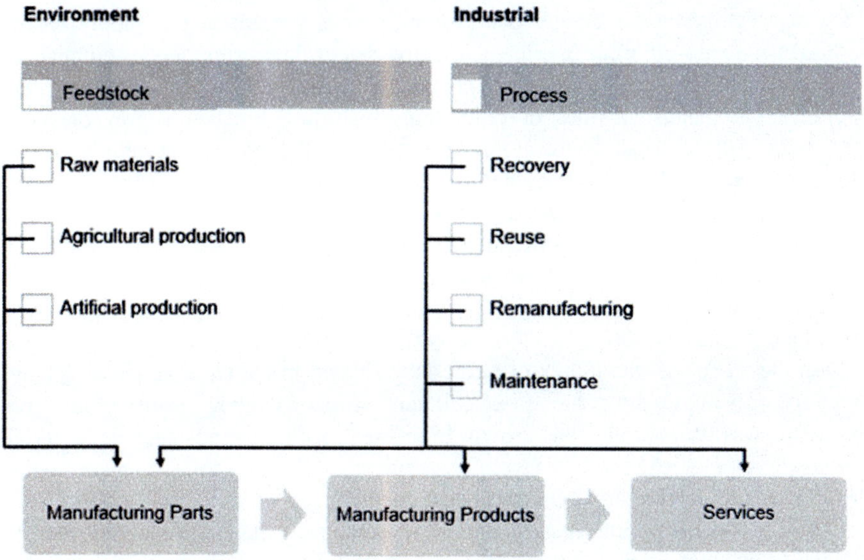

Fig. 2 Circular economy. (Source: [11])

environmental effects and expand business competitive advantage [16]. Therefore, in order to manage environmental management, it is necessary to measure its performance, but this assessment is complex because it involves systemic factors in the production chain.

The flow of resources can come from natural materials or industrial processing. In both sources there are specific situations due to the form of extraction, preparation, transport and industrialization. In addition, for specific recovery, recycling or processing for remanufacturing, as can be seen in Fig. 2:

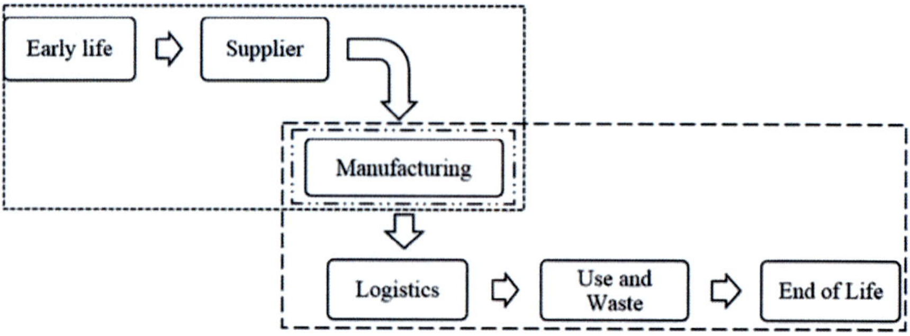

Fig. 3 System boundaries

The data are also associated with aspects of externalities, such as market demand that interferes with the volume of productivity or linked to the properties of the materials. In addition, like the principles present in the Circular Economy [11], the idea of sustainability is added to the organizations present in the supply chain. Companies responsible for intermediate deliveries or direct suppliers of companies or industrial parks generate integrated technical interests with environmental responsibilities, by means of rules or regulations regarding procedures for logistical processes. All of these connections exist because the product system involves different values and functions in the product's life cycle, as shown in Fig. 3:

At each stage in system, data associated with the product, function, process and results achieved can generate structural and economic effects, and can be considered in its evaluation [2, 27] to reduce environmental pollution, optimize the use of natural resources and increase productivity, energy efficiency and offer new sources for economic growth [30].

For these aspects, the number of criteria evaluated in the environmental dimension, the prevention and reduction of pollutant emissions and in cleaner production processes and technologies has grown. In this approach, methods such as Material Flow Analysis (MFA) and Life Cycle Assessment (LCA) are applied to measure the level of resource reuse and the effects generated [7].

LCA is considered the main technique for analyzing the environmental performance of products and processes [33]. The method can represent variation of the results according to the procedure and criteria adopted in the evaluation of the impacts and the limitation of the database available for the study. LCA presents an assessment of the static condition, and does not represent variations that may originate from specific factors [6]. Dynamic systems are mathematical models that can associate variables that are subject to variation. The model presents a systemic evaluation because it can contain data referring to values of geographic, temporal, technical references of product or connectivity with other factors allowing representative and adequate interpretation of the sample. These data generate complexity in the evaluation because they add economic and environmental aspects with an interrelated hierarchical structure [4].

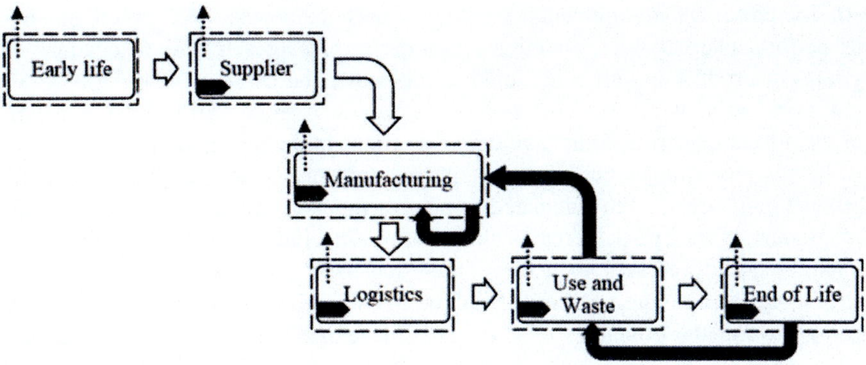

Fig. 4 Product system split

The factors associated with the materials can also influence other elements of the system. The properties of the resources are particular characteristics that determine the resilience of the materials and determine, how and for how long, their characteristics will be changed and may influence the system. This aspect is the resilience property existing in the ecosystem, which relates the changes that the system can change, its degree of self-organization or its capacity to adapt to the interference received [38]. One way to model the assessment is to consider the system's ability to absorb influences without changing its characteristics, known as ecological resilience [17]. The variables considered for the assessment of resilience involve the condition of living with changes, insertion of diverse externalities and the possibility of rearrangement in cross scales related to time and space [18, 39]. Another way, in a mathematical context, involves modeling by dynamic systems that involve alternative state spaces, feedback mechanisms and interactions at different scales that associate the flows of biosystems in the form of mass or energy balance with their specific properties. Therefore, the evaluation of each process cycle and its interactions, is the most adequate form of measurement to avoid a static analysis that may not be repeated in other processing steps.

Although there are numerous interference variables, the evaluation that integrates the internal factors of the industrial system, is the model of greatest application because it considers the factors directly related to the industrialization processes [2, 19, 22, 33]. The environmental performance of industrial processes involves the time of industrialization, number of operations and the data of inputs and outputs of the processes. The production stages can also be subdivided into operational and non-operational phases, if reactions occur in processes that generate environmental effects [40]. As it is not possible to predict these reactions, it is recommended to subdivide the processes present in the system, so that the evaluation is delimited for each stage, being able to analyze and identify the results of each production phase. The Fig. 4 shows the partial assessment of the productive system.

The division into productive stages (- -) demonstrates the inflows of inputs (primary resources or by-products) in the processes (➡) and the emissions generated

(↑). Therefore, the environmental behavior of each subprocess is observed, relating its performance variables, avoiding allocation (a precept that defines locations or groups of product system conditions) as recommended by ISO 14040: 2006 [23], thus avoiding consequences that a process can cause over another and the modeling of the effects is more specific, avoiding distortions in the results [12].

In this conception, the form of modeling measures partial efficiencies of the production process, often calculated as rates of inputs (resources) required per unit of product. Thus, partial efficiencies measure specific situations and allow the monitoring of processes to achieve pre-established goals. For this reason, the allocation principle must be considered to obtain efficient use of resources without going beyond the principles of sustainable development. In this conception, the focus on eco-efficiency, such as end-of-life management, used in closed cycle models, represents the best productive and environmental performance in reprocessing inputs than discarding by-products or exploring primary resources [33, 35]. Therefore for this characteristic, the closed-loop production system has aroused the interest of industries to reduce the consumption of raw materials and energy sources.

In the remanufacturing process, the reprocessing capacity is different. The specific coefficient of contribution to production is altered by the property of the by-product material being different from the primary process. This characteristic is a volatile and complex measurement parameter because each material has particular properties and the productive potential changes according to the installed technological structure. Therefore, the remanufacturing cycle is the determining factor for operations [42]. The way to assess the closed-loop process is appropriate to the allocation conditions and is assessed by the subsequent number of production cycles to assess environmental impacts [24, 25].

Thereby the rate of emissions is related to the number of times in the reprocessing cycle (n) [19]. This conception allows to consider the materials present in the closed loop operations cycle, which represent the inventory of data necessary for the evaluation of this phase of the life cycle. The life cycle inventory (LCI) is a process of quantifying energy and virgin materials required and the atmospheric, liquid and solid waste generated in the life cycle of a product, process or activity [38]. The principle of the first law of thermodynamics is also preserved by conserving the balance of energy flow in the system. In this way, it is possible to identify process points that can be improved through the collection of material flow data [7, 8]. This form of modeling is adequate to sustainable standards and generates simplification for not exploiting the resources of nature and for not generating waste disposal.

The relationship between the level of pollutant emissions and the number of closed-loop production cycles can also be assessed from the perspective of a mathematical model. The model considers the flow of inputs (E) with the outputs by the transfer function of the primary process (G) and the closed-loop in the relationship of the recovered material (R) and its recovery rate (T) and the specific emissions coefficient (C) in Eq. (1). Starting from the ideal condition of lower emission rate with the total use of the recovered material, it means, in scalar values, the lowest value of for the highest value of (Eqs. 1 and 2).

This principle is considered with the concepts of optimization of engineering processes by mathematical functions of maximization and/or minimization [20] and in the condition of the optimal production cycle associated with the generated environmental effects, as provided by ISO 14044 and ISO 14049 [24, 25]. In this sense, considering the execution of all material recovered in the remanufacturing cycle, it means, the complete processing for all R and n admitted. Considering the mathematical axiom of function minimization to identify the relationship with its assigned variable, for Eq. 1, Min (Y) for the admitted value, we will have the lowest emission value generated, applied in Eq. (1) [29]:

$$_{\text{Min}}Y = \left[\left(E(s).G^{-1} + \left[\sum\nolimits_{i=1}^{n}\left(\sum\nolimits_{k=1}^{n} R_{i,k}.C_i\right)\right]\right.\right. \tag{1}$$

How there is interrelation between the variables and this axiom presents the real condition in the best condition of productive operation. This approach considers that the performance is related to the mass and energy flow with the reprocessing cycles to evaluate the optimization of the system in relation to its operations [20]. Therefore the frequency of remanufacturing cycles is evaluated using Eq. 2 to identify the production condition in the closed loop:

$$n = Log_R\left(\frac{T \cdot R}{E} - \frac{R}{E} + T + 1\right) \tag{2}$$

The frequency of closed loop cycles will present the lowest volume of atmospheric emissions for the condition of productivity. From the perspective of eco-efficiency, this state allows the evaluation of the flows of materials in transformation related to the best use of resources in a closed cycle. This form of modeling considers that the environmental effects are associated with the devices of the productive system due to the existing correlation between the technological infrastructure and the productive and environmental performance. For other types of industrialization and products, studies must be carried out according to their specific data because the applicability factors are different [1].

The mathematical axiom of the equation enables the structured measurement of closed loop production systems and the specifications for:

- The transfer rate of the closed loop process represents the resilience of the End of Life theory (EOL) remanufacturing system [41];
- Mass balance of the flow of materials and energy involved in the process [7, 34];
- Thermodynamic conditions of the energy flow of the processes [8, 9];
- The specific coefficient of the recovered material represents the condition of ownership of this material as to its new condition of reprocessing [9];
- "n" represents the number of closed-loop, to determining factor for assessing the emissions generated [24, 25].

3 Material and Methods

To explore the context of eco-efficient manufacturing with closed-loop production, wood pellets were chosen because, they are a clean energy source and highlighted for their energy and renewable potential. Pellets are excellent environment-friendly energy products, because it has “carbon neutral” characteristic, and all CO_2 emitted in their burning is recovered throughout its life cycle, from tree growth to the reuse of forest residues in industries [32].

Regarding the amount of CO_2 emission per kWh, along its supply chain, it is about six times smaller than that of fuel oil, when compared to residential heating [3]. Due to these positive aspects, the world production of wood pellets has had strong growth in recent years. In 2010 it reached more than 18 million tons, in 2015 it was 28 million and it is estimated to reach about 40 million in the next ten years, due to the high growth of the energy market [31].

The research was carried out in a pellet industry in Brazil (Plant BR). Its production serves the national and international market, mainly in Europe. The industry has ENplus A1 certification, which ensures high quality standards in the entire pellet production chain, using reforested wood and certified by the Forest Stewardship Council (FSC).

The reference process considered the database process called “industrial wood waste pelletizing, wood pellets A1”, which considers industrial waste processing in accordance with DIN EN 14961-2 (2010) [13]. Although Brazil produces and exports pellets, the country does not yet have a specific standard on the product. Brazilian companies adapt their industrialization process to the European standards of the European Committee for Standardization (CEN) to be able to serve the foreign market with DIN and ISO quality standards, in the ENplus® standard [14].

Pellets are produced from pinus wood residues in accordance with the ENplus® standard [14]:

- Diameter: 6 ± 1 or 8 ± 1 mm
- Length: $3.15 < L < 40$ mm
- Moisture: $\leq 10\%$ w
- Ash: $\leq 0.7\%$ w (A1)
- Mechanical Durability: $\geq 98.0\%$ w
- Temperature: ≤ 40 °C
- Net Calorific Value: ≥ 4.6 kWh/kg (≥ 16.5 MJ/w)
- Bulk Density: $600 \leq BD \leq 750$ kg/m^3

The product system is delimited from “gate to gate” (LCA boundaries), considering the ICV of pellet processing operations in plant BR. The study considers the industrialization processes of the preparation (moisture), grinding, drying, pelletization, cooling and packaging, which are within the product system shown in Fig. 5 (flow chart). The patrimonial infrastructure data related to the construction processes such as the use of concrete or steel from the factory to exclusively evaluate the industrialization processes.

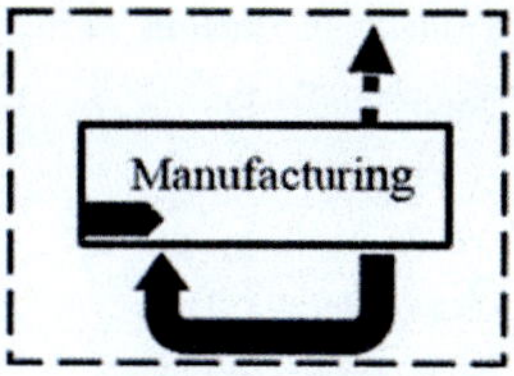

Fig. 5 Boundaries system of manufacturing

The industrialization ICV considers the resource flows of inputs, closed-loop reprocessing and outputs, following subtitles the Fig. 4. The accounting for specific material properties follows the recommendations of the German OpenLCA® application by Green Delta, which operates on the Ecoinvent® database that presents internationally recognized information [10].

The experiment was carried out in stages, at hourly data collection intervals, under regular operating conditions of the industrial plant. The functional unit (UF) of production presented average results of 3325 t/h. For the analysis, the value of 19,950 t was considered, which represents the annual production volume, counted in 6000 h of work, to present more significant results in terms of unit of analysis. In the closed-loop process, it is not necessary to have a statistical sample of process information because the representative data collection is equivalent to the limit number of remanufacturing cycles (n).

The main raw material, pinus waste, is received dry within the initial moisture specification of 8.0–16.4%. The equipment used in the pelletizing process is the Pelletizer press 45-1250, serial number 26.212, with motor power of the pelletizer 315 kW and matrix with holes diameter of 6.5 mm. Cooling is carried out gradually, with thermal exchange with the environment and use of auxiliary fans.

At the end, the pellets are packed in 1 t big bags and 15 kg small bags. The energy consumption of the weight scale is very small, therefore, the information was unrecognized in the evaluation. The Table 1 presents the data (UF) considered from the production processes:

The evaluation showed results similar to other research on environmental impacts generated in the industrialization of pellets [3, 26, 31, 32]. The total emissions generated in the primary processes are higher because they represent more than 90% of the entire production process. In relation to the energy flow for the manufacture, if we observe the emissions from the processes separated by the primary and closed-loop phases, the results show that the remanufacturing contributions generate greater environmental impacts than in the primary process (Fig. 6).

In addition to causing discharges to the environment, the energy flow of the production system interferes with the production condition, mainly in the pelletization and drying processes, which also have effects on the properties of the pellets. Atmospheric emissions are generated in the furnace of the drying machine. In this process, thermodynamic heat exchange occurs to dry the materials and reduce the moisture content. The drying gas flow goes to filters to reduce emissions, featuring the use of "end of pipe" technology, demonstrating the condition for the development of improvements to the drying process or installation of a new drying system (greenfield project).

Table 1 Production process data

Input – materials		
Variables	Data	Unity
Wood waste (pinus)	3.65	t
Water (additive)	100.32	l/t
Dry sawdust (additive)	0.33	t
Green sawdust (additive)	0.16	t
Process		
Press pressure	70.00	bar
Hydraulic pressure	168.75	bar
Matrix temperature	89.75	°C
Oil temperature	69.75	°C
Energy (pelletizing)	67.30	Kwh
Energy (cooling)	14.71	Kwh
Temperature pellet	101.35	°C
Fines	6.80	%
Output – emissions		
Steam water	0.09120	kg water
CO_2 eq. total – primary process	1,923.00795	kg CO_2 eq.
CO_2 eq. total – closed-loop process	74.98825	kg CO_2 eq.
Specific Properties		
Productivity rate (pellet/madeira)	0.9009	Kg
Net calorific value – Pinus wood	19.78	MJ/kg
Bulk density	0.64525	Kg/l
Moisture	6.28	%
Pellet Durability Index (PDI)	98.80	%

Sources: Process material data: Plant BR; Specif property of processes: (GreenDelta, 2013)

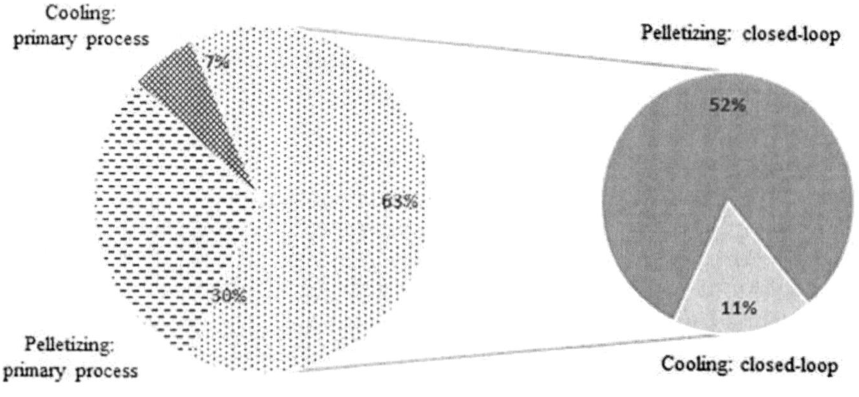

Fig. 6 Chart of CO_2 emissions

4 Discussion

The characteristic of the raw material of wood, in terms of its moisture content and homogenization of the blended, has an important influence on the pelletizing and cooling processes of the pellets. These processes require greater energy consumption and cause the greatest environmental impacts, confirming a similar result with other research, regardless of the industrial sector under analysis. The closed-loop production process improves industrial productivity in the concept of eco-efficiency because it reduces the need for natural resources while also increase the economic aspect of production.

The homogenization of the blended pick up the pressing process and reduces the friction in the matrix holes. This condition improves the pelletizing process and reduces energy consumption. The density of the product is also increased for the best performance of this process. The volume of fines that feeds the process is processed with the same characteristics as the primary process.

The evaluation demonstrated that one remanufacturing cycle, would be the ideal condition of closed-loop production from an environmental perspective, mainly due to the small volume of material recovered. This result is directly related to technical specifications established in the requirements of the European standard EN 15210-1, which refers to the general ISO 3310-2 standard on the characteristics of the mill sieve to pelletizing. The volume can also be reduced because the ENplus® standard indicates fines content of less than 1%.

5 Conclusions and Future Perspectives

This paper presents a model for measuring the number of production cycles in a closed loop to achieve the best production and environmental performance. The results demonstrate the possibility of analyzing the use of primary raw material and the supply of materials at the end of life, in order to maximize the rate of recovery of waste material. However, this form of analysis is in conflict with the standards of the international ENplus standard. This observation can be considered critical because the standards regulate different aspects related to products and processes.

In general contributions, the research highlighted the importance of reusing materials, in economic, productive and environmental aspects, increasing energy efficiency in the production system. The work also contributes to new companies seeking to develop sustainable manufacturing through closed-loop production.

Acknowledgments The authors acknowledge Fundação para a Ciência e a Tecnologia (FCT-MCTES) for its financial support via the project UIDB/00667/2020 (UNIDEMI).

References

1. Allacker, K, Mathieux, F, Pennington, D, Pant, R (2017) The search for an appropriate end-of-life formula for the purpose of the European Commission Environmental Footprint initiative. Int J Life Cycle Assess (2017) 22:1441–1458
2. Carlson, R (2009) Eco-Efficiency – The conceptual model and an operational data structure. Center of Environmental Assessment of Product and Material Systems (CPM). Chalmers University of Technology. Goteborg, Sweden
3. Cavallet, O, Slettmo, SN, Cherubini, F (2018) Energy and Environmental Aspects of Using Eucalyptus from Brazil for Energy and Transportation Services in Europe. Sustainability 2018, 10, 4068
4. Cinelli, M, Cole, SR, Kirwan, K (2014) Analysis of the potentials of multi criteria decision analysis methods to conduct sustainability assessment. Ecological Indicators, 46, 138–148. https://doi.org/10.1016/j.ecolind.2014.06.011
5. Center Information Brazil of United Nations (UNIC) (2019) Transforming Our World: The 2030 Agenda for Sustainable Development. Rio de Janeiro. https://sustainabledevelopment.un.org
6. Chertow, M, Ehrenfeld, J (2012) Organizing self-organizing systems – toward a theory of industrial symbiosis. Journal of Industrial Ecology, v.16 (1), 13–27
7. Coelli, T, Lauwers, L, Huylenbroeck, G (2007) Environmental efficiency measurement and the materials balance condition. Journal of Productivity Analysis, 28 (1–2), 3–12
8. Dakpo, KH (2016) On modeling pollution-generating technologies: a new formulation of the by-production approach. SMART – LERECO N^0 1606. Rennes, France
9. De Jong, E, Jungmeier, G (2015) Chapter 1 – Biorefinery Concepts in Comparison to Petrochemical Refineries. In: Pandey, A., Höfer, R., Taherzadeh, M., Nampoothiri, K. M. & Larroche, C. (eds.) Industrial Biorefineries & White Biotechnology. Amsterdam: Elsevier
10. Ecoinvent, Jungbluth. N, et al. (2007) Life Cycle Inventories of Bioenergy. Data v 2.0 (2007). Ecoinvent report 17, Swiss Centre for Life Cycle Inventories, Dubendorf, CJ
11. Ellen Macarthur Foundation (2013) Towards the circular economy. Vol. 1: na economic and business rationale for na accelerated transition. Disponível em: https://www.ellenmacarthurfoundation.org/publications
12. Ekvall, T, Weidema, BP (2004) System Boundaries and Input Data in Consequential Life Cycle Inventory Analysis. International Journal Life Cycle Assessment, 9, (3), 161–171
13. European Pellet Council (2014) The new European standard for wood pellets (EN 14961-2) and its implementation through certification ENplus. http://www.cebio.net/web/images/stories/PellCert_flyer.pdf
14. European Pellet Council (2015) European Biomass Association. EN plus Manual, part 3 - Pellet Quality Requirements. Bélgica, V 3.0
15. Federation of Industries São Paulo State (FIESP) (2017) Climate Change - Assessment of the effects of emission reduction targets on the Brazilian economy and industry. São Paulo: FIESP, 2017
16. Fraccascia, L, Giannoccaro, I, Albino, V. (2017) Rethinking Resilience in Industrial Symbiosis: Conceptualization and Measurements. Ecological Economics 137 (2017), 148–162
17. Folke, C, Colding, J, Berkes, F (2003) Building resilience and adaptive capacity in social-ecological systems. In: Berkes, F.; Colding, J.; Folke, C. (Ed.). Navigating social-ecological systems. Cambridge: Cambridge University Press, p. 352–473
18. Folke, C, Carpenter, SR, Walker, B, Scheffer, M, Chapin, T, Rockstrom, J (2010). Resilience thinking: integrating resilience, adaptability and transformability. Ecology and Society [online], vol. 15, no. 4. http://www.ecologyandsociety.org/vol15/iss4/art20
19. Geyer, R, Kuczenski, B, Zink, T, Henderson, A (2015) Common Misconceptions about Recycling. Journal of Industrial Ecology, 1–9
20. Grossmann, IE, Aguirre, PA, Barttfeld, M (2000) Optimal synthesis of complex distillation columns using rigorous models. Computers and Chemical Engineering 29 (2005) 1203–1215

21. Institute for Applied Economic Research (IPEA) (2018) Agenda SDG 2030 – National Sustainable Development Goals. Brasília, DF
22. Intergovernmental Panel on Climate Change (IPCC) (2014) Climate Change 2014. Energy Systems – Chapter 7. Working Group III - Mitigation of Climate Change
23. International Organization for Standardization (ISO) ISO 14040: 2006 Environmental management – Life Cycle Assessment – principles and framework. https://www.iso.org/standard/37456.html
24. International Organization for Standardization (ISO) ISO 14044: 2006 Environmental management – Life Cycle Assessment – requirements and guidelines. https://www.iso.org/standard/57110.html
25. International Organization for Standardization (ISO) ISO/TR 14049: 2012 Environmental management – Life cycle assessment – Illustrative examples on how to apply ISO 14044 to goal and scope definition and inventory analysis. https://www.iso.org/standard/57110.html
26. Jawahir, IS, Bradley, R (2016) Technological Elements of Circular Economy and the Principles of 6R-Based Closed-loop Material Flow in Sustainable Manufacturing. 13th Global Conference on Sustainable Manufacturing - Decoupling Growth from Resource Use. Procedia CIRP 40 (2016) 103 – 108
27. Johannsdottir, L (2014) Transforming the linear insurance business model to a closed-loop insurance model: a case study of Nordic non-life insurers. Journal of Cleaner Production 83 (2014), 341–355
28. Nations Research Council (NRC) (2010) Real Prospects for Energy Efficiency in the United States. National Academy of Sciences/National Academy of Engineering. https://www.nap.edu/resource/12621/energy_054844.pdf
29. Numata Junior, F, Navas, H (2019) Evaluation off the Productive and Environmental Potential of Closed-Loop Manufacturing Processes. Proceedings ICPR 2019 25th International on Conference Production Research. Manufacturing Innovation: Cyber Physical Manufacturing. Chicago, Illinois
30. Organisation for Economic Co-operation and Development (OCDE) (2013) Why New Business Models Matter for Green Growth. OCDE, Paris. http://www.oecd.org/greengrowth/oecdworkongreengrowth.htm
31. Paletto, A, Bernardi, S, Pieratti, E, Teston, F, Romagnoli, M (2018). Assessment of environmental impact of biomass power plants to increase the social acceptance of renewable energy technologies. Heliyon 5 (2019) e02070
32. Reed, D, Berman, R, Kim, J-W, Taylor, A, Harper, D, Jones, D, Knowles, C, Puettmann, ME (2012) Cradle-to-Gate Life-Cycle Inventory and Impact Assessment of Wood Fuel Pellet Manufacturing from Hardwood Flooring Residues in the Southeastern United States. Forest Prod. J. 62(4):280–288
33. Rebitzer, G, Ekvall, T, Frischknecht, R, Hunkeler, D, Norris, G, Rydberg, T, Schmidt, W-P, Suh, S, Weidema, BP, Pennington, DW (2004) Life cycle assessment Part 1: Framework, goal and scope definition, inventory analysis, and applications. Environment International 30, (5), 701–720
34. United Nations Environment Programme/Society for Environmental Toxicology and Chemistry (UNEP/SETAC) (2011). Towards a Life Cycle Sustainability Assessment – making informed choices on products. UNEP, 1–86
35. United Nations Environment Programme/Society (UNEP) (2010) Waste and climate change: Global trends and strategy framework. Osaka, Japan: International Environmental Technology Center, United Nations Environmental Program (UNEP). www.unep.org/ietc/informationresources/publications/wasteandclimatechange/tabid/79353/default.aspx.
36. United Nations Sustainable Development Goals Platform (UN/SDG) (2019) https://sustainabledevelopment.un.org
37. United Nation World Commission on Environment and Development. Report of the World Commission on Environment and Development (WCED) (1987) Our Common Future – Brundtland Report. Oxford University Press, 1987.

38. US Environmental Protection Agency (US EPA) (2006) Life cycle assessment: Principles and practice. US EPA, Washington, D.C. http://www.epa.gov/nrmrl/std/lca/lca.html
39. Scheffer, M, Carpenter, SR, Dakos, V, Van Nes, EH (2015). Generic Indicators of Ecological Resilience. Annual Review of Ecology Evolution and Systematics, 46, 145–167
40. Shokravi, S, Smith, AJR, Burvill, CR, Maheswararajah, S (2012) Industrial environmental performance evaluation – Comparison and sensitivity analysis. iEMSs 2012 – Managing Resources of a Limited Planet: Proceedings of the 6th Biennal Meeting of the International Environmental Modeling and Software Society. Leiping, Germany, 1037–1044
41. Yazan, DM, Romano, VA, Albino, V (2016) The design of industrial symbiosis: an input-output approach. Journal of Cleaner Production 129, 537–547
42. Zink, T, Geyer, R, Startz, D (2015) A Market-based framework for quantifying displaced production from recycling or reuse. Journal of Industrial Ecology, 20(4), 719–729

Logistics for Disaster Waste Management: A Case Study of the 2019 Oil Spill in Brazil

Luiz Fernando Netto, Alex Igor Sanghikian, Hugo Tsugunobu Yoshida Yoshizaki, and Irineu de Brito Junior

Abstract During the response and recovery phases of a disaster, an important issue is waste management, which includes generation, transportation, transfer, processing, recycling, and reuse of debris. In this paper, we analyzed the logistical process related to the management of 5400 metric tons of waste from the 2019 oil spill in Brazil, considering environmental impacts and cleaning efforts. The analyses show the quantities handled by the state governments responsible for the operation and the main destination of the residues, which was the cement industry.

Keywords Logistics · Disaster waste management · Brazilian oil spill

1 Introduction

A disaster can be defined as an event that results in destruction, ecological or human impacts, suffering, and health deterioration in proportion to the need for an extraordinary response from outside the community zone [1]. Regardless of the magnitude or repercussion of a crisis, one of the most important stages in the disaster

L. F. Netto (✉)
Chico Mendes Institute for Biodiversity Conservation, Pirassununga, Brazil
e-mail: luiz.netto@icmbio.gov.br

A. I. Sanghikian
Department of Production Engineering, Sao Paulo State University, Sao Paulo, Brazil

H. T. Y. Yoshizaki
Department of Production Engineering, Sao Paulo State University, Sao Paulo, Brazil

Graduate Program in Logistics Systems Engineering, Sao Paulo University, Sao Paulo, Brazil
e-mail: hugo@usp.br

I. de Brito Junior
Graduate Program in Logistics Systems Engineering, Sao Paulo University, Sao Paulo, Brazil

Environmental Engineering, São Paulo State University, Sao Jose dos Campos, Brazil
e-mail: irineu.brito@unesp.br

J. Vargas Florez et al. (eds.), *Production and Operations Management*, Springer Proceedings in Mathematics & Statistics 391,
https://doi.org/10.1007/978-3-031-06862-1_20

management process is the recovery phase [2]. This stage is defined as the process of restructuring the community or area affected after the disaster until the situation returns to normal [3].

Several studies involving a literature review were conducted on the topic of post-disaster waste management [2]. Emergency response can be divided into two stages: pre-event and post-event [4]. Pre-event actions consist of predicting and analyzing potential hazards and necessary developments for action plans aimed at mitigating the risks. The post-event response phase consists of actions while the disaster is in progress. At this stage, the challenge is to locate, allocate, coordinate, and manage available resources. In this management phase, an effective response plan must integrate both stages from the perspective of a single objective, as separating the stages can lead to inferior solutions [4].

An important issue in the disaster response and recovery process is waste management, associated with waste generation, transportation, processing, recycling, reuse, and disposal [2], including the collection and transfer of material to temporary storage locations and then to its final destination [5]. In some cases, the federal government reimburses response & recovery operational costs of State and local governments and nonprofit organizations, as performed by the US Federal Emergency Management Agency – FEMA [6].

The oil spill that hit northeastern Brazil in 2019 is considered the largest oil spill in the country [7] or even in a coastal tropical region [8]. The first news of the emergence of oil pellets dates to August 30, 2019, on the coast of Paraíba, followed by Sergipe and Pernambuco [9], and the contingency actions extended until March 19, 2020 [10]. Finding the oil was quite difficult due to the characteristics of the oil, which remained subsurface in the water column. The containment actions before its arrival on the Brazilian coast were inefficient, and the focus of the operation was on collecting the waste that reached the coast [9].

It is estimated that the oil reached more than 3000 km off the Brazilian coast [9], and it is estimated that approximately 870,000 people were affected by the oil spill (mainly employed in artisanal fishing and local tourism) [11]. The oil affected approximately 55 protected areas [7, 8] and caused exposure to oil-related disturbances in several marines and coastal habitats and at least 27 threatened species [11]. Despite the aforementioned social and environmental impacts, there was no record of fatal victims directly related to the oil spill. The origin and location of the spill remain unidentified to date [10].

Environmental issues of the contingency were the responsibility of the Brazilian Institute for the Environment and Natural Resources (Portuguese acronym: IBAMA). IBAMA is a federal agency responsible for implementing environmental policies, and it is a permanent member of the Monitoring and Evaluation Group (MEG), which is part of the crisis management framework established in the National Contingency Plan for Incidents of Oil Pollution in Waters under National Jurisdiction (NCP) [12]. Thus, the agency was directly responsible for waste management throughout all stages of the crisis, supervising and advising from the collection of waste to its final destination.

Considering the described scenario, this article contributes to the knowledge of waste management processes in a disaster, with a focus on logistics during and after a crisis event involving waste management. It is intended to answer the following research question: How was the management of the waste generated in the contingency actions of the oil spill that hit the Brazilian coast in 2019, from its collection in the environment to its destination?

The main objective is to present logistical difficulties and solutions and perform an analysis on the management of waste generated in contingency actions concerning the management process of logistics during and after the crisis. To this end, a case study was conducted involving the occurrence of the oil spill that hit the coast of northeastern and part of southeastern Brazil, which began in September 2019.

2 Material and Method

The methodology used was the case study, according to the classification adopted by Nakano [13] for studies in production engineering. It consisted of the analysis of documents generated by IBAMA related to the management of the oil spill crisis off the Brazilian coast in 2019, focusing on data related to the management of waste generated during and after contingency actions.

To carry out the study, access to IBAMA documents related to the oil spill disaster in the northeast in 2019/2020 had been formally requested. Among the available documents, the main source of information on solid waste management consisted of the "ICS-209 Incident Status Report Formulary". The documents contained centralizing data provided by various sources and agencies. Emphasis was placed on information related to waste generated and destined by contingency actions, seeking the information that provides relevant data.

Data referring to the following information were selected: the 1-Total amount of waste collected, by state and offshore, in metric tons; 2-Affected locations reported, 3-Daily number of personnel-related to federal agencies involved in the action (including the volunteers they manage), 4-Transport and final destination of the waste managed. Data from the period between September 14, 2019, and March 19, 2020, were considered.

The values related to location consist of extensions equivalent to 1.0 km of the coast or fraction of this length, which had been affected by oil, in any degree of magnitude. In this way, a beach, municipality, or region can include more than one location.

It is important to emphasize that the values include not only oil but also contaminated materials and waste, such as personal protective equipment (PPE), tarpaulins and sand, and all contaminated material that required proper disposal.

Only data related to federal agencies and institutions were considered. The values include all human resources related to federal agencies, such as analysts, agents, and technicians, but also support and support personnel, drivers, pilots, and crews of vessels and aircraft and drone pilots.

3 Results and Discussion

Federal agencies involved in crisis management and provision of equipment and personnel include IBAMA, Chico Mendes Institute for Biodiversity Conservation - ICMBio (the Federal Agency responsible for managing protected areas), Brazilian Navy, Brazilian Army, National Agency of Petroleum, and National Civil Defense, and two consultants from ITOPF - International Tanker Owners Pollution Federation Limited. The number includes all human resources related to federal agencies, such as analysts, agents, technicians, support personnel, drivers, pilots, the crew of vessels and aircraft, and drone pilots.

IBAMA, as a member of the MEG and responsible for environmental issues in the management of the crisis, coordinated the collection actions, its temporary storage, and articulated the transport and final disposal of waste, in addition to advising the coordinators or local leaders established to monitor the incident. In the waste collection stage, MEG was also responsible for mobilizing resources and distribution of equipment such as drums, big bags, personal protection equipment, tarpaulins to cover the ground and make temporary coverings, among other materials necessary for the proper storage of the waste.

Local agents performed the monitoring of the temporary storage through inspections of storage structures, mainly IBAMA or Civil Defense, whose registration was carried out via a form on the JotForm platform, which the inspector fed with information such as location, the form of packaging of waste, the local structure (floor, roofing, insulation, etc.), deadline for transport, destination, and photographic record. IBAMA compiled this information to compose the SCI-209 Forms. The frequency of issuance of these forms was daily, and its regularity began to decrease at the end of the response management stage in mid-February 2020, as the situation was controlled and stable.

3.1 Impact Magnitude

Until November 2019, all states from the Brazilian northeast region were affected (from north to south: most of the coast of Maranhão, Piauí, Ceará, Rio Grande do Norte, Paraíba, Pernambuco, Alagoas, Sergipe, and Bahia), as well as Espírito Santo and the north of Rio de Janeiro in the southeast region, totaling 11 Brazilian states affected in different magnitudes. By January 29, 2020, 130 municipalities were reached, totaling 1009 localities.

However, even though oil fragments were identified in the State of Rio de Janeiro and the state was officially considered affected by the disaster, it reported the

Table 1 Waste managed (metric tons - t) by affected state (ordered by geographic position, from North to South)

State	Coast (km)	Waste (t)	Waste %	Localities* affected (qtt)	Waste/ Coast (t/km)	Waste/ Location* (t)	Location*/ Coast (%)
MA	640	13.69	0.25	47	0.02	0.29	7.34
PI	66	10.46	0.19	21	0.16	0.50	31.82
CE	573	39.76	0.74	51	0.07	0.78	8.90
RN	410	35.18	0.65	80	0.09	0.44	19.51
PB	117	0.85	0.02	21	0.01	0.04	17.95
PE	187	1676.26	31.16	55	8.96	30.48	29.41
AL	229	2564.58	47.67	128	11.20	20.04	55.90
SE	163	569.35	10.58	105	3.49	5.42	64.42
BA	932	459.49	8.54	376	0.49	1.22	40.34
ES	392	6.26	0.12	123	0.02	0.05	31.38
RJ	636	0	0.00	2	0.00	0.00	0.31
Total	4345	5379.76	100.00	1009	1.24	5.33	23.22

Coastline length data (km) for each state, Localities affected, Residues (metric tons.) by Coastline (km), Residues (metric tons.) by Locality and estimated percentage of the affected coast represented by the Locality reached value (km) by Coastline (km)
*The "Locality or Location" is equivalent to 1.0 km or fraction of coastline hit by oil
Brazilian States: *MA* Maranhão, *PI* Piauí, *CE* Ceará, *RN* Rio Grande do Norte, *PB* Paraíba, *PE* Pernambuco, *AL* Alagoas, *SE* Sergipe, *BA* Bahia, *ES* Espírito Santo, *RJ* Rio de Janeiro

collection of only 320 grams of oily material by the end of the incident; therefore, the state was excluded from the analyses.

Table 1 presents the data compiled from the ICS-209 forms, broken down by the state reached, as well as the coastline values for each of them, in addition to presenting relationships between the values found.

The volume shows the magnitude of the environmental accident, as the waste was dispersed across 11 states along the Brazilian coast. In total, 5379.76 tons were collected and destined. The beginning of November 2019 was the critical period for the removal of waste, and there was a sharp increase in the number of materials handled. Figure 1 illustrates the total waste, in tons, that was managed by each state.

The state that had to manage the largest amount of waste was Alagoas, with 2564.58 tons, representing 47.67% of the total managed waste, followed by Pernambuco, with 1676.26 tons (31.16%), Sergipe with 569.35 tons (10.58%) and Bahia with 459.49 tons (8.54%) (Table 1 and Fig. 1).

However, if we consider the values of affected locations, the state of Bahia is now in the first place, with 376 locations touched by oil (37% of registered locations), with Alagoas in second place, with 128 locations (12.69%), followed and tied by Espírito Santo, with 123 locations affected (12.19%) and only in the fourth place was the state of Sergipe, with 105 (10.41%) (Table 1). These values can provide an estimate of the dispersion of oil along the coast, and thus, considering that the greater number of locations affected by oil demands greater effort to manage its collection

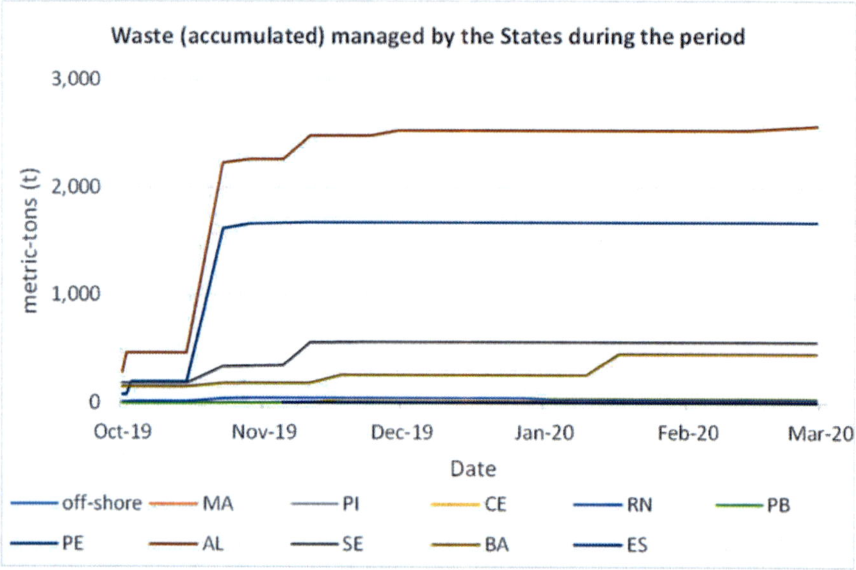

Fig. 1 Residuals (metric tons (t)) managed by the state during contingency

and destination, considering the distribution of PPE, displacement of personnel, transport to temporary warehouses and the final destination.

To estimate the percentage of the affected coast, the locality values related to the coastline were used, giving an estimate of the reached coast. Initially, the data indicate that 23.22% of the coast was touched by oil to some degree of magnitude (Table 1). However, it must be considered that several affected locations were unregistered, either due to the absence of human presence or difficulty of access, which has led to an underestimation of the number of locations, especially in states with an extensive coastline, such as Bahia and Maranhão.

When each state was individually observed, these data show that Sergipe was the one with the highest percentage of its coast affected (64.42%), followed by Alagoas (55.9%), Bahia (40.34%), and tied in fourth place, the states of Piauí (31.82%), Espírito Santo (31.38%) and Pernambuco (29.41%) (Table 1). However, it is possible to observe that even though Sergipe was the state with the highest percentage of coastline reached, it was not the state that suffered from the largest volume of oil (inferred by the amount of waste), in this case, the state of Alagoas.

It is important to mention that the state of Piauí has only 66 km of coastline, which has influenced the high proportion of the affected coastline (Location/ Coast = 31.82%) (Table 1). Considering that it is geographically located between the states of Maranhão (Locality/Coast = 7.34%) and Ceará (Locality/Coast = 8.9%), it appears that the value for Piauí differs not only from the average between the last two but also does not follow the trend of increasing impacted coastline, which has its peak in Sergipe (Table 1). One possible explanation is that, as the coastline is smaller, the state could monitor more efficiently and, in its entirety, not by sampling or waiting for notifications from the population, which would indicate that the monitoring methodology should be revised; however, this is not conclusive.

The relationship between the amount of waste managed by each state and its coastline can provide an idea, not only of the effort directed toward waste management but also of the impact on the population and the marine/coastal environment and resources. In this parameter, Alagoas appears again in the first place, with 11.2 tons of waste managed per kilometer of the coast, followed by Pernambuco (8.96 t/km) and then Sergipe (3.49 t/km) (Table 1).

However, if we estimate the average of the waste generated by the location reached, relating values for waste (metric tons.) by Location, giving an estimate of the amount of oil per location that each state had to deal with, the positions between the first two places are inverted, with Pernambuco in the first place, with an average of 30.40 t/location of waste managed by each location affected, followed by Alagoas (20.04 t/location), Sergipe (5.42 t/location) and Bahia (1.22 t/location) (Table 1).

3.2 *Contingency Management Effort*

To illustrate the general effort involved in the response work during contingency, attention was paid to the daily values of the number of federal agents involved in the response work. These data include employees from IBAMA, ICMBio (including volunteer programs), Navy, Army, ANP, ITOPF, and National Civil Defense. This quantity provides only an idea of the effort involved in the response action and does not include the movement of personnel related to state and municipal employees, NGOs, volunteers, civil society organizations, and collective efforts organized by local communities.

On November 22, 2019, the day on which 11,257 federal agents were deployed can be considered the heyday of the crisis. The drastic reduction in personnel mobilization occurred from January 15, 2020, indicating the beginning of the demobilization phase. On February 1, approximately a thousand agents were working, largely related to the monitoring of demobilization and waste management.

It draws attention that in the two periods in which there was a decrease in personnel, both in December 2019, the second period was more pronounced and occurred during the second half of December 2019. A possible explanation for these two reductions in staff numbers may be related to the personal change due to the long period of performance at the event, boosted by the second reduction with issues related to the end-of-the-year festivities. The issues of changing personnel in long-term emergencies are important within crisis management and should be considered in future analysis and worked on preventively in the preparation phase for possible future occurrences.

3.3 Responsibility for Waste Management

According to Brazilian environmental law, the polluter must answer for the damage caused to the environment, bearing the burden of its recovery [14]. However, given the nonidentification of those responsible for the oil spill, the government was responsible for providing all contingency actions, including the management and disposal of waste generated.

The MEG guided and supervised the coordinators responsible for locally managing those responsible for the collection, temporary storage, transport, and final destination of waste. The MEG defined that the responsibility for managing the waste generated in response to emergencies was the responsibility of the municipal authorities, which should act locally in the transport and final disposal of waste, as recommended by the National Solid Waste Policy, according to which it is the responsibility of the municipalities the integrated management of solid waste generated in their territories, with the support of the states [15]. Thus, the performance of MEG was in the sense of this guidance, play as an articulator and organizer of information on waste.

However, according to the Brazilian Federal Constitution (Article 23, VI) [16], the protection of the environment and monitoring of pollution is common competence for the entities of the federation. Thus, the effort to respond to the oil spill was a shared responsibility between the federation, the states, and the municipalities. Thus, in places where local authorities were unable to act, MEG coordinated efforts to optimize resources, concluding the proper disposal of waste.

Some states carried out the transport and disposal of waste by themselves, often under the responsibility of the State Environmental Agency (SEA), or even the city halls. In these cases, the MEG monitored the periodic information to provide integrated management of efforts. Where it was impossible to carry out waste management by the state itself, the MEG coordinated the transport and disposal of waste, carried out by Petrobras, through the activation of the PNC, which carried out adequate transport of waste from the temporary storage locations until the destination.

For the contingency actions, resources were available according to the capacity of each institution in a staggered manner, as established by a flowchart of logistical needs decisions. Thus, when an institution did not have the resources to act, it requested logistical support from the higher level.

In this way, the provision of resources was established, passing to the next instance according to the logistical incapacity of each actor or administrative sphere: (1) Claiming body; (2) Municipal Agency; (3) State Agency; (4) MEG, which directed the demand to the (5) Executive or Support Committee of the NCP, and finally, if there was no logistical resource available, the demand was forwarded to (6) Petrobras, through activation provided for in the NCP, which finally managed the availability of the logistical resource for the temporary packaging, transport and final destination of waste, which must be reimbursed by the Federal Government at a later time.

It is worth highlighting the situation in which the state of Alagoas declared, on November 10, 2019, that it was unable to manage the collection and disposal of waste, with these assets being under the competence of the Federal Government from then on. It is worth noting that Alagoas was one of the states most affected by the disaster, which had a considerable impact and likely reached the limits of its resources.

3.4 Waste Destination

Seeking an alternative for the final destination of the waste generated, MEG led the articulation of entities related to cement industries, such as the National Union of the Cement Industry and the Brazilian Cement Association, in the search for cement factories that were interested in receiving waste. These companies use the residues for co-processing in their furnaces, burning to generate energy for use by the industry itself. Figure 2 illustrates the final destinations and temporary storage sites in the affected region.

That was the main strategy in the disposal of the oil waste generated on the contingency of the disaster, with 68.43% of the waste being the destination. Approximately 31.2% of the waste was sent to landfills, and almost 4% of the destination corresponded to waste treatment centers (Table 2).

Due to the small amount of oil collected in the state of Rio de Janeiro (320 g), the destination of the waste was "sent for analysis", without further details. Waste collected offshore was disposed of according to the state of origin.

Petrobras was responsible for the temporary storage and transport to the final destination of 59.15% of the managed waste (3179.57 metric tons), and the affected states were responsible for the transport of 40.85% (2196.31 metric tons) (Table 2). which demonstrates that its participation and mobilization via NCP was fundamental in contingency actions and waste disposal.

4 Conclusion

Given the nonidentification of the source and the magnitude of the incident, the public administration was responsible for the contingency actions and management of waste generated, most of the time, providing resources to enable the actions of waste destinations.

The role of the federal government was decisive as a central coordination and integrated management of available resources, with the federal agencies also playing a decisive role in waste management, dealing with the monitoring of the temporary storage locations, articulating with the cement industries the destination of waste and often enabling the transport of contaminated material.

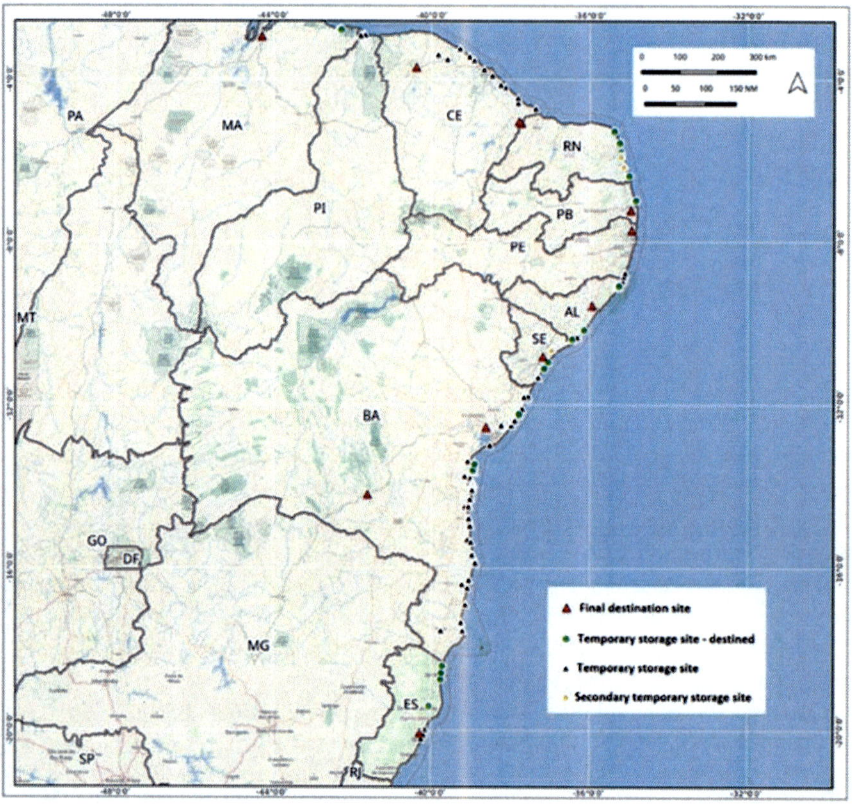

Fig. 2 Final destination (red triangles) and temporary storage sites (green dots, black and yellow triangles) for waste. (Source: Map modified from IBAMA [9])

Despite the success of the operation to dispose of the immense amount of waste generated by the contingency actions, it is evident that the assessment and prior preparation for waste management are essential to mitigate the impacts of the event. In the case study in question, the prior articulation of partnerships with cement representatives, with knowledge of their logistics chain and prior establishment of protocols and procedures, would have streamlined the crisis management and subsequent recovery and reconstruction of the affected society, in particular from the perspective of socio-environmental aspects, as well as the pressure on the demand for human resources to manage the contingency, leaving availability to others.

The environmental accident showed that waste management requires a well-structured, pre-established logistics chain with specific actors involved in each phase and that incorporates issues related to environmental management in all stages of the process. In this way, through the activation of the National Contingency Plan, Petrobras also played an important role by properly transporting waste from temporary storage locations to destinations in cement industries.

Table 2 Destination and quantity of collected waste, broken down by states, responsible for the temporary storage and transport and places of destination

State	Total (t)	Temporary storage and transport	Destination
AL	2564.58	Alto Jericó waste management center, Petrobras, Carmópolis, SE	Votorantim cement, Laranjeiras, SE
BA	459.49	State: Environmental agency and the private company contracted	InterCement
CE	39.76	State: Private company contracted	Apodi cement plant. Quixeré, CE.
ES	6.26	State: Private company contracted	Industrial waste management center
MA	13.69	State: Environmental agency and the private company contracted	Titara waste management center. Rosário, MA
PB	0.85 t	State	Class I landfill. João Pessoa-PB
PE	1676.26	State: Private company contracted	Licensed landfill, Igarassu-PE
PI	10.46	Petrobras	Votorantim cement. Sobral-CE
RN	35.18	Petrobras	Mizu cement, Baraúna-RN
SE	569.35	Waste Management Center of Alto Jericho, Petrobras. Carmópolis-SE	Votorantim cement. Laranjeiras, SE
Total	5379.76		

States, *MA* Maranhão, *PI* Piauí, *CE* Ceará, *RN* Rio Grande do Norte, *PB* Paraíba, *PE* Pernambuco, *AL* Alagoas, *SE* Sergipe, *BA* Bahia, *ES* Espírito Santo, *RJ* Rio de Janeiro

Thus, this article contributed to the registration of the best waste management practices, involving oil spills, detailing the stages and actors of the supply chain involved in the humanitarian logistics process to minimize environmental impacts. As limitations, the article did not go into detail about the social and environmental impacts of the oil spill and waste management, an important topic to be explored in future research, as well as more articles dealing with case studies for a broader spectrum of analysis of this type of phenomenon.

Acknowledgments The authors acknowledge the support of Coordination for the Improvement of Higher Education Personnel (CAPES) [88887.387760/2019-00]. we also thank Intituto Brasileiro do Meio Ambiente e dos Recursos Naturais Renováveis - IBAMA for allowing access to documents, as well as Instituto Chico Mendes de Conservação da Briodiverdidade – ICMBio for supporting this research.

References

1. World Health Organization. Risk reduction and emergency preparedness: WHO six-year strategy for the health sector and community capacity development, 2007. http://www.who.

int/hac/techguidance/preparedness)/emergency_preparedness_eng.pdf. Accessed: Nov 11 2019.
2. Boonme, M., Arimura, M., Asada, T., Facility Location Optimization Model for Emergency Humanitarian Logistics, Int. J. Disaster Risk Reduction. https://doi.org/10.1016/j.ijdrr.2017.01.017. 485–498, 2017.
3. Leiras, A., de Brito Jr, I., Peres, E. Q., Bertazzo, T. R., Yoshizaki, H. T. Y. Literature review of humanitarian logistics research: trends and challenges. J. Humanit. Logist. Supply Chain Manag. 4, 95–130 (2014). doi: https://doi.org/10.1108/JHLSCM-04-2012-0008.
4. Tufekci, S., Wallace, W.A. The Emerging Area of Emergency Management and Engineering. IEEE Transactions on Engineering Management 45(2), 103-105 (1998)
5. Sharif, M.S., Pishvaee, A, Aliahmadi, A., Jabbarzadeh. A bi-level programming approach to joint network design and pricing problem in the municipal solid waste management system: a case study, Resour. Conserv. Recycl. 131, 17–40, 2018.
6. FEMA. 'Public assistance debris management guide'. Federal Emergency Management Agency. http://www.fema.gov/pdf/government/grant/pa/demagde.pdf. Accessed Dec 25, 2017.
7. Ladle, R.J., Malhado, A.C., Campos-Silva, J.V., Pinheiro, B.R., Brazil's mystery oil spill: an ongoing social disaster. Nature 578 (37) (2020). https://doi.org/10.1038/d41586-020-00242-x.
8. Soares, M.O., Teixeira, C.E., Bezerra, L.E., Rossi, S., Tavares, T., Cavalcante, R.M., Brazil oil spill response: time for coordination. Science 367 (6474), 155–156 (2020). doi: https://doi.org/10.1126/science.aaz9993.
9. IBAMA. Source: Manchas de óleo no Nordeste. www.ibama.gov.br/manchasdeoleo. Accessed: Aug 24, 2020.
10. Zacharias, D. C., Gama, C. M., Fornaro, A. Mysterious oil spill on Brazilian coast: Analysis and estimates. Marine Pollution Bulletin. Volume 165, 112125, April (2021).
11. Magris, R. A., Giarrizzo, T. Mysterious oil spill in the Atlantic Ocean threatens marine biodiversity and local people in Brazil. Marine Pollution Bulletin 153 110961 (2020).
12. Brazil. 2013. Federal Decree No. 8,127. National Contingency Plan for Incidents of Oil Pollution in Waters under National Jurisdiction. October 23 (2013).
13. Nakano, D. Capítulo 4 - Métodos de Pesquisa Adotados na Engenharia de Produção e Gestão de Operações. In: CAUCHICK, P. M. (org.) Metodologia de Pesquisa em Engenharia de Produção e Gestão de Operações, Rio de Janeiro: Campus (2009).
14. Brazil, 1981. Federal Law No. 6,938. National Environmental Policy. September 2 (1981).
15. Brazil. 2010. Federal Law No. 12,305. National Solid Waste Policy. August 3, (2010).
16. Brazil. Constitution. Constituição da República Federativa do Brasil. Brasília, DF: Senado Federal: Centro Gráfico (1988).

Process Optimization in a Peruvian Cheese Microenterprise Through the Synergy of Lean Manufacturing and Ergonomic Tools

Tania Sthefany Gamboa Rojas, Patricia Gianella Sánchez Huallpa, Yosy Staicy Trejo Cacha, and Shakira Malionof Mamani Bonifacio

Abstract Small and medium-sized companies have been affected by the economic contraction generated by covid-19, so they must seek greater competitiveness in their internal operations. That is why this research articulates lean manufacturing tools and ergonomics in the performance of a Peruvian cheese company.

Keywords Lean manufacturing · Value stream mapping · Ergonomic factors

1 Introduction

In 2019, the participation of MIPYMES (micro, small and medium-sized enterprises) in Peru reached 99.6%, this being the economic segment that generates the highest income in the country. Among these MSMEs, companies in the manufacturing sector dedicated to the production of food and beverages have a participation of 16.4% [1]. By 2020, the COVID 19 pandemic makes it difficult to import logistics for products from other countries, including the cheese industry. In the absence of this product, the national cheese production in Peru has increased. Thanks to the increase in demand, the levels of cheese sales at the national level increased by 25% at the end of 2020 and according to the director of MINAGRI (Ministry of Agriculture) it is expected that the commercialization of this product will continue to grow at 8% by the end of 2021 [2]. However, an increase in sales also requires producers to increase the manufacturing capacity of the plant, buy new production equipment, hire more labor and even acquire more raw materials to produce the necessary volume of cheese in the shortest time possible [3]. An alternative to this problem is the application of Lean Manufacturing and ergonomics in order to optimize the production process of the cheese factories in time and costs because it would able to

T. S. Gamboa Rojas (✉) · P. G. Sánchez Huallpa · Y. S. Trejo Cacha · S. M. Mamani Bonifacio
Pontificia Universidad Católica del Perú, San Miguel, Peru
e-mail: tania.gamboa@pucp.edu.pe; patricia.sanchez@pucp.edu.pe; yosy.trejo@pucp.edu.pe; shakira.mamani@pucp.edu.pe

J. Vargas Florez et al. (eds.), *Production and Operations Management*, Springer Proceedings in Mathematics & Statistics 391,
https://doi.org/10.1007/978-3-031-06862-1_21

reduce the number of wastes, to standardize and organize the work areas and guarantee the well-being of people through quality and sustainable standards of working training for the pandemic [4]. In this research we will study the case of a manufacturing company located in Peru that offers a wide group of products made with milk: cheese, yogurt and butter; whose main product is fresh cheese, due to its quality, flavor and demand. The company uses batch production, has its own plant, recurring suppliers and indirect clients such as institutional concessionaires.

2 Literature Review

The participation of MSMEs (micro, small and medium-sized companies) generates high annual sales figures that are constantly growing. Likewise, it represents a high percentage of the participation of the EAP (47.7%). However, most of them, despite their great potential, still operate in scenarios of low competitiveness that limit their growth. Faced with the commemoration of Peru's bicentennial, MSMEs face different challenges, mainly to reactivate the economy, for which they must improve their productivity and thus achieve sustained growth. To reactivate the economy, it's important to improve the productivity of the MSMEs in order to achieve sustained growth.

3 Methods

3.1 Lean Manufacturing

This tool is based on the Lean philosophy, which was first developed in the automotive industry and is currently being developed in other sectors [5] such as the healthcare sector where organizations have tried to incorporate Lean principles [6]. This philosophy is focused on achieving a manufacturing system free of waste and reducing activities that do not add value, such as unnecessary transportation, excessive production, wasting time, among others [7].

3.2 Ergonomics Factors

Ergonomics is a science that studies the needs, capacities and abilities of human beings [8]. With the intention of increasing productivity, in a continuous work environment, the best experience for the worker comes from improving the design [9]. The design is applicative since it shows the characteristics of a productive operation and the facilities and unnecessary spaces in the work area appear visually [10]. It is a qualitative method to indicate the importance of the location of the areas

[11]. On the other hand, the effectiveness of the regulations develops an experience approach in good labor practices in the operators and strengthens hygiene and health at work, avoiding worker fatigue and exhaustion to promote human productivity [12].

4 Case Study

The microenterprise has been in charge of producing various dairy products since 2010. Among the variety of its products, it has a higher demand for producing fresh cheese due to the added value for price, quality and the unique flavor that it delivers to the customer. The plant massively offers fresh cheese to concessionaires, institutions and universities, but they have not yet reached a natural consumer, which they would like to enter, but they do not know how to start, since they have the uncertainty of competing with cheese industries already established in the Peruvian market with 50% participation. Which have been made to the plant to determine how the current situation is in order to propose and implement industrial improvements in its processes.

4.1 Current Situation

The processes to produce fresh cheese include the following sequence: reception of the main material, pasteurizing, standardizing, curdling, molding, pressing, straining, packing, weighing, and storing. The current status of the process to be optimized (See Fig. 1).

The current process that this research aims to help (Table 1).

The current design is shown in Fig. 2 with the order of operations. There is evidence of many interactions between operations due to the inefficient layout of workstations. On the other hand, there are greater distances traveled between the following operations: standardized, reception and pasteurized to curd. In summary the Distance walked was 34.7 m and time Invested was 1.875 h (See Fig. 2).

In the graph of the mass balance, it is observed that there are losses such as curdling and whey, the amount reduced represents approximately 220 cheeses per month; therefore, the yield of the process is 90% (Fig. 3).

In the following diagram, the main activities that coincide in the value of the product can be identified, as well as the main entities involved so that the production of cheese can be carried out, which are the suppliers and customers. Among the

Fig. 1 Flow diagram process

Table 1 Current process

DOP	Susceptible Process	Description of current situation	Illustration of current situation
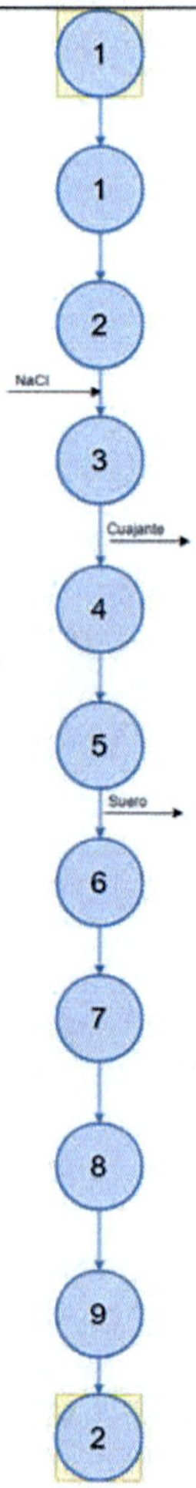	Curding (3)	There is muscle fatigue when the operator bends over to cut the curd in the tank. Inappropriate postures and movements, bent posture and standing for hours, improper handling of hot curd results in the operators involved burning their hands and arms.	
	Molding (4) Pressing (5)	There are many molded cheeses on the molding table because it is not enough for its capacity. Molding containers are diverse and of non-standard size for all production. The pressing is manual because the operators must put another container and leave it resting on the receiving container.	
	Packing (7) Weighting (8)	After cooling, packing and weighing. Weighing is not standardized due to the different molds used, so they have to be weighed one by one. This operation must be repeated until the expected weight is reached. The required weights range from 1 kg to 3.5 kg, depending on the packaging.	
	Stored (2)	There is no access between chamber 1 and chamber 2, forcing the operator to walk a long distance rather than go straight. If there is access between cameras, operators will reduce their operating time.	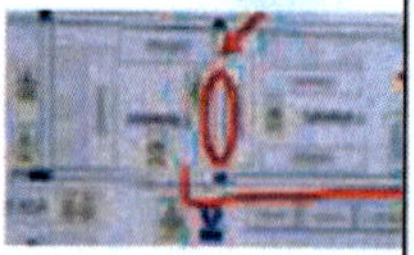

activities, one of those that achieves longer cycle times is molding, since it is done manually and with molds that are not identical, which means that the time in the cheese mold can be increased. Another activity is packaging, due to the imperfect

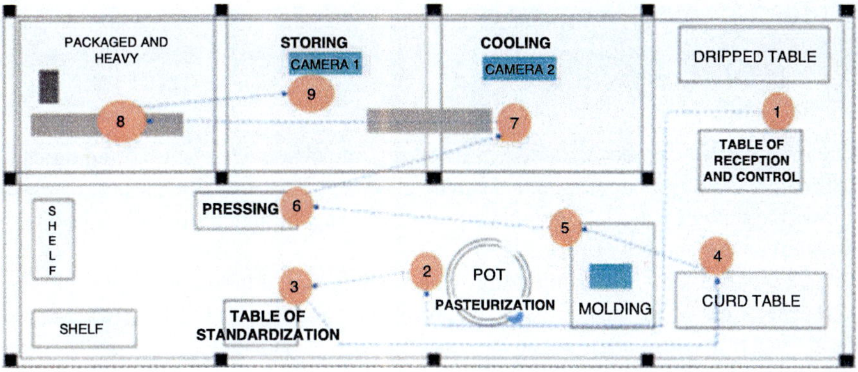

Fig. 2 Current situation layout

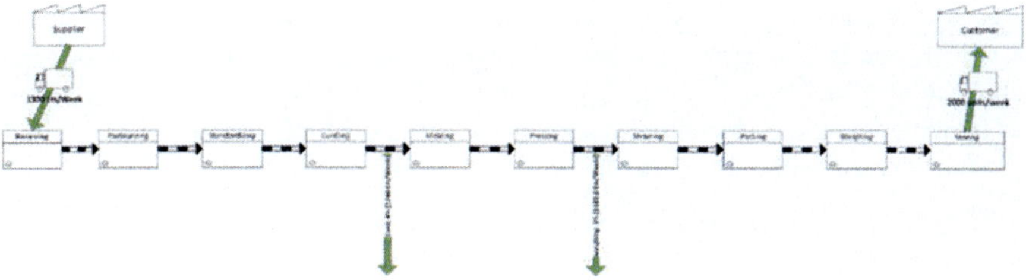

Fig. 3 Mass balance

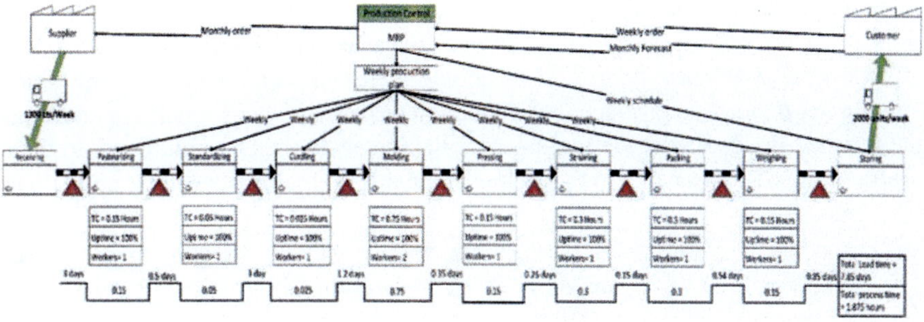

Fig. 4 Value stream mapping

shape of the cheese that makes the task of adapting them to the available packages not standardized and requires more time and packaging materials. Finally, weighing is another relevant activity with respect to cycle times because the molds do not reach similar weights and have to be separated by the closest similarity, which makes this activity tedious when it should be easy to perform (See Fig. 4).

5 Improvement Proporsals

5.1 *Lean Manufacturing Tools*

The improvement proposal started from asking ourselves. Are resources being used optimally? Is there an operation that can be grouped? Therefore, the proposed design allows the flow of materials and linear processes to be without reprocessing and with less movement and travel (See Fig. 5).

The proposal arises from the improvements mentioned in the proposed situation to reduce or eliminate the causes of the problems in the current situation. On the one hand, they propose to standardize the molding processes since there are unnecessary and overexposed activities on the part of the operator. In other words, the operator must use his criteria to shape the fresh cheeses, and this generates inconveniences due to the defective appearance of the fresh cheese. Therefore, the Distance walked was 23.5 m and it represents an improvement of 47.66% and the time Invested was 1.375 h and it also represents an improvement of 36.36%. In conclusion, the application of the layout of Fig. 3 will reduce problems of current processes such as molding and the overstrain of operators so that in the future it will generate good productivity indicators and generate added value for the company.

5.2 *Ergonomics Applications*

Some problems can be identified, they are related to ergonomics and the lack of standardization of processes, other problems that affect not only production, but also affect operators because their level of fatigue increases and, therefore, many times they do not manage to produce the production requested by customers. In addition, the fact that the operators are multi-functional increases the ergonomic risk to their future health. It may be due to the lack of standardized molds since a big problem is

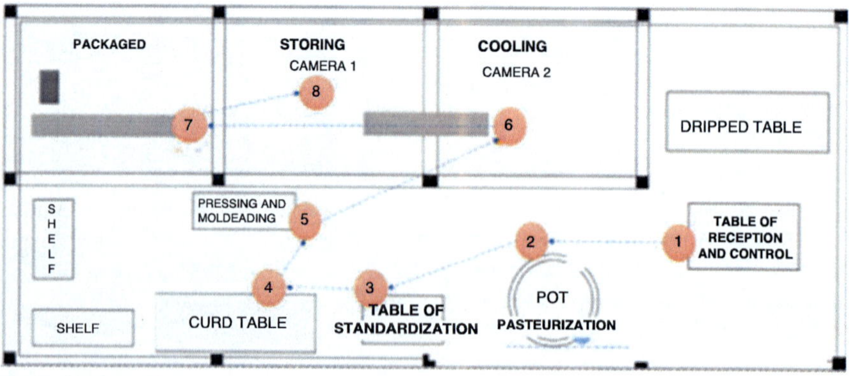

Fig. 5 Proposed situation layout

Table 2 Ergonomic risk evaluation matrix

	ERGONOMIC RISK EVALUATION MATRIX OWAS METHOD						
Campus:1	Lima, Peru						
Date test:	16/11/2019						
Produced by:	Tania Gamboa Rojas, Yosy Trejo, Patricia Sanchez, Shakira Mamani						
Area	**Exposed charges**	**Chores**	**Risk Assessment method**	**Time occurred by the day (hours)**	**Duration of the Labor Day (hours)**	**Evaluation of positions**	**Global assessment of risk**
Administration	Production manager	Oversees production and process control	OWAS	8	8		1
Reception	Cargo Assistant	Move everything needed in the plant and stores	OWAS	6	8		2
Industrial processes	Operators	They carry out all the activities of making cheese	OWAS	6	8		4
Stored	Operators	They seal the package and return it to refrigerate	OWAS	2	8		3

often the products that are delivered with more weight to the customer, this generates waste and loss of the main raw material, milk, which could be used to produce more cheese or other products of the company. On the other hand, there are unnecessary transfers due to the inadequate location of the equipment and machines in the plant, which may be due to the fact that it was not planned from the beginning where they should be located, this was mainly due to the misuse of the available space in the plant. The proposed improvement in ergonomics is to allow operator satisfaction, therefore, it must have standardized implements such as molds from 1 kg to 3.5 kg. As well as clothing that supports the temperature of pasteurized milk and the cooling of fresh curd, avoiding serious burns and injuries due to poor posture.

Likewise, according to the risk assessment matrix (See Table 2), the possible causes of inappropriate ergonomic practices and postures in the personnel working in the plant were evidenced in the company. Following the OWAS method (Ovako Working Analysis System) [13], an inspection of each personnel participating in the plant was carried out, for which the matrix was prepared with interpretation of results (See Table 3).

Table 3 Interpretation of results

Interpretation of results		
In the results, the color code shown in the following table is used to classify the risk of the postures adopted. Each color indicates one of the four risk levels defined by the OWAS method.		
Risk level	**Posture effects**	**Required action**
1	Normal and natural posture without effects on the musculoskeletal system.	No action required.
2	Posture with the possibility of causing damage to the musculoskeletal system.	Corrective actions are required in the near future.
3	Posture with harmful effects on the musculoskeletal system.	Corrective actions are required as soon as possible.
4	The load caused by this posture has extremely damaging effects on the musculoskeletal system.	Corrective actions are required immediately.

The following will explain the evaluation criteria according to the posture codes used in the OWAS method and determining which one should be audited by SIG personnel (See Table 4).

It is concluded that the positions of the operators and the personnel are adequate if the codes of positions are used properly supervised by the safety personnel at work, since the positions in manual transfer are always in constant change, in the force they apply for the loads of full trays and continuous transfer of heavy materials to press the cheese mold. It is proposed to train and train constantly through frequent exercises to avoid fatigue and future illnesses.

6 Results

- It is expected that the company can begin to sell its product to direct consumers and increase 30% of its profits from the sale of fresh cheese and achieve low production costs with the proposed improvements of the layout applying Lean Manufacturing and ergonomics within 1 year.
- Plant capacity will increase by 19% of actual capacity due to waste disposal and proper handling of materials such as cheese molds.
- The company is expected to see a 14% increase in staff satisfaction surveys during normal business hours.
- Among the proposals to improve the layout design is the reduction of the distance between the cold room and the packaging, this would be done by buying a conveyor belt, therefore, it is necessary to tear down a wall that is not an essential part of the plant architecture to transport the trays of 15 cheeses with the action of placing the trays and they will transit by themselves on the belt.

Table 4 Table of postures according to OWAS

CDG	BACK	CDG	LEGS	CURRENT	IMPROVEMENT
1	Right	1	Sitting	Back: 1: Right movement 2: Movement and transfer of materials. 3: Spin to collect items 4. To move the kettle in a circular way	Back: 1: Regular posture 2,3,4: Exercises and active 5-minute breaks
2	Bent forward, or backward	2	Standing with both legs straight		
3	Turned or bent to the side	3	Standing with the weight on one leg	Legs: 2: Standing throughout the cheese-making process, in addition to moving objects standing up all the time..	Legs: 2: Posture exercises and active breaks every 5 minutes for the legs. Chairs outside the process area.
4	Girada y doblada hacia adelante o un costado	4	Standing or squatting with both knees bent		
CDG	ARMS	5	Parado o de cuclillas con una rodilla doblada	Arms: 1: When storing fresh cheeses they should not raise their arms above their shoulders	Arms: Exercises in the arms so as not to fracture the muscles. But to improve the design of the storage area of the molds.
1	Both arms below the level or above the shoulders	6	Kneeling on one or both knees		
2	One arm at or above the shoulders	7	Walking or moving		
CDG	CHARGE OR USE OF FORCE			Load: 1: Weight that they load when they are moving the molds one by one 2: Weight that presses when moving the cart with all the molds 3: When they must load the tray of 15 cheeses that weigh between 1 kg to 3.5 kg in each transfer of the trays.	Load: 1: Exercises to avoid muscle fatigue Buy standard molds and a presser to not exert manual pressure but automatic pressure in the molding area.
1	Weight or force required less than 10 Kg.				
2	Weight or force required exceeds 10 kg but is less than 20 Kg.				
3	Weight or force required is greater than 20 Kg.				

7 Recommendations and Conclusions

- The application of OWAS (Ovako Working Analysis System) will reduce the fatigue and inappropriate postures that are generated in susceptible processes such as the molding and weighing of cheeses.

- It is important and valuable to take small and medium-sized companies as a case study, since this research has shown that they can improve their performance by applying engineering tools.
- The proposal of the improvement in ergonomics is to allow the satisfaction of the operator, therefore, it must have standardized implements such as molds from 1 kg to 3.5 kg, avoiding serious burns and injuries due to poor posture.
- Plant representatives are suggested to be able to sell their products to end customers and not just to distributors.
- It is recommended that the company carry out an economic evaluation of setting the equipment automatically and ergonomically.
- Compliance with Law 29,783 on Occupational Health and Safety Law must be ensured, to guarantee the right of workers to safety and health at work in the face of epidemiological and sanitary risks

References

1. Produce. https://ogeiee.produce.gob.pe
2. Diario Gestión. https://gestion.pe/economia
3. Kohtamäki M, Heimonen J, Parida V (2019) The nonlinear relationship between entrepreneurial orientation and sales growth: The moderating effects of slack resources and absorptive capacity. Journal of Operations Management, vol 100: 100–110. https://doi.org/10.1016/j.jbusres.2019.03.018
4. Ali A (2021) How Can Lean Manufacturing Lead the Manufacturing Sector during Health Pandemics Such as COVID 19: A Multi Response Optimization Framework. Computers, Materials & Continua 66, vol 2: 1397–1410. https://doi.org/10.32604/cmc.2020.013733
5. Rajadell M, Sánchez J (2010) Lean Manufacturing: La evidencia de una necesidad. Díaz de Santos, Madrid.
6. Salim G, Newbold D, Li M (2013) Lean principles in Healthcare: an overview of challenges and improvements. IFAC Proceedings Volumes 46, vol 24: 229–234. https://doi.org/10.3182/20130911-3-BR-3021.00035
7. Ealde. https://www.ealde.es/principios-lean-direccion-de-proyectos/
8. Fachal C, Motti V (2016) Ergonomics and the workplace. Sociedad Ecuatoriana de Seguridad y Salud Ocupacional, vol 5: 2–8
9. Del Prado J (2016) Ergonomics and its influence on work quality. Sociedad Ecuatoriana de Seguridad y Salud Ocupacional, vol 5: 12–13
10. Neumann W, Winkel J, Magneberg R., Mathiassen S (2006) Production system design elements influencing productivity and ergonomics: A case study of parallel and serial flow strategies. International Journal of Operations & Production Management 26, vol 8: 904–923. https://doi.org/10.1108/01443570610678666
11. Golabchi A, Guo X, Liu M, Han S, Lee S, AbouRizk (2018) An integrated ergonomics framework for evaluation and design of construction operations. Automation in Construction, vol 95: 72–85. https://doi.org/10.1016/j.autcon.2018.08.003
12. Smolander J, Louhevaara V (n.d) Trabajo Muscular. Enciclopedia de Salud y Seguridad en el Trabajo, vol 3: 29
13. Kivi P, Mattila M (1991) Analysis and improvement of work postures in the building industry: application of the computerised OWAS method. Applied Ergonomics 1, vol 22: 43–48. https://doi.org/10.1016/0003-6870(91)90009-7

Part III
Defense, Healthcare and Humanitarian Logistics

Defense Offsets: Propositions and Different Perceptions

Fernando de Almeida Silva and Rodrigo Antônio Silveira dos Santos

Abstract This paper presents a systematic literature review to analyze propositions related to offset practices in the international defense market. 73 propositions were identified from the material obtained by means of a systematic review according to Tranfield, Denyer and Smart method, published in 2003. As a result, 46.58% of the identified propositions are directly related to the interest of the Buyer State. This reflects the weight of this actor on the defense offset practice. Buyers' perspectives are responsible for 16.44% of the identified propositions, while sellers' perspective represents 21.92% of the propositions. Besides, the seller State point of view is present on 15.07% of the identified propositions. Regarding offset policies, it is observed that some countries impose mandatory defense offsets, however, without an adequate study if the gain brought by the mandatory offset would outweigh the advantages related to the lower prices that could have been negotiated. It was observed that the subordination of the institutions that make the purchases and negotiate the agreements to the superior demands of the States can weaken their position during the negotiations, taking away their bargaining power. This weakness can make the traded product more expensive, although it politically strengthens the decision to purchase selected defense products. In this way, the work contributes to the control and continuous improvement of processes and dynamics related to offset agreements already in progress, as well as future agreements that take place in a perspective of alignment with the interests of national policy and strategies.

Keywords Defense offset · Systematic literature review · Industrial cooperation

F. de Almeida Silva (✉) · R. A. Silveira dos Santos
Universidade da Força Aérea, Rio de Janeiro, RJ, BRAZIL

J. Vargas Florez et al. (eds.), *Production and Operations Management*, Springer Proceedings in Mathematics & Statistics 391,
https://doi.org/10.1007/978-3-031-06862-1_22

1 Introduction

The practice of offset agreements, or defense offset, has become commonplace in the international trade of defense products. The theme is quite complex and knowledge on the subject is dispersed among different forms of publication. In order to make research on this subject, it is necessary to search and organize different sources of information that can be used to deepen the theme.

The concept of offset derives from the understanding of a non-monetary compensation, in which a purchasing government requires that an exporting company provide, as a condition for the sale, a percentage of investments as a form of return to the purchasing country. Offsets can be directly related to the object of the contract, known in the literature as direct offset, or benefit other themes and industries, different than the contracted object, cases defined as indirect offset.

The practice of offset is quite recurrent in the international trade of defense products, but this dynamic still needs to be better understood. This practice, although common, is not standardized and results from intense negotiations between seller and buyer. Although defense compensation is mainly linked to international tenders, the practice has been used by the Brazilian government also in defense contracts with national companies that have foreign suppliers involved in the manufacture of contracted defense products.

Thus, the present work has as general objective to produce a systematic literature review to identify, in a single study, the main propositions about the practice of off-set existing in the specialized literature, from the perspective of the different actors that take part in this practice. It is important to realize that the proposal of the work is really to synthesize extensive primary research articles in a single text, providing different insights, through a method that has the due methodological rigor and, in order to meet this proposal, the Systematic Literature Review showed as the most suitable method.

The justification for the research is the need for control and continuous improvement of processes and dynamics related to offset agreements already in progress, as well as future agreements that take place in a perspective of alignment with the interests of the National Defense Policy and Strategy. Thus, this literature review configures the first stage of a broader research, which will allow the deepening and contact with different perspectives on the same subject.

This article is divided into six sections. The first presents the concept and context of offsets related to defense products. The second section exposes the theme of offsets to a necessary theoretical review. Next, the systematic literature review methodology used in the work is shown. The fourth part of the article demonstrates the result of the search made through different databases. Subsequently, a brief analysis of the identified propositions is provided, within the appropriate perspectives. The last section concludes the results of the systematic review analysis and suggests further research, highlighting the limitations of the study.

2 Theoretical Review

An offset agreement is a legal instrument that formalizes the commitment and obligations of a foreign supplier to offset the expenses incurred with an import, containing a set of clauses which include the object or objects of compensation [1]. Offset arouses the interest of governments and military forces as it is seen as a way of reducing technological dependence, as it allows critical knowledge to be transferred to the domain of users and maintainers of the acquired systems [1].

The practice of offsets related to defense products involves the participation of the Defense Industrial Base of the country, which makes the purchase in various modalities that aim to support the operation and evolution of the contracted object, highlighting the co-production for the supply of materials and equipment, the technological and industrial updating of companies and universities, and the nationalization of materials and services [1]. While some countries seek offsets with the purpose of creating jobs or to make commercial effects favorable, in Brazil technology transfer is the central argument for the demand for compensation related to purchases of defense products [2].

Offsets related to defense products focus different interests, from different perspectives. Hanna, Willen and Zuazua [3] list the following actors as responsible for conflicts of interest in offset negotiations: the government of the contracting country; the Armed Forces of the contracting country; the contracted company; and the government of the country of the contracted company. In the present study, the perspectives described can also be read as the State of buyers, buyers, sellers, and the perspective of the State of sellers.

The government of the contracting country insists on the practice of offset basically aiming to obtain strategic technologies; minimize the costs of the country's balance of trade and increase the capacity of local industries [1]. The Armed Forces of the contracting country seek, through offset, mainly to minimize the life cycle cost of the contracted object and minimize the risks of delivery of the contracted object [1]. The contracted company basically aims at maximizing its profits. The government of the contracted company's country is primarily concerned with protecting its strategic technologies, and maintaining investments and jobs in the country [1].

3 Systematic Literature Review Methodology

The method chosen for conducting this study was the Systematic Literature Review, as published by Tranfield, Denyer and Smart in 2003 [4]. In the development of the review, the researcher maps and evaluates relevant intellectual content, in order to deepen the research that will be developed [4]. Systematic reviews differ from other, even more traditional forms of review in that they adopt a replicable, scientific and transparent procedure, that is, a detailed technology, which aims to minimize bias through exhaustive searches in the literature by providing an audit trail of the reviewers' decisions, procedures and conclusions [4].

The central point for choosing a Systematic Literature Review is the possibility of synthesizing knowledge from reliable bases and in a structured, systematized and replicable way. A good systematic review should make it easier for the researcher to understand the research by synthesizing extensive primary research articles from which it was derived, as systematic search starts with identifying keywords and search terms, which are constructed from a defined scope [4].

The process used for a systematic review increases methodological rigor and reduces bias, distortions and subjectivity, in addition to defining the research limits, as it follows pre-defined criteria [4]. A literature review can even use complementary methods such as the snowball technique, which consists of looking for other works from the references of the articles themselves, to enrich the content, however, this is not mandatory. The use of the snowball method can expand the search too much, causing the researcher to enter an indefinite spiral, increasing the possibility of adopting subjectivity and decreasing impartiality in the selection of these new articles.

Since the objective of the Systematic Literature Review is to provide collective in-sights through a theoretical synthesis in fields and subfields, the model proposed by Tran-field, Denyer and Smart [4] will allow the adequate examination of the propositions found in order to achieve the general objective proposed by the article of identifying, in a single study, the main propositions about the practice of offset existing in the specialized literature, from the perspective of different actors.

Conducting a Systematic Literature Review provides the gathering of knowledge dispersed from different sources, as well as the understanding of the evolution of a topic, given the chronology of the material found. This form of literature review al-lows us to identify the methodological paths that have already been developed, so that the main paradigms that guide research already carried out can be outlined, as a researcher who ignores previous research runs the risk of proposing something trivial or exhaustively debated in his work [5]. Rowe [6] states that a literature review pursues the following points: (1) Synthesis of knowledge of a topic or domain of interest; (2) Identification of biases and knowledge gaps in the literature; and (3) Proposition of corresponding future research directions.

Literature on offset deals in Portuguese is very scarce. The work of obtaining sources through systematic literature review was developed through access to the SCOPUS database, accessed through the CAPES journal portal. Tranfield, Denyer and Smart [4] didactically describe a Systematic Literature Review and divide it into three stages: (1) planning, (2) execution, and (3) elaboration and dissemination. Thus, we sought to establish a structured, logical and distinct approach to researching offsets related to defense products [4].

The development of the research plan, in the planning stage, comprised the sources and parameters of interest, as well as the alignment with the proposed research objective. The development of the systematic review protocol did not take place immediately. Initially, it was necessary to know the terms to be searched in view of the im-portance of correctly directing research. The purpose of the research is to understand the dynamics of compensation agreements in the area of defense products. Therefore, the choice of search terms was made trying to gather some "set intersections" that would allow a good survey of bibliographic material. The search terms were searched in the English language, given the greater existence of articles in this language and for reaching a greater number of scientific publications. The terms used for the search derive from the practice of offsetting, aiming to expand the search results, in English: Offset agreement, Defense Offset (American expression), Defence Offset (British ex-pression), Defense economics, Defence economics (British expression), Aerospace industry, Industrial cooperation, Industrial participation and arms trade.

The protocol was designed to carry out the search through the SCOPUS portal, with the search terms already defined, and directed the search to articles that have been published in scientific journals (journals), without limitation of publication date. Inclusion and exclusion criteria include semantic relevance and the possibility of accessing the full text. The choice of Boolean operators that composed the equation sought terms that allowed the intersection of "sets" (AND), as well as the union of these results with other subsets (OR).

Thus, the following search equation was established: (("Defense Off-set"AND"Aerospace industry")OR("Defence Offset"AND"Aerospace industry")OR("Defense Offset"AND"Defense economics")OR("Defence Off-set"AND"Defence economics")OR("Defense Offset"AND"Industrial cooperation")OR("Defence Offset"AND"Industrial cooperation")OR("Defense Off-set"AND"Industrial participation")OR("Defence Offset"AND"Industrial participa-tion")OR("Offset agreement"AND"Aerospace industry")OR("Offset agree-ment"AND"Defense economics")OR("arms trade"AND"offset")), with filter for DOCTYPE options – ARTICLE and SOURCE TYPE – JOURNAL.

Once the research protocol (Table 1) is defined, the first stage of the Systematic Literature Review process is concluded. The second stage, the execution stage, started with the identification of the research and the consequent implementation

Table 1 Research protocol

Language	English-only
Date range	No data range
Search fields	Search terms were applied to titles, abstracts and keywords
Search terms	Keywords used on SCOPUS: *Offset agreement, defense offset, Defence offset, defense economics, Defence economics, aerospace industry, industrial cooperation, industrial participation* e *arms trade*
Document type	Article
Source type	Journal
Exclusion criteria	Semantic relevance; access to the full text
Criteria	Relevance to the research problem

Source: Created by the authors

of the protocol. On July 9, 2021, the result showed 14 articles, which were gathered in a Microsoft Excel table for the correct organization of the results.

With the organization of the selected articles, the exploratory reading [7] of the articles for the first selection filter was started. In the exploratory reading of aca-demic articles, reading the abstracts is enough to verify the alignment between the research and the material. At this stage, an article was excluded, according to the criterion of semantic relevance, as it was not related to the research developed. Subsequently, a selective reading of the material was started [8], aiming to select the parts of the material that would really interest the research. In this examination, another article was excluded due to the impossibility of obtaining the pdf file by the means available for the research.

The analytical reading [8] of the material started next. In this new exam, the articles were read in more depth, seeking the answer to questions such as:

- What is the purpose of the study?
- What concepts are discussed?
- How do the author(s) discuss the topic of their research?
- What important definitions appear in the text?
- What is the method, the amount of data collected and the type of survey?
- What are the main results or contributions of the study?

In this analytical reading stage, the answers found during the reading were gathered in Analytical Reading Sheets, assembled by the author himself, which also included a table to separate the identified propositions.

The analysis of the sources did not identify specific aspects related to defense contracts with national companies that have foreign suppliers, which highlights a knowledge gap to be filled, as well as reinforcing this review as a basis for future studies. A thorough reading of the articles allowed the identification of 73 important propositions about offsets related to defense material purchases, as well as the identification of different perspectives. Therefore, this information was organized within a 6 W matrix, identifying aspects such as Who (Author), When (Year), Where (Article/Journal), What (propositions), Whose perspective (perspective of the most relevant actor) and how (positive or negative classification of the aspect).

4 Results

The result of the search carried out on the SCOPUS portal revealed 14 articles, according to the parameters listed. Of the 14 articles, one was eliminated due to the impossibility of accessing the full text and the other two according to the criterion of semantic relevance. As a result, 46.58% of the identified propositions are directly related to the interest of the Buyer State. This reflects the weight of this actor on the defense offset practice. Buyers' perspectives are responsible for 16.44% of the identified propositions, while sellers' perspective represents 21.92% of the propositions. Besides, the seller State point of view is present on 15.07% of the identified propositions. The table containing the results will be available in the Appendix section of the text.

5 Discussions

Regarding the identified perspectives, it is important to note that the same item can be approached from more than one perspective. When an item can be evaluated from more than one perspective, the framing and analysis will be delimited according to the institutional relevance of the person responsible for the perspective. For example, the country of the seller or buyer will have greater relevance than the company selling the equipment or the subordinate state organization (Ministry, Armed Forces and other institutions).

The propositions found were observed from the perspective of buyers (B), State of buyers (BS) of sellers (S) and from the perspective of State of sellers (SS), in order to better understand the aspects. The classification of aspects in positive (P) and negative (N) refers respectively to the interests, benefits and disadvantages that "B", "BS", "S" and "SS" have in view of the practiced dynamics.

5.1 Buyers' Perspective

From the perspective of buyers, 12 proposals were listed, 16.44% of the total, with 6 negatives and 6 positives. In essence, it is possible to infer that offsets end up making the purchase of defense materials more expensive, in addition to removing flexibility in negotiating with sellers [9]. An interesting point for the buyer concerns the competition between sellers that offset makes possible [9], allowing them to seek to combine the best equipment with the best conditions. Furthermore, the buyers' perspective does not normally include indirect offset, as these do not relate to the object of the contract (Table 2).

5.2 Perspective of the Buyer State

The buyer's State perspective represents 46.58% of the total of 73 proposals identified, 28 of which are characterized positively and 6 negatively. As a rule, this perspective overlaps that of the person who buys the material, be it a ministry or an armed force, as this subordinate relationship with the central government is the key element in the matter.

Briefly, some aspects stood out more from the perspective of the buyers' state. An interesting point observed reinforces that offsets contribute more to promoting sales of defense products than to improving the quality of purchases or to price

Table 2 Buyers' perspective propositions

What (Propositions)	Whose perspective	hoW (P/N)
Offset prerequisite makes the item more costly and removes the Buyers' capacity to negotiate lower prices [1]	B	N
Offsets might expand the expense of arms bought under primary arrangements [1]	B	N
Under the established offsets rules, it is oriented to look for "in kind" bundle improvements rather than equivalent price discounts [1]	B	N
By denying the adaptability to arrange the most beneficial arrangements, including profound pricing discounts, the compulsory offsets prerequisites don't appear to serve the material purchaser's general benefits [1]	B	N
The seller enjoys information asymmetry over the buyer and, the application of obligatory offset requirements is likely to put the buyer in more disadvantageous bargaining position [1]	B	N
Offsets normally constitute serious conflicts of interest and may represent challenges to democratic accountability [2]	B	N
There is no question that the appeal of offset bundles has become a critical standard in the determination of bidders [3]	B	P
Offsets work as a market reality for companies competing for global defense trade [1]	B	P
Imports and offsets have become a method by which to understand its technological independence aspiration [4]	B	P
Offset uninhibitedly offered by the merchant can be viewed as a benefit offered by the dealer to win a market dispute [5]	B	P
Offset is seen as a criterion used by a defense purchaser for choice [6]	B	P
Improved logistic support is a motivation for compensation [7]	B	P

Source: Created by the authors

competition among suppliers [16]. Another important point concerns the degree of technology transfer obtained, which is usually lower than the expectation generated [17].

Politically, the State demonstrates an expectation of compensating the taxpayer's money spent on such materials by creating jobs and attracting investments to the country, in the medium and long term, whether through direct or indirect offset [18]. From the point of view of the buyers' state, offset presents itself as a way to gain access to technologies that the country would not normally have [9].

An offset agreement is usually linked to a main commercial contract. Some States have offset policies that demand compensation for international purchases, in specific areas such as defense, which can reach more than 100% of the commercial contract value [9]. In Brazil, for example, the requirement is 100% of the commercially contracted value for imports of defense products with a total value above 50 million US dollars [19] (Table 3).

Table 3 Buyer State's perspective propositions

What (Propositions)	Whose perspective	hoW (P/N)
Benefits are risky and some may never eventuate – Given the changing nature of defense contingencies and the pace of technological change, while, in other hand, many costs are reasonably certain [1]	BS	N
A recipient country's absorptive capacity for offset deliverables can only be determined with any precision after all the process, being hard to assess it before [1]	BS	N
There is some kind of difficulty in verifying the economic effects of offset, especially indirect offsets [5]	BS	N
It might be more limited than expected, the degree of technology transfer for defence-oriented local companies [8]	BS	N
Offsets are better at promoting arms sales than at improving quality or furthering price competition [9]	BS	N
Offset effects like technological spin-off to military industries tend to be exaggerated [10]	BS	N
Normally including design, blueprints, skills and manufacturing capability, technology transfer represents the most important variable within an offset package [3]	BS	P
Technological emulation is the most important objective of an offset arrangement. Because of this, most states make huge efforts for improving the absorptive capacity and make offset work [3]	BS	P
Offset represent an opportunity for the importing nation to engage in activities, which would otherwise find either impossible or too costly to engage in, acquiring technologies which would not otherwise have been accessible [1]	BS	P
Building their own skills and developing important relationships for their companies are some opportunities for individual member states that offsets offer [1]	BS	P
Believing that more local subcontractors will be engaged by foreign companies, that a wider range of new product and process technologies will be transferred to them, and that their export opportunities will be enhanced, defense offsets tend to be mandated by material-importing nations [1]	BS	P
Offset gives entry to developing technology that, in other way, the country could not have individually [4]	BS	P
Offset offers job creation by allowing the opportunity to participate in the production of the system that is being bought [4]	BS	P
Offset reduce the financial impact of the defense procurements [11]	BS	P
Offsets allow retaining their foreign currency reserves [11]	BS	P
Offsets allow grabbing valuable technology and manufacturing know-how [11]	BS	P
Offsets allow increasing domestic workforce skills and capabilities, especially in the defense industrial base [11]	BS	P
Offsets bring opportunity to create new jobs and preserve already existing domestic employment [11]	BS	P
Offsets allow strengthening their defense industries, including targeted industrial and other regional sectors [11]	BS	P

(continued)

Table 3 (continued)

What (Propositions)	Whose perspective	hoW (P/N)
Offsets allow enhancing bid-offers among companies as they compete to tender the most compelling offset packages [11]	BS	P
Offsets make procurements of defense foreign goods more politically palatable to their taxpayers [11]	BS	P
A procuring country establishes different multipliers to particular activities that the government believes to be highly valuable or desirable, such as technology transfers [11]	BS	P
The demand for compensation from aircraft manufacturers is favored by the competitive nature of the market and large sums involved in negotiations [5]	BS	P
Major customers can impose purchasing conditions, like technology transfer and other offset transactions, that aircraft manufacturers cannot ignore or they will be out of the dispute [5]	BS	P
Offsets may condition macro-level development policy decisions because of they can have an impact well beyond their projected costs [8]	BS	P
Not expected benefits might in time flow from the offset project [8]	BS	P
Offsets are viewed as a path to country development by stimulating local employment and acquiring advanced technologies [9]	BS	P
Offsets intend to reduce the net capital outflow from the buyer state [6]	BS	P
Offsets disseminate impacts to the general economy [10]	BS	P
Offsets allow spin-off effects on human capital (military training, education and modernization) [10]	BS	P
The technology transfer obtained by offset increase investment productivity [10]	BS	P
Transfer of technology is the most important motivation for offsets [7]	BS	P
Defence industrialization is another motivation for offsets [7]	BS	P
Learning and spillover effects motivate offset arrangements [7]	BS	P

Source: Created by the authors

5.3 *Seller's Perspective*

Of the 73 proposals identified, the perspective of sellers was characterized by 21.92%, with 9 negative aspects and 7 positive. In summary, the analysis of the points showed the concern of defense material exporting companies with regard to feeding their competition with their technology through offset [11]. At the same time, the importance of offset as opening doors to new world markets is also recognized, which allows companies a greater capacity for competition and compliance with requirements imposed by buyers [7]. It is also noteworthy that a positive aspect relevant to the perspective of sellers concerns the possibility of reducing costs through the development of new suppliers in countries that receive the offsets, which is undeniably advantageous in terms of savings in the logistics chain [11] (Table 4).

Table 4 Seller's perspective propositions

What (Propositions)	Whose perspective	hoW (P/N)
Offset-based technology transfers may outcome in the displacement of domestic producers or products unable to oppose increased global competition by helping rival foreign suppliers [1]	S	N
Offsets may undermine its economic and national security interests, but it is a reality of the marketplace in order to compete for international arms trade [11]	S	N
Offset arrangement may result in firms contributing to the evolution of their competitors [11]	S	N
Offset can impact the aircraft manufacturer loose of control of the supply chain structure [5]	S	N
Offset as a procurement obligation might be viewed as a non-duty barrier for certain organizations [5]	S	N
Offsets represent an added restriction that limits the acquirement of allocative efficiency and deflect trade [9]	S	N
Sellers are concerned about technology leak from their defense industrial bases [7]	S	N
Defense companies worry about illegal sales of their products to other countries without authorization [7]	S	N
Offsets may create, sows the seeds, its own future international competition [7]	S	N
Offsets may create other associated benefits for the exporting country (offsets provider) by inducing new exports [1]	S	P
Establishing new international cooperative agreements and/or joint ventures, offset may reduce production costs by developing new suppliers in foreign countries [11]	S	P
The production of components, subsystems or systems by companies in buyer's country is the common way to offset be put into practice [5]	S	P
Offset works as a strategic alliance between companies in the buyer country and the selling company [5]	S	P
Costs reduction and increase of competitiveness are motivation for offsets [7]	S	P
Access to new markets works as motivation for offsets [7]	S	P
Production continuity is a motivation for offsets [7]	S	P

Source: Created by the author

5.4 *Perspective of Seller State*

The sellers' country point of view cannot be overlooked. The United States, as a hegemonic power and the world's largest arms exporter, is very concerned about the loss of internal production capacity for components that end up being produced in other countries due to offset agreements [18]. Another aspect that concerns the powers that manufacture weapons is the possibility of losing control of the manufacture of military and related equipment [16]. In a way, the possibility of transferring technology to countries that do not share the same values as Western society [21] is a concern not only to central countries.

However, for the State of sellers, offsets do not only represent the possibility of loss of domain and risk to national security, they can also represent benefits in the field of strategic international relations. By using offsets as an important foreign policy tool, the sellers' State expands its influence, allows for standardization and helps the development of the industrial technological park for the defense of allied countries [18].

In the analysis carried out in this work, it was verified that the sellers' country perspective is responsible for 15.07% of the relevant aspects identified here, with 9 negative aspects and only 2 identified as positive aspects (Table 5).

5.5 *Analysis*

The percentages generated from the different perspectives of the 73 proposals identified allow an interesting analysis on the subject. Of the total, 46.58% of the identified points are directly related to the interest of the State of buyers and this reflects the weight of this actor on the practice of offsets. Buyers are responsible for 16.44%, sellers for 21.92% and the state of sellers for 15.07% of the identified proposals. In the negotiation of offsets related to defense products, despite the weight and interest of the buyer's State, it does not negotiate directly with the seller and this has a great impact on the negotiation. The demand of an external actor to a negotiation ends up reducing the subordinate party's bargaining power, generating negative consequences for the weakened side.

In terms of offset policies, it is observed that some countries impose mandatory defense offsets. However, sometimes, without having an adequate study if the gain brought by the obligatory offset would outweigh the advantages related to the lower prices that could have been negotiated [9]. In Brazil, there is an obligation of offset related to purchases and contracting of defense products, which imply in importation above 50 million US dollars. The requirement is applied not only to foreign companies, but also to Brazilian companies that carry out imports related to the purchase or contracting of defense products [19].

It is important to emphasize that this work achieved the proposed objective of identifying the main propositions about the practice of offset existing in the specialized literature, from the perspective of different actors. And in this way, the research managed to synthesize extensive primary research articles in a single text, providing distinct insights, through a method that presented the due methodological rigor [4].

Table 5 Seller State's perspective propositions

What (Propositions)	Whose perspective	hoW (P/N)
US government prohibits any of its agencies from stimulate, take part, or committing US firms to any offsets agreements in connection with the arms trade and services to foreign government, because considers that offsets are economically inefficient, deform trading [1]	SS	N
US Government tries to protect US defense exporter by hostiling any offset efforts [1]	SS	N
Decrease business for U.S. suppliers, decreased employment for skilled labor in those sectors, and the transfer of capability from the U.S. industrial base to the foreign country [11]	SS	N
If the offsets result in the loss of critical skills in the defense industry workforce, that ones that cannot later be replaced or expanded quickly, it may be a threat to the national security [11]	SS	N
The practice of offsets increase the number of arms suppliers over time as technologies are transferred, and this also exacerbates the difficulty in forming arms control regimes [9]	SS	N
Offset agreements involving unapproved technology transfers may undermine U.S. national security, and also negatively impacts the U.S. defense industrial base and its workforce capabilities [11]	SS	N
The defense industrial basis may be declined and disrupted with critical parts of the defense supply chain, as offsets may undermine national security interests [11]	SS	N
The chances that leading edge weapons and the technology for producing them may go to countries that represent a threat to U.S. national security interests may be multiplied by offsets arrangements [11]	SS	N
Offsets could represent transfers of substantial resources to authoritarian governments under conditions of near total loss of control, or unaccountability [2]	SS	N
Offset may be an opportunist vehicle to lock client states into long-term defense industrial and technology partnerships and it may be used for some states [3]	SS	P
Offsets can be used to increase political influence serving as important foreign policy and national security objectives of the United States, such as increasing the industrial capabilities of allied countries, standardizing military equipment, and modernizing allied forces [11]	SS	P

Source: Created by the authors

6 Conclusion

Reading the selected articles allowed the identification of 73 propositions distributed by the perspectives of the buyer, buyer's state, seller and seller's state, as well as their classification in positive and negative aspects. The perspective of the State of buyers reflects 45.95% of the identified points, that of buyers 16.44%, that of sellers 21.92% and that of the State of selling companies 15.07%. This statistic ends up reflecting the weight existing in the specialized literature for the perspective exercised by each of the actors identified in this work.

During the offset negotiations related to defense products, it is observed that the submission of the agencies that make the purchases and negotiate the agreements to the superior demands of the States weakens them and takes away their bargaining power. This weakness ends up making the traded product more expensive, but politically strengthens the decision to purchase selected defense products. The situation creates the expectation of compensating the taxpayer's money spent on such materials by creating jobs and attracting investments to the country, while allowing access to technologies that the country would not normally have.

From the perspective of sellers, the propositions found highlight the concern of defense material exporters with respect to feeding their technology through offsets to the competition. While the importance of offset for opening doors to new markets was also recognized. The state of sellers also presented a unique perspective, revealing concern about the loss of internal production capacity for components that end up being produced in other countries due to offset agreements. However, even from the perspective of the States of the sellers, offsets do not only represent risks and threats to security. Vendor States can use the offset mechanism related to sales of defense products as an opportunity to develop, standardize allied countries and, consequently, expand their influence.

Regarding offset policies, it is observed that some countries impose mandatory compensation related to the purchase of defense products. However, sometimes, this obligation is made without an adequate study if the gain brought by the obligatory offset outweighs the advantages related to the lower prices that could have been negotiated. It is verified in Brazil the obligatory requirement of offset related to purchases and contracting of Defense Products, for imports above 50 million US dollars. The requirement is made not only in relation to foreign companies, but also to Brazilian companies that need to carry out imports to meet the supply of purchases or contracts for Defense Products.

It is important to emphasize that this work achieved the proposed objective of identifying the main propositions about the practice of offset existing in the specialized literature, from the perspective of different actors. And in this way, the research managed to synthesize extensive primary research articles in a single text, providing distinct insights, through a method that presented the due methodological rigor. The impact analysis of the raised propositions, as well as the correlation between these propositions and the negotiations of Defense products, appear as new possibilities for future studies in relevant debates on the possibility of technology absorption in the Defense area.

The study did not identify specific aspects related to defense contracts with national companies that have foreign suppliers, which highlights a knowledge gap to be filled, a new possibility for study. The information gathered by the study is important and directly contributes to the control and continuous improvement of processes and dynamics related to offset agreements already in progress, as well as future agreements that take place in a perspective of alignment with the interests of national policy and strategies.

Appendix

Search results

No.	Author, Year	Article title	Journal -	H index	Selected for analysis?
1	Christensen (2021) [22]	The morality of substitution intervention: The case of Yemen	Politics	x	Discarded by the semantic relevance criterion
2	Perlo-Freeman (2020) [23]	Red flags for arms trade corruption	Economics of Peace and Security Journal	1	Discarded. No full text file access.
3	Arifin et al. (2020) [24]	Countertrade mechanism of global arms trade: Case study of Indonesia	International Journal of Financial Research	7	Discarded by the semantic relevance criterion
4	Li and Matthews (2017) [11]	"Made in China": An emerging brand in the global arms market	Defense and Security Analysis	12	Yes
5	Markowski and Hall (2014) [9]	Mandated defence offsets: Can they ever deliver?	Defense and Security Analysis	12	Yes
6	Hagelin (2012) [12]	Into the black box? Technology sharing in major arms transfers and beyond	Defense and Security Analysis	12	Yes
7	Guerra (2011) [13]	Systems integration model in the aerospace industry: Motivating factors	Gestao e Producao	16	Yes
8	Petersen (2011) [18]	Defense and commercial trade offsets: Impacts on the U.S. Industrial base raise economic and national security concerns	Journal of Economic Issues	46	Yes
9	Marshall (2009) [10]	Money for nothing? Offsets in the U.S.-Middle East defense trade	International Journal of Middle East Studies	43	Yes
10	Haines and Hosking (2005) [17]	A bridge too far? The arms deal, the Coega IDZ, and economic development in the Eastern Cape	Society in Transition	x	Yes
11	Sandler (2000) [16]	Arms trade, arms control, and security: Collective action issues	Defence and Peace Economics	38	Yes
12	Catrina (1994) [14]	Main Directions of Research in the Arms Trade	The ANNALS of the American Academy of Political and Social Science	102	Yes

(continued)

No.	Author, Year	Article title	Journal -	H index	Selected for analysis?
13	Soesastro (1994) [20]	Military Expenditure and the Arms Trade in the Asian-Pacific Region	Asian-Pacific Economic Literature	21	Yes
14	Matthews (1991) [15]	Offset to decline for Britain's defence-industrial base?	RUSI Journal	20	Yes

Source: Table organized by the authors

References

1. Brustolin, V., Oliveira C., Senna C.: Análise das Práticas de Offset nos Contratos de Defesa no Brasil. Rev da Esc Guerr Naval 22, 169–196 (2016).
2. Perlo-Freeman S.: Offsets and the development of the Brazilian arms industry. In: Dune, J., Brauer, J. Arms Trade and Economic Development: Theory, Policy, and Cases in Arms Trade Offsets, pp. 185–200. Routledge Taylor & Grancis Group, London (2005)
3. Hanna, J., Willen, B., Zuazua, M.: GCC Defense Offset Programs: The Trillion- Dollar Opportunity, 10 p. (2013).
4. Tranfield D, Denyer D, Smart P. Towards a Methodology for Developing Evidence-Informed Management Knowledge by Means of Systematic Review*. 2003; 14: 207–222.
5. Godoi, C., Bandeira-De-Mello, R., Silva, A.: Pesquisa Qualitativa em estudos organizacionais: Paradigmas, Estratégias e Métodos. 2nd edn. Saraiva, São Paulo (2010).
6. Rowe, F.: What literature review is not: diversity, boundaries and recommendations. Eur J Inf Syst, 23: 241–255 (2014).
7. Gil A.: Como elaborar projetos de pesquisa. 4th edn. Editora Atlas S. A., São Paulo (2002).
8. Gil A. C. Como elaborar projetos de pesquisa. 6th edn. Editora Atlas S. A., São Paulo (2017).
9. Markowski, S., Hall, P.: Mandated defence offsets: Can they ever deliver? Def Secur Anal 30, 148–162 (2014).
10. Marshall, S.: Money for nothing? Offsets in the U.S.-Middle East defense trade. Int J Middle East Stud 41, 551–553 (2009).
11. Li, L., Matthews, R.: 'Made in China': An emerging brand in the global arms market. Def Secur Anal 33, 174–189 (2017).
12. Hagelin, B.: Into the black box? Technology sharing in major arms transfers and beyond. Def Secur Anal 28, 163–175 (2012).
13. Guerra, J.: O modelo de integração de sistemas da indústria aeronáutica: fatores motivadores. Gest e Prod 18, 251–264 (2011).
14. Catrina, C.: Main Directions of Research in the Arms Trade. Ann Am Acad Pol Soc Sci 535, 190–205 (1994).
15. Matthews, R.: Offset to decline for Britain's defence-industrial base? RUSI J 136, 58–63 (1991).
16. Sandler, T.: Arms trade, arms control, and security: Collective action issues. Def Peace Econ 11, 533–548 (2000).
17. Haines, R., Hosking, S.: A bridge too far? The arms deal, the Coega IDZ, and economic development in the Eastern Cape. Soc Transit 36, 1–23 (2005).
18. Petersen, C.: Defense and commercial trade offsets: Impacts on the U.S. Industrial base raise economic and national security concerns. J Econ Issues 45, 485–492 (2011).
19. Brasil. Portaria Normativa N° 61/GM-MD, de 22 de outubro de 2018 – Imprensa Nacional. Brasília: Ministério da Defesa, https://www.in.gov.br/materia/-/asset_publisher/Kujrw0TZC2

Mb/content/id/46673332/do1-2018-10-23-portaria-normativa-n-61-gm-md-de-22-de-outubro-de-2018-46673171, last accessed 2021/08/20.

20. Soesastro, H.: Military Expenditure and the Arms Trade in the Asian-Pacific Region. Asian-Pacific Econ Lit 8, 27–47 (1994).
21. Huntington, S.: The Clash of Civilizations? Foreign Affairs, 72(3), 22–49. doi:10.2307/20045621 (1993).
22. Christensen, J.: The morality of substitution intervention:The case of Yemen. Politics. Epub. doi: https://doi.org/10.1177/02633957211014694 (2021).
23. Perlo-Freeman, S. :Red flags for arms trade corruption. Econ Peace Secur J 15. Epub. doi: 10.15355/epsj.15.1.5 (2020).
24. Arifin, Z., Suman, A., Khusaini, M: Countertrade Mechanism of Global Arms Trade: Case Study of Indonesia. Int J Financ Res 11. Epub. doi: 10.5430/ijfr.v11n1p307 (2020).

Scaling Operations to Address Forced Migration Flows: The Case of Venezuelan Immigration

Luiza Ribeiro Alves Cunha, Adriana Leiras, and Paulo Gonçalves

Abstract The ongoing political crisis in Venezuela is the cause of one of the most substantial migratory movements in recent history, attracting the attention of academics interested in social-impact. This study analyses the Brazilian Federal Government's Operational response to receive and assist Venezuelans migrating into Brazil. We develop a process model and a causal loop diagram to explore the relations across factors affecting the Operation and its scalability. The findings highlight the central role of planning and investment for scalability, contributing to reasoned theoretical and practical discussions.

Keywords Operations' scalability · Process modelling · Causal loop diagram

1 Introduction

The number of international migrants has sharply grown over the past 15 years [1]. In Latin America, migration flows have followed a similar trend, gaining international repercussions [2]. The growth in migration flows attracted academic attention, leading to studies that encompass refugee mental health [3, 4], vaccination campaigns [5], refugee camps or shelters management – including waste management and inventory challenges [6–8], regularization of immigrant documentation [9], refugee policies [10], refugee labor market integration [11], and criminalisation of migration and migrants [12].

Nation-states have classified migrants into specific categories [13]. An international migrant is any person who has moved across an international border away from his/her habitual place of residence regardless of the cause, legal status, and length of the stay [1]. A refugee is a person fleeing conflict or persecution defined

L. Ribeiro Alves Cunha (✉) · A. Leiras
Pontifical Catholic University of Rio de Janeiro, Rio de Janeiro, Brazil

P. Gonçalves
Universita della Svizzera Italiana (USI), Lugano, Switzerland

J. Vargas Florez et al. (eds.), *Production and Operations Management*, Springer Proceedings in Mathematics & Statistics 391,
https://doi.org/10.1007/978-3-031-06862-1_23

and protected by international law. Refugees should not be expelled or returned to situations where their lives and freedom are at risk. Unlike refugees, who cannot safely return home, migrants do have the freedom to return home and continue to receive their government's protection [1].

One of the most recent migratory movements, and most notorious, as grew exponentially in the past 5 years [12] is Venezuela's ongoing political crisis. A massive migration to other countries has marked the Venezuelan most significant population mobilization [14], the largest external displacement crisis in Latin America's, and the second largest in the world [12]. Over 5 million Venezuelans were forced to leave their homeland over the past 2 years [15]. Venezuela's political crisis is responsible for a massive migration movement to neighboring Latin American countries, such as Colombia, Peru, Chile, Ecuador, and Brazil. In this research, we analyse the Venezuelan migration to Brazil and the Brazilian government's operational actions to host these immigrants.

Created in 2018 to receive with dignity Venezuelan immigrants and refugees, the Brazilian Federal Government's "Acolhida" Operation (Welcome Operation) goals include receiving; identify; sort; immunize; shelter; and internalize immigrants in a vulnerable situation, resulting from the migratory flow caused by a humanitarian crisis.

The number of temporary residence and refugee applications shows that Venezuelan immigration to Brazil has grown from 2015 to 2019. Records show that in 2019 the number of Venezuelans crossing the border was over 800 people a day [16]. However, since mid-March 2020, borders have been closed due to the COVID-19 pandemic, including for Venezuelans [17]. As a result, the contingents previously mobilized to act in the frontier order were dismissed or relocated, the shelters decreased their population as Venezuelans continued to be interned, and the operations procurement strategy needed to be revised [18]. With the opening of the borders after the pandemic, a suppressed flow of Venezuelans is expected. Nevertheless, the Operation went the opposite way, to demobilize its contingent due to the COVID-19.

In this scenario of constant fluctuations in Venezuelans flow to Brazil, our work seeks to answer the research question: Which factor(s) influence the scalability of the Brazilian Operation in response to Venezuela migration? Scalability is the ability to scale the supply chain's capacity to meet the changing demands in a humanitarian context. Therefore, scalability may adapt to humanitarian supply chains that can increase or decrease their capabilities when there is a change in demand [19].

To answer the research question, we adopt the case study methodology, process mapping, and a causal loop diagram (CLD), an essential tool from System Dynamics (SD) method [20], to present the main results. SD is a promising tool to be used in modeling humanitarian operations [21], besides being the most predominantly simulation technique in disaster management (DM) context [22]. Researches in the field of DM encompass vehicle fleet management [23], road-rush-repairs after a disaster [24], trade-offs between a provision of relief assistance and capacity building in humanitarian organizations [25], and investments in disaster management capabilities and pre-positioning of inventory [26]. More recently, studies applying system dynamics to research refugee crises have also gained attention [9, 27, 28].

The paper contribution is justified given the challenges faced by decision-makers in migratory movements response operations, besides the strong social impact that the subject involves. Therefore, through processes mapping, development of a CLD and analysis of the main interactions concerning the scalability of the Brazilian federal government operation, insights are generated for decision making in similar contexts operations. Among the insights, the need for financing for physical infrastructure constructions, allocation of human resources and supplies is emphasized.

This paper is organised as follows. We detail the research method in Sect. 2, followed by a Section presenting and analysing the results and findings. Section 4 brings the conclusions and recommendations for future research.

2 Methodology

We adopt a six-step case study methodology proposed by [29]: (1) plan, (2) project, (3) preparation, (4) data collection, (5) data analysis, and (6) sharing.

1. Plan:

This step identify relevant aspects of "Operação Acolhida" (Welcome Operation). In January 2018, the Brazilian Federal Government signed three Provisional Measures (n° 823/2018, n° 857/2018, and n° 860/2018) that allocated a total of R$280.3 million (> USD 87 million) to assist Venezuelan refugees and migrants through the establishment of Operação Acolhida. The response operation included (i) expanding the supply of documentation, housing, protection of women's rights, children, adolescents, people with disabilities, and indigenous Venezuelans; (ii) voluntary internment to other Brazilian states and host communities; and (iii) the provision of infrastructure and sanitation. A year later, through Provisional Measure No. 880/2019, an additional R$223.8 million (> USD 57 million) was released by the Brazilian government to provide emergency humanitarian assistance to Venezuelans [15]. The Brazilian Federal Government coordinates the actions of international organizations - such as the United Nations High Commissioner for Refugees (UNHCR) and the International Organization for Migration (IOM) -, governmental bodies and civil society entities.

Operação Acolhida has three main pillars [30]:

A. Border ordering Involving the regularization as well as social and medical treatment of Venezuelan immigrants upon arrival in Brazil;
B. Shelter involving the shelter of vulnerable immigrants in Pacaraima and Boa Vista, daily distribution of food, water and hygiene products; and
C. Internalization involving the socio-economic integration of Venezuelan immigrants to Brazilian society as a whole.

2. Project:

In the project step, we defined the research question and design our unique case study focusing on the massive influx of Venezuelan immigrants in Brazil. Previously the Operação Acolhida border control installation, thousands of unregulated Venezuelan migrants have settled in Pacaraima and Boa Vista, overwhelming local public services, contributing to crime and homelessness. With the installation of the Operation, not only did the ordering of the border take place, but also the sheltering and internalization of vulnerable immigrants, drawing Venezuelans attention and increasing the flow to Brazil (reaching the mark of 800 people crossing the border per day). Given these challenges, our study focused on understanding the factors that can enable or hamper the scalability of the operation. Thus, answering the research question: Which factor(s) influence the scalability of the Brazilian Operation in response to Venezuela migration?

3. Preparation:

In the preparation step, we developed a protocol for our case study to increase reliability during data collection and analysis. The protocol included scheduling specific dates for field visits in lodgings, shelters, and military installations (during 2020), and face-to-face interviews with key stakeholders at the operation site (in Boa Vista and Pacaraima).

4. Data Collection:

The data collection step involves the content and direct observations gathered during interviews, field visits, as well as review of open-access documents and internal documents from Operação Acolhida. Data triangulation is guaranteed through multiple means of data collection [31], increasing robustness and validity of the case study. Key interviewees included military personnel (e.g., the Operation's Coordinator, shelter managers, military installations coordinators, internalization personnel, military officers that accompanied the field visits), and representatives from different humanitarian organizations (HO) (e.g., IOM, UNHCR, UN Women, MSF (Doctors Without Borders), and UNICEF (United Nations International Children's Emergency Fund)).

Field visits included visits to several shelters located in two different cities: Pacaraima (close to the border) and Boa Vista (capital of Roraima state). In Pacaraima we visited two shelters: the BV-8 pass-through accommodation and the indigenous shelter. In Boa Vista, we visited three shelters, two spontaneous occupations (e.g., unregulated immigrant occupations within the city) and military installations of Operação Acolhida (e.g., interagency base, Reception and Support Desk (PRA), identification post (PRI), and internalization and triage post (PI Trg)).

Therefore, the interviews and field visits carried out served for a better understanding of the Operation and, consequently, to map the processes involved in the three pillars. The open-access and internal documents, in turn, served to validate the processes seen and explained by the military involved. Data referring to the number of military personnel involved in the operation, the number of Venezuelans entering and remaining in Brazil, the number of daily distributed food and personal hygiene products, the number of Venezuelans internalized, are also among the data collected. These data support the development of the CLD, once human resources and supplies are among the resources needed to scale the Operation.

5. Data Analysis:

The data analysis step consists of examining, categorizing, and tabulating evidence to draw conclusions based on what has been obtained empirically [32]. In this step we attempt to shed light on the mechanisms enabling or hampering the scalability of the Operação Acolhida through the establishment of causal interlinkages across key factors gathered during the data collection step [28]. From the data collected, we map the main processes in the Operação Acolhida, and develop a comprehensive causal loop diagram capturing the key relationships across different pillars of the Operation. Thus, we organize and structure key aspects of the fragmented data collected into a process map and a causal loop diagram (CLD) capturing key feedback processes.

6. Sharing:

The current manuscript represents one aspect of the sharing step.

3 Results and Findings

This section details each of three pillars of the operation and presents a process map in Figs. 1 and 2, and a CLD in Fig. 3.

Border ordering constitutes the first pillar of Operação Acolhida. It involves the regularization of Venezuelan immigrants arriving in Brazil. Prior to its inception, Brazil did not have any control of the number of Venezuelans crossing its borders each day. To gain control, Brazil built a Reception and Identification Station (PRI) in Pacaraima, and a Post of Internalization and Screening (PI Trg) to assist Venezuelans declaring their intention to stay in Brazil.

The fraction of Venezuelans who intend to stay in Brazil, but do not have the financial means to support themselves, are considered unassisted. Those unassisted

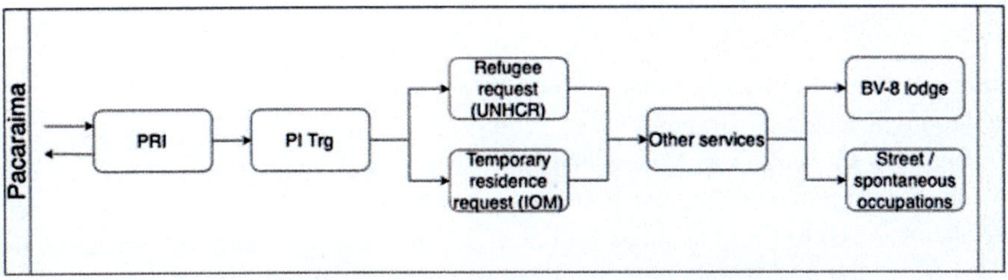

Fig. 1 Process map of Operação Acolhida in Pacaraima

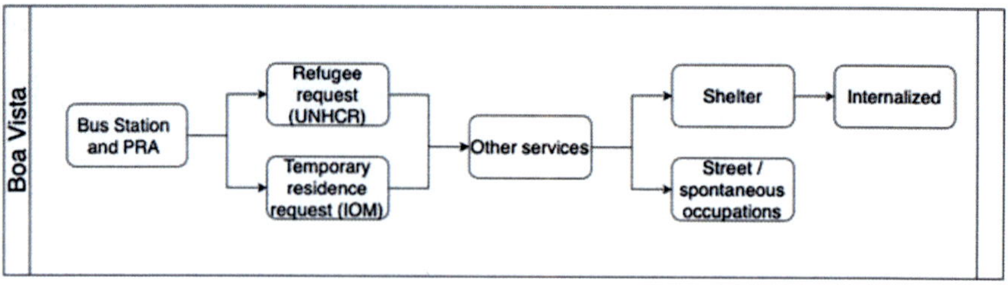

Fig. 2 Process map of Operação Acolhida in Boa Vista

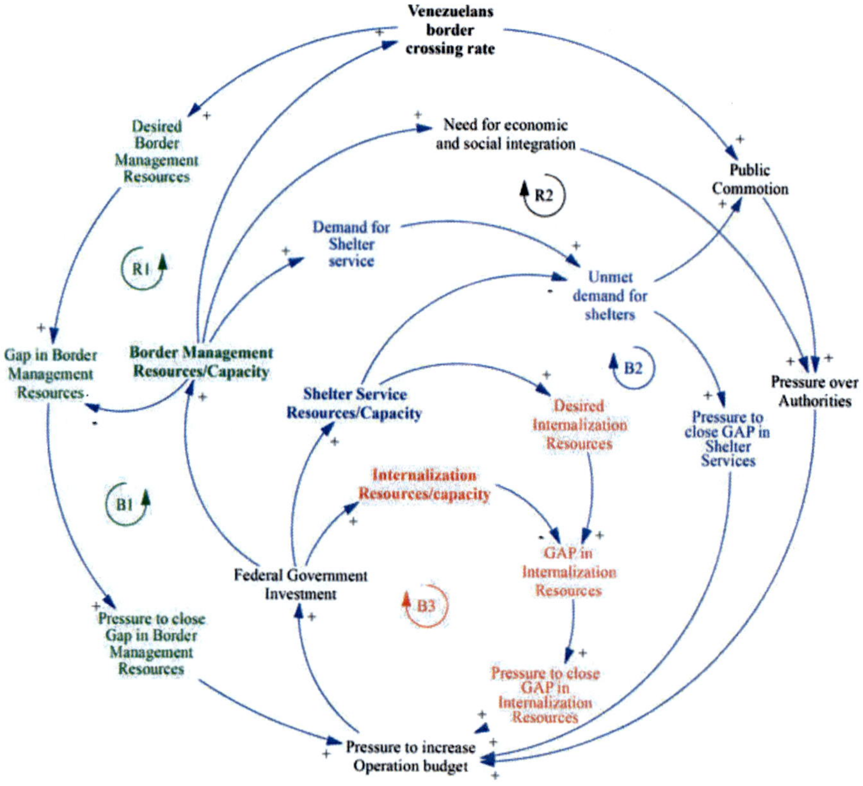

Fig. 3 Causal Loop Diagram capturing key interactions among pillars of Operação Acolhida

immigrants require humanitarian support from Operação Acolhida until their socio-economic insertion in Brazil.

Figure 1 shows that as Venezuelans arrive in Brazil, they go to the PRI. Afterwards, the migrant declares whether he/she has the intention to stay in Brazil. If so, he/she proceeds to the next stage, PI Trg. If not, he/she may return to Venezuela. In the PI Trg, the immigrants must define whether he/she wants to apply for refugee status (with support from UNHCR), or request temporary residence (with support from IOM). Regardless of their choice other services (e.g, vaccination, medical care, and psychological care) are provided for all immigrants. Finally, depending on the vacancies available and the vulnerability of the immigrant, they are allocated in the BV-8 shelter (i.e., the accommodation available in Pacaraima). Immigrants sheltered in the BV-8 facility are eventually transferred to shelters in Boa Vista, according to availability. Some of the immigrants settling on the streets, or spontaneous occupations, in Pacaraima, can eventually go on their own to Boa Vista. Therefore, the ***shelter pillar*** consists in sheltering Venezuelans who do not have the financial means to support themselves, considered unassisted, also offering food, products and means for personal hygiene.

Similar processes take place in Boa Vista. Venezuelan immigrants can reach the city by bus and find support (overnight stay, cafeteria, restrooms) at the Reception and Support Desk (PRA). In Boa Vista, they can also apply for refuge or temporary residence status (with support from either UNHRC or IOM) and have access to other services. Finally, they are either allocated to shelters (according to availability and vulnerability), or not. Sheltered immigrants are eventually internalized to other states in Brazil (***internalization pillar***).

Figure 3 portrays a Causal Loop Diagram (CLD) capturing key interactions among the three pillars of Operação Acolhida. First, the variable Venezuelans border crossing rate captures the flow of Venezuelans entering Brazil. As the operation grew and reception efforts became more structured, Brazil became a migration option for more Venezuelans, which in turn increased the number of daily immigrants further (represented by the reinforcing loop R1). The balancing loop (B1) describes the pressure to increase the budget allocated to Border Management, capturing the causal relationships referring to the Venezuelan ***Border ordering pillar***. As more Venezuelans enter Brazil, the greater the amount of resources required to manage the border. Considering the operation's reception capacity and the country's willingness to receive immigrants, there is a gap in borders management resources. This gap creates pressure on Brazil's Federal Government to increase investment to receive Venezuelans.

The border ordering pillar directly influences the next pillar of the Operation, the shelter of vulnerable Venezuelans, characterized by the balancing loop (B2). The more unassisted Venezuelans enter Brazil, the greater the number of immigrants in need of shelter. As the number of Venezuelans requiring shelter increase, against the installed capacity to receive Venezuelans, the operation faces an increase in unmet demand for shelters. As Brazil fails to serve all immigrants in need, Spontaneous Occupations emerge. In turn, the unmet immigrant demand and the growth of Spontaneous Occupations increased the pressure for further investment in the ***Shelter pillar***.

Finally, the main variables referring to the Venezuelans' internalization to other states in Brazil are captured by the balancing loop (B3). The government identified that it would not be feasible to create an infinite amount of installed capacity in shelters in the state of Roraima without an option for the integration of these immigrants in other states in the medium and long term. Such realization was followed by the need to internalize Venezuelans to other states in the country.

Therefore, as the number of Venezuelans sheltered increased so did the need to internalize these immigrants and open a vacancy to assist new unassisted people. Once again, the actual internalization rates are lower than the demand for internalization, which generates pressure on the government to make more significant investments in the ***Internalization pillar*** of the operation. The reinforcing loop (R2) highlights the need for socio-economic integration of Venezuelan migrants that permeate the operation.

4 Conclusion

The present research focuses on the scalability of the Welcome Operation. To answer the research question, we present some external factors that drive the need for the Operation scalability, and factors internal to the Operation that make it liable to be scaled.

Therefore, Venezuela's ongoing political and economic instabilities, further accentuated by the COVID-19 pandemic that exacerbates hunger and despair in Venezuela, are examples of external factors. The borders closure due to the pandemic, is another external factor that generated a repressed flow of Venezuelans wanting to migrate to neighbouring countries in search of better living conditions. Those external factors cannot be directly influenced by the operation. In contrast, our CLD captures operational actions that can directly influence public opinion and can pressure authorities to pursue specific actions.

To understand the internal factors driving the scalability of the operation, we first map its key processes to understand the dynamic interactions among its three pillars. Then, we develop a CLD to understand how such interactions can lead to specific bottlenecks to scale the operation. Our CLD highlights the central role of investment in the three pillars of the operation (e.g. border ordering, shelter, and internalization). In addition to federal funds, other forms of investment (e.g. services provided by humanitarian organizations, civil society, or donations from private companies) could assist in the scalability of a humanitarian operation focused on receiving, sheltering, and internalizing migrants. The investment here highlighted as the main influential factor in the Operations scalability serves to close the gaps in border management, shelter services, and internalization resources. These gaps consider: (i) supplies such as food, medicines, water, and personal hygiene products; (ii) space and physical structure; (iii) specialized and trained personnel. The government investment explicitly influences these three resource groups; however, they affect the entire Operation since, without them, the Operation stagnates.

The proposed CLD serves as a basis for developing a System Dynamics simulation model to define policies to allocate resources among the three pillars to ensure optimal Operation performance in future work. Other research gaps identified through the case study are the coordination between the military and humanitarian organizations to avoid duplicate work, the adoption of measures to avoid xenophobic manifestations, the distribution logistics for non-food items (NFIs) and WASH (water, sanitation and hygiene) resources, staff turnover, and lack of labor for Venezuelans in the region.

Acknowledgments The authors acknowledge the support of Coordination for the Improvement of Higher Education Personnel (CAPES) [88887.199861/2018-00 – Finance Code 001].

References

1. Sweileh, W. M., Wickramage, K., Pottie, K., Hui, C., Roberts, B., Sawalha, A. F., Zyoud, S. H.: Bibliometric analysis of global migration health research in peer-reviewed literature (2000–2016). BMC public health 18(1), 777 (2018).
2. Mougenot, B., Amaya, E., Mezones-Holguin, E., Rodriguez-Morales, A. J., Cabieses, B.: Immigration, perceived discrimination and mental health: evidence from Venezuelan population living in Peru. Globalization and Health 17(1), 1–9 (2021).
3. Löbel, L. M.: Family separation and refugee mental health – A network perspective. Social Networks 61, 20–33 (2020).
4. de Graaff, A. M., Cuijpers, P., Acarturk, C., Bryant, R., Burchert, S., Fuhr, D. C., McDaid, D.: Effectiveness of a peer-refugee delivered psychological intervention to reduce psychological distress among adult Syrian refugees in the Netherlands: study protocol. European Journal of Psychotraumatology 11(1), 1694347 (2020).
5. Chin, T., Buckee, C. O., Mahmud, A. S.: Quantifying the success of measles vaccination campaigns in the Rohingya refugee camps. Epidemics 30, 100385 (2020).
6. Salem, M., Raab, K., Wagner, R.: Solid waste management: The disposal behavior of poor people living in Gaza Strip refugee camps. Resources, Conservation and Recycling.153, 104550 (2020).
7. McCoy, J. H., Brandeau, M. L.: Efficient stockpiling and shipping policies for humanitarian relief: UNHCR's inventory challenge. OR spectrum 33(3), 673–698 (2011).
8. Karsu, O., Kara, B. Y., Selvi, B.: The refugee camp management: a general framework and a unifying decision-making model. Journal of Humanitarian Logistics and Supply Chain Management 9(2), 131–150 (2019).
9. Losoncz, I., Marlowe, J.: Regulating Immigrant Identities: the Role of Government and Institutions in the Identity Construction of Refugees and Other Migrants. Journal of International Migration and Integration 21(1), 117–132 (2020).
10. Lavenex, S.: The Europeanization of refugee policies: Normative challenges and institutional legacies. Journal of Common Market Studies 39(5), 851–874 (2001).
11. Seidelsohn, K., Flick, U., Hirseland, A.: Refugees' labor market integration in the context of a polarized public discourse. Qualitative Inquiry 26(2), 216–226 (2020).
12. Freier, L. F., Pérez, L. M.: Nationality-Based Criminalisation of South-South Migration: the Experience of Venezuelan Forced Migrants in Peru. European Journal on Criminal Policy and Research, 1–21 (2021).
13. Garcia-Zamor, J. C.: The global wave of refugees and migrants: complex challenges for European policy makers. Public Organization Review 17(4), 581-594 (2017).
14. Rodríguez-Morales, A. J., Suárez, J. A., Risquez, A., Villamil-Gómez, W. E., Paniz-Mondolfi, A.: Consequences of Venezuela's massive migration crisis on imported malaria in Colombia, 2016-2018. Travel medicine and infectious disease 28, 98 (2019).
15. R4V (Response for Venezuelans plataform) Homepage, https://r4v.info/, last accessed 2021/01/15.

16. Silva, G. J., Cavalcanti, L, Oliveira, T, Macedo, M.: Refúgio em Números, 5th edn. Observatório das Migrações Internacionais; Ministério da Justiça e Segurança Pública/ Comitê Nacional para os Refugiados. Brasília, DF: OBMigra, 2020. Available at: https://www.justica.gov.br/seus-direitos/refugio/refugio-em-numeros
17. ACNUR Brasil (2020) Homepage, https://r4v.info/es/situations/platform/location/7509, last accessed 2021/03/28.
18. Lamenza, A. A. S., Fontainha, T. C., Leiras, A.: Purchasing strategies for relief items in humanitarian operations. Journal of Humanitarian Logistics and Supply Chain Management 9, 151–171, 2019.
19. Tabaklar, T.: Scalability and Resilience in Humanitarian Supply Chains. Helsinki, Finland: Hanken School of Economics (2017).
20. Sterman, J. D.: Business Dynamics: Systems Thinking and Modeling for a Complex World. McGraw-Hill, Irwin (2000).
21. Sopha, B. M., Asih, A. M. S.: Human resource allocation for humanitarian organizations: a systemic perspective. In MATEC Web of Conferences, vol. 154, pp. 01048 (2018).
22. Mishra, V., Sharma, M. G.: Understanding Humanitarian Supply Chain Through Causal Modelling. South Asian Journal of Business and Management Cases 9(3), 317–329 (2020).
23. Besiou, M., Stapleton, O., Van Wassenhove, L. N.: System dynamics for humanitarian operations. Journal of Humanitarian Logistics and Supply Chain Management 1(1), 78–103 (2011).
24. Zhang T. Z., Lu Y. M.: Study on simulation and optimization of the road rush-repair model after disaster. Applied Mechanics and Materials 50(2), 298–303 (2011).
25. Gonçalves, P.: Balancing provision of relief and recovery with capacity building in humanitarian operations. Operations Management Research 4(1-2), 39–50 (2011).
26. Kunz, N., Reiner, G., Gold, S.: Investing in disaster management capabilities versus pre-positioning inventory: a new approach to disaster preparedness. International Journal of Production Economics 157, 261–272 (2014).
27. Yang, J., Dong, H.: A prediction based migration route evaluation method for refugees. In: 10th International Congress on Image and Signal Processing, BioMedical Engineering and Informatics (CISP- BMEI), pp. 1–5 (2017).
28. Dolezal, O., Tomaskova, H.: System dynamics in migration modeling. In: Proceedings of the 31st International Business Information Management Association Conference (IBIMA) 2018: Innovation Management and Education Excellence through Vision, pp. 5056–5060 (2018).
29. Yin, R. K.: Case Study Research: Design and Methods, 5th edn. Sage Publications, Los Angeles (2013).
30. Operação Acolhida (2020) Homepage, https://www.gov.br/acolhida/historico/, last accessed 2021/03/28.
31. Voss, C., Tsikriktsis, N., Frohlich, M.: Case research in operations management. International journal of operations and production management 22, 195–219 (2002).
32. Godoy, A. S.: Estudo de caso qualitativo. Pesquisa qualitativa em estudos organizacionais: paradigmas, estratégias e métodos. Saraiva, São Paulo (2005).

Optimizing Human Resources: The Case of Venezuelan Migration in Lima, Peru

Irineu de Brito Junior, Renato Quiliche, Mariana Moyano, and Mario Chong

Abstract In 2021, Peru received 1,049,970 Venezuelan inhabitants due to the country's crisis, which caused inflation of more than 1000% and an unemployment rate of 250% from 2013 to date. This paper presents a new perspective on the social problem shown above. A systemic approach from logistic theory allows observing the Venezuelan anthropogenic crisis and the migratory process as an uncontrolled flow of human resources. Thus, if migration to Peru is represented as a supply chain model whose flows are human resources, a theoretical framework can be applied to optimize them and help incorporate them into Peruvian society. The reason is to improve both their quality of life and the economic activity of the recipient country. A survey focused on knowing the profile of the Venezuelans who arrived in Peru was carried out. The importance of characterization will help propose solutions such as creating migration policies and future logistical and consumption impacts that can affect certain residence districts and, finally, get to know their future expectations in our country. Adaptation to new migrants is a growth opportunity for the city and a generator of business opportunities that help improve the city and economy of the country.

Keywords Humanitarian logistics · Labor markets · Adaptation · Migration · Policies

I. de Brito Junior (✉)
São Paulo State University, São Paulo, Brazil
e-mail: irineu.brito@unesp.br

R. Quiliche
Brasil Pontificia Universidad Católica do Rio de Janeiro, Rio de Janeiro, Brazil
e-mail: renato.quiliche@aluno.puc-rio.br

M. Moyano · M. Chong
Facultad de Ingeniería, Universidad del Pacífico, Lima, Peru
e-mail: md.moyanom@alum.up.edu.pe; m.chong@up.edu.pe

J. Vargas Florez et al. (eds.), *Production and Operations Management*, Springer Proceedings in Mathematics & Statistics 391,
https://doi.org/10.1007/978-3-031-06862-1_24

1 Introduction

In recent years, Venezuela's political and economic instability (International Monetary Fund 2018) has intensified dramatically to the point of triggering an anthropogenic crisis (Smilde and Ramsey 2019). This situation has forced nearly 1.5 million Venezuelans to leave their country to seek a better quality of life. Due to the good economic conditions of Peru, a large number of Venezuelans have opted for this country as their new home (Economic Commission for Latin America and the Caribbean 2019). Until February 2021, 1,049,970 Venezuelans moved to Peru, which means that Peru is the second-largest destiny for Venezuelan migrants (Plataforma de Coordinación Interagencial para Refugiados y Migrantes de Venezuela 2021). Many of them are students, professionals, or parents who, seeing their future truncated by the conditions of their country, migrated to other cities in search of new opportunities. However, joining a completely different environment means starting from scratch (Bhugra 2004). This refers to a set of effects and processes that consequence human decisions.

Although it is usual for a humanitarian crisis to seek to take aid to the place where the event was triggered, in the Venezuelan case, migration has made the aid focus on reaching those affected in the area where they are now. In addition, we will seek to characterize the profile of these migrants to identify their primary needs and find the appropriate logistics that must be behind in search of satisfying their necessities. Specifically, the gap in the quality of labour between the current situation of Venezuelans and the conditions under which they lived in their homeland will be contrasted. We hypothesize that Venezuelans travelled for economic reasons in need of a different source of income to maintain their families. However, labour markets in Peru offered low skilled jobs, and Venezuelans with a high educational level had to take what labour markets provided. These hypotheses will be discussed in the function of empirical insights. This work presents the results of a survey applied to the Venezuelan population in Lima. This represents an opportunity for research to find results that contribute to creating strategies to favor the economic adaptation of migrants.

The advantages of conducting this type of research lie mainly in generating reliable information about the socio-economic and living conditions experienced by migrants in the country and their perception of how their lives have changed in this new country. From the former perspective, the recipient country must respond to the increasing demand for the required job vacancies. Thus, the research questions are as follows: What are the principal reasons to emigrate? What changes were the most important in terms of labour? What policies can be proposed to improve adaptation and meet Venezuelans' needs according to their plans? We will test the hypotheses described above with survey data from Venezuelan immigrants.

The following article will be divided into five sections. First, the actual state of relevant literature for addressing the humanitarian crisis in Venezuela will be delimited. Second, the methodology used to collect data for the profile of migrants and the hypotheses testing methods will be described. Third, the analysis of the results and their interpretation will be carried out. Finally, conclusions and recommendations on the study will be presented.

2 Literature Review

Since the early years, humans have been in constant movement for different reasons. According to the International Organization for Migration (IOM), human mobility can be defined as the mobilization, voluntary or forced, of people from one place to another to exercise their right to free movement. It is a complex process and is motivated by various reasons. Recently, migrations around the world have increased by 22% (Portal de Datos Mundiales sobre la Migración 2019). In 2019, the number of migrants reached 272 million people compared to 220 million in 2010 (ONU, DAES 2019).

An economic collapse is a breakdown of a national, regional, or territorial economy. It occurs at the onset of an extreme version of an economic contraction, depression, or recession. It can last any number of years, depending on the severity of the circumstances. When an economic collapse causes massive migration or human mobility, we can define the phenomenon as a humanitarian crisis. The Venezuelan exodus is a humanitarian crisis. It produces human suffering in several ways; it affects people's livelihoods to the extent that people have to move to another country to survive hunger and extreme poverty (Gedan 2017).

Venezuela is the second-largest cause of high migration around the world. Although people usually leave a country after a violent war or a systemic political collapse (Aburas et al. 2018), Venezuelan mass migration began after a systemic collapse during the golden oil decade (2004–2014), where crude oil prices exceeded $100 per barrel. The systemic collapse was followed by the country's violation and destruction of economic rights; furthermore, an extraordinary increase in indebtedness caused secure access to credit in international markets (Nelson 2018; Seelke 2018). Figure 1 illustrates, in short, the Venezuelan crisis.

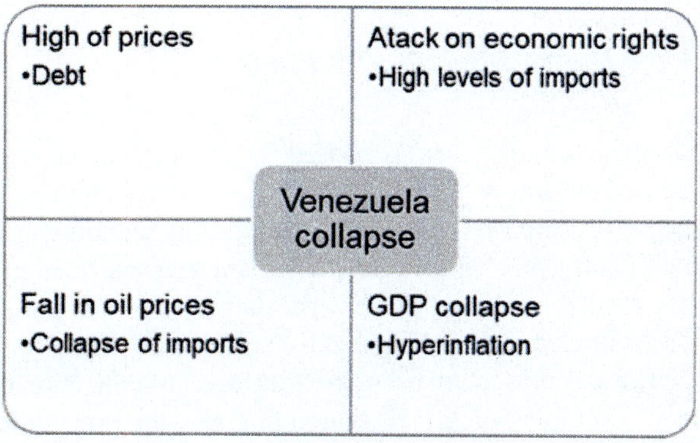

Fig. 1 Summary of the Venezuelan crisis according to Hausmann and Neffke (2019)

A wide variety of literature about Venezuelan migration has been written. The literature focused on the impacts of Venezuelan migrants covers two critical topics: economic impacts (Denisse and Morales 2020; Olivieri et al. 2020) and health impacts (Carroll et al. 2020; Tuite et al. 2018). Most of the literature emphasizes the effects of migrants on resident population economic and health dynamics, such as employment, salaries (Denisse and Morales 2020; Olivieri et al. 2020) or disease spread (Tuite et al. 2018). There is scarce literature that emphasizes impacts on migrants, such as their mental health (Carroll et al. 2020) or other aspects of their adaptation to a new environment. This paper contributes to the literature that focuses on impacts on the population of Venezuelan migrants. Furthermore, this paper covers an aspect of economic impact that includes a comparison with the previous socio-economic status of migrants before they move to the country of residence: Perú.

3 Methodology

A survey was developed in function of six thematic axes based on the report "Status of foreign migrants in Peru and their access to social services, health services, and education" (Sphere Association 2018; DTM 2019). Online forms were made in the Survey Monkey online web tool (es.surveymonkey.com) and were sent to the objective population via e-mail. The accurate population was selected randomly based on a database or registry of Venezuelan migrants. The survey results may be limited to Venezuelans who have access to the internet. Still, they provide interesting insights that could be used for inference, at least for the population with access to the internet.

The data collection methods will be explained; specifically, the construction of questions for the applied survey will be depicted. Then, the main hypotheses of this study will be presented. These hypotheses are based on the literature review and contribute to the empirical literature on the impacts of Venezuelan migrants on the migrant population.

3.1 Data Collection Methods

Six thematic axes were proposed for the survey. The first, called "Personal Data", focused on knowing slightly more about the migrant without much detail about identity. Therefore, the main interest of this section was to know when they entered the country and what were their main reasons to migrate to understand whether migration was related to the Venezuelan crisis and which were the most important reasons to leave their homeland.

The second section asks about the level of education or instruction the immigrant has. From this point, it is desired to check the preparation of Venezuelans to see if

Table 1 Description of variables/parameters and values

Variable/ parameter	Description	Value
N	Population size, retrieved from R4 V (2021) data	1,049,970
Z_a	Constant value obtained from standard normal distribution at 95% of confidence	1.96
p	Proportion of females in sample	0.5
q	Complement of p	0.5
d	Desired sample error	0.057

they are genuinely underqualified and their talent is being wasted with the activities they perform. In this way, reasonable strategies could be established to insert them into the Peruvian labour market.

The family theme is presented as a third point of the survey. In this regard, the survey asks about the number of children the respondents could have and if they live with them in Peru. The objective of these questions is to know the number of families that have been separated as a result of this crisis and find alternative solutions to bring them back together in the country where they want to develop.

The fourth point focuses on the plans of Venezuelans to see their interest in returning to their country or staying in Peru. To do this, they are asked if they want to carry out any study in the country, what role and in which sector they would like to perform in their following jobs. This goes hand in hand with the fifth section of the survey, which emphasizes the current activities in the country. The purpose of these questions is to identify the gap between their reality and their aspirations, to understand this gap and propose action plans to reduce it.

Finally, the sixth block of questions is based on what Peru has been able to offer them in their time of permanence, despite the country's deficiencies in its public administration. The objective of these questions is to know the migrants' perspective on whether the government is helping them with their development.

Regarding the sample, since we approximate the population size to 871,000 of the Venezuelan population in Peru, we opt for the following formula (Gómez and Requena 2018), considering a 95% confidence level for our inferences based upon the sample characteristics:

$$n = \frac{N.Z_a^2.p.q}{d^2.(N-1) + Z_a^2.p.q} \quad (1)$$

Table 1 shows the description of variables/parameters and their values.

The resulting sample size has 151 observations. In our survey, we collected information on 154 people of Venezuelan nationality, so we met the minimum sample size condition for inference. The sample was conducted to characterize the Venezuelan population that lives in the city of Lima.

3.2 *Hypothesis and Testing*

The central hypothesis of this paper is that Venezuelans did not get a job in Peru that corresponds to their level of education and past work experience in Venezuela. This economic impact on the migrant population is significant because it measures economic adaptation. The insights help to propose public policies oriented to improve the adaptation of Venezuelans. This study also includes other hypotheses, such as Venezuelans having more children than maintaining or staying and working in Peru in the future. According to these insights, we will propose public policies oriented to improve adaptation and optimize Venezuelan migrants as essential human resources.

The most relevant questions are: What is your actual level of education? What is the kind of job you did in Venezuela? What is your occupation in Peru? What are your plans for the future? These questions will be cross-tabulated, and some graphics will be added to generate insights for the central hypothesis of the paper. Some descriptive statistics and tables are constructed regarding the rest of the hypotheses. The survey results will serve to test the hypotheses of this research quantitatively.

4 Results and Discussion

According to the survey results, 62.9% of the individuals declared that they moved to Peru mainly for work and 15.6% for political reasons. This result suggests that the search for paid work is the main reason behind the displacement. The need for earnings in a different currency motivated Venezuelans to travel to Lima. In most cases, Venezuelans send money, from country to country, to their families. Figure 2 shows the reasons that motivated the migration of Venezuelans.

Another survey result was that 46.8% of Venezuelans had at least one child, but only 9.7% reported that he/they lived in Peru with their children. This result suggests that, at least for this case, Venezuelan migration is based on the work search and not on bringing the children or moving with them to Lima. In this sense, migration is economic. Due to the lack of opportunities in the country of residence, Venezuelans

Fig. 2 Reasons behind Venezuelan migration

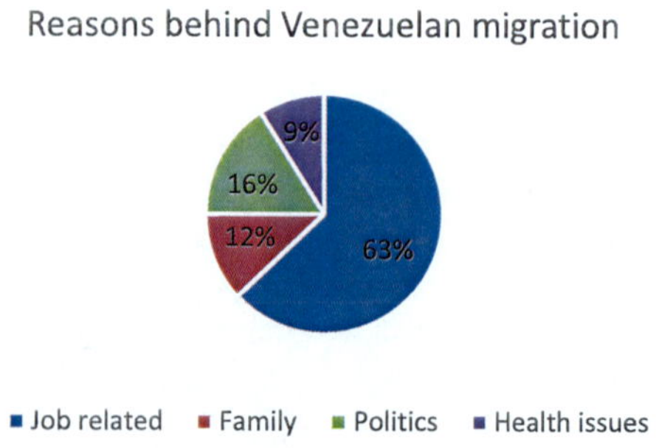

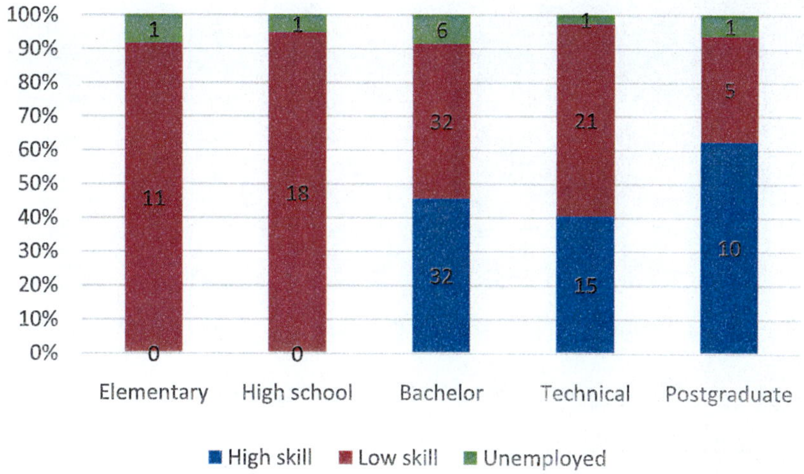

Fig. 3 Job skill in Venezuela by study degree

moved to another city searching for better opportunities, specific job opportunities. Nevertheless, one of the main problems addressed in this paper is that the jobs in which Venezuelans become involved are not the jobs they must have according to their educational level. This is why most Venezuelans decide to start a business or do any economic activity regardless of what they have studied for or their level of education. Figure 3 shows the level of skill of the job in Venezuela by level of education:

Additionally, it is worth mentioning that 45.5% of Venezuelan migrants have a bachelor level of education, 24.0% have a superior technical education, and 10.4% have postgraduate education. Thus, one critical insight was that, in comparison to Peru, the migrant population is composed of people with a high level of education. Then, we observed that high-skill jobs such as economists, professors, engineers, police officers, medics, and psychologists, among others, are mainly carried out by postgraduates. On the other hand, high skill jobs are not done by elementary or high school Venezuelans. This figure proves that the classification of high skill and low skill jobs makes sense, and it may be correct. Figure 4 shows the job skill in Peru by study degree:

Next, we observed higher levels of unemployment, showing severe difficulty in finding a job in Peru. Due to this harder difficulty, most Venezuelans have changed their positions to low skill jobs, such as street vendors, cab drivers, security guards, and chefs. There are few, most of them with a postgraduate level, Venezuelans, that still have high skill work. This insight shows that Venezuelans have obstacles to adapting to a new environment, the Peruvian environment. Figure 5 shows the plans for Venezuelans by their level of skill in their jobs:

Concerning Venezuelan's plans, its composition is uniform among the categories except for the study. Few Venezuelans plan to keep studying (13.6%). High-skilled workers plan to undertake and start a business or invest their money, or travel

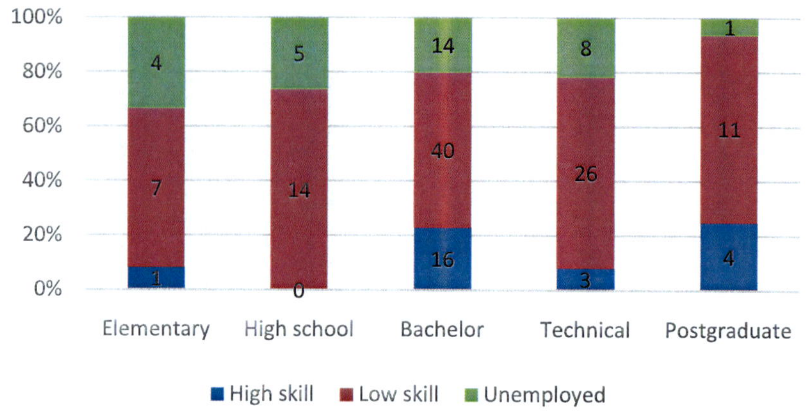

Fig. 4 Job skill in Peru, by study degree

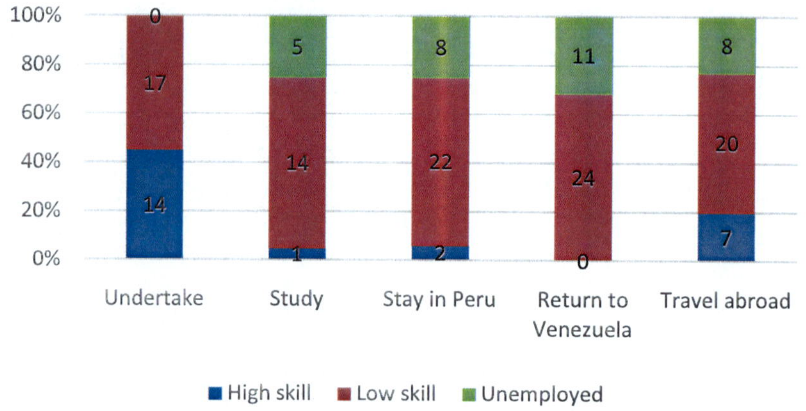

Fig. 5 Future plans of Venezuelans by job skill in Peru

abroad. The rest of the workers, who were low-skilled or unemployed, plan to study, stay in Peru, return to Venezuela or travel abroad. No highly skilled workers plan to return to Venezuela, and only a few plan to keep studying or stay in Peru. These insights suggest that Venezuelans prefer to undertake, especially high-skill workers, given the difficulties of finding jobs or adapting to the local labour markets. Figure 6 shows how plans vary across the level of study of Venezuelans:

We observed that bachelors and technicians plan to become involved in high-skill jobs. Postgraduates mainly plan to travel abroad, regardless of whether they have a high skill job in Peru. Elementary and high school Venezuelans plan to stay in Peru or return to Venezuela. These insights suggest that more educated Venezuelans plan to work and undertake. In contrast, less educated Venezuelans plan to remain in Peru and work, as it is an essential source of income for their families.

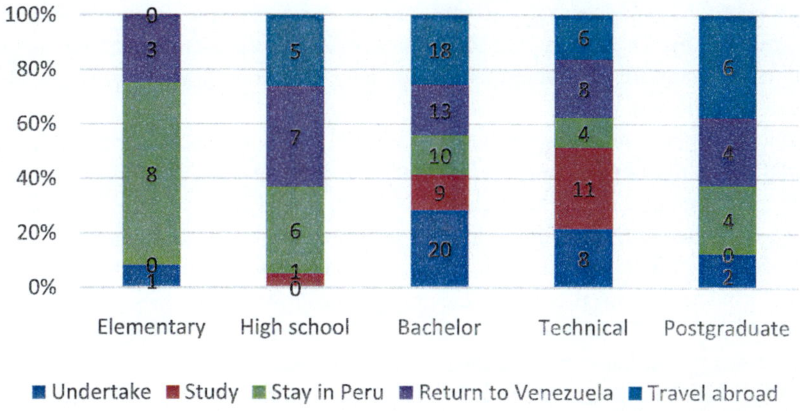

Fig. 6 Plans of Venezuelans by the level of education

Given these results, the optimal policy should consider the intentions of Venezuelans and their current situation in Peruvian labour markets. On the one hand, policymakers need to support adaptation to labour markets by creating jobs for Venezuelans that they can take according to their level of education or encouraging private investment that creates the required job opportunities. In practice, creating jobs can be complex, so a short-run solution may be to create these jobs from public investments. Another short-run policy can be implemented as a particular credit program for Venezuelan entrepreneurs considering Venezuelan plans.

5 Conclusions

This research has provided an overview of the reasons behind massive Venezuelan migration. As Peru holds the second-largest case of Venezuelan migration, it is essential to know what kind of human resources are moving to the country (Moulin & Magalhães 2020) and what policymakers can do to support them. To optimize the human resources, the insight suggests that there is a severe difficulty of adaptation for Venezuelans regarding the local labour markets; most of them have low skill jobs regardless of their educational level. The 79.9% of Venezuelans who have arrived at Lima have superior education, so policymakers need to think about how the city can optimally use these human resources. Also, considering the Sustainable Development Goals established in the 2030 UN Agenda, it is essential to highlight that regards such as respecting labour rights of migrant workers, reducing the costs of transferring remittances and providing legal identity for all (United Nations 2017) should be taken into account by policymakers. Thus, the proposal for short-run policies is to simplify the steps to access legal documents like immigration cards or temporary residence permits. This is given that, in the case of Peru, the multiplicity of migration categories generates confusion and uncertainty in Venezuelans (Blouin and Freier 2019). Moreover, in some cases, it hinders the labour insertion of Venezuelans because employers commonly request these documents, and, as there is a high demand for obtaining them, it can take several months for Venezuelans to have a legal permit.

On the other hand, long-run policies should promote the transition to a more formal economy. For this purpose, it is mandatory to develop a comprehensive and integrated strategy cutting across a range of areas and involving institutional and civil society actors (International Labour Organization 2018). For example, regarding the plans of Venezuelans, policymakers should pay attention to those who plan to start a business, it is crucial for the country of destiny that Venezuelans do not face many obstacles at the moment of starting a business, so there could be some facilities to micro and small enterprises or also, with more prominent companies, encourage private investment that creates jobs and meet labour demand. Following these recommendations, the Peruvian government can provide an excellent economic environment for Venezuelans, which will, in turn, improve the economic environment for Peruvians and create more development opportunities.

References

Aburas, R., Najeeb, A., Baageel, L., & Mackey, T. K. The Syrian conflict: a case study of the challenges and acute need for medical humanitarian operations for women and children internally displaced person. Bmc Medicine, 16, 1. (2018).

Bhugra, D. Migration and mental health. Acta Psychiatrica Scandinavica, 109(4), 243-258. https://doi.org/10.1046/j.0001-690X.2003.00246.x (2004).

DTM. Monitoreo de flujo de población venezolana en el Perú, Reporte 6. https://data2.unhcr.org/es/documents/download/71522 (2019).

Economic Commission for Latin America and the Caribbean [ECLAC]. Atlas of migration in Northern Central America. (2019).

Gómez, C., & Requena, M. El Panel de Hogares y la toma de decisiones comerciales. Madrid: Difusora Larousse - Ediciones Pirámide. (2018).

Hausmann, R., & Neffke, F. M. The workforce of pioneer plants: The role of worker mobility in the diffusion of industries. Research Policy, 48(3), 628–648 (2019).

International Monetary Fund. World economic outlook. Washington, DC: International Monetary Fund. (2018).

ONU International migrant stock. Departamento de Asuntos Económicos y Sociales, 2019 https://www.un.org/en/development/desa/population/migration/data/estimates2/estimates19.asp (2019).

Portal de Datos Mundiales sobre la Migración. Una Perspectiva Global: Número total de migrantes internacionales 2019. https://migrationdataportal.org/es?i=stock_abs_&t=2019 (2019).

Nelson, R. M., & Library of Congress. Venezuela's economic crisis: issues for congress. (CRS reports (Library of Congress. Congressional Research Service)) (2018).

Seelke, C. R., & Library of Congress. Venezuela: Background and U.S. Relations. (CRS reports (Library of Congress. Congressional Research Service)) (2018).

Smilde, D. & Ramsey, G. El difícil camino hacia adelante: Venezuela y el grupo de contacto internacional. Institución Fundación Carolina. (2019).

Sphere Association. The Sphere handbook: Humanitarian Charter and minimum standards in humanitarian response. (2018).

R4 V Plataforma de Coordinación Interagencial para Refugiados y Migrantes de Venezuela. Data base of Venezuelan migrants in LATAM countries. https://www.r4v.info/es/refugiadosymigrantes (2021).

Olivieri, S., Ortega, F., Carranza, E., & Rivadeneira, A. The Labor Market Effects of Venezuelan Migration in Ecuador. https://doi.org/10.1596/1813-9450-9336 (2020).

Moulin, C., & Magalhães, B. Operation shelter as humanitarian infrastructure: material and normative renderings of Venezuelan migration in Brazil. Citizenship Studies, 00(00), 642–662. https://doi.org/10.1080/13621025.2020.1784643 (2020).

Denisse, M., & Morales, F. Adjustments in the Labor Market. (2020).

Gedan, B. N. Venezuelan Migration: Is the Western Hemisphere Prepared for a Refugee Crisis? SAIS Review of International Affairs, 37(2), 57–64. https://doi.org/10.1353/sais.2017.0027 (2017).

Carroll, H., Luzes, M., Freier, L. F., & Bird, M. D. The migration journey and mental health: Evidence from Venezuelan forced migration. SSM - Population Health, 10, 100551. https://doi.org/10.1016/j.ssmph.2020.100551 (2020).

Tuite, A. R., Thomas-Bachli, A., Acosta, H., Bhatia, D., Huber, C., Petrasek, K., Watts, A., Yong, J. H. E., Bogoch, I. I., & Khan, K. Infectious disease implications of large-scale migration of Venezuelan nationals. Journal of Travel Medicine, 25(1), 1–8. https://doi.org/10.1093/jtm/tay077 (2018).

United Nations, Department of Economic and Social Affairs, Population Division. International Migration Policies: Data Booklet (ST/ESA/SER.A/395) (2017).

Blouin, C., & Freier, L. F. Población venezolana en Lima: entre la regularización y la precariedad. L. Gandini, F. Lozano Ascencio y V. Prieto (coords.) Crisis y migración en la población venezolana. Entre la desprotección y la seguridad jurídica en Latinoamérica. México: UNAM. (2019).

International Labour Organization. Informal Economy (Decent work for sustainable development). https://www.ilo.org/global/topics/dw4sd/themes/informal-economy/lang%2D%2Den/index.htm (2018).

An Analysis of Public Hospital Services and Technologies 4.0: A Conceptual Framework for Health Management

Annibal Scavarda, Douglas Markonne, Gláucya Lima Daú, Ana Isabel Sousa Magalhães Guerra, and Rabea Qassim Nafil

Abstract The health 4.0 coming from the fourth industrial revolution technologies has influenced several segments in the private and public sectors. This research study analyzes the disruptive technologies and how these technologies can contribute to public health institutions, creating a conceptual management model. The research was developed through a literature review in the Web of Science database, where 75 articles were found based on search equations and pre-established filters. In total, 62 abstracts were selected for reading, five being excluded for not meeting the objectives, leaving 57 articles. Of these, 35 articles were read in full and the ten that met the scope were selected for analysis. The research period was between February and August 2021. A summary table of the articles found presents the results, helping to compose the findings, identifies disruptive technologies, and their contributions to hospital management. In a public health system patients are constantly admitted in the same health institutions. It is discussed how disruptive technologies can provide practical and innovative solutions that meet the desires of interested parties to maintain a constant and effective flow of care in a way that overcrowding does not occur. In the conclusion, the integration between health institutions and disruptive technologies can bring improvements in the work process and in the performance of professionals in the public hospitals. This integration can support theoretical elements for the preparation of project proposals and new management models based on health 4.0.

Keywords Public hospital · Hospital administration · Disruptive technology

A. Scavarda · D. Markonne · G. L. Daú · A. I. S. M. Guerra (✉) · R. Q. Nafil
Universidade Portucalense Infante D. Henrique, Porto, Portugal
e-mail: ana.guerra@upt.pt

J. Vargas Florez et al. (eds.), *Production and Operations Management*, Springer Proceedings in Mathematics & Statistics 391,
https://doi.org/10.1007/978-3-031-06862-1_25

1 Introduction

In the public health management field, there are many challenges faced by hospital administration. The disruptive technologies as innovative proposals for solutions, coming from industry 4.0, have contributed over time. These contributions are the promotion of improvements in the logistical processes, the identification of events, and the actions of standardization in the management of institutions. The disruptive technologies embedded in healthcare services provide opportunities and challenges. These opportunities and challenges are the difficulty of absorbing concepts that are still under theoretical investigation, predicting the effects and measuring the impacts, resulting from the introduction of these technologies in healthcare work environments and processes.

This research study seeks to answer how disruptive technologies contribute to the emergence with a new conceptual model of management in public health institutions as a working tool and research source for researchers, managers, and health professionals. Currently, there is a tendency to integrate different knowledge in health with disciplines focused on technological areas. This focus brings to provide an acceptable level of safety for medical care, to promote the quality of patient care, and to help of the professionals in the development their critical care and administrative activities.

Therefore, the study becomes relevant insofar as the European Commission has joined efforts to support action plans for innovation that - highlighted is the eHealth project 2012–2020. In this project, the main objectives are to invest in research seeking excellence in the medicine of the future, to improve the quality and availability of care, and to reduce healthcare costs. Celdrán et al., [1].

In Germany, the engagement in the eHealth project and started its government strategies for industry 4.0 are highlighted. It worked on the advancement of innovative concepts and mergers of different technologies, in specific of disruptive technologies. The motivation elements for the development of this study were the promising potential of the theme, the numerous possibilities for studies, and the various forms of applicability.

It is believed that integrating theoretical models of innovation with the findings identified on disruptive technologies in the analyzed papers. The study results can significantly contribute to the emergence of new understandings of integrated technology management, which applicable in public hospital organizations can successfully impact all stakeholders interested in the proposal.

Therefore, it is proposed to analyze how disruptive technologies can foster a conceptual management model. In the next section, named Innovation and disruptive technology in healthcare institutions, are explained concepts and fundamentals of how disruptive technologies are related to healthcare institutions.

2 Innovation and Disruptive Technology in Healthcare Institutions

A study by Herrmann [2] on digital innovation and disruptive technologies in healthcare management reports the possibility of verifying that there is a willingness of some corporations to invest in the sectorial digital transformation. In this sense, it is believed that this technological trend of digitization of labor processes can contribute to the qualification of the management of public hospitals. In addition, it is able to mitigate challenges faced by health managers in the face of the strong bureaucratization of the political system of countries and the centralization of state governance, which are striking and structuring characteristics of public administration.

In this context of traditional governance, the public hospital administration has been challenged to maintain efforts to meet people's health expectations and solve the problems of patients affected by various diseases. Their healthcare costs have been increasing progressively. The public hospital administration receives few integrated solutions with sustainable and scalable properties that can meet the balance equation between service supply and demand.

Returning to the understanding of the digital transformation in the health sector, Herrmann [2] states that there are great benefits in the insertion of disruptive technologies in administration. These technologies improve the performance of the healthcare professionals' care, reduce operating costs, improve the quality of care, and promote an environment of collaboration between people, organizations, and business models.

The research developed by Bonilla-Asalde et al., [3] in a health institution in Peru, addressed the determination of barriers and perspectives of synergy in the implementation of a public hospital oriented towards administrative and care results. It was found in the results that an analysis of the hospital management and that a favorable attitude towards changes prevailed among professionals. The barriers were identified in all areas of exploration, the most relevant being inadequate forms of planning and budget management, deficiencies in personnel management and control mechanisms, dysfunctions in organizational culture, and the political context.

It is understood that health technologies meet these objections in order to promote the rethinking of conceptual paradigms that underlie isolated management models. However, they need new business models that allow the integration of several problem-issues, administrative, and technological tools, professionals and organizations in a single objective. In this regard, the innovations promoted in hospital management by disruptive technologies can support new methods and management models. These resolve questions arising from strategic and budgetary planning, technical skills of employees, service flows, monitoring and control of people and processes, and finally the cultural and organizational transformation that innovative technologies have the potential to offer.

3 Methodology

The study was developed through a integrative literature review. According to Whittemore et al., [4], the integrative literature review is a broad method that allows the inclusion of theoretical and empirical literature for analysis.

The literature search was carried out in the Web of Science database, the searches were based on a mathematical algorithm expression where a specific code was used, the TS descriptor (the symbol represents that the search was to find keywords in the following places in the platform: topic, title, abstract and keywords). Thus, the search equations of this study were "TS = (hospital administration and disruptive technologies)" where 33 texts were found and "TS = (hospital and disruptive technologies)" where 75 texts were found. The inclusion criteria were only complete papers and available electronically, peer-reviewed, in English, having been published in the period from 2016 to 2021, and contemplating the proposed object with updated papers. The editorials, letters to the editor, works published in event proceedings, incomplete texts, papers that appear one more time were excluded.

Thus, 62 abstracts were included for floating reading, of which three were excluded for being conference proceedings and two from the year 2015. For full reading, 35 papers were selected and ten papers of these remained, that met the scope of the work and composed the synthesis of analysis. The study period took place from February to August 2021.

The selection process of the findings was performed by reading the abstracts in their entirety with the aid of Rayyan software (text manager). The integration of the papers was on Medley software (citation and bibliography manager) so that ten were for the final selection papers that met the criteria guided by the PRISMA flowchart.

4 Results and Discussion

The discussion section seeks to analyze how the applicability of disruptive technologies can occur in public and private health institutions and their implications for ways to foster a conceptual management model 4.0 (Table 1).

In a public health system people can be admitted constantly and regularly admitted to the same hospitals. The use of information and communication systems, artificial intelligence, and care automation robots integrated with cloud computing can encourage them to become actors active social workers and co-participants in their treatments. So, in order to help they make decisions collectively and in a unique experience that brings security and significance to everyone, patients, family members, managers, and technical professionals [5].

It is believed that these disruptive technologies can provide their health workers and political agents with specific and specialized knowledge; collaborating in the monitoring of labor processes in order to advise the prioritization of investments, allocation of human, and material resources.

Table 1 The synthesis of papers analyzed – conceptual model of public hospital management 4.0

N°	Title	Authors	Disruptive technology	Contribution to management in health institutions
01	Strategies for technology-based service encounters for patient experience satisfaction in hospitals.	Lee, Donhee.	Medical information and communication technology (ICT), biotechnology, internet of things (IoT) and sensors, intelligent self-service devices, artificial intelligence (AI) and robots.	It encourages patients to participate in the treatment process, positively affect satisfaction due to patients' experience. It provides new insights to hospital administrators, regarding the allocation of the strategic investments in technology.
02	Avoidance of disruptive health technology: A revealed causal mapping (RCM) approach [11].	Samhan, Bahae; Joshi, KD.	Information systems, electronic hospital record (REH)	The research source for secondary research. Optimizes data handling and management of patient and hospital information.
03	Changing the practice of pediatric surgery due to the emergence of connected health technologies.	Riikka Niemeläf, Minna Pikkarainena, Mari Ervastib, Jarmo Reponena.	Scanning, mobile apps (app), hospital computer system.	It promotes an exchange of internal (hospital) and external (home) communication between doctors and parents. It offers technological support to support shared decisions about the child's treatment.
04	Development and innovation of system features to optimize patient care [10].	Johnson, Thomas J.; Brownlee, Michael J.	Hospital electronic record (REH), 3D printing, drones, telehealth, integration of multi-hospital health systems.	It leverages the standardization of systems, a support resource for optimizing test results, improves the ways in which medications are administered, and improves patient access to pharmaceutical services.
05	Exploring Mobile work in healthcare: Clinical perspectives on the transition to a first Mobile work culture	Nisha Shah, guy Martin, Stephanie archer, Sonal Arora, Dominic king, Ara Darzi	Technologies and mobile devices.	Support the daily clinical and administrative practices in the hospital.

(continued)

Table 1 (continued)

N°	Title	Authors	Disruptive technology	Contribution to management in health institutions
06	The role of informatics in patient-centered care and personalized medicine [9]	Matthew G. Hanna, Liron Pantanowitz.	Information technologies, electronic medical records, image digitization, and big data	The computerized support for pathologies to meet precision medicine, modify laboratory processes and workflows, promote interaction between technical teams, integrate medical reports to support clinical decisions.
07	Diagnostics 4.0: The medical laboratory in digital health	Neumaier, Michael	Cloud computing, artificial intelligence (AI), health bots, and scanning	It promotes predictive medicine, improves the visibility of the patient-physician-laboratory relationship, efficiently contributes to medical dialogue and support for clinical decisions.
08	Towards Smart, Collaborative eHealth IoT: From Device to Mist and Cloud [8]	Bahar Farahani, Mojtaba Barzegari, Fereidoon shams Aliee, Khaja Ahmad Shaik	Internet of things (IoT), artificial intelligence (AI), cloud computing	It allows continuous monitoring of patient health, provides additional knowledge in real time, improves decision-making power
09	A mobile device app to reduce medication errors and time to drug delivery during pediatric cardiopulmonary resuscitation simulation: Multicenter, randomized, controlled, crossover trial.	Johan N. Siebert; Frederic Ehrler; Christian Lovis; Christophe Combescure; Kevin Haddad; Alain Gervaix; Sergio Manzano.	Mobile device app	It reduces the occurrence of errors in the calculation of dosages, preparation time and administration of pediatric drugs in the hospital
10	Sustainable protection of cyber medical systems for the health of the future	Alberto Huertas Celdrán, Manuel Gil Pérez, Félix J. García Clemente, Gregorio Martínez Pérez.	Information and communication technology (ICT), cloud computing, intelligent health systems, and cyber security	The monitors, processes and makes autonomous decisions without the need to involve health professionals.

By applying 3D printing technologies in the hospital environment, the digitization of imaging and laboratory exams and Big Data are strategic mechanism to structure large volumes of health data. It intends to collect, organize, and interpret information from the care registered in the fixed and mobile electronic devices performed by health professionals; and provide excellent opportunities to replace small parts of specific materials, lost or broken, through the creative capabilities of additive manufacturing. Thus, it is believed that the use of disruptive innovations in order to support a conceptual and practical model of hospital management can:

1. To support health informatics in a standardized way and work processes;
2. To improve medication disposition and administration;
3. To improve availability of access and exam results;
4. To modify externally assisted patient flows, bed management, and patient transfers;
5. To leverage the organization with clinical information;
6. To promote interactions between professionals from different hospitals;
7. To integrate medical reports in the same database;
8. To provide means of digital support for secondary research.

According to Neumaier, [6], the association between disruptive artificial intelligence technologies, digitization, and the internet of things can promote in the hospital environment, particularly in the care support sectors, a precision medicine. The technical support departments, as examples laboratories and pharmacy, can contribute in order to enhance the dialogues and to specialize written communication of the physicians. Furthermore, it provides simplified, coded and additional real-time knowledge and continuous monitoring of vital patient data, improving the visibility of diagnoses, medical treatments, and nursing care.

In a research developed by Siebert et al., [7] about the creation of a mobile application that helps nursing teams to calculate the dosages of the pediatrics medications, reducing probable human errors and administration time. Thus, it was found that this disruptive technology promotes communication between doctors and parents, provides technological support for shared decisions and support for the clinical practices of nursing professionals involved in care in emergency rooms.

4.1 A Conceptual Framework for Health Management

A conceptual framework for health management (blue box) is presented in the Fig. 1. In the organizational structure are inserted the technologies and the public and private institutions with the technical, administrative, and assistance characteristics of their governance. The blue arrow indicates the transition from the traditional model of the health management to health 4.0 management promoted by disruptive technologies. The traditional health management (grey box) faces some challenges in finding scalable solutions. The disruptive technologies (orange box) can influence and contribute (indicated by the black arrow) in the traditional management of health

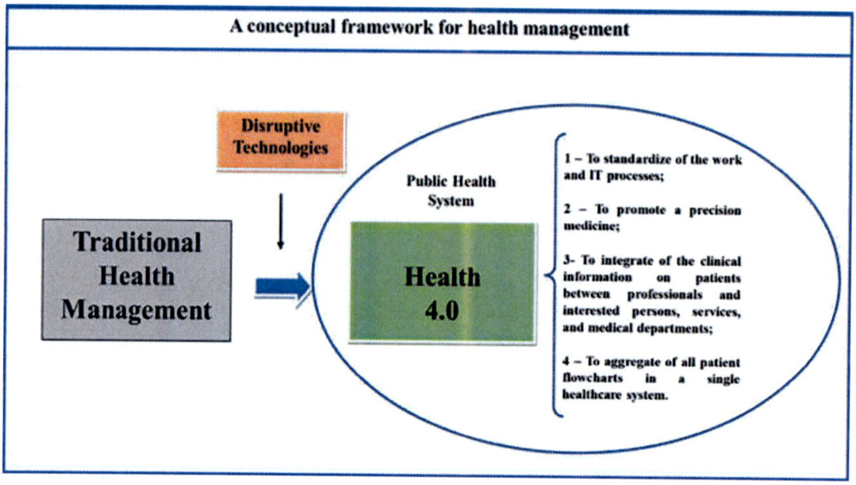

Fig. 1 A conceptual framework for health management. (Source: The authors)

institutions transition, due to sustainable investment proposals. The new conceptual management model called health 4.0 (green box) exposes distinct elements of the management mentioned above. It can be expressed by some characteristics:

1. To standardize of the work and IT processes;
2. To promote a precision medicine;
3. To integrate of the clinical information on patients between professionals and interested persons, services, and medical departments;
4. To aggregate of all patient flowcharts in a single healthcare system.

5 Conclusion

The result of the literature review, carried out by the authors of this research study, responded to the objective of how disruptive technologies can foster a conceptual model of management in public hospital institutions. The disruptive technologies can be exemplified by medical information and communication technology, biotechnology, internet of things and smart sensors, artificial intelligence, health robots, self-service devices, hospital electronic records, telehealth, 3D printing, drones, mobile communication devices, systems integration computerized, image typing, big data, cloud computing, and cyber security. All these technologies collaborate in a specific way for the qualification of managerial and care work for health workers and hospital administrators. In addition, they cooperate to support a management model based on health 4.0.

The main contributions of this paper are in the valuation of health care centered on the satisfaction of hospitalized patients. Other contributions are that the disruptive technologies can modify processes and improve traditionally ineffective concepts of public hospitals and in a new practical approach that express needs that can be accommodated in public hospitals with digital technologies.

The authors of this study suggest expanding the scope of investigation and search sources for future researches to promote advances in the possibilities of finding relevant outcomes that occur in the occurrence of how disruptive technologies impact people, work processes, and the organizational structure of the hospital setting.

References

1. Celdrán, AH, Pérez, MG, Clemente, FJG and Pérez, GM (2018). Sustainable protection of cybernetic medical systems for the health of the future. Sustainable Computing: Informatics and Systems, 19, 138–146.
2. Herrmann, M., Boehme, P., Mondritzki, T., Ehlers, JP, Kavadias, S., & Truebel, H. (2018). Digital Transformation and Disruption of the Health Sector: an Internet-Based Observational Study. Journal of medical internet research, 20 (3), e9498.
3. Bonilla-Asalde, CA, Adrianzen, E., Jáuregui, J., Quiroz, J., Camacho, E., & Rivera-Lozada, O. (2020). Results-oriented hospital administration: barriers and perspectives for synergies in a public hospital in Peru. Pak. J. Med. Health Sci, 14(2), 846–852.
4. Whittemore, R., & Knafl, K. (2005). The integrative review: updated methodology. Journal of advanced nursing, 52 (5), 546–553.
5. Lee, D. (2018). Strategies for technology-based service encounters for patient experience satisfaction in hospitals. Technological forecasting and social change, 137, 118–127.
6. Neumaier, M. (2019). Diagnostics 4.0: the medical laboratory in digital health. Clinical Chemistry and Laboratory Medicine (CCLM), 57(3), 343–348.
7. Siebert, JN, Ehrler, F., Combescure, C., Lovis, C., Haddad, K., Hugon, F., ... & Juzan, A. (2019). A mobile device app to reduce medication errors and time to medication administration during sham pediatric cardiopulmonary resuscitation: a multicenter, randomized, controlled, crossover study. The Lancet Child & Adolescent Health, 3 (5), 303–311.
8. Farahani, B., Barzegari, M., Aliee, FS and Shaik, KA (2020). Towards smart, collaborative eHealth IoT: from device to fog and cloud. Microprocessors and Microsystems, 72, 102938.
9. Hanna, MG and Pantanowitz, L. (2017). The role of informatics in patient-centered care and personalized medicine. Cancer cytopathology, 125 (S6), 494–501.
10. Johnson, TJ and Brownlee, MJ (2018). Development and innovation of system features to optimize patient care. The Bulletin of the American Society of Hospital Pharmacists, 75 (7), 465–472.
11. Samhan, B. and Joshi, KD (2019). Avoidance of Disruptive Technology in Healthcare: A Revealed Causal Mapping (RCM) Approach. International Journal of Healthcare Information Systems and Informatics (IJHISI), 14 (2), 28–48.

The Emergency Care Unit Operations Supply Chain Management: An Analysis of the Healthcare Service Challenges and Opportunities

Lídia Santos Silva, Annibal Scavarda, Ana Dias, Zdenek Uherek, and Miguel Sellitto

Abstract The Emergency Care Unit (UPA) has the purpose of being a place of rapid assistance to primary and medium cases. However, during the covid-19 pandemical period, the administration of inputs and supplies represented challenges for management. The shortage of equipment and health professionals may have been caused by the pandemic, butit could also have been influenced by management. This paper analyzes how the problems with the supply chain in UPA interfere in health services. This article seeks to analyze, through qualitative research of CNES data, whether there were problems with the supply chain in the UPAs and whether supply chain risk management can contribute to better management (SCRM) of emergency healthcare establishments.

Keywords Emergency care · Management · Healthcare · COVID-19

L. S. Silva (✉) · A. Dias
Federal Center for Technological Education Celso Suckow da Fonseca-CEFET/RJ, Rio de Janeiro, Brazil

A. Scavarda
Federal Center for Technological Education Celso Suckow da Fonseca-CEFET/RJ, Rio de Janeiro, Brazil

UniRio, Rio de Janeiro, Brazil

Z. Uherek
Charles University, Prague, Czechia

M. Sellitto
Universidade do Vale do Rio dos Sinos – Unisinos, São Leopoldo, Brazil

J. Vargas Florez et al. (eds.), *Production and Operations Management*, Springer Proceedings in Mathematics & Statistics 391,
https://doi.org/10.1007/978-3-031-06862-1_26

1 Introduction

The World Health Organization, on March 11, 2020, officially declared the COVID-19 pandemic, which has drastically intervened in all areas of the world, including the supply chain [1]. With that, there was a need for the management of the supply chain to become resilient and face the challenges that arose [2]. Companies and researchers have started to pay more attention to SCRM, which is remarkably triggered by the frequency and intensity of catastrophes, disasters and crises that are increasing on a global scale [3]. However, health is an area where SCRM is still beginning to act [4], and Brazilian public health has some administrative steps [5] that can generate more problems for managers of emergency care units. Thus, the objective of this work is to find out whether there were problems in the supply chain in public emergency care units in Brazil during the pandemic, and whether there is space for supply chain risk management in the health area. To try to answer these questions, this work was divided into four parts: literature review, methodology, result and discourse, and conclusion. Where the Brazilian public health scenario was contested, in the health supply chain area, it was explained where the studied data are located and how we analyzed them, the result and the conclusion we found.

2 Literature Review

Starting this work, dividing the framework into three parts: Supply Chain, Public Health in Brazil, and UPA Management. First, showing how medical care is divided on Brazil's public health, according to complexity. Then contextualizing the concept of the supply chain in the hospital area. And finally, how the management of the UPA and administration of inputs is carried out, which is the focus of this article.

2.1 Public Health in Brazil

According to the Brazilian constitution, health is a right for all and a duty of the State [6] and, thinking of serving everyone, the unified health system, the SUS, was created. The SUS is a public and universal health system financed by the government and is structured into three complementary hierarchical levels of health care primary, medium and high complexity care [6]. When it comes to urgency, each of these components of the care network must participate, respecting the limits of its complexity and resolution capacity. It is expected that the population that need of care can be welcomed at any level of establishment and referred to other levels when the complexity of the care required exceeds the local service's capacity for assistance. However, in the hope of receiving assistance, and also because of the different times that an emergency can occur, a large number of patients used to go to the hospital' emergency unit, causing overcrowding and long lines [7].

In order to reduce the huge and slow-moving lines of hospital emergencies, the Regulatory Ordinance No. 1863, of September 29, 2003, launched a package of national policies and emergency care [7]. Dividing healthcare among units according to their degree of complexity [8]. In this scenario, the Emergency Care Units (UPA) emerge as one of the strategies of the National Emergency Care Policy for better organization of healthcare, articulation of the services, and defining flows [9]. This strategy appears as one of the resolving initiatives for the problem of overcrowding in hospital emergencies [10]. UPAs, unlike the primary health care and the Family Health Strategy, must have work 24 h a day, performing risk classification screening, providing resolutive care to patients affected by acute or acute chronic conditions, cases of low complexity, at night and in weekends, in this way, also, a stabilizing of the critical patient so that the transfer to a hospital can be carried out [9].

To manage the beds and that there is also no overcrowding in the UPA care beds, Article 7 of Ordinance No. 342 of the Ministry of Health provides that if the patient does not obtain clinical improvement within 24 h, the patient must be transferred to a Hospital [8]. In this way, the UPA becomes a unit for emergency care and quickly, having its main function the complete care of emergencies of primary and intermediate complexity, and stabilize of intermediate patient that will require a hospitalization. Carrying out this screening, which were previously the responsibility of hospital emergency unit. With the creation of the UPA, the hospitals did not stop responding to emergencies of low and medium complexity, but to ensure the efficiency of the project of Ordinance No. 1863, the buildings of the UPAs were made in a way to serve society, in strategic locations to minimize the social costs [11]. If the location of facilities occurs ineffectively, it can create problems of productivity or the provision of services [12].

2.2 *Supply Chain in the Healthcare*

Supply chain risk (SCR) is an area that has been reinventing itself over the years since its inception. Some researchers define SCR as an area that acts when there is a failure in the supply and demand relationship [13, 14], while other authors define supply chain management as the integration and organization of business chain processes, from suppliers and manufacturers to the end-user [15, 16]. Supply chain management can also be defined as the management of all material, both information and financial, of an organizational network that produces and delivers products or services to customers [17], and there are authors who say that there is still no consensus on the definition of supply chain risk management (SCRM) [3]. However, it can be said that SCR is a resilient area [2]. SCRM can be diversified, depending on the area in which it is inserted. The healthcare supply chain (SC) is a differentiated segment, as its main objective is to save lives and not profit [4]. The production of health is related to improving the quality of life, which can be seen as a form of social progress [18]. In public health, cost management must be effective to make good use of the money, which comes from tax contributions [4], and for each UPA have an investment amount that only varies with the number of beds, not monthly expenses [19], which emphasizes the need for an SCRM. Although the health sector is not mentioned by the authors of academic literature as one of the segments of application of the SCRM [4], but in practice, the need for your application is seen.

With the pandemic, SC suffered from a lack of workers, for reasons of health and safety, and sanitary barriers to contain the spread of the COVID-19 virus [1]. However, interruptions in the supply chain can lead to financial losses [14], and when health it is the subject, the interruption in CS can represent a loss of life. Especially because the pandemic has caused a shortage, even for the purchase of personal protective equipment (PPE), which compromised both the safety of health professionals and the entire population [20]. The crisis in CS caused by the pandemic also came about because of the period and place where it started. The first cases of COVID-19 appeared in China, which is the largest manufacturer of PPE, and the production of safety materials was already facing a seasonal low due to the New Year celebrations [20]. And after the official WHO decrees declaring a pandemic period [1], many countries restricted the export of PPE and hospital supplies, due to the great internal demand of each one of them [20]. Within this scenario, SC needed to be resilient, to overcome its vulnerability [21]. Resilience in healthcare offers the opportunity for CS to return to its starting position or even a more ideal position than in the past [2]. The concept of resilience can be defined as the ability of a system, community, or even an individual to transform a problem into a device to reduce negative impacts of future crises, becoming an effective strategy in general health CS for imminent crises [22, 23].

2.3 UPA Management

According to article 8 of ordinance No 342 [19], it is incumbent to the manager responsible for the UPA: to implement the Reception process with Risk Classification, in a specific environment, considering the identification of the patient in need of immediate treatment, with the establishment of potential risk, health problems or degree of suffering, to prioritize care according to the degree of suffering or the severity of the case; establish and adopt compliance with clinical care protocols, risk classification, and administrative procedures; and ensure technical and logistical support for the proper functioning of the unit [19]. To define the staff and the amount of public investment to be designated, the UPAs are classified into Sizes I, II and III [11]. In the Table 1, presents the classification of the 24-hour UPA by size, population, physical area, number of planned visits, number of doctors per shift, and minimum number of rooms for care. Among the functions established for the 24-hour UPA manager is the management of people, defining strategies to ensure the presence of medical, nursing, technical and administrative support personnel 24-hour a day and every day of the week, holidays, enabling the first assistance and stabilization of patients affected by any type of emergency [19].

Table 1 Definition of sizes applicable to 24-hour UPA

UPA 24H SIZES	Population of the coverage area of the UPA	Minimum physical area	Medical care in 24 hours	Doctors from 7 am to 7 pm	Doctors from 7 pm to 7 am	Minimum of observation rooms
SIZE I	50.000 to 100.000 inhabitants	700 m^2	Average of 150 patients	2 doctors	2 doctors	7 rooms
SIZE II	100.001 to 200.000 inhabitants	1.000 m^2	Average of 250 patients	4 doctors	2 doctors	11 rooms
SIZE III	200.001 to 300.000 inhabitants	1.300 m^2	Average of 350 patients	6 doctors	3 doctors	15 rooms

In Brazil, the purchase of inputs from public agencies is a different process from what is seen in private corporations. The Brazilian Constitution establishes that any corporation that is directed, directly or indirectly, by the government must purchase inputs and equipment through a bidding process [5]. Bidding is an administrative process that aims to ensure equal conditions for everyone who wants to enter into a contract with the Government, and is divided into 7 modalities: Competition; Price taking; Invitation card; Contest; Auction; Reverse Auction; Competitive Dialogue. Competition is the modality for contracting special goods and services or common works [5]. Price taking is the modality where the proposals are already registered in the database and will be chosen through this [5]. The invitation is the modality the public administration invites at least 3 interested parties to present their proposals [5]. The contest is the modality for hiring or awarding people for public office [5]. Auction is the modality for the sale of movable or immovable property without value or legally seized [5]. The reverse auction is similar to an auction, but for all types of goods and where the winner is the bid with the lowest value or the highest discount [24]. And in April 2021, Law No. 14,133 came into force to make the bidding process faster and more virtual, extinguishing the Price taking and Invitation modalities, but it was presented in this work since these modalities are still valid until 2023, and institutionalized the modality Competitive Dialogue, which occurs when the public agency exposes its problems in the public notice and the participants present proposals that they deem to be efficient for the presented problem [25]. This bidding process is necessary to ensure the good use of public investment, but it represents another challenge for public health SCRM. In view of this, during the pandemic the Brazilian government implemented Provisional Measure No. 926 [26] in Law No. 13,979 [27] which dispensed with bidding and other formalities for works and purchases of goods and services aimed at combating the coronavirus.

3 Methodology

The present work carried out a qualitative analysis of the UPA data that are available on the CNES (National Register of Health Establishments) website,[1] in reference to June 2021.

To carry out the preparation of the data, the software Rstudio version 4.0.2 was used. First, the CNES database was downloaded, and the bases referring to establishments, complementary establishments, professional activities, equipment, types of rooms, the reason for deactivation, main activity, and managing state were selected. All data were compiled into a single file based on the reference code of health units. After this step, only establishments that had "UPA" or "emergency service unit" in their name were filtered. Afterwards, a second filter was applied, to select only establishments that were registered as emergency care as their main activity, which were public agencies and which had no registered reason for deactivation. At the end of the filter, had a total of 786 establishments. A summation formula was applied to obtain the total of equipment, doctors and rooms per UPA. After that, the UPA's architectural project[2] was consulted, which establishes the minimum amount of medical equipment necessary for the UPA's functioning, and was selected 5 equipment that was registered in the CNES database and that could provide some assistance for patients with suspected or diagnosed COVID, which were: Respirator, infusion pump, oxygen generator, defibrillator and electrocardiogram monitor. After obtaining the total of selected equipment, the number of beds and doctors, per unit of the UPA, verified whether it met the minimum quantities established in the project and Ordinance No. 342 (Table 1), and was considered that all UPA was size I, that is the smaller unit of UPA. The minimum quantities of equipment, doctors and beds were gathered in Table 2. We created a categorical variable with a "yes" or "no" answer for the units that reached the minimum amount necessary for operation, and for those UPAs that did not, respectively, as established by the Ministry of Health of Brazil. Finally, a descriptive analysis of the data was performed using the SPSS software.

Table 2 The minimum quantities necessary based on UPA's architectural project

Equipament	Min. quantity
Oxygen	1
Infusion pump	1
Desfibrilator	2
Electrocardiograph	2
Respirator	2
Doctors	2
Room	7

[1] http://cnes.datasus.gov.br/pages/downloads/arquivosBaseDados.jsp

[2] https://antigo.saude.gov.br/images/pdf/2018/janeiro/26/ PROGRAMAARQUITETONICO - MINIMO-UPA-24-H-VERSAO-2.0-2018.pdf

4 Result and Discussion

Senna et al. (2021) [4] expressed in his work that there is a lack of empirical studies on SCRM and the approaches being used to identify and mitigate health supply chain risks. Given this, sought to carry out a qualitative survey of the CNES data to verify whether the SCRM is being carried out, and if not, if there is a need for its application. Analyzing Fig. 1, sees that in June 2021, of the786 establishments analyzed, in none of the variables studied, the frequency of 100% was reached. Emphasizing that this work sought to analyze the minimum amount for operation [19], and not the amount needed to meet the demand for service in the region where the UPA is located. Was noticed that more than half of the establishments managed to reach the minimum amount of mechanical respirator, defibrillator and electrocardiograph, but this represents that 24.05% (n = 189) of the UPAs do not have at least 2 mechanical ventilators, and that 37.53% (n = 295) and 30.03% (n = 236) of the establishments, which are for emergency care, also do not have at least 2 defibrillators and electrocardiograph, respectively. The inverse perspective further aggravates the problem of managing these units, as 63.87% (n = 502) of the UPAs do not have

Respirator

	Frequency	%
no	189	24.05
yes	597	75.95

Electrocardiograph

	Frequency	%
no	236	30.03
yes	550	69.97

Oxygen

	Frequency	%
no	502	63.87
yes	284	36.13

Doctors

	Frequency	%
no	758	96.44
yes	28	3.56

InfusionPump

	Frequency	%
no	159	20.23
yes	627	79.77

Room

	Frequency	%
no	652	82.95
yes	134	17.05

Defibrillator

	Frequency	%
no	295	37,53
yes	491	62,47

Fig. 1 Descritive analysis of UPAs data

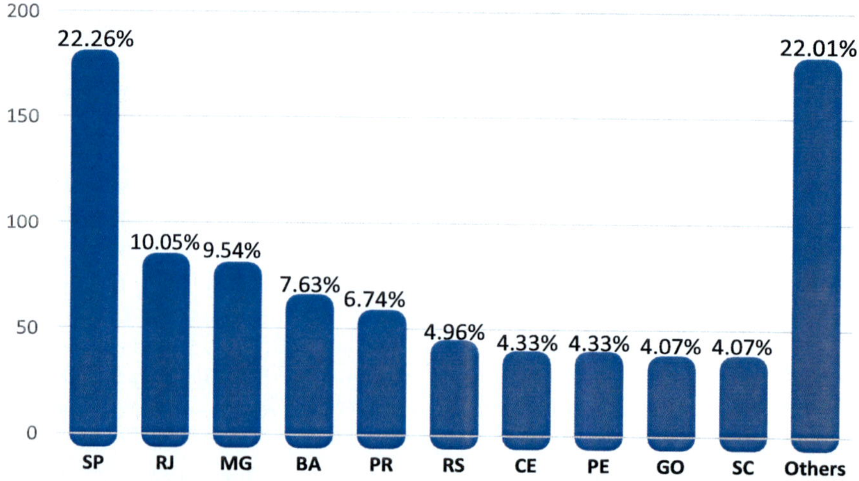

Fig. 2 UPA establishment by state

at least one oxygen pump, 82.95% (n = 652) did not reach the minimum number of beds. For the service to be carried out, the UPA project provided for at least 2 doctors for7 beds per establishment, but looking at the June 2021 data from the CNES, 134 establishments offered at least 7 beds, but only 28 establishments had a minimum of 2 doctors registered in the system, this represents 3.56% of the establishments studied.

Of the selected equipment, only 5 establishments had the minimum amount of all of them, and looking at the relationship between doctors and rooms, 9 establishments had the minimum necessary combination, and these 9 UPAs are located in the states of Minas Gerais and Paraná. However, none of the studied UPAs presented the minimum amount of equipment, beds and doctors registered in the CNES. When performing this analysis by state, in Fig. 2, sees that the states of São Paulo and Rio de Janeiro, two states that are among the largest economically, are the ones with the most establishments registered in the CNES. 79 UPAs are located among 92 municipalities in Rio de Janeiro, and 175 UPAs in São Paulo, are located among 645 municipalities in the state. In the "Others" category are all the states that had less than 4% of the total UPAs (Acre; Alagoas; Amapá; Amazonas; Distrito Federal; Espírito Santo; Maranhão; Mato Grosso; Mato Grosso do Sul; Pará; Paraíba; Piauí; Roraima; Rondônia; Rio Grande do Norte; Sergipe; Tocantins). The lack of equipment, beds and doctors could have an explanation in the great demand for these supplies during the beginning pandemic. This shortage led to the lack of health professionals, including for health reasons [28, 29]. But after a year of pandemic, this problem could be resolved if a resilience measure of SC had been applied.

Looking from another perspective, within the scope of management, these data may represent another lack within management, that could be a space to SCRM act. According to ordinance No. 1646 [30], the Ministry of Health institutes the registration and updating of data in the CNES as mandatory and attributes this function to the manager of the health establishment. It also establishes the responsibility of the municipalities to verify the information of the establishments. As relationships between companies become more complex, such as those of public institutions in Brazil, the need for accurate information becomes crucial. Even if the good in question is health and because it is a public institution, which does not aim for profit, the efficient management of SCRM would guarantee a greater quantity of care and quality. And for this to happen, an important factor is the adoption of a transparent performance management process [31]. Emphasizing that was analyzed all establishments as if they were of Size I, the most basic size, since that was found in the database the limitation that did not allow us to analyze each UPA with its proper size.

5 Conclusion

The objective of this work was to carry out a qualitative analysis of data from the UPAs, which were collected at CNES. It was found that none of the establishments contained the minimum amount of equipment, beds and doctors necessary for operation, according to current legislation. This information may represent that there was a management problem in the logistics, where the manager was not able to guarantee the minimum amount needed, or it could also represent a problem in the management administration, which did not fulfill its obligation to record its data. Any of these aspects can result in negative ways in public health services in Brazil, and other countries. It is suggested that further studies be carried out to understand the management problems encountered and to see how this interfered with care during the pandemic. Another encouraged future study is a case study in one or more health units, in order to improve and design a structure that presents the constructs.

As relationships between companies become more complex, such as those of public institutions in Brazil, the need for accurate information becomes crucial. Even if the good in question is health and because it is a public institution, where profit is not measured in money, but in the quantity of quality care. Quality management of the supply chain would increase public healthcare. CS solutions have become an important management tool for organizations seeking resilience. Applying SCRM concepts in public health in Brazil could change this scenario, and ensure an efficient and effective system, with quality care for all people in need.

References

1. Nagurney, Anna. "Supply chain game theory network modeling under labor constraints: Applications to the Covid-19 pandemic." European journal of operational research 293.3, 880-891 (2021).
2. Bastani, Peivand, et al. "Strategies to improve pharmaceutical supply chain resilience under politico-economic sanctions: the case of Iran." Journal of Pharmaceutical Policy and Practice 14.1 (2021): 1-14.
3. Sodhi, ManMohan S., Byung-Gak Son, and Christopher S. Tang. "Researchers' perspectives on supply chain risk management." Production and operations management 21.1, 1-13 (2012).
4. Senna, Pedro, et al. "A systematic literature review on supply chain risk management: is healthcare management a forsaken research field?." Benchmarking: An International Journal (2020).
5. Constituic¸ao Federal, http://www.planalto.gov.br/ccivil03/leis/l8666cons.htm. Last accessed 25 August 2021.
6. Constituic¸ao Federal, http://www.planalto.gov.br/ccivil03/leis/L8080.htm. Last accessed 25 August 2021.
7. O'Dwyer, Gisele. "A gestao da aten¸c˜ao às urg^encias e o protagonismo federal". Ci^encia saúde coletiva 15, 2395–2404 (2010).
8. Machado, Cristiani Vieira, et al." Gestao do trabalho nas Unidades de Pronto Atendimento: estrategias governamentais e perfil dos profissionais de saúde." Cadernos de Saúde Publica 32 (2016).
9. Oliveira, Saionara Nunes de, et al." Unidade de Pronto Atendimento-UPA 24h: percepc˜ao da enfermagem." Texto Contexto-Enfermagem 24 (2015): 238–244.
10. Bittencourt, Roberto Jośe, and Virginia Alonso Hortale." Interven¸c˜oes para solucionar a superlota¸c˜ao nos servi¸cos de emerg^encia hospitalar: uma revisa˜o sistem'atica." Cadernos de Saúde Pública 25 (2009): 1439–1454.
11. Briozo, Rodrigo Amancio, and Marcel Andreotti Musetti. "Metodo multicriterio de tomada de decis˜ao: aplica¸cao ao caso da localiza¸c˜ao espacial de uma Unidade de Pronto Atendimento–UPA 24 h." Gesta˜o Produ¸c˜ao 22 (2015): 805–819.
12. Randhawa, Sabah U., and Thomas M. West. "An integrated approach to facility location problems." Computers Industrial Engineering 29.1-4,261-265 (1995).
13. Norrman, Andreas, and Ulf Jansson. "Ericsson's proactive supply chain risk management approach after a serious sub-supplier accident." International journal of physical distribution logistics management (2004).
14. J uttner, Uta, Helen Peck, and Martin Christopher. "Supply chain risk management: outlining an agenda for future research." International Journal of Logistics: Research and Applications 6.4, 197-210 (2003).
15. Croxton, Keely L., et al. "The supply chain management processes." The international journal of logistics management 12.2, 13-36 (2001).
16. Lambert, D. M., and T. L. Pohlen. "Supply Chain Metrics International Journal of Logistics Management.", Vol. 12 No. 1, 1-19 (2001).
17. Tang, Christopher S. "Perspectives in supply chain risk management." International journal of production economics 103.2, 451-488 (2006).
18. Illich, Ivan. "A expropria¸cao da saúde." Ne^mesis da Medicina. Rio de Janeiro: Nova Fronteira (1975): 33.
19. Ministerio da Saúde, https://bvsms.saude.gov.br/bvs/saudelegis/gm/2013/prt034204032013.html. Last accessed 25 Aug
20. Burki, Talha. "Global shortage of personal protective equipment." The Lancet Infectious Diseases 20.7 (2020): 785-786.
21. Peck, Helen. "Drivers of supply chain vulnerability: an integrated framework." International journal of physical distribution logistics management (2005).

22. Van De Pas, Remco et al. Interrogating resilience in health systems development. Health policy and planning, v. 32, n. suppl 3, p. iii88–iii90, (2017).
23. Yaroson, Emilia Vann, et al. "Resilience strategies and the pharmaceutical supply chain: the role of agility in mitigating drug shortages." Pharmaceutical Supply Chains Medicines Shortages. Springer, Cham, 249–256 (2019).
24. Constituic¸ao Federal, http://www.planalto.gov.br/ccivil03/leis/2002/l10520.htm. Last accessed 27 August 2021.
25. Constituic¸ao Federal, http://www.planalto.gov.br/ccivil03/ato2019−2022/2021/lei/L14133.htm. Last accessed 27 August 2021.
26. Constituic¸ao Federal, http://www.planalto.gov.br/ccivil03/ato2019−2022/2020/lei/l13979.htm. Last accessed 27 August 2021.
27. Constitui¸cao Federal, http://www.planalto.gov.br/ccivil03/ato2019−2022/2020/lei/l13979.htm. Last accessed 27 August 2021.
28. Johnson SU, Ebrahimi OV, Hoffart A. PTSD symptoms among health workers and public service providers during the COVID-19 outbreak. PLoS One. 2020 Oct 21;15(10):e0241032. https://doi.org/10.1371/journal.pone.0241032. PMID: 33085716; PMCID: PMC7577493.
29. Kader, N., Elhusein, B., Chandrappa, N.S.K. et al. Perceived stress and posttraumatic stress disorder symptoms among intensive care unit staff caring for severely ill coronavirus disease 2019 patients during the pandemic: a national study. Ann Gen Psychiatry 20, 38 (2021). https://doiorg.ez108.periodicos.capes.gov.br/10.1186/s12991-021-00363-1
30. Constitui¸cao Federal, https://bvsms.saude.gov.br/bvs/saudelegis/gm/2015/prt164602102015.html. Last accessed 27 Aug
31. Scheidl, H. A., A. T. Simon, and F. C. De Campos. "Conhecimento Strategic and Operational Information Management: The Logistic Operators Relationship Management from the Perspective of Business Intelligence and Knowledge Management." Espacios, vol. 37, no. 11(2016).

A Telemedicine and Telehealth Conceptual Managerial Framework: Opportunities, Challenges, and Trends in the Healthcare Promotion

Lídia Santos Silva, Annibal Scavarda, Ana Dias, Edgar Ramos, and Sofía Esqueda

Abstract The combination of technology and medicine presents new opportunities to transform the provision of health services more effectively and sustainably. Telemedicine emerges as a tool to face the challenges of access to health, but it also faces challenges to be popularized among health professionals and the population. From an economic point of view, telemedicine can be seen as a way to increase the rate of preventive medicine, to reduce the high rate of hospital occupancy and worsening of diseases, reducing spending on hospitalizations and surgeries. And from a social perspective, it can democratize access to health services, offering more health care to reach more remote areas. The purpose of this article is to make a framework of the literature found at Scopus that relates the themes Telemedicine, Telehealth, and Latin America to analyze the participation of these concepts in health promotion. In view of the articles found, there was a low rate of publication in global journals, but several articles talking about the insertion of these concepts in the education of health professionals. Another important factor observed is how the technological structure and the socioeconomic vulnerability found in Latin America are points that need government assistance to be successful in the provision of health. And that the concepts of telemedicine and telehealth can help the country's development by reaching people and places where health care is scarce.

L. S. Silva (✉) · A. Dias
Federal Center for Technological Education Celso Suckow da Fonseca-CEFET/RJ, Rio de Janeiro, Brazil

A. Scavarda
Federal Center for Technological Education Celso Suckow da Fonseca-CEFET/RJ, Rio de Janeiro, Brazil

UniRio, Rio de Janeiro, Brazil

E. Ramos
Universidade Peruana de Ciencias Aplicadas, Lima, Peru

S. Esqueda
Instituto de Estudios Superiores de Administración IESA, Caracas, Venezuela

J. Vargas Florez et al. (eds.), *Production and Operations Management*, Springer Proceedings in Mathematics & Statistics 391,
https://doi.org/10.1007/978-3-031-06862-1_27

Keywords Telemedicine · Telehealth · Health assistance · Social development · Latin America

1 Introduction

There has been much debate about the use of technology in several areas. In several countries, concepts such as telemedicine are common, but in Latin America, it encounters structural and cultural challenges. When relating technology to medicine, we are faced with conservative fronts that see this union as the dehumanization of medicine [17]. However, telemedicine expands medical care and contributes to the continuity of the patient's treatment with the same professional who has already been accompanying his condition. In a way, we can suggest that the greater frequency provided by virtual consultations can intensify the doctor-patient relationship, and thus also contribute to the greater acceptance of patients in this new service modality. However, doubts about the capacity of Latin American countries to implement telemedicine are recurrent, as it is necessary health structures and policies to contribute to this service, a reform in the health system in Latin America [19]. Due to this, the purpose of this article is to review the work on the use of telemedicine and telehealth in health care in Latin America, to assess the evolution and challenges of implementing these concepts in the Latin American health system.

2 Literature Review

The pandemic, in addition to the concern about being contaminated with COVID19, contributes to anxiety to take better care of health. Mainly because it is a virus that has a greater effect on people with some pre-existing diseases that, for many, their care was neglected, such as hypertension and diabetes arising from eating habits and lack of physical activities [15]. However, at the same time, social isolation and the high rate of contagion pose more obstacles for people to consult with doctors, psychologists, nutritionists, and other health professionals. Fear of contagion caused many consultations and therapies to be postponed, but it doesn't prevent or relieve pain and injury, anxiety attacks, and worsening of depressions. In this context, the concept of telemedicine resurfaces as an outlet for these people. Telemedicine can be understood as the practice of medicine, or health care services, through interactive technologies of audiovisual communication to offer health care [20]. Although telemedicine became more popular during the pandemic, its application already existed and had been trying to conquer a place in the health market along with another concept, that of telehealth. The concept of telehealth represents the use of medical information shared from one location to another through information and communication technologies, to improve and promote health care [4].

The telemedicine faced a lot of prejudice on the part of doctors, being justified that this would bring the dehumanization of medicine [17], however, the pandemic helped to popularize this concept. Having to adapt to this market on a compulsory basis, many who were against it, saw a promising and advantageous market. Some specialties are easier to adapt to telemedicine, such as psychology and psychiatry, but specialists in digital health in Latin America have been working with the World Organization to promote the development of new specialties and other projects related to telemedicine [3]. It knows that the good quality internet and efficient applications and software are prerequisites for telemedicine [22], and the pandemic contributed to this aspect, anticipating the release of some updates to these programs and increasing the demand. There is a lack of internet signal structure and the reach of telephone companies in some places compromises the service of some people [1], for this reason, that for the dissemination of telehealth, a partnership between the public and private sectors is necessary [2]. On the other hand, telemedicine is an advantage for those regions where it is more difficult to find specialists in certain areas, or even clinical healthcare professionals sufficient for the demand. With the current level of technology, it is already possible to do some health actions at a distance without compromising the quality of the service.

Another concept that telemedicine is interconnected is that of healthcare 4.0, which uses artificial intelligence to assist health, in which attached to the telemedicine electronic medical record, manages to report for certain exams, using large data libraries, Big Data. Klauser and Albrechtslund describe Big Data as the collection, management, and automated analysis of software-based data. This collection can be done in several ways, even through our access to electronic pages. And when we talk about health, this data collection contributes to a large database that can help us to identify diseases [14]. Some health specialties, such as ophthalmology, already used telehealth as a large database of results and magazine work to assist in the report of new exams [1]. The Latin American continent is a group of countries that are in development, but with gaps that leave part of the socioeconomically vulnerable population, which compromises its human development index (HDI) [13]. These gaps contribute to the inequality in the health supply between incomes, races, and genders [5]. Cultural and structural aspects also hamper equal health promotion. Some authors argue how telemedicine can democratize access to health in vulnerable areas due to a lack of adequate infrastructure. And they also highlight the unpreparedness of health professionals in Latin America to exercise this type of care [18, 19].

3 Methodology

A systematic review of the literature was carried out because the work seeks to answer a guiding question about the concepts of Telemedicine and Telehealth in Latin America. The method was chosen because it sought to carefully select and evaluate the studies selected for data interpretation [21]. The work followed the criteria of the Cochrane Handbook and CDR Report publications: Key question; definition of the database and location of the documents; critical evaluation; data collection; analysis and interpretation of data; and presentation of data [21]. To organize the review, we used the following guiding question: What is the participation of telemedicine and telehealth in health promotion in Latin America? The search was carried out in the Scopus database, with no publication deadline and without cutting the type of article. About the keywords, the following were used: "Telemedicine"; "Telehealth"; "Latin American". The Boolean operator used was the "and" in the search that occurred in the title, abstract, and keywords. In the first moment, the searches presented 15 articles containing one of these words in the main subject. After that, a filter was made, by reading the title and the abstract, for adequacy and pertinence to the guiding question.

In this filter, 2 articles were removed from the selection. A complete reading of the articles was carried out, to analyze in detail and their contribution to the proposed theme.

4 Result

The list of articles used in the literature review can be found in the table (Table 1).

Of the 13 articles found, after the exclusion process, we noticed articles published from 2010 to 2021, an average of 0.85 articles published per year on this subject. Certainly, as the database is Scopus and it presents academic papers indexed in global publishers that dominate this market, articles published in national journals in Latin American countries are not included in the list. However, we are talking about a continent that is still developing and there is not so much investment in technologies [13], so it is important to increase the discourse on this topic in global magazines to attract partnerships and learn from experiences from other countries [3].

Table 2 shows that the articles were divided equally between the places of publication, with 3 conference and review articles, 1 editorial and 6 articles (Fig. 2). Regarding the authors, there was also a regular distribution of the works, only Ph.D. Humberto Jośe Alves participated in two articles, the others only had one publication. This theme has been more addressed in the Medical areas and about health professionals, as shown in Fig. 1, with only 5% of the work being carried out by the social areas, a science that studies the socioeconomically vulnerable population, which was the population that showed itself open to this new modality of medical care [1, 2].

The documents were also separated by their central information and were divided into three groups: Practical studies; Generalist Analysis; Professional. The articles selected as a practical study are those that in their methodology selected an application or a health clinic to evaluate the use of telehealth. Generalist analyses are the articles that study the importance of the topic in general, and the professional group represents the articles that speak of the application of the concepts of telemedicine and telehealth in the training of health professionals. As shown in Fig. 3, most of the articles were for practical applications, around 53% of the database. Consequently,

Table 1 List of article

Title	Author	Year
Implementing Telemedicine Visits in an Underserved Ophthalmology Clinic in the COVID-19 Era	Adeli Mona and William R. Bloom	2021
Telemedicine and the current opportunities for the management of oncological patients in Peru in the context of COVID-19 pandemic	Paola Montenegro, Luis Pinillos, Frank Young, Alfredo Aguilar, Indira Tirado-Hurtado, Joseph A. Pinto and Carlos Vallejos	2021
Building capacity and training for digital health: Challenges and opportunities in Latin America	Walter H Curioso	2019
Telehealth in Latin America: A review of the studies registered in clinicaltrials.gov	Quispe-Juli CU, Moquillaza-Alćantara VH and Arapa-Apaza KL	2019
Digital health, gender and health equity: Invisible imperatives	Chaitali Sinha and Anne-Marie Schryer-Roy	2018
MHealth intervention to improve the continuum of maternal and perinatal care in rural Guatemala: A pragmatic, randomized controlled feasibility trial	Boris Martinez, Enma Coyote Ixen, Rachel Hall-Clifford, Michel Juarez, Ann C. Miller, Aaron Francis, Camilo E. Valderrama, Lisa Stroux, Gari D. Clifford and Peter Rohloff	2018
The effectiveness of e- & mHealth interventions to promote physical activity and healthy diets in developing countries: A systematic review	Andre Matthias Müller, Stephanie Alley, Stephanie Schoeppe and Corneel Vandelanotte	2016
Telemedicine services in the Chilean public sector; A first quantitative study	Antonio Rienzo and Ćesar Galindo	2016
Providers Expectations on Telemedicine: A Qualitative Research in a Large Healthcare Network of Latin America	Giussi Bordoni, Plazzotta F., Sommer J., Benítez S., Garćıa G., Luna D. and González B De Quiŕos	2015
Telehealth distance education course in Latin America: Analysis of an experience involving 15 countries	Alaneir de F´atima dos Santos, Humberto Jośe Alves, Janaina Teixeira Nogueira, Roso´alia Moraes Torres, and Maria do Carmo Barros Melo	2014
An overview of telehealth initiatives in Latin America	Alaneir de Fátima dos Santos, Marcelo D'Agostino, Mauŕicio Simon Bouskela, Andŕes Fernandez, Luiz Ary Messina and Humberto Jośe Alves	2014
Experiences in mHealth for chronic disease management in 4 countries	John D. Piette, Joaquin A. Blaya, Ilta Lange and Juan B. Bru Sanchis	2011
E-health's promise for the developing world	Susan Dentzer	2010

the articles that had the most citations were those of practical applications, which had a total of 90 citations until March 2021. In all, the articles had 119 citations over that period, the most cited being the article on the importance of using applications health to promote well-being.

Table 2 Places that published on the topic and year

Place of publication	Documents	Year
Journal of primary care & community health	1	2021
Critical Reviews in Oncology/Hematology	1	2021
Journal of Medical Internet Research	1	2019
Revista Cubana de Informacion en Ciencias de la Salud	1	2019
Journal of Public Health (United Kingdom)	1	2018
Reproductive Health	1	2018
International Journal of Behavioral Nutrition and Physical Activity	1	2016
CHILECON 20155	1	2016
Studies in Health Technology and Informatics	1	2015
Telemedicine and e-Health	1	2014
Pan American Journal of Public Health	1	2014
ACM International Conference Proceeding Series	1	2011
Health Affairs	1	2010

Martinez et al. (2018) reported that the introduction of health support technologies for home births is feasible and may improve complication detection and referral to health care if delivery worsens [6]. Andŕe Muller (2016) showed that most studies on telemedicine interventions had effective results in promoting physical activity and healthy diets in developing countries, but recommended investigating the cost-effectiveness and scope of interventions and focusing on emerging technologies such as smartphone apps and handheld activity trackers [7]. Rienzo and Galindo's paper analyzed data on health facilities in Chile and showed that specialties such as dermatology, cardiology, and ophthalmology were in great demand for teleconsultations. This rapid advance of telemedicine in Chile represents a great adhesion by the population and a good alternative to complement the other health programs already operating in the country [8]. And despite being utopian for some and facing several challenges [16], medical assistance over the phone is already achieving results, even to help control patients' medications [12]. However, for telemedicine to have positive results, Santos et al. (2014) warns about the importance of including distance care in the training of health professionals, so that professionals are able to offer the same quality service, or better, than in person [10]. Telemedicine in diabetes control in Chile has been shown to have more effective results than patients treated face-to-face in a clinic during a 15-month study the assisted patients virtually stabilized their glycemic control, significantly reducing emergency room visits and access to care ambulatory improved [11]. However, the expansion is limited due to the lack of incentives from managers in the preparation of health professionals [11]. And action by the managers is also necessary to develop a training plan and get the professionals in place [9].

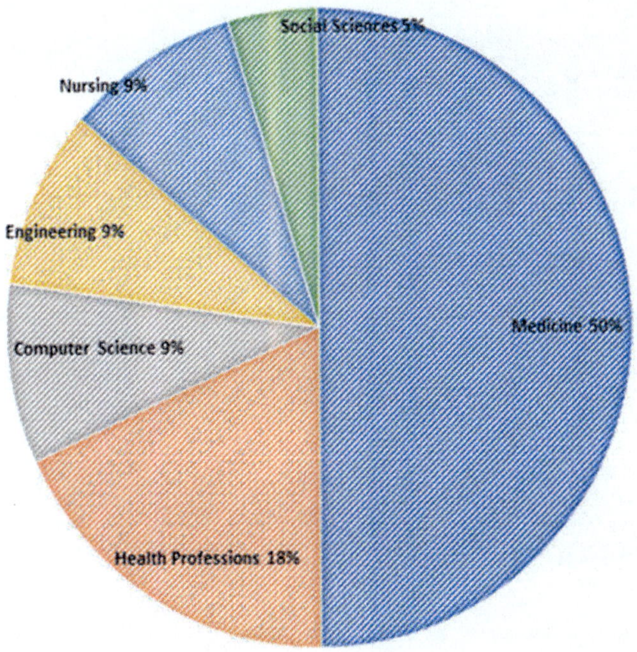

Fig. 1 Subject area

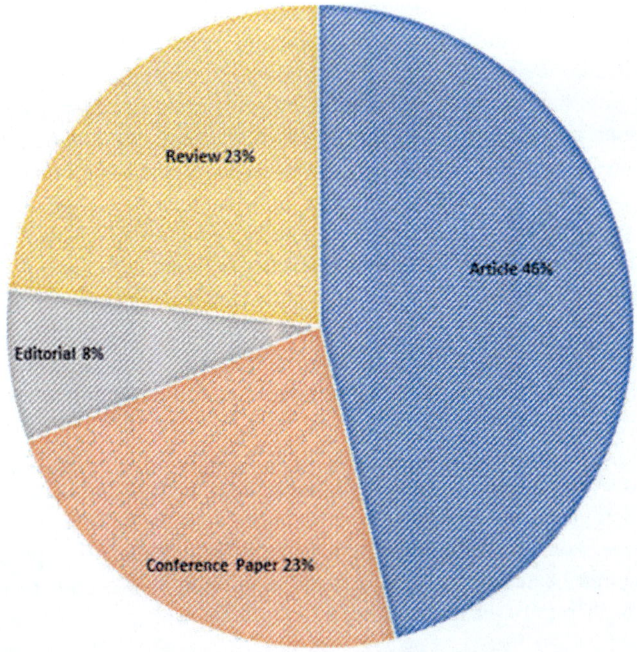

Fig. 2 Type of documents

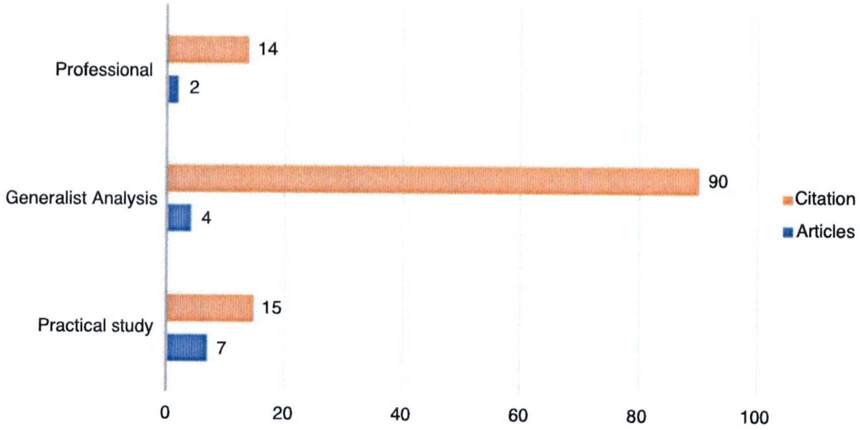

Fig. 3 Theme groups and Citation

5 Conclusion

The academic studies analyzed show that telemedicine and telehealth represent a great ally to promote the increase in health supply in the countries of Latin America and other continents. However, it is still a premature subject in the practical aspect, there is limited evidence of the applications of the concepts of telemedicine and telehealth in the market public and private. Despite the limited number of articles, he found that they were published in journals with a global impact on related subjects. There should be an incentive for more publications on these topics in global journals.

The systematic review of the literature allowed to verify the evidence of training of health professionals for this new market. Which shows the acceptance on the part of professionals in this new type of consultation and assistance, and this qualification will cause the paradigm of dehumanization of telemedicine to be broken. However, an action by managers is necessary to train professionals who have already graduated and improve their telehealth management. The method applied also contributed to observe that patients are willing to use this service, and not only that, they see in this aspect a way of increasing the health supply, which for many was scarce due to the situation of vulnerability of the place where they lived. And also the need for investment in technologies, as they can cooperate to reduce the inequalities in access to health found in Latin America, being essential tools for promoting knowledge and innovation.

Certainly, that telehealth was indirectly performed for a specific target audience by parts of doctors and other health professionals. Through phone assistance and messaging applications. The analysis of the articles also raised points such as the need to define work to standardize assistance. From an economic point of view, telemedicine can be seen as a way to increase the rate of preventive medicine, to reduce the high rate of hospital occupancy and worsening of diseases, reducing spending on hospitalizations and surgeries. And from a social perspective, it can democratize access to health services, offering more health care to reach more remote areas. A management operation is important to ensure equipment and training for professionals, monitor the main problems reported by patients, and monitor the evolution of telehealth care.

References

1. Adeli, Mona, and William R. Bloom. "Implementing telemedicine visits in an underserved ophthalmology clinic in the COVID-19 era." Journal of Primary Care Community Health 12 (2021): 2150132721996278.
2. Montenegro, Paola, et al. "Telemedicine and the current opportunities for the management of oncological patients in Peru in the context of COVID-19 pandemic." Critical Reviews in Oncology/Hematology: 103129 (2020).
3. Curioso, Walter H. "Building capacity and training for digital health: Challenges and opportunities in Latin America." Journal of medical Internet research 21.12: e16513 (2019).
4. Quispe-Juli, Cender Udai, V́ıctor Hugo Moquillaza-Alćantara, and Katherine Linda Arapa-Apaza. "Telehealth in Latin America: A review of the studies registered in clinicaltrials. gov." Revista Cubana de Informaci'on en Ciencias de la Salud (ACIMED) 30.4: 1–12 (2019).
5. Sinha, Chaitali, and Anne-Marie Schryer-Roy. "Digital health, gender and health equity: invisible imperatives." Journal of Public Health 40.suppl2: ii1 ii5(2018)
6. Martinez, Boris, et al. "mHealth intervention to improve the continuum of maternal and perinatal care in rural Guatemala: a pragmatic, randomized controlled feasibility trial." Reproductive health 15.1: 1–12 (2018).
7. Müller, Andre Matthias, et al. "The effectiveness of e mHealth interventions to promote physical activity and healthy diets in developing countries: a systematic review." International Journal of Behavioral Nutrition and Physical Activity 13.1 (2016): 1–14.
8. A. Rienzo and C. Galindo, "Telemedicine services in the chilean public sector; a first quantitative study," 2015 CHILEAN Conference on Electrical, Electronics Engineering, Information and Communication Technologies (CHILECON), pp. 215–218, 2015, https://doi.org/10.1109/Chilecon.2015.7400378.
9. Bordoni, Giussi, et al. "Providers Expectations on Telemedicine: A Qualitative Research in a Large Healthcare Network of Latin America." Studies in health technology and informatics 216: 890–890 (2015).
10. Santos, Alaneir de Fatima dos, et al. "Telehealth distance education course in Latin America: analysis of an experience involving 15 countries." Telemedicine and e-Health 20.8: 736–741 (2014).
11. Piette, John D., et al. "Experiences in mHealth for chronic disease management in 4 countries." Proceedings of the 4th International Symposium on Applied Sciences in Biomedical and Communication Technologies (2011).
12. Dentzer, Susan. "E-Health's promise for the Developing World.", Health Affairs 29(2): 229 (2010).
13. Tosoni, German Alarco, and Ćesar Castillo Garćıa. "Indice de desigualdad y crecimiento ecońomico en America Latina." Investigacíon Economica 79.314: 106–134 (2020).
14. Klauser, Francisco R., and Anders Albrechtslund. "From self-tracking to smart urban infrastructures: towards an interdisciplinary research agenda on Big Data." Surveillance Society 12.2: 273–286 (2014).
15. Ferreira, Ricardo Bruno Santos. "V́ıctimas preferidas de COVID-19 en diferentes páıses según raza/color de la piel." Revista Cubana De Enfermeŕıa 36 (2020).

16. dos Santos, Alaneir de Fatima, et al. "An overview of telehealth initiatives in Latin America/ Uma visao panoramica das acoes de telessaude na America Latina." Revista Panamericana de Salud Pública 35.5: 465–471 (2014).
17. Gadow, Sally. "Touch and technology: Two paradigms of patient care." Journal of Religion and Health 23.1: 63–69 (1984).
18. Li, Benjamin, et al. "Impact of a SBRT/SRS longitudinal telehealth training pilot course in Latin America." Critical reviews in oncology/hematology (2020): 103072.
19. Litewka, Sergio. "Telemedicina: un desafio para America Latina." Acta bioethica 11.2 (2005): 127–132.
20. Maldonado, Jose Manuel Santos de Varge, Alexandre Barbosa Marques, and Antonio Cruz. "Telemedicine: challenges to dissemination in Brazil." Cadernos de saude publica 32 (2016).
21. ROTHER, ET. "Revisão sistematica x Revisão narrativa. Rev." Acta Paulista de Enfermagem, São Paulo 20.2.
22. Silva, Angélica Baptista, and Ilara Hammerli Sozzi de Moraes. "O caso da Rede Universitaria de Telemedicina: análise da entrada da telessaúde na agenda política brasileira." Physis: Revista de Saúde Coletiva 22 (2012): 1211–1235.

A Sustainable Development Managerial Analysis of the Integration Among Healthcare, Safety, Ergonomics, and Environment

Geraldo Assis Cardoso, Annibal Scavarda, Ve Adamu, Miranda Harizaj, and Miguel Afonso Sellitto

Abstract In the contemporary scenario presents a high prevalence of accidents and occupational diseases with impacts on health and safety in the work environment. This study, through a narrative review of the literature, proposes an integrated health, safety, ergonomics and environment management policy. This integrated management allows for the improvement of workers' health and safety in line with sustainable development policies. An integrated management model allows for more global actions in relation to workers' health, not dividing the risks and therefore not segmenting prevention actions. Organizations need to evolve from a management based on simple compliance with legislation, such as occupational health and safety and prevention of environmental accidents, to sustainable management, with proactive positioning and actions in relation to eco-efficient product projects.

Keywords Ergonomics · Management · Sustainability · Safety · Worker's health

---

G. A. Cardoso (✉) · A. Scavarda · M. Harizaj
Federal University of the State of Rio de Janeiro, Rio de Janeiro, RJ, Brazil

Polytechnic University of Tirana, Tirana, Albania

Universidade do Vale do Rio dos Sinos, Porto Alegre, RS, Brazil
e-mail: geraldo.cardoso@foa.org.br

V. Adamu · M. A. Sellitto
Polytechnic University of Tirana, Tirana, Albania

Universidade do Vale do Rio dos Sinos, Porto Alegre, RS, Brazil

Euclid University, Sukuta, Gambia

J. Vargas Florez et al. (eds.), *Production and Operations Management*, Springer Proceedings in Mathematics & Statistics 391,
https://doi.org/10.1007/978-3-031-06862-1_28

1 Introduction

Occupational Health is a branch of Public Health and its primary object of study is the relationship between work and health and the objective of promoting and protecting workers' health, through the development of actions to monitor the risks present in the environments and in working conditions and injuries resulting from exposure to them [1].

Historically, the most important record in the analysis of the relationship between work and health dates back to the release of the book De Morbis Artificum Diatriba, in 1700, by the Italian physician Bernardino Ramazzini, who based his work on the study of 54 professions known at his time, listing their exercises with the consequent diseases, proposing adequate treatment and prevention for these disorders [1, 2]. His lessons remained the basic text of preventive medicine for nearly two centuries [1–3]. In addition to these contributions, he expanded the view of the working environment to the factory's surroundings, reporting the deleterious effects on the health of the population that inhabits the environments close to the factories, as well as describing the environmental conflicts that occurred at the time, similar to those that occur currently [4].

About 200 years after the publication of Ramazzine's work, the industrial revolution emerged with the increase in serial production, making increasingly evident the fragility of the worker in the face of machines and modern means of production, alarmingly increasing the number of invalid workers and deaths from work accidents. In this scenario, occupational medicine emerges with the vision of placing the right man in the right place, whose main characteristic was the placement of the doctor inside the factories to attend to the sick worker, keeping this workforce productive. During this period, the first laws for the protection of workers also appeared, in England, in 1802 and later in other European countries in the following years, until arriving in Brazil, through Legislative Decree no. 3724, of January 15, 1919 [1].

Over the years, it was realized the need to go beyond the to go beyond simple medical care within companies. It was necessary to identify the causal factors for the effective prevention of the health of the workers' community. Then comes the contribution of Engineering through Occupational Hygiene and, later, Ergonomics, whose multidisciplinary analysis includes the participation of physiologists, psychologists, architects, physicians and engineers.

In this scenario, the "Occupational Health" stage begins in the mid-twentieth century. The concept of health was expanded with the creation of the World Health Organization – WHO – in 1946 and Brazil expanded the standards of occupational safety and medicine, instituting the Specialized Services in Safety Engineering and Occupational Medicine – SSSEOM – and the Internal Accident Prevention Commissions – IAPC. In 1978, the Ministry of Labor published the consolidation of occupational safety and medicine standards, through Ordinance n. 3.214. During this period, the worker and the union movement became concerned with working conditions, actively intervening in the health and safety conditions of the work environments. The era of Worker's Health begins [5].

In Brazil, the Federal Constitution (CF) of 1988 was a fundamental milestone in the evolution of this new concept of worker's health. Health was considered a social right, and workers were guaranteed the reduction of risks inherent to work, through health, hygiene and safety standards. The FC establishes the competence of the Union to take care of the worker's safety and health through the actions developed by the Ministries of Labor and Employment, Social Security and Health [6].

In Brazil, regarding the legislation that regulates and directs the actions of companies, in the field of worker health and in the structuring of Occupational Safety and Medicine services in companies inserted in the formal market, we highlight the Regulatory Standards (NR) of the Ministry of Labor, in which we highlight NR 4 which establishes the competences of the SSSSEOM; NR 7, which underlies the Medical Control and Occupational Health Program – OHMC and NR 17 – Ergonomics, which establishes parameters that allow the adaptation of working conditions to the psychophysiological characteristics of workers, in order to provide maximum comfort and safety for the workers [5, 7].

Importantly, the interrelationships production/work, health and environment, determined by the mode of production and consumption in a given society, are the main references to understand the living conditions, the profile of illness and death of people, the vulnerability of certain social groups and environmental degradation. The knowledge of this reality is fundamental for building change alternatives capable of guaranteeing life and health for the environment and the population, including the working population [8].

The labor legislation as it is structured, according to the researched literature, dichotomizes the actions of the NR and the National Occupational Health Policy itself, causing the Occupational Health services to act in accordance with these standards, but in a dissociated manner, where the Management of the Medical Control Program in Occupational Health occurs without proper integration with the Ergonomics Programs, as well as with environmental policies. Companies, to meet legal requirements, implement different types of programs with different methodologies, with different purposes, but without integration and not always reaching the main objective, which is the health and well-being of workers, as well as the development of a policy of sustainability [3, 5, 9].

This study, through a narrative review of the literature, analyzes the legislation that supports worker health in Brazil and sustainability policies, and proposes an integrated health, safety, ergonomics and environment management policy. This integrated management can allow the improvement of workers' health and safety in line with sustainable development policies.

2 Legal Basis for Workers' Health Actions in Brazil

2.1 Brief History

The Industrial Revolution that took place in Europe, between the years 1760 and 1850, had a great impact on people's lives and health. Evidencing the fragility of the worker in the unfair fight with the machine, causing a frightening increase in the number of dead, mutilated, sick, orphans and widows. It was during this period that the stage of "Occupational Medicine" emerged, whose main characteristic was the placement of a doctor inside the factories to assist sick workers and keep the workforce productive. The first laws regarding work-related accidents also appeared in England, Germany and other countries in Europe [1, 2]. The creation of the International Labor Organization – ILO – by the Treaty of Versailles increased the production of preventive norms, so much so that, already at its first meeting in 1919, six conventions were adopted, which directly or indirectly aimed at protecting health, as well-being and physical integrity of workers, as they dealt with the limitation of working hours, unemployment, maternity protection, night work for women, minimum age for admission of children and night work for minors [1, 3].

In Brazil, in 1943, Decree-Law no. 5452, created the Consolidation of Labor Laws, CLL, which established standards of safety, hygiene and occupational medicine, in Chap. V. In 1972, Ordinance no. 3237, from the Ministry of Labor, determined the existence of SSSEOM. Law no. 6514, from 1977, empowered the Ministry of Labor to regulate, through Ordinances, the matters of Safety, Hygiene and Occupational Medicine Services in companies, and this year, Ordinance no. 3.214, editing the Basic Regulatory Standards [1, 3].

2.2 Brief The National Occupational Safety and Health Policy

It is highlight, among these legal instruments, the National Policy on Occupational Safety and Health (2004), which should be developed in an articulated and cooperative manner by the Ministries of Labour, Social Security and Health, with a view to ensuring that work, basis of social organization and fundamental human right, is carried out in conditions that contribute to improving the quality of life, personal and social fulfillment of workers and without prejudice to their health, physical and mental integrity [7].

The purpose of this policy is to promote the improvement of the quality of life and health of workers, through the articulation and integration, in a continuous manner, of Government actions in the field of production relations. It proposes a National Occupational Safety and Health Policy – PNSST, seeking to overcome the fragmentation, disarticulation and overlapping of the actions implemented by the Labor, Social Security, Health and Environment sectors. For the purposes of this Policy, all men and women who carry out activities to support themselves and/or their dependents, whatever their form of insertion in the labor market, in the formal or informal sector of the economy, are considered workers [7].

2.3 *The Regulatory Standards of the Ministry of Labor*

NR 01 – General Provisions and Occupational Risk Management
This NR 01 was sanctioned by SEPRT Ordinance No. 6.730 of March 9, 2020 and should enter into force in January 2022. General Provisions and Occupational Risk Management" [1, 3, 7, 8].

The purpose of this Standard is to establish the general provisions, field of application, terms and definitions common to the NR relating to occupational health and safety and the guidelines and requirements for the management of occupational risks and prevention measures in Safety and Health at Work – SHW.

This standard states that the organization must implement, by establishment, the management of occupational risks in its activities. Occupational risk management must constitute a Risk Management Program – RMP. At the organization's discretion, the RMP may be implemented by operating unit, sector or activity. The RMP can be served by management systems, as long as they comply with the requirements set out in this NR and in legal provisions on safety and health at work. The RMP must contemplate or be integrated with plans, programs and other documents provided for in the occupational health and safety legislation [1, 3, 7, 8].

This standard also recommends that the Occupational Risk Inventory must include, at least, the following information: characterization of work processes and environments; characterization of activities; description of hazards and possible injuries or harm to workers' health, with identification of sources or circumstances, description of risks generated by the hazards, with indication of groups of workers subject to these risks, and description of implemented prevention measures; data from the preliminary analysis or monitoring of exposures to physical, chemical and biological agents and the results of the ergonomics assessment in terms of the risk assessment, including classification for the purpose of preparing the action plan; and criteria adopted for risk assessment and decision making [1, 3, 7, 8].

NR4 – Specialized Services in Safety Engineering and Occupational Medicine
Regulatory Standard No. 4 aims to indicate the implementation in all companies with employees governed by the CLL of Specialized Services in Safety Engineering and Occupational Medicine, aimed at promoting health and protecting integrity [1, 3, 5]. The implementation of these Services varies according to the number of employees the company has and the degree of health risk [1, 3, 5].

The degree of risk of companies is based on the National Classification of Economic Activities, and ranges from 1 to 4.

The SSSEOM service must be composed of a team composed of Occupational Physician, Occupational Safety Engineer, Occupational Safety Technician, Occupational Nurse and Occupational Nursing Assistant.

The professionals of this service are responsible for applying knowledge to the environment and components; determine when necessary the use of Personal Protective Equipment; collaborate, when requested, in the projects and implementation of new physical and technological facilities in the company; be responsible for the guidance and collection of what is presented in the NR's; maintain a relationship with IAPC by supporting, training and serving it; promote awareness, education and guidance activities for workers to prevent accidents and occupational diseases; analyze and record in a specific document, all accidents that occurred, with or without victims and all cases of occupational disease; monthly register updated data on accidents at work, occupational diseases and unhealthy conditions and send a map with annual assessment of the data referred to the Department of Safety and Occupational Medicine; keep the records at the Specialized Services headquarters or in some accessible place, ensuring the understanding of the content, for a period longer than 5 years; the work must be preventive, and emergency action is not prohibited when necessary, and the control of catastrophes, firefighting or any other type of accident must also be their responsibility [1, 3, 5].

All SSSEOM must be registered with the Ministry of Labor, and must include the name of the components, registration number, number of degree of risk and number of employees, specification of shifts and working hours [1, 3, 5].

NR 7 – Occupational Health Medical Control Program
The object is to establish mandatory preparation and implementation, by all employees and institutions that admit workers as employees, the OHMC, aims to promote and preserve the health of all its workers [1, 3, 5]. The OHMC is an integral part of the company as a whole in the area of worker health, so it should have the character of prevention, tracking and early diagnosis of work-related health problems, in addition to the existence of cases of occupational diseases or irreversible damage to the health of the workers [1, 3, 5].

According to item 7.2.4 of this NR, the OHMC must be planned and implemented based on the risks to the health of workers, especially those identified in the assessments provided for in the other NR's [1, 3, 5]. For the execution of the OHMC, a coordinator must first be indicated among the physicians of the SSSEOM of the company, and in cases where the company is not obliged to maintain occupational physician, according to NR 4, the employer must appoint an occupational physician, a non-company employee, to coordinate the OHMC [1, 3, 5].

According to item 7.3.1.1.3 of the NR 4, by determination of the DRT (Regional Labor Office), based on the conclusive technical opinion of the competent regional authority in matters of occupational safety and health, or as a result of collective bargaining, the companies they may be required to appoint a coordinating physician, when their conditions represent a potential serious risk to workers.

The OHMC must include, among others, the performance of mandatory medical exams, such as: admission, periodic, return to work, change of function and dismissal. These exams include: clinical evaluation (occupational anamnesis and physical and mental exam) and complementary exams [1, 3, 5].

According to item 7.4.4 of the NR 4, for each medical examination performed, the physician will issue the Occupational Health Certificate (OHC) in two copies. The OHC must include: the worker's full name, identity, function, specific occupational hazards that exist according to technical instructions issued by the SSST (Secretary of Safety and Health at Work), indications of the medical procedures that the worker was submitted, name of the coordinating physician and the respective Regional Council of Medicine (RCM), definition of "CAPABLE or UNABLE" of exercising a specific function that the employee will perform, name of the physician in charge of the examination [1, 3, 5].

In item 7.4.5 of this NR, the data obtained in the medical exams including clinical evaluation and complementary exams, the conclusions and the measures applied must be registered in an individual clinical record, which will be under the responsibility of the coordinating physician of the OHC. After the employee leaves the company, their clinical records must be kept for 20 years and if there is a change of doctor, the notes and medical records must be forwarded to their successor.

The OHMC must comply with a plan in which the health actions that will be carried out during the year are foreseen, and these actions must be the object of an annual report. After the annual report has been assembled, it must be presented to the IAPC, if the company has it, and this report can be kept in the form of a computerized file, provided it is kept easily accessible by the agent of the labor inspection [1, 3, 5].

In this standard it is observed that ergonomic hazards and occupational accidents are not mentioned in it, but can be characterized and registered, in a preventive manner, at the discretion of the coordinating physician of the OHMC. According to NR 7, OHMC must have a preventive character, so the inclusion of ergonomic risks could be a great tool for the prevention of musculoskeletal diseases [1, 3, 7].

NR9 – Environmental Risk Prevention Program – PPRA

NR9 is the regulatory standard responsible for the Environmental Risk Prevention Program, also known as PPRA. NR9 determines the mandatory protection necessary to ensure the physical and mental health of workers in unhealthy environments. For the purpose of this NR, environmental risks are considered to be physical, chemical and biological agents existing in work environments that, due to their nature, concentration or intensity and exposure time, are capable of causing damage to the worker's health. It also establishes the important stages of the Risk and Accident Prevention Program, which are: Anticipation and recognition of risks. Evaluation and control priorities and goals. Implementation of control measures and evaluation of their effectiveness.

NR 17 – Ergonomics

The seventeenth regulatory norm came to establish criteria in order that the adaptations of working and psychophysiological conditions can provide maximum comfort, safety and performance, has its legal existence ensured at the level of legislation through articles 198 to 199 of the CLL. With this set of criteria comes ergonomics.

According to item 17.1, “this NR aims to establish parameters that allow the adaptation of working conditions to the psychophysiological characteristics of workers, in order to provide maximum comfort, safety and efficient performance” in order to allow managers to quantitative data control [1, 3, 9].

This standard refers to working conditions where aspects related to lifting, transporting and unloading materials, furniture, equipment and environmental conditions of the workplace are reported, and the organization of work itself.

To assess the adaptation of working conditions to the psychophysiological characteristics of workers, it is up to the employer to carry out an ergonomic analysis of the work, which should address, at least, the working conditions, as established in this NR [1, 3, 5, 9].

2.4 The Federal Constitution

The execution of actions aimed at workers’ health is the responsibility of the Brazilian Unified Health System – BUHS, prescribed in the Federal Constitution of 1988 and regulated by the Organic Health Law – LOS. Article 6 of this law gives the national management of the System the responsibility to coordinate the worker’s health policy. According to paragraph 3 of article 6 of the LOS, worker health is defined as “a set of activities that are intended, through the actions of epidemiological surveillance and health surveillance, to promote and protect worker health, as well as aiming at the recovery and rehabilitation of workers subjected to risks and injuries arising from working conditions” [1, 6, 10].

In addition to the Federal Constitution and the LOS, other federal instruments and regulations guide the development of actions in this field, within the scope of the Health sector, among which the Ordinance/MS no. 3.120/1998 and Ordinance/MS no. 3908/1998, which deal, respectively, with the definition of basic procedures for the surveillance of workers’ health and the provision of services in this area. The operationalization of activities must take place at the national, state and municipal levels, to which different responsibilities and roles are assigned [11–13].

2.5 National Occupational Health Policy

Integrated with the National Occupational Safety and Health Policy, the National Occupational Health Policy, instituted in August 2012, aims to define the principles, guidelines and strategies to be observed in the three spheres of management of the BUHS – federal, state and municipal, for the development of comprehensive care actions to Occupational Health, with an emphasis on surveillance, aimed at promoting and protecting the health of workers and reducing morbidity and mortality resulting from development models and production processes. It is aligned with the set of health policies within the BUHS, considering the transversally of workers’ health actions and work as one of the determinants of the health-disease process. Among its objectives, we can highlight: strengthening Occupational Health Surveillance and integration with the other components of Health Surveillance; promote health and healthy work environments and processes; ensure comprehensive care to the worker's health; expand the understanding that Worker's Health TS must be conceived as a transversal action, and the health-work relationship must be identified at all points and instances of the care network; incorporate the category of work as a determinant of the health-disease process of individuals and the community and joint planning actions between surveillances, with the election of common priorities for integrated action, based on the analysis of the health situation of workers and the population in general and in the mapping of productive activities with potential environmental impact in the territory; promoting health and healthy work environments and processes, which presupposes, among other actions, the strengthening and articulation of health surveillance actions, identifying environmental risk factors, with interventions both in work environments and processes, as well as in the surroundings, in view of the quality of life of workers and the surrounding population [14].

3 Worker's Health in the International Plan

At the international level, since the 70s, documents from the WHO – World Health Organization, such as the Alma Ata Declaration and the proposal of the Health for All Strategy, have emphasized the need to protect and promote health and safety at work, through the prevention and control of risk factors present in work environments. More recently, the topic has received special attention in the focus of health promotion and the construction of healthy environments by the Pan American Health Organization [1, 7, 15].

The International Labor Organization (ILO), in Convention/ILO no. 155/1981, adopted in 1981 and ratified by Brazil in 1992, establishes that the signatory country must institute and implement a national policy on safety and the environment of work. Brazil, as a member of the ILO, has already ratified several conventions related to safety, health and the environment at work. In fact, the ILO has been promoting the international standardization of Labor Law, in order to provide a harmonious evolution of the norms of worker protection and achieve the universalization of social justice and decent work for all [3].

3.1 British Standard Standard – BS 8800

The British standard BS 8800 was considered a first attempt to implement a safety, health and environment management system. It is widespread and implemented in several countries. Its objective is to continuously improve the conditions of the

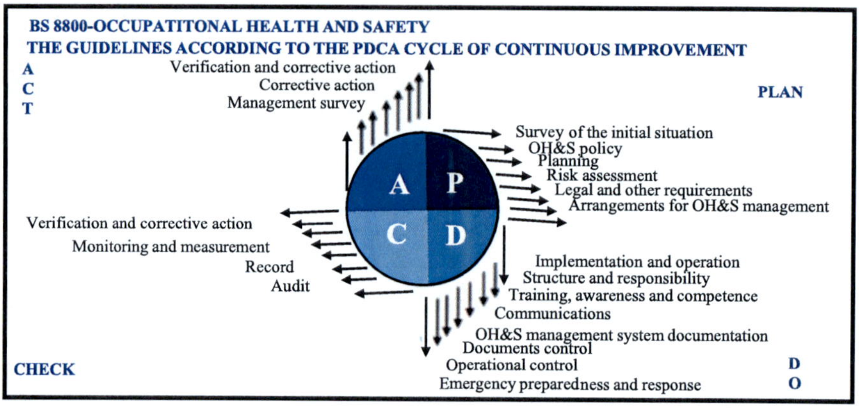

Fig. 1 PDCA applied to Occupational Health and Safety management (BS 8800). (Source: Adapted from Quelhas et al. [22])

working environment. It is important to emphasize that the principles of this standard are in line with the guidelines of the ISO 9000 series (Quality System) and ISO 14000 series (Environmental Management) standards. The BS 8800 standard can be used as a guide that allows organizations to implement occupational health and safety management, enabling the protection of their employees, who may have their health and safety status affected by the activities carried out by the organization [12, 21, 22].

The standard is structured into four chapters, which describe the following topics: guidance on the development of occupational health and safety management systems and the links with other management systems standards; references to publications, which may be consulted in addition. This standard was developed based on the PDCA cycle (Plan, Do, Check, and Act). This cycle of continuous improvement in management and the way in which it is integrated into the global system, guarantees the continuous improvement and maintenance of the organizations' routine, taking into account all stages of implementation, as shown in Fig. 1 [12, 21, 22].

The PDCA cycle brings the guidelines that companies must follow, based on the ISO 14000 standard for their organization. The contents presented in the standard are considered, by several authors, essential for an effective management system [12, 21, 22].

3.2 *Occupational Health and Safety Assessment Series – OHSAS 18001*

The OHSAS Standards for Occupational Health and Safety Management – SHW, aim to provide organizations with elements of an effective SHW management system, which can be integrated with other management requirements and help them achieve their SHW goals and economical. Developed from the experiences of the BSS 8800, by a selection of the main trade bodies, international standards and certification bodies. The OHSAS 18001 standard was developed based on the PDCA cycle [11, 12].

This standard specifies requirements for an SHW management system to enable an organization to develop and implement a policy and objectives that take into account legal requirements and information on SHW risks. It also allows an organization to control its risks of accidents and occupational diseases and improve its SHW performance.

The objectives of this standard must be communicated to all employees, so that they become aware of their individual obligations in relation to SHW; Employee involvement and commitment are vital to successful SHW management.

It is necessary to make employees aware of the effects of SHW management on the quality of their own work environment. It is recommended that they are encouraged to actively contribute to SHW management.

Employees (at any level, including Administration levels) will hardly be able to contribute efficiently to the management of SHW, unless they understand their responsibilities and are competent to perform the required tasks. This requires the organization to clearly communicate its SHW policies to employees; in order to provide them with a framework with which to measure their own SHW performance [11, 12].

4 Ergonomics and Worker Safety and Health

4.1 Concept and Areas of Expertise

According to Iida [13] ergonomics is the study of adapting work to man, where work has a broad meaning, covering not only work performed with machines and equipment, used to transport materials, but also every situation in which men are related to a productive activity, thus involving not only the physical environment, but also the organizational one. With a long view, ergonomics encompasses planning and design activities, which take place before the work is performed, and control and evaluation taking place during and after the work. All these aspects are necessary for the work to achieve its desired goals. ABERGO (Brazilian Ergonomics Association) defines ergonomics as:

> Ergonomics is understood as the study of people's interactions with technology, organization and the environment, aiming at interventions and projects that aim to improve safety, comfort, well-being and effectiveness in an integrated and non-disassociated way of human activities [14].

Internationally, the International Egonomics Association (IEA) approved a definition in 2000, conceptualizing ergonomics and recognizing its specializations: "Ergonomics (or Human Factors) is the scientific discipline, which studies the interactions

Table 1 Ergonomics goals

Goals	Favorable working conditions
Worker's health	It is maintained when the demands of work and the environment do not exceed their energy and cognitive limitations, in order to avoid situations of stress, risk of accidents at work and occupational diseases
Safety	Safety is achieved with the projects of the workstation, environment and work organization, which are within the worker's capabilities and limitations, in order to reduce errors, accidents, stress and fatigue
Satisfaction	It is the result of meeting the needs and expectations of the worker. Satisfied workers tend to adopt safer behaviors and are more productive
Efficiency	Consequence of good planning and organization of work, which provides health, safety and satisfaction to the worker

between human beings and other elements of the system, and the profession that applies theories, principles, data and methods, to projects aimed at improving human well-being and performance global systems" [15].

Within the discipline, according to the IEA, there are three main areas of specialization that represent competences in specific human characteristics, among them, physical ergonomics which is related to posture at work, handling of materials, repetitive movements, disorders work-related musculoskeletal disorders, job design, safety and health; [15].

The main objective of ergonomics is always the well-being, health, comfort and consequent increase in worker productivity, providing favorable working conditions [13]. The Table 1 defines each of these points aimed at by ergonomics and its benefits for worker health and safety.

The role of ergonomics in industries has great contributions such as increasing efficiency, reliability and quality of operations, however these gains can be obtained through three ways: the improvement of the human-machine-environment interaction, the organization of the work process and the improvement of working conditions [9].

Ergonomics is a strong field of action that is focused on organizational aspects of work, with the aim of reducing fatigue, repetitive work and the lack of worker motivation due to little participation in the decisions of their work. Another important point is the analysis of environmental working conditions, such as temperature, noise, vibrations and gases [9, 13, 14].

The systematic performance of ergonomics in factories and industries aims to identify more serious anti-ergonomic situations, where some factors stand out, such as occupational diseases, accidents, high number of errors at work, employee turnover and absenteeism. The causes of these problems are varied, such as workers' inadequacy of the work instruments, failure in the work organization process, environmental discrepancies, among others that lead to these situations [9, 13, 14].

4.2 *The Ergonomics Integrated into the Occupational Health Medical Control Program – PCMSO*

The proposed implementation of the ergonomics program integrated with OHMC is based on NR 4, 7 and 17 of the Ministry of Labor and Employment and on the references of Iida and Grandjean on ergonomic bases [13, 16].

According to the guidelines of OHSAS 18001, the SHW services of organizations must work in an integrated management, aiming at the continuous improvement of health and safety and with the active participation of employees [12]. Oliveira [3] proposes the inclusion of ergonomic risks in OHMC, thus increasing their preventive dimension.

Within this context, we report the experience of a large steel company in the State of Rio de Janeiro, Brazil, where the ergonomics service is linked to occupational medicine, whose OHMC management is articulated with the ergonomics management ergonomics come from occupational medical examinations, direct request from the work areas, situations identified by the ergonomics committees and related work accidents. After identifying a possible situation that requires an ergonomic assessment, an epidemiological study of the diseases detected in periodic examinations and medical dismissals due to musculoskeletal disorders is carried out. These collected data are discussed in a team, together with the occupational physician responsible for the area, the OHMC coordinator, the ergonomics coordinator and the ergonomics consultant to verify if the data collected from the epidemiological survey are valid. After these studies, ergonomic evaluations are initiated by the ergonomics team, together with the ergonomics committees of the areas involved.

With the assessment completed, a meeting is held with the ergonomics team, the ergonomics coordinator and the OHMC to reach a consensus on what was raised and thus propose the improvements that will be relevant to better adapt the work position. Once these conclusions have been reached, a subsequent meeting should be scheduled with the area's management and supervision to clarify the facts verified, demonstrate the ergonomic solutions that were proposed and discuss the feasibility of their implementation.

5 Environmental Health and Worker's Health

In Brazil, the National Environmental Policy, Law 6938, of August 31, 1981 [17], aims to: reconcile economic and social development with the preservation of environmental quality and ecological balance; establishment of criteria and standards of environmental quality and norms related to the use and management of environmental resources; the development of national research and technologies aimed at the rational use of environmental resources; the dissemination of environmental management technologies, the dissemination of environmental data and information, and the formation of public awareness of the need to preserve environmental quality and ecological balance; the preservation and restoration of environmental resources with a view to their rational use and permanent availability, contributing to the maintenance of the ecological balance conducive to life; the imposition, on the polluter and predator, of the obligation to recover and/or indemnify the damage caused and, on the user, the contribution for the use of environmental resources for economic purposes. As principles we can highlight: maintenance of ecological balance; rationalization, planning and inspection of the use of environmental resources; protection of ecosystems; control of potentially polluting activities; between others.

According to the literature, Brazil, in recent years, has evolved towards a movement towards the environmental issue and its relationship with health. A gradual, theoretical and practical approach between the fields of Environmental Health and Workers' Health, in an integrative and transdisciplinary perspective. The relationships between environment and health are widely recognized, where human health is highly dependent on society's ability to manage the interaction between the physical, biological and human activities [17, 18].

The Basic Health Care, of the Brazilian public health system, BUHS, is an advanced theme towards the integration of worker and worker health in the health system, as well as workers' users of basic health units. In 2011, Decree no. 7508, which regulates the Organic Health Law No. 8080/90, highlights the importance of Health Surveillance, and its fields of action are the promotion of population health, surveillance, protection, prevention and control of diseases and health problems, which should be organized from the articulation of the Epidemiological, Sanitary, Environmental and Occupational Health Surveillance [4].

According to Dias [4], health surveillance actions in BUHS Primary Health Care, related to productive activities, still remain unsystematic and discontinuous, but according to the author, "it is possible to recognize the growing understanding that the construction of health takes place beyond the spaces and practices of health units and services, covering everyday life". This empowerment of people's daily lives allows knowledge of production processes and the dynamics of life in cities and the countryside, which is reflected in interventions on the determinants and conditioning factors of health, including aspects of worker health and the environment.

On the path to the integration of worker's health with the environment, the 3rd National Conference on Worker's Health of 2005, includes in its agenda the thematic axis 2: How to incorporate the health of workers in health policies sustainable development in the country? Among the deliberations related to this axis, we highlight: To oblige multinationals to obey, at least, the same standard of protection for workers and the environment that is given in the country of origin; Obliging companies to provide a list of the substances they use and their risks; Elaboration of a government policy, ensuring that technological advances take into account the need to preserve health; Obligation of information, by the employee, to the worker, their families and the community, regarding the risks to which they are subject due to indirect and environmental contamination resulting from the company's activity; That deforestation, installation of dams and agro-industries be controlled and endorsed by the community and local entities; Articulation with a technology import policy that ensures the mandatory adoption of safety criteria according to the most rigorous and efficient principles recognized by the international community; Guarantee that all public works contracts include in their clauses the obligation of companies to maintain the safety of work environments [19].

6 The Integrated Management of Health, Safety and Ergonomics and the Challenge of Integration with Environmental Health

In Brazil, the concept of Occupational Health as a practice related to sustainable development has been growing since the beginning of the twenty-first century, through the implementation of this field as a health policy. In the same direction, throughout this period, health legislation, and also that of workers' health, has been proposing and articulating some actions that aim to promote changes in the practices of health organizations, aiming at the quality of life of workers in territories impacted by large projects and works that are essential for the country's growth [20].

In this sense, the concept of workers' health is related to a public health policy aimed at promoting and protecting the health of workers, which has been built over the last few years. In this way, we can state that worker health actions can be understood as a set of actions aimed at the realization of sustainable development considering the socio-environmental conditions related to work [20].

In Brazil, health and safety management is well established in formal market companies, where we have workers governed by the Consolidation of Labor Laws – CLL, based on Regulatory Standard number 4, which advocates the management of SSSEOM, which is part of the Standard Regulatory number 7, which advocates the Medical Control Program in Occupational Health – OHMC [3].

Ergonomic actions, NR 17, are commonly developed outside these two standards. The new NR 1, which is not yet in force, calls for an integrated management, through the Risk Management Program – RMP [8].

According to this standard, organizations must develop actions in occupational health of workers integrated with other prevention measures in Occupational Health and Safety – SHW, according to the risks generated by the work. The control of employee health must be a planned, systematic and continuous preventive process, in accordance with the classification of occupational risks and in accordance with the other Regulatory Standards. The RMP must contain some documents that are fundamental for an integrated management, such as: risk inventory and action plan. The documents that make up the RMP must be prepared under the responsibility of the organization, in compliance with the provisions of other Regulatory Standards. They must always be available to interested workers or their representatives and to the Labor Inspectorate.

Data from hazard identification and occupational risk assessments must be consolidated into an occupational risk inventory that must include, at least, the following information: characterization of work processes and environments, as well as the activities performed; description of hazards and possible injuries or harm to workers' health, with identification of sources or circumstances and description of implemented prevention measures; data from the preliminary analysis or monitoring of exposures to physical, chemical and biological agents and the results of the ergonomic assessment pursuant to NR-17 [8].

According to Vitoreli [21], there is a growing need for organizations to meet market requirements for international certifications and this has led to the increasingly frequent adoption of different health, safety, ergonomics and environment management systems, where each one covers some minimum requirements for meeting a given objective. Among these certifications we highlight: ISO 9001, ISO 14001 and OHSAS 18001. According to this author, "with the increase in the adoption of different management systems by organizations, difficulties arise related to the parallel management of these systems".

Thus, integration is seen as a way to generate greater efficiency in several aspects, alleviating these difficulties. Standardized management systems are those based on norms, of a national or international character, elaborated due to the need of organizations to meet legal and market demands. Vitoreli also states that "due to the difficulties of parallel management of these systems, their integration into a single Integrated Management System has been seen as a way to generate several benefits". The literature points to some benefits, such as cost reduction and management improvement [21].

According to Quelhas [22], "increasingly, the concerns of the government, businessmen and unions in improving safety, health and the conditions of the work environment are highlighted". This author also highlights "many organizations in Brazil still have a restricted view in relation to safety, occupational medicine and occupational health. The treatment of these issues is restricted to the collection of statistical data, reactive actions to work accidents and responses to labor claims". Literature points out that occupational health and safety begins as a management system through international standards such as SHWAS 18001, respecting national standards and legislation and evolving with the integration of these management systems [22].

7 The Integrated Management Systems: Occupational Health Safety, Ergonomics and Environment

According to Vitoreli [21], an Integrated Management System can be defined as "a set of interrelated processes that divide a set of human, financial, material resources, in addition to infrastructure and information, in order to achieve a set of objectives related to the satisfaction of stakeholders". When thinking about health, safety, ergonomics and the work environment, this management system can be structured in a set of initiatives, based on policies, programs, procedures and processes that integrate the organization's activity in order to optimize compliance with legal requirements in line with the organization's philosophy and culture, conducting its activities within the principles of ethics and social responsibility [21].

A safety and well-being management system in occupational environments must go beyond a simple risk management tool. It must be considered an operational challenge for organizations, structured in a system for evaluating and controlling the risks arising from their activities, aiming at a continuous improvement in performance and in the constant development of best practices. Another important objective of an integrated management system is the sharing of procedures, processes and practices adopted by an organization, so that it can implement its strategic planning, policies and guidelines, aiming to achieve its goals, goals and the development of health programs, safety, environment and ergonomics. These integrated management systems also aim at meeting regulatory and legal requirements, contributing to the continuous improvement of the organizations' sustainable, ethical and responsible performance [21, 27].

According to the researched literature, the OHSAS 18001 standard presents fundamental requirements for the management of occupational health and safety in the organization, filling a gap due to the lack of an international occupational health and safety standard. It was developed based on the PDCA cycle, and its requirements can be related to each of the stages of this cycle [21, 22, 26]. Another issue raised in the researched literature is the difficulties of parallel management of management systems in health and safety, including the integration of the management of some international quality standards such as ISO 9001 and 14,001. The integration in a single Management System has seen as a way to ensure several benefits for organizations, such as cost reduction and optimization in the management process. The management system of the OHSAS standard does not include the inclusion of ergonomic risks, so we think that it still leaves a gap. Based on this issue, we propose an integrated management model with the inclusion of ergonomic risks, as shown in Fig. 1 [21].

According to Fig. 2, within the stages of an Integrated Management Cycle, based on the PDCA cycle, in the Plan stage, the planning requirements must include the development, by the organizations, of an occupational health and safety policy, with identification of the hazards and risk assessment of the work environment, including ergonomic risks and also a planning of improvements through objectives and goals of safety and health that must be settled by the organizations. Within the "do" stage of the cycle, related requirements include the implementation and implementation of the controls and preventive measures that were identified in the first phase of the cycle.

Also within this stage, it is essential to carry out training for employees so that activities are carried out in accordance with health and safety policies and also so that they are aware of the importance of the safety and health management system within the company. In the "verify" stage of the PDCA cycle, organizations can assess and monitor the performance of their Occupational Health and Safety Management System – SGSSO. In the last stage of the cycle, "acting correctively", organizations

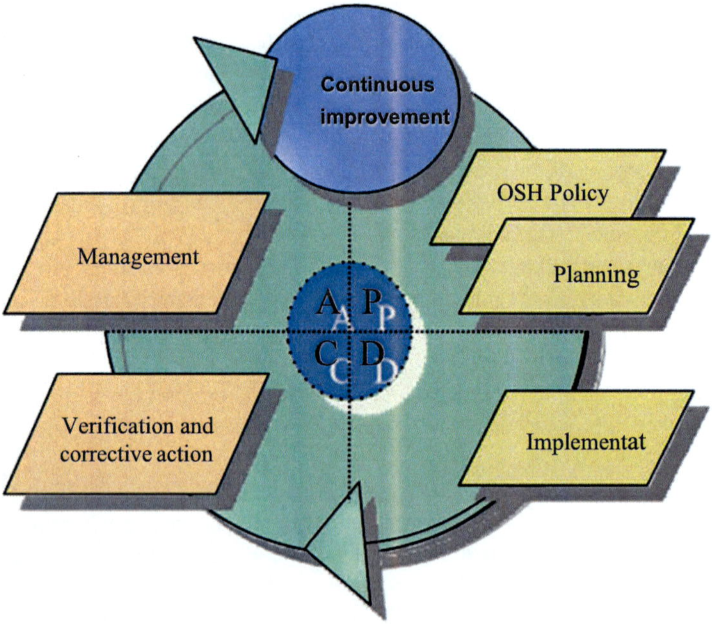

Fig. 2 Management model in safety, worker health, ergonomics and the environment

must, through a critical analysis of management, generate actions to improve the performance of their integrated health and safety management system [21, 22].

The Chart 1 shows the stages of the Integrated Safety, Occupational Health, Ergonomics and Environment Management Cycle based on PDCA.

8 Results and Discussion

Environmental Health and Occupational Health are fundamental fields of Public Health in the social determination of the health-disease process. They seek to transform the reality of production-consumption-environment and health relations, in order to make them favorable to life with quality and social justice. The social responsibility of an organization consists of the decision to participate more directly in community actions in the region in which it operates and to reduce possible environmental damage resulting from the type of activity it carries out.

During the development of this study, we can highlight that in Brazil, health and safety management is more established, still consolidating, in companies in the formal market, where they walk in an attempt to comply with the Regulatory Standards – NR, especially in the NR 4 standards, which advocates the management of SSSEOM, which is part of NR 7, which advocates the Medical Control Program

Chart 1 Stages of the integrated safety, occupational health, ergonomics and environment management cycle

P – Plain	D – Do	C – Check	A – Corrective Act
Legal requirements	Implementation of the controls and preventive measures identified in the first step	Monitor established controls and the performance of the Occupational Health and Safety Management System (SGSSO), ergonomics and the environment	Management review
Federal Constitution	Employee Training	Records	Actions to improve SGSSO, ergonomics and the environment
Regulatory Standards – NR1, NR4, NR7, NR9 and NR17	Management System Documentation	Audits	
Environmental legislation	Documents control		
Occupational health and safety policy	Operational control		
Identification of hazards and risk assessment of the work environment (including ergonomic risks	Emergency Preparedness and Response		
Improvement planning			

in Occupational Health – OHMC. Ergonomic actions, NR 17, are commonly developed outside these two standards [3].

Sustainable Management must be articulated with Organizational Social Responsibility and must be recognized as a continuous and ethical commitment of organizations, aligned with economic development. It must promote the improvement of the quality of life of workers, their families and the local community. Sustainable management, integrated with the management of health, safety and ergonomics in companies, is still under construction and needs more inspection with regard to environmental damage [23–25].

The current trend of sustainable management is to go beyond the control actions of pollutants or waste generated in the production process. Prevention should be prioritized in the structural phase, with an emphasis on prior project analysis. An action that would start before the construction and operation of factories, examining the standards of environmental protection, together with safety and health at work, aiming to establish the necessary modifications in the production process, avoiding the generation of contaminants and defining effective management programs of risks.

The Brazilian environmental legislation recommends that a large part of the projects be submitted to the environmental licensing procedures, requiring the presentation of technical documentation containing a series of requirements, such as the description of the production process, operational techniques, (raw materials and products used in the production process, possible contaminants generated, control measures and the foreseen environmental impacts in this production process.

An evaluation of the effectiveness of this policy is necessary, aiming at the integration of health, safety and environmental actions and optimization of results in line with the vast legislation On this path we can see that the working population has been empowered by this issue, with regard to health, safety and sustainability, see the discussion and reports of the 3rd Occupational Health Conference, which need to be included in the priority of the discussion agenda of country authority [19, 23].

According to the researched literature, we understand that Brazilian companies should evolve to an integrated management model, based on the guidelines of international standards, BSS and OHSAS 18001, where the OSH services of organizations must work in an integrated management, aiming at continuous improvement health and safety, ergonomics and environment, as well as the adaptation to the new NR 01 [21, 22, 26, 27].

9 Final Considerations

Initially, we found throughout this work that, historically, worker health has evolved over several centuries. Starting from Ramazzini, passing through the industrial revolution, where the first laws to protect workers in Europe appeared, starting the period of occupational medicine there, which evolved over the years to occupational health with dissemination of health and safety standards at work by the International Labor Organization. Despite the relative normative progress, work-related illnesses and accidents continued to severely affect the working class, especially due to the rapid process of industrialization. The concept of occupational medicine only evolved into worker health when workers started to demand better health and safety conditions. From this evolution, workers assume the role of actors, of subjects capable of thinking and thinking themselves, producing their own experience.

Workers, individually and collectively in organizations, are considered subjects and participants in health actions, which include: the study of working conditions, identification of technical intervention mechanisms for their improvement and adequacy, and control of the health services provided. Even with this normative evolution in Brazil, worker health actions are still dissociated, especially with regard to the Regulatory Norms of the Ministry of Labor and environmental protection laws. Based on this context, the companies' occupational health services need to work in articulation with occupational hygiene and safety, constituting their SSSEOM, which in turn need to promote joint actions with the ergonomics services and likewise integrate with the principles of an environmental management. A proposal to include ergonomic risks in the OHMC can increase its preventive dimension, following the determinations of NR 1.

The study also emphasize that companies must follow the guidelines of international standards, such as OHSAS 18001, where the SHW services of organizations must work in an integrated management, where the inclusion of ergonomic risks has been proposed, in this management system (Fig. 2), aiming at the continuous improvement of health and safety and with the active participation of employees, who must know this company's health and safety policy. An integrated management model allows for more global actions in relation to workers' health, not dividing the risks and therefore not segmenting prevention actions. Organizations need to evolve from a management based on simple compliance with legislation, such as occupational health and safety and prevention of environmental accidents, to a sustainable management with proactive positioning and actions in relation to eco-efficient product projects.

References

1. Saúde do trabalhador e da trabalhadora no Brasil: uma abordagem holística e integrada / Organizado por Tatiliana Bacelar Kashiwabara et al. – Diamantina: UFVJM, 249 (2021).
2. Ramazzini B.: De Morbis Artificum Diatriba (Tratado sobre as Doenças dos Trabalhadores), Módena, 1700. Tradução Raimundo Estrela. São Paulo: Fundacentro (2000).
3. Oliveira, S.G.D.: Estrutura Normativa da Segurança e Saúde do Trabalhador no Brasil, Rev. Trib. Reg. Trab. 3ª Reg., Belo Horizonte 45 (75), 107–130 (2007).
4. Dias, E. C., Silva, T. L., & Almeida, M. H.: Desafios para a construção cotidiana da Vigilância em Saúde Ambiental e em Saúde do Trabalhador na Atenção Primária à Saúde. Cad Saúde Colet, 20 (1), 15–24 (2012).
5. Araújo, G.M.: Normas regulamentadoras comentadas. Rio de Janeiro. 6ª Ed. GVC, NR17 (2007).
6. Brasil, Constituição da República Federativa do Brasil. Promulgada em 5 de outubro de (1988).
7. Arcuri, A. S. A.: A Política Nacional de Segurança e Saúde do Trabalhador. INTERFACEHS, 2 (4) (2007).
8. Brasil. Ministério do Trabalho. Norma Regulamentadora n.° 01 – disposições gerais e gerenciamento de riscos ocupacionais. Portaria SEPRT n.° 6.730, de 09 de março de (2020).
9. Mesquita, D. F.: Ergonomia na era do Teletrabalho: Impactos para a Saúde e Segurança do Trabalho (2020).
10. Brasil. Lei Federal n° 8.080, de 19 de setembro de 1990. Dispõe sobre as condições para a promoção, proteção e recuperação da saúde, a organização e o funcionamento dos serviços correspondentes e dá outras providências (1990).
11. International Organization for Standardization.: iso Concept Database (2010).
12. BSI, a, OHSAS 18001 – : Especificação para Sistemas de Gestão da Segurança e Saúde no Trabalho, Reino Unido (2007).
13. Iida, I., Buarque, L.: Ergonomia: projeto e produção. 3. ed São Paulo: Blucher (2018).
14. ABERGO – Associação Brasileira de Ergonomia (2018).
15. International Ergonomics Association (IEA).: Human Factors/ Ergonomics HF/E (2020).
16. Grandjean, E. Manual de ergonomia. Porto Alegre: Artes Médicas (2005).
17. Ministério da Saúde (BR). Subsídios para construção da Política Nacional de Saúde Ambiental. Ministério da Saúde, Conselho Nacional de Saúde. Brasília: Ministério da Saúde, 56 (2007).

18. Rigotto, R. M.: Saúde Ambiental & Saúde dos Trabalhadores: uma aproximação promissora entre o Verde e o Vermelho. Revista Brasileira de Epidemiologia, 6, 388–404 (2003).
19. Brasil. Ministério da Saúde. 3.ª Conferência Nacional de Saúde do Trabalhador: 3.ª CNST: "trabalhar, sim! adoecer, não!": coletânea de textos / Ministério da Saúde, Ministério do Trabalho e Emprego, Ministério da Previdência e Assistência Social. – Brasília: Ministério da Saúde, 214 (2005).
20. Silva, J. M. D., Santos, M. O. S. D., Augusto, L. G. D. S., & Gurgel, I. G. D.: Desenvolvimento sustentável e saúde do trabalhador nos estudos de impacto ambiental de refinarias no Brasil. Saúde e Sociedade, 22, 687–700 (2013).
21. Vitoreli, G. A., & Carpinetti, L. C. R.: Análise da integração dos sistemas de gestão normalizados ISO 9001 e OHSAS 18001: estudo de casos múltiplos. Gestão & Produção, 20, 204–217 (2013).
22. Quelhas, O., Alves, M., & Filardo, P.: As práticas da gestão da segurança em obras de pequeno porte: integração com os conceitos de sustentabilidade. Revista Produção Online, 4(2), (2004).
23. Augusto, L. G. S. A construção do campo da saúde do trabalhador e da saúde ambiental. In: AUGUSTO, L. G. S. (Org.). Saúde do trabalhador no desenvolvimento humano local: ensaios em Pernambuco. Recife: Editora Universitária UFPE, 17–47 (2009).
24. Organização Mundial da Saúde. Atención Primaria de Salud: Informe de la Conferencia Internacional sobre Atención Primaria de Salud. Ginebra: Organización Mundial de la Salud, 91 (1978).
25. WHO – World Health Organization. Health impact assessment as part of strategic environmental assessment. Geneva (2001).
26. da Silva, E. H. D. R., Daniel, B. H., & de Oliveira, D. B.: Os sistemas de gestão em segurança e saúde no trabalho em auxílio à prevenção de acidentes e doenças ocupacionais. Revista de Gestão em Sistemas de Saúde, 1(2), 157–172, (2012).
27. Ruella, N. C., & sa-petrobras, p. b. (2004). Processo de implementação de sistemas de gestão integrada com base nas ISO 9001, ISO 14001, OHSAS 18001, BS 8800, SA 8000 e OIT SGSST 2001. In II Congreso latinoamericano de calidad en la industria del petroleo y gas, (2004).

Demand Estimation for Humanitarian Aid Due to Earthquakes in Lima's Cliff Area Using Simulation

Gianmarco Raymundo, Jorge Vargas Florez, and Christian Cornejo-Sanchez

Abstract Humanitarian logistics requires the number of victims estimated. "Costa Verde" runaway is a highly vulnerable coastal strip due to its traffic and cliffs. This work simulates vehicular flows in this place, to estimate the number of victims due to a large earthquake in Lima, it is estimated 3825 victims on average due to a large earthquake to push Lima Coastal.

Keywords Demand estimation · La Costa Verde · Humanitarian aid · Pedestrian and vehicular flow · Simulation · Traffic · Cliffs · Earthquakes

1 Introduction

The humanitarian logistics operations required the estimation of the number of people affected to define the number of resources to be deployed. Since large-scale disasters are random events, traditional forecasting models are not useful for estimating the demand, which is why this work opts for the simulation to approach the complex system formed by the incoming and outgoing flows of vehicles on "La Costa Verde" fast lane. Which is a highly vulnerable coastal strip due to cliffs under an Earthquake.

This work seeks to estimate the number of people vulnerable to a seismic event and subsequent tsunami on the Costa Verde in the district of Miraflores. First, we begin by reviewing the literature and case studies carried out in different areas belonging to the Ring of Fire, which will be used for a comparison with the real situation described in the second part. After, a qualitative and quantitative analysis of the state of cliffs, fast lane, bridges, etc. during a tsunami warning is presented. Finally, the simulation model to estimate the demand is proposed.

G. Raymundo (✉) · J. Vargas Florez · C. Cornejo-Sanchez
Pontifical Catholic University of Peru, Lima, Peru
e-mail: gianmarco.raymundo@pucp.pe; jorge.vargas@pucp.edu.pe; cscornejo@pucp.edu.pe

J. Vargas Florez et al. (eds.), *Production and Operations Management*, Springer Proceedings in Mathematics & Statistics 391,
https://doi.org/10.1007/978-3-031-06862-1_29

2 "La Costa Verde" Cliff Vulnerability

The beaches and cliffs that the city of Lima has, are called " *la Costa Verde* ", this name is also used to refer to the expressway that extends over the lower part of the cliffs. The mentioned cliffs have a natural presence of vegetation due to the presence of groundwater or high groundwater [1]. *La Costa Verde* include the following districts: Callao, San Miguel, San Isidro, Miraflores, Barranco and Chorrillos. This work focuses on the part of the coastal district of Miraflores, the adjacent beaches are of the sandy type, although in recent years there has been a loss of this. According to the National Center for the Estimation, Prevention and Reduction of Disaster Risk [2], 192,141 inhabitants of the aforementioned districts are in a high and very high-risk zone in the event of a seismic movement. An investigation carried out by the Directorate of Hydrography and Navigation [3], warns and identifies flood zones if an earthquake occurs with magnitudes 8.5 and 9.0 Mw.

In [4] was reported that the *Costa Verde* has 8 entrances from the city to the circuit. It also points out that, in summer, the area receives around 350,000 people daily. Teves points out that in recent years clearing landfills have been carried out on the coast. In 1971, Teves says that the *Consorcio Corporación Peruana de Ingeniería* (CORPEI) and Aramburú Menchaca Asociación carried out a study called "Proyecto Costa Verde" [5], where the areas of the cliffs with risk of landslide in case of earthquakes were established. Also in that report, Teves indicates that a geotechnical safety line was established at a distance of 100–200 m from the edge of the cliffs, on which buildings should not be built due to the danger of existing faults in the area. According to Tavera [6], faced with an 8 MW earthquake in Lima, this would generate a tsunami of waves of more than 7 m in Callao, with an arrival time to the coast of 18 min. Despite all the aforementioned, the nearby urban projects do not respect the technical recommendation. Santa Cruz [7] mentions that Miraflores has a smaller population compared to other districts of Lima, however, this varies considerably in the months of January and February, as a result of the summer season and, in particular, which increased mass circulation in the Costa Verde vehicle network, this route being chosen for its tourist attraction and, above all, for being a travel alternative avoiding the transit of the metropolis.

On the other hand, escape routes on the *Costa Verde* are scarce, both for floating people (this is the name given to people who transit through the area to a temporary destination [8]), as well as for the vehicles that transit. To find a possible solution to this problem, it will be necessary to study the behavior and traffic flow in the area; likewise, it will be necessary to present various scenes; for example, lane blocking caused by car accidents or falling to the ground; in order to place ourselves in the worst scenario and estimate the number of vulnerable people. The information obtained will help to choose solutions in the face of disruptive scenarios such as earthquakes: escape routes, traffic flow control, etc.

3 Casualties Forecasting in Disaster

According to EM-DAT [9] between 2011 and 2020 decades, on average natural disasters affected more than 150 million people each year; the magnitude of that impact suggests the complexity of the challenges facing humanitarian operations management planning. On the other hand, one of the starting points to improve the planning of these operations is the forecast of the number of affected, which has among other purposes to determine the amount of humanitarian aid and then plan the distribution in order to attend to the disaster's victims. Nevertheless, in the context of this research, the casualties forecasting refers to the estimation of the number of pedestrian and people inside the car that circulate on the *Costa Verde* at different times of the day, to later determine whether the access and exit infrastructure to that coastline it is suitable to evacuate people in a 15–30 min period, which according to the Instituto Geofísico del Perú [10] is the time interval that the near-field tsunami (the one that originates from an earthquake off the coast) arrives at the coastline, in addition, according to [11], the near-field tsunamis are devastating because arrive at the coast in a short time after the earthquake.

In relation to the above, this research is inserted in the disaster risk management, specially through the traffic simulation model on the Costa Verde, it involves the civil infrastructure in that coastal area as well as the estimation of casualties. In addition, the results of the model contribute to three of the principles recommended by [12] to mitigate the negative impacts of this infrequent but rapid-onset hazard: (1) "know your community's tsunami risk, hazard, vulnerability, and exposure"; (2) "take special precautions in locating and designing infrastructure and critical facilities to minimize tsunami damage" and (3) "plan for evacuation"; in order to improve the response capacity that consists of people arriving in a safe area when receiving a tsunami warning signs [13].

4 Simulation Models for Estimate Tsunami Casualties

In the scientific literature there are studies to determine the casualties due to tsunamis using simulation [11] with an agent-based simulation modelling applied to the study of Seaside city, Oregon, determined that the mortality ratios increase as the quantity of evacuees who used automobile because of the traffic congestion; in addition, they conclude that the walking speed "has significant impacts on the number of casualties in the horizontal evacuation on foot".

On the other hand, [14] with a GIS (Geographical Information System) and agent-based discrete-event simulation framework applied in several scenarios of large-scale disaster in the city of San Francisco in California. They state that it is not necessary to increase the number of shelters, but to use them to their maximum capacity and that it is more important to reduce traffic congestion during evacuation.

With an optimization-based algorithm and agent-based approaches, [15] model a pedestrian evacuation for near-field tsunamis applied in a Puerto-Rican municipality and estimate the tsunami casualties at 5 min intervals, between 5 and 15 min and more than 15 min. Also, with an agent-based approach y and considering the level of education and the effectiveness of the warning as important factors of human behaviour, [16] develops a model for the estimation of casualties due to tsunamis, and [17] estimate casualties in Kamakura, Japan with the aim of estimating tsunamis casualties and providing mitigation measures. The researchers simulated several scenarios with different starting times of the evacuation and they found that the congestion was not as severe as expected, that is to say, the extent of congestion varies according to how people are evacuated and particularly by the starting time.

Numerical calculation and GIS-based framework was used by estimate the tsunami human damage in maximum inundation areas with tsunami in USA town, Japan [18]. The numerical calculation for estimate the human damage consider several factors such as map data, distribution of inhabitants, starting time of the evacuation, evacuation speed, and a grid on the map designed with time of arrival to the shelter and the distance from the grid, routes, and evacuations places, as well as other factors. One of the outstanding conclusions of this study is that human damage decreases due to the effect of early evacuation.

An investigation focused on the coastline of La Punta, an inhabited area located at the sea level in Callao, Perú, simulates tsunami evacuation and estimates the casualties. An integrated simulation was designed take into account aspects of human behaviour and [19] simulated horizontal and vertical evacuations. The researchers determined that the number of casualties in vertical in evacuation casualties is lower than horizontal, likewise, vehicles can cause difficulties in evacuation operations due to traffic.

5 Methodology

The methodology is divided into three stages; (1) Literature review, (2) Information and data collection from the subject of study, highway called Costa Verde, Miraflores area, (3) model formulation. To integrate the concepts of resilience, vulnerability, urban mobility, and transportation flows within a disruptive scenario, it is proposed to use a macrosimulation model based [20]. The authors suggested that single-sector vulnerability analysis is not recommended. There is the possibility of higher risks when the sectors are analyzed together.

The case study requires the use of vehicle flow measurements based on the vehicle count. The collection methodology will be based on historical information and taking a sample according to the recommendations [21, 22] to obtain a statistical analysis for a short-term model. A short-term model considers a small sample of the traffic flow to forecast. The authors analyze the annual behavior of the city of Duisburg, Germany through historical data, also, they mention the causes that generate long holidays and school vacations. These analyzes were taken into account in the methodology to collect the data. There are several methodologies to forecast vehicular flow, [23] establish three (1) random walks forecast, (2) historical average forecast and (3) deviation from historical average forecast. This case study will forecast based on a probability function using *Arena Software*.

6 Transport Model on a Cliff Area

The following section presents a proposal for estimating the demand in a cliff area in three sections: geographic information, highway information and transport modeling.

6.1 Geographic Information

The vulnerability of each area depends of the geographic status. In a cliff area, is very important to get information about the risk and what areas could have more probabilities to collapse. In the Fig. 1, the red zone indicates the flood area by a tsunami after an earthquake (8.5 Mw). According to DHN [3], the entire highway will be flooded. It is estimated that the time when the waves crush the area will be in 15 min after the earthquake.

According to the Geophysical Hazard Assessment Report in the Miraflores district [8], there are 77 critical points in the cliff area. The Table 1 presents the resume:

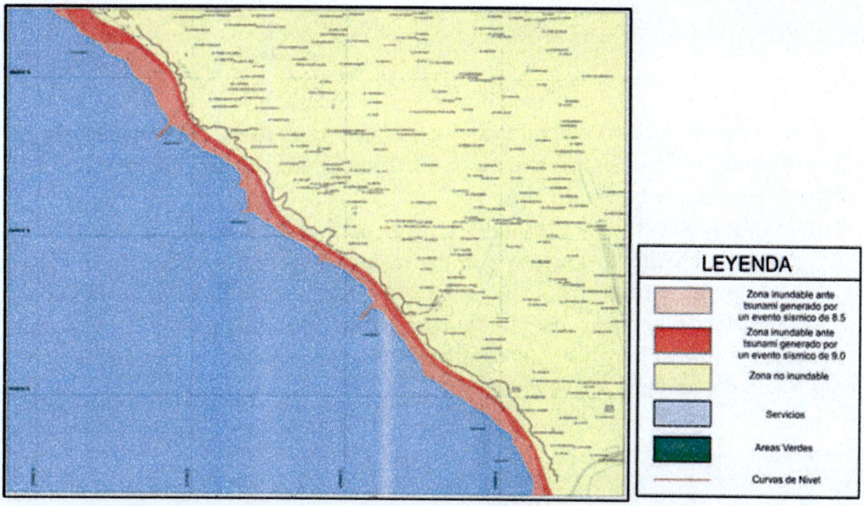

Fig. 1 Flood letter-Costa Verde-Miraflores [3]

Table 1 Critical point Cliff Area-Miraflores [8]

Critical points	Amount
Uncoated mesh slope	5
Deteriorated geogrid	10
Eroded geogrid	2
Unstitched geogrid	28
Rocks suspended on hillside	2
Falling geogrid	1
Rockslide and rock fall deposit	1
Deteriorated protection wall	1
Dangerous fence	1
Probable ground displacement	23
No Fence	1
Spike	2
Total	77

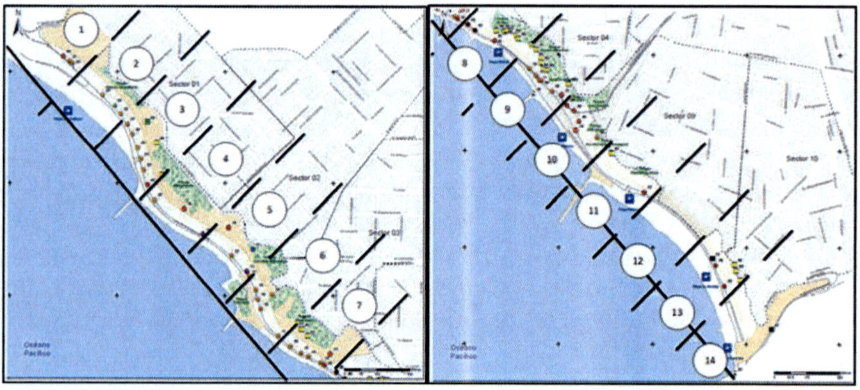

Fig. 2 Cliff Costa Verde-Miraflores (IGP, 2019)

The Fig. 2 presents the cliff area divided in 14 areas.

To determine the most vulnerable area it was necessary to use the weighted-factor rating method. It was assigned from 0 to 10. Table 2 presents risk factors and their criterial weight for each area.

The analysis shows that the most critical area is area 9 which means it is extremely probable that this area will collapse during an earthquake.

To sum up, the entire zone is vulnerable after a huge earthquake at Lima's coast and people in the highway has 15 min to escape. According to IGP, there are 77 critical points and they can generate more vulnerable areas. The more vulnerable zones, the more amount of people die.

Table 2 Weighted-factor rating method

Risk factors	Criteria weight	1	2	3	4	5	6	7	8	9	10	11	12	13	14
Geographic criticality	0.35	3	3	7	6	3	3	8	10	10	6	10	0	10	7
Proximity to buildings	0.20	3	0	0	0	0	10	0	0	3	4	0	9	0	9
Slop without geogrid	0.20	3	0	0	10	10	0	10	0	10	10	0	0	9	0
Location and condition of bridge	0.15	0	7	0	8	0	5	0	5	0	0	0	0	0	0
Geogrid in poor condition	0.10	4	6	6	3	2	6	8	7	10	9	3	0	6	0
Total weighted score	1.00	2.7	2.5	3.1	5.6	3.3	4.4	5.6	5.0	**7.1**	5.8	3.8	1.8	5.9	4.3

6.2 Highway Information

According to Jin [24], the maintenance of the road is a relevant factor in Risk Management. There are three main highways in the area: *Costa Verde* highway, *Paseo de la Republica* Highway and *Bajada Armendariz* Highway. To get a good condition, it is necessary to get clean, there has to not get potholes. During an inspection of the Municipality of Lima, these three highways are in a good condition.

There are 5 bridges, they connect the beach with the high cliff area. From the 5 bridges, 3 of them are in a good condition, the rest is inoperative. In the evacuation map, these 3 bridges are used to evacuate from the area. The zone with more people is in the Bajada Balta, because it is a touristic zone. There is 1 bridge close and the time to evacuate is 7 min approximately. The Fig. 3 presents the resume of the highway.

There are 4 entrances (San Isidro, Bajada Balta, Bajada Armendariz, Barranco) and 5 exits (San Isidro, Bajada San Martin, Bajada Balta, Bajada Armendariz, Barranco). It has 5 km long approximately.

6.3 Transport Modeling

To get the vehicular flow, the amount of vehicle entries and exits was counted. The worst scenario after an earthquake in the cliff will occur when people could not escape from the area. This will happen in the peak time, which is between 7:00 am to

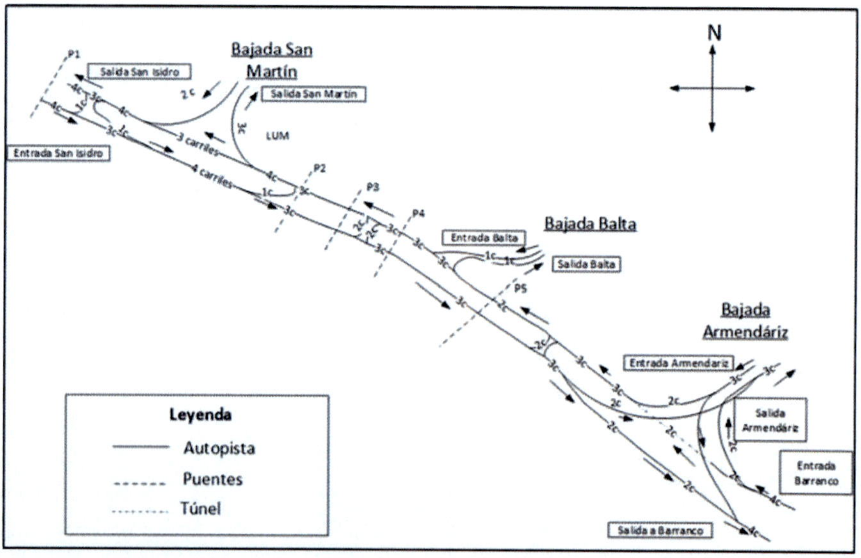

Fig. 3 Highway Costa Verde-Miraflores. (Source: Own elaboration)

Table 3 Expression of input variables

Variable	Expression
Interval time San Isidro	WEIB (0.00957, 12.4)
Interval time Bajada Balta	0.12 + GAMM (0.00653, 3.23)
Interval time Bajada Armendariz	0.02 + 0.09 * BETA (1.27, 1.53)
Interval time Barranco	0.02 + 0.05 * BETA (0.711, 0.955)

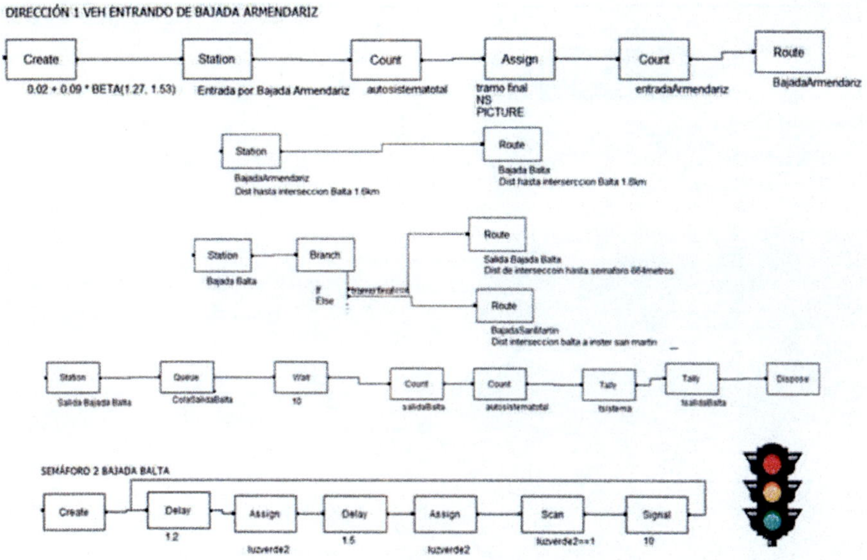

Fig. 4 Transport model Bajada Armendariz-Bajada Balta. (Source: Own elaboration)

9:00 am. The field research started to count from 7:45 am to 9:15 am, the interval between was 15 min. In total, it was counted 31,830 vehicles in the entire area. The Table 3 presents the 4 inputs for the model using *input analyzer*.

Arena Simulation software v.14 is chosen to modelling the traffic flow and to determine the amount of people. This software is used to model discrete events and operational processes. The appeal of using ARENA versus other software comes from the fact that it is versatile, and processes can be modeled using a number of different methods [25].

This model is non-terminal system, and the full-time simulation was 4,780,000 min. Assumptions:

- Vehicles won't back to the original station when they are at the system
- The time between stations is an average
- The time in the traffic lights between green and red is constant.
- The model doesn't consider crushes between vehicles.
- The model doesn't consider the driver's behavior.

The Fig. 4 presents the vehicular flow form Bajada Armendariz to Bajada Balta.

Table 4 Origin-destination distribution

Origin/destination	San Isidro	San Martin	Balta	Armendariz	Barranco
San Isidro	0	0.45	0	0.15	0.4
Bajada Balta	0.65	0.35	0	0	0
Bajada Armendariz	0.45	0.4	0.15	0	0
Barranco	0.25	0.4	0.35	0	0

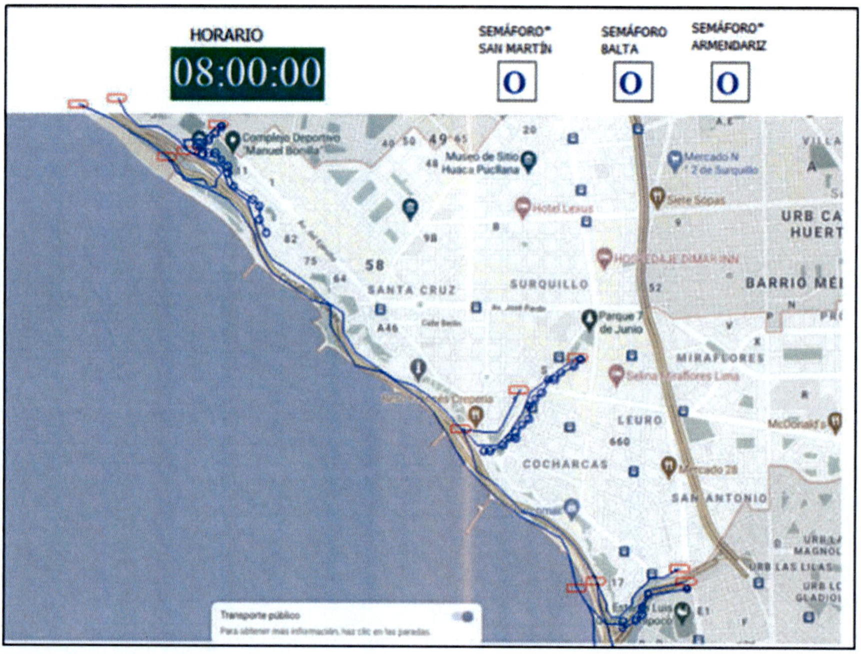

Fig. 5 Vehicular flow map. (Source: Own elaboration)

The figure shows how vehicles enter from Bajada Armendariz to the system. Then, they are assigned one attribute: "tramo final", this attribute indicates their final destination. Furthermore, it is necessary to create internal entities for traffic lights when the vehicles are coming out. The model shows the block Wait with a Signal of 10 and a Scan Block changing the variable "luzverde2", when "luzverde2" is 1, it means that traffic light's color is green (1.2 min) and when "luzverde2" is 0, the color is red (1.5 min).

The attribute "tramo final" requires a distribution of the final destination of the vehicles. Therefore, the Table 4 indicates the distribution based on the vehicle count.

The following Fig. 5, presents the entry and exit stations of the model using the animation software. Also, it indicates the traffic lights. Cero when is red, it means there will be a queue on the road and one when the color is green. The longest stretch is 5 km approximately.

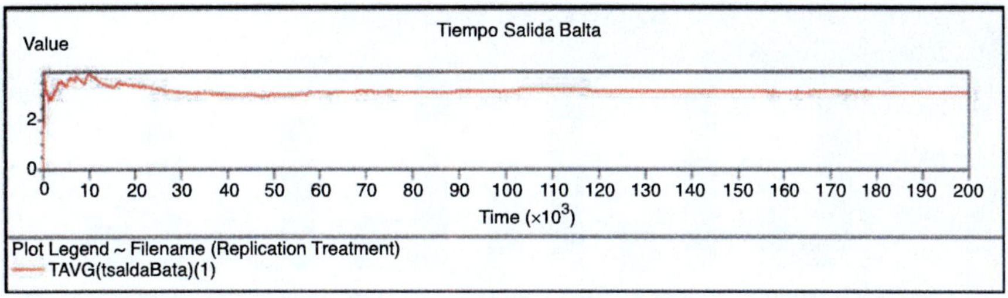

Fig. 6 Exit time Balta. (Source: Own elaboration)

After stablishing the model, it is necessary to define our outputs. The outputs of the model are the number of cars in each queue, the time of exit in each Bajada, the time in the system and the number of vehicles in the system. The first-time simulation was 200,000 min. After using *output analyzer* for each variable, the system is stable after 100,000 min (Fig. 6). Then, to identify the number of batches, the long-time simulation were 4,000,000 min. The number of batches were 400 and the number of observations were 130. It was chosen the observation number 24, because its correlation was under 0.10.

In resume, to stablish a transport modeling, is necessary to analyze three important points. First, get information about the risk in the geographical zone. Isolate possible cliff collapses. Second, highway and bridge maintenance. It is useful to have the map of the zone and identify bridges, the raceway geometry and speed limit. Finally, have information about the car's behavior that enter and leave the area to establish their distribution income.

7 Results

The next results were taking from fieldwork from 7:45 am to 9:15 am. It shows the behavior of each Bajada in Miraflores.

Bajada San Martin has the largest number of vehicles entering the system. On average, 1570 vehicles enter to the system in 15 min, it means 104 vehicles inter per minute from San Isidro. In the case of Bajada Balta, there are more vehicles coming out of the system, but the amount is not more than Bajada Armendariz. In the last Bajada, the geometry of the roads was more complex. The Fig. 7 indicates the most of the vehicles come from Barranco than Armendariz.

According to the model simulation, the amount of vehicle in the system was 1598 vehicles. Using the factor of 2.4 person per vehicle in average, the amount of the people that will be vulnerable after an earthquake is 3825 people approximately.

This research doesn't consider other districts in the coastal area of Lima, they are also in a vulnerable area. It is recommended to know the behavior of those districts. Furthermore, this research just takes a sample from the morning. The second peak time is at night, from 6:00 pm to 8:30 pm. It will be important the differences

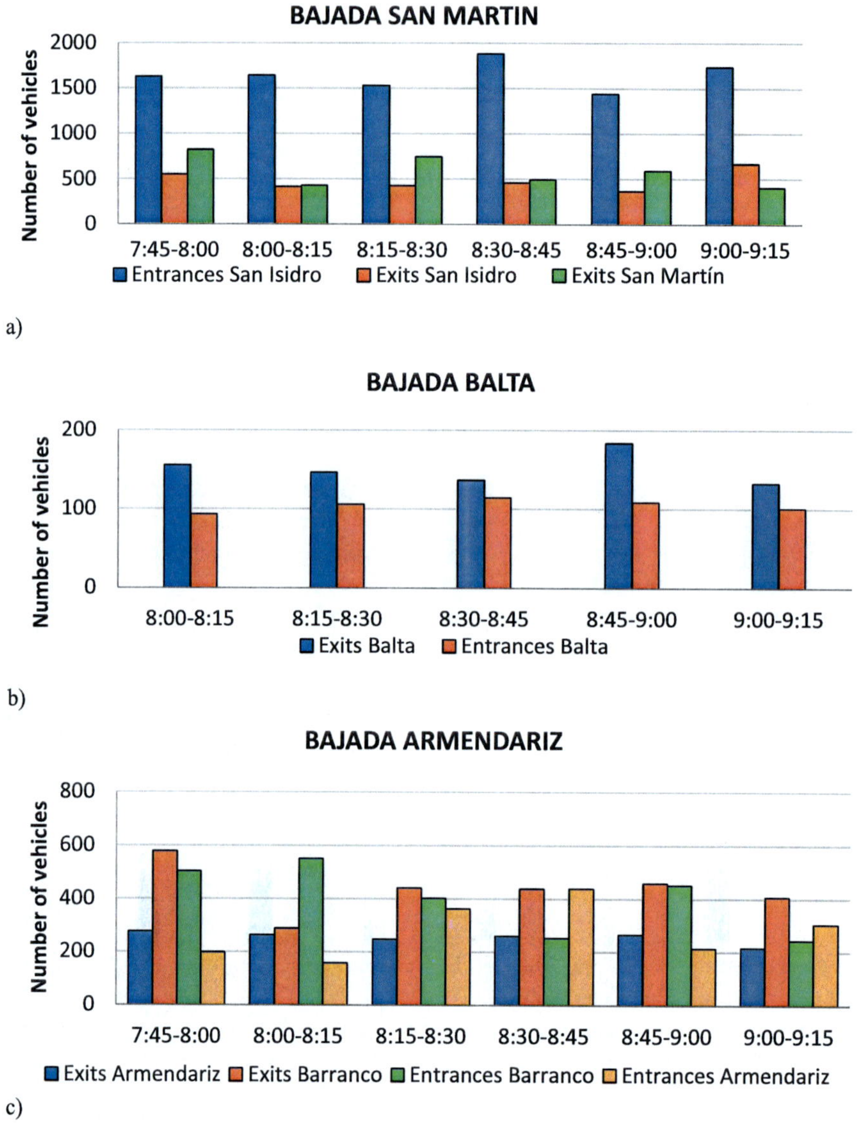

Fig. 7 Number of vehicles entrances and exits-Miraflores. (**a**) number of exits and entrances at San Martin. (**b**) number of exits and entrances at Balta. (**c**) number of exits at Armendariz

between day and night. Finally, estimating the number of vulnerable people in the entire *Costa Verde* after an earthquake will help to plan the humanitarian support needed in each district.

References

1. Teves, N.: Revista del Colegio de Ingenieros de Lima 75, 8–9 (2016).
2. CENEPRED: Escenario de riesgo por sismo y tsunami para Lima Metropolitana y la provincia constitucional del Callao. Centro Nacional de Estimación, Prevención y Reducción del Riesgo de Desastres (2017).
3. Dirección de Hidrografía y Navegación Homepage, https://www.dhn.mil.pe/secciones/departamentos/oceanografia/apps/cartastsunamis/tsunamis_prevencion/tsunamis_inundacion.htm, last accessed 2021/01/07.
4. Municipalidad de Lima: Proyecto Costa Verde. Mi Costa Verde Blog, Link https://micostaverde.blogspot.com/2011/02/autoridad-del-proyecto-costa-verde.html, last accessed 2019/07/10.
5. Santa Cruz, W.: Territorios Fragmentados El Caso de la Costa Verde. Pontificia Universidad Católica del Perú, Lima. (2018).
6. TAVERA, H. "Escenario de sismo y tsunami en el borde occidental de la región central del Perú". Lima, Perú (2014).
7. Gándara, C., Padilla, F., Gutiérrez, P.: Población flotante y ciudad desde una perspectiva socioespacial: revisión de estudios recientes. Revista de estudios transfronterizos. https://scielo.conicyt.cl/pdf/ssa/v20n1/0719-0948-ssa-20-01-103.pdf (2020).
8. Instituto Geofísico del Perú.: Evaluación de peligros geofísicos en el distrito de Miraflores. Instituto Geofísico del Perú, Lima (2019).
9. EM-DAT The International Disaster Database Homepage, https://public.emdat.be/, last accesed 2021/08/12.
10. Instituto Geofísico del Perú.: Reporte técnico La ciencia y la gestión de tsunamis en el Perú. Instituto Geofísico del Perú (2012).
11. Wang, H., Mostafizi, A., Cramer, L., Cox, D. Park, H.: An agent-based model of a multimodal near-field tsunami evacuation: Decision-making and life safety. Transportation Research Part C 64, 86–100 (2016).
12. Eisner, R.: Planning for tsunami Reducing future losses through mitigation. Natural Hazards 35, 155–162 (2005).
13. Post, J. Wegscheider, S., Mück, M., Zosseder, K., Kiefl, R., Steinmetz, T., Strunz, G.: Assessment of human inmediate response capability related to tsunami threats in Indonesia at a sub-national scale. Natural Hazard and Earth System Sciences 9, 1075–1086 (2009).
14. Suk Na, H., Banerjee, A.: Agent-based discrete-event simulation model for no-notice natural disaster evacuation planning. Computers & Industrial Engineering 129, 44–55 (2019).
15. Faucher, J-E., Dávila, S., Hernández-Cruz, X.: Modeling pedestrian evacuation for near-field tsunamis fusing ALCD and agent-based approaches: A case study of Rincón, PR. International. Journal of Disaster Risk Reduction 49, (2020).
16. Yeh, H.: Tsunami hazard and casualty. In 10th Proceedings of the National Conference in Earthquake Engineering. Earthquake Engineering Research Institute, Anchorage, US (2014).
17. Takabatake, T., Shibayama, T., Esteban, M., Ishii, I.,: Advanced casualty estimation based on tsunami evacuation intended behavior: Case study at Yuigahama Beach, Kamakura, Japan. Natural Hazards 92, 1763–1788. (2018).
18. Sugimoto, T., Murakami, H., Kozuki, Y., Nishikawa, K., Shimada, T.: A Human Damage Prediction Method for Tsunami Disasters Incorporating Evacuation Activities. Natural Hazards 29, 585–600 (2003).
19. Mas, E., Adriano, B., Koshimura, S.: An Integrated Simulation of Tsunami Hazard and Human Evacuation in La Punta, Peru. Journal of Disaster Research 8(2), 285–295 (2013).
20. Murray, A., Matisziw, T., y Grubesic, T. "A methodological overview of net-work vulnerability analysis". Growth and Change. Vol 39. pp. 573–592. (2008).
21. Leon, L., Y. Kibangou, A. and Canudas de Wit, C. "Adaptive Kalman Filtering for Multi-step ahead Traffic Flow Prediction (2013).

22. Chrobok, R., Kaumann, O., Wahle, J. and Schreckenberg, M. "Different methods of traffic forecast based on real data". European Journal of Operational Research 155, 558–568 (2004).
23. M. William, B., Asce, M., A. Hoel, L. "Modeling and Forecasting vehicular traffic flow as a Seasonal ARIMA Process: theoretical basis and empirical results" Journal of Transportation Engineering 129 (6): 664–672 (2003).
24. Jin, C., Wang, J., Chen, K., Su, J., Bi, M.: Research on Monitoring and Emergency Response Project of the Urban Road Network. In: 11th International Conference of Chinese Transportation Professionals. ASCE Library (2011).
25. Davis, K. "Simulating traffic flow for emergency evacuation in Manhattan, KS using rockwell arena". Kansas State University (2010).

Rescue Robot Against Risks in Natural Disasters Using Arduino

Ana Luna, Mario Chong, Pilar Hidalgo, and Aldo M. Panfichi

Abstract Nowadays, nobody can deny that advanced technology is widespread in almost every aspect of our daily lives. Our main objective in this research is to provide a tool that is complementary to the work of rescuers in the face of a natural disaster that minimizes the risks of loss of life during the SAR (Search And Rescue) process. The main function of the designed prototype, named "Rescue Bot", is the search of missing persons after collapses of large-scale structures of any kind, where it is difficult to locate people caught in a landslide. The rescue bot has several advantages: it is lightweight, it has small dimensions, it is cheap, and it uses a simple and low-level programming language through the free hardware platform Arduino. Its mission is to reduce the number of fatalities and rescue time and take the risks that rescuers face in every natural disaster. We have used several sensors and an infrared camera as indispensable accessories in the Rescue Bot. The information collected is sent to the control centre in real-time.

Keywords Arduino board · Relief operations · Low-cost sensors · Rescue terrain robot

1 Introduction

The geography of Peru is prone to damage of large magnitude due to natural phenomena such as rain, frost, landslides, floods, earthquakes, and droughts, which generate large human and infrastructure losses, as well as economic losses that are barely recoverable in the short term. Alongside these, we must also consider anthropogenically caused disasters. According to INEI (the National Institute for

A. Luna (✉) · M. Chong · P. Hidalgo
Universidad del Pacífico, Lima, Peru
e-mail: ae.lunaa@up.edu.pe

A. M. Panfichi
Pontificia Universidad Católica del Perú, Lima, Peru

J. Vargas Florez et al. (eds.), *Production and Operations Management*, Springer Proceedings in Mathematics & Statistics 391,
https://doi.org/10.1007/978-3-031-06862-1_30

Statistics and Informatics), Peru has experienced 36,882 natural disasters and 12,355 anthropogenic disasters in the last decade [1]. The UN has reported that around 60% of Peruvians are not prepared to face natural disasters; more than half of Peruvians are vulnerable to most disasters. Therefore, it is crucial to be prepared for them. One of the critical issues at the time of a natural and anthropogenic disaster is how to quickly resort to an efficient rescue in the shortest possible time to safeguard many injured people.

In many cases, both the lack of equipment and rescuers facing these catastrophes must also be considered. As such, the post-disaster stage must have a clear protocol of action, and the use of technology is a tool that could be of great help in these types of events, specifically during the rescue process. Some robots are already implemented in urban search and rescue tasks that include autonomous exploration of disaster sites and the recognition of victims and other objects of interest. However, in Peru, this new concept has not yet been implemented.

On the other hand, the desire to cultivate higher education interest in resolving social problems has motivated universities to focus on social responsibility in professional training. According to [2] and the UN Global Compact,[1] social responsibility is the commitment to the impact generated by an agent in society due to exercising his functions, be it academic, business, industrial, or anything else. In this context, the states advance if their institutions circumscribed in their territories promote and follow a direct route based on the current legislation.

The Universidad del Pacífico in Lima, Peru, has integrated into its engineering program a focus on developing projects as a learning strategy aimed at solving social problems in local and global spheres. In this vein, the Physics course as part of the Business and Information Engineering curriculum complements theory classes with laboratory sessions where concepts can be put into practice in developing various prototypes and projects. Robots are designed here to fulfil a business model, but their educational value also considers their use to accomplish social responsibilities.

This article describes the prototype designed and built at the Universidad del Pacífico in a Business Engineering degree course. We use a free hardware platform, Arduino. The Rescue Bot device is economical and can be used as a complement or replacement for rescuers' functions. The current research goals are to identify, analyze, and interpret a real social problem and find a solution.

This research article is organized as follows: Sect. 2 describes the state of the latest developments in rescue robots. Section 3 presents the methodology, and Sect. 4 describes the experimental setup of the Rescue Bot. Section 5 discusses opportunities to improve the project and learning strategies. Finally, Sect. 6 outlines the conclusions and future research.

[1] Available on: https://www.unglobalcompact.org/

2 Literature Review

The authors of Kohlbrecher et al. [3] developed an open-source software module for a system capable of autonomous exploration of disaster sites and recognition of victims in Urban Search and Rescue (USAR) environments that includes mapping in a degraded urban area. This was one of the first modules of open source software for urban search and rescue robots, and many have followed up with similar work in recent years.

Other researchers Bernard et al. [4] have used aerial robots, made up of three small helicopters, to help victims, particularly injured people with minimal mobility during rescue operations. We learned from this work the value and use of cooperative detection mechanisms, using several different types of sensors. However, the use of wireless devices is not always feasible since they are sensitive to winds or gusts and even to storms and visibility conditions, limiting their use.

In the research work of Dixit et al. [5], the authors showed the design and implementation of a wireless robot that detects living bodies with the help of a PIR (Pyroelectric Passive Infrared) sensor and permanent internet connection. The robot has a camera that moves horizontally and vertically and records what it sees. The robot's movements are controlled from a web page corresponding to the user interface, providing an improved view of the surroundings. Although the use of the Internet does not bring range limitation, the big problem is that usually, in natural disaster events, internet access is not necessarily available.

In the work of Papoutsidakis et al. [6], a remote-controlled robotic platform was designed. This project consists of sending data and images to one or several central control stations. Potential applications would be in remote monitoring and telemetry in places that are difficult to access or dangerous for humans. More research should be done to improve this prototype in the field of robotics. A lesson from this work is that their code tracks were driven by a modular construction approach to use them as separate parts for different projects. It could also be set up to be semi-automatic or fully autonomous, depending on the application's needs. This work sets a precedent for the latest efficient platforms when working with remote control systems.

The author of [7] proposed using a drone to help with the search and rescue activities over disaster scenarios. The advantages that researchers highlight about drones are that they provide a temporary communication structure, create updated maps of the affected region and search for hot spots where the rescue teams may have more chances of finding victims. However, drones can perform these functions with high-resolution sensors and 3D cameras, among others. On the other hand, it is unlikely that people in the terrain will have the human resources to control drones and ensure perfect full coverage of a given area. In addition, there is a legal vacuum in most countries regarding the licenses necessary for the drones' flight since they are not fully autonomous and the human operator always decides their trajectories and functions. The ideal in a rescue situation would be to have a fleet of drones. But this architecture is costly compared to robots, which are electro-mechanical machines, which could fulfil similar functions to drones without the disadvantages they present.

Mobile Robots are being widely used, and another team of researchers has presented the design and implementation of a robot with obstacle avoidance capability for remote sensing and monitoring (Sunehra et al., [8]). The system enables the user, located at a base station, to send commands to the remote station (mobile robot). They also used Global Positioning System (GPS) technology to provide the user with the robot's location. The system also provided the user with real-time video monitoring of the remote area using an internet-enabled device. This work was particularly motivating and inspiring for the project that we will present in the following sections.

The author of [9] describes that robot construction and educational robotics are practical tools for project-based learning in STEM, coding, computational thinking, and engineering skills, all wrapped up in one project, alongside soft skills including communication, collaboration, presentation skills, and patience. However, this work does not mention the relationship between these skills and raising awareness of social responsibility.

Similarly, Araújo et al.'s [10] work have as its principal contribution to the development of open-source tools that allow the use of Arduino platforms in Robot Operating Systems (ROS) as a starting point towards the use of other Arduino-based controllers. It also points out that the main advantage of this integration is the reduction in construction time due to the utilization of code and an extensive library of tools. Despite this, the research does not show how the use of integrated platforms leads to strengthening soft skills and social awareness.

In [11], Osborne et al. mention the development of a program for undergraduates to mentor teams of local middle and high school students on robotics projects, applying in this manner scientific, mathematical, and computational basics through robotics, as well as designing algorithms. The authors document an overview of the program and some evaluation results. The program proposed here mentions the soft skills of cooperative work; however, social responsibility for the community is not a variable taken into account when supervising and grading these robotics projects.

An approach to project-based learning is studied in [12], which shows how building robots as projects helps to reinforce in students some concepts acquired at school through direct application of those concepts and acquired knowledge in a valuable way to solve real-life problems:

> When students faced real situations and problems where they can use the mathematics, science and physics theories as well as concepts learned at school, they realize that knowledge can be a useful tool in life. The teamwork developed in this activity has an important impact on the general behavior of students in society; due to students always interacting inside real social situations.

This work has a casual and underdeveloped conclusion on the social impact of these projects; however, it refers to their importance within education in sciences and mathematics.

With this background in mind, we propose and show a project carried out by undergraduates that solve real social problems using technology and programming in robotics. In addition, they improve upon the concepts studied in their courses.

3 Methodology

The execution of projects, such as the one presented in this work, is the final learning achievement of the Physics course of the second year in the Engineering degrees offered by the Universidad del Pacífico. Through this project, students are expected to implement new solutions to existing problems and needs, designing, analyzing, and explaining in detail the functionality of a device based on its properties and physical principles. The course itself makes students aware of the importance of an accurate description of the elements of a system and the analysis of how they are interrelated. In this way, the logical and rational studies required to explain physical phenomena act as the basis for any engineering diagnosis of processes or technology.

The course consists of 16 weeks, four theoretical hours plus 2 h of laboratory per week. In the latter, techniques for the design and construction of robots that perform basic operations and specific jobs are addressed, guiding the student to the understanding and behaviour of this type of technology. In addition, the physical fundamentals are taught through the use of electronic sensors, and the student is simultaneously introduced to programming using the free access Arduino software. In the theoretical classes, the course covers topics of Mechanics, Electrostatics, and Electrodynamics. The dynamics of the classes seek to permanently exploit students´ curiosity, questioning what they think about the world that surrounds them and giving explanations of the various phenomena to which they are usually exposed.

To guarantee the achievement of the final learning achievement of the course, professors used, within the evaluation system, not only traditional exams (a midterm and a final exam) but also the Immediate Feedback and Evaluation Technique (IF-AT cards) to review the fundamental physical concepts of the course [13–17]. Professors' choice to use this methodology is primarily to prioritize the collaborative aspect of the IF-AT cards and the development of questions that support peer instruction that is potentially highly beneficial strategies in the learning process and within the context of team learning (TBL). Finally, the students have two optional weekly hours for general consultations.

As mentioned in Sect. 1, Physics competencies for the Engineering program include a focus on the concept of social responsibility. In the first stage of the prototype construction, the student, having previously acquired the basic knowledge in the theoretical-practical course, must recognize a social problem and design a technological solution that also includes developing a code in Arduino that executes the desired functions. Then, the undergraduate proposes a possible solution to their advisor, who guides the student in conducting and analyzing the project's feasibility. Once the final project is approved, the student presents it orally and receives additional feedback before the final delivery. Next, the student has one extra week to carry out the improvement aspects observed.

The two main competencies evaluated within the Physics course were: a) The originality of the prototype and how it contributes to the solutions proposed by the student. Creativity and innovation are also taken into account. b) Communication and teamwork, looking forward to presenting a robot that uses technology for real solutions highlighting the knowledge acquired during the course.

4 Experimental Setup

In this section, we present the design of one of the best students' projects, not only taking into account priority competencies but also their costs compared to commercial prototypes that address solutions to similar problems [18].

The name of the robotic vehicle project is the Rescue Bot, and its function is to monitor disaster areas that are not easily accessible by people. The robot is provided with a camera module to obtain real-time images of the area, temperature and humidity sensors to measure and monitor the changes of those parameters in the surrounding area and the immediate open space around the robot, without the necessity of human intervention. A GPS device and communication register location information occurs through the Wi-Fi module.

An Arduino MEGA board (Fig. 1) is used to control all the accessory devices. Arduino is a free hardware platform based on a board with a microcontroller and a development environment through software. The advantages of Arduino are several: it is an open-source platform, it is low cost, its programming is simple (a simplified version of C++), and it is highly flexible and can be combined with other platforms, among other advantages. We used Arduino to write and upload the computer code to the board using a USB cable in this particular case.

Fig. 1 Arduino MEGA. (Source: https://images.app.goo.gl/Bt52PTbCc2Qf1PUg7)

Fig. 2 Sensors (From left to right: HC-SR04, DHT11). (Source: https://www.makerelectronico.com/producto/dht11-sensor-temperatura-humedad/)

The sensors used are four ultrasonic sensors (HC-SR04) and a temperature and humidity sensor (DHT11) (Fig. 2). The first allows the robot to move through the structure without hitting any rock or wall. Its principal function measures the distance to the closest objects present in its trajectory through echo and trigger signals. The trigger sends an ultrasonic signal (340 m/s), and the echo measures the distance to a nearby object concerning the sensor in a particular time frame. This distance is then easily calculatable with Eq. (1):

$$range = \frac{return\ time}{2} \times 340\ m/s \tag{1}$$

One of the limitations of these sensors is that their detection range is between 2 and 400 cm within a 30-degree field of view.

The other sensor is used for continuous monitoring of two variables that are critical in rescue events. DHT11 temperature and humidity sensors have a calibrated digital signal output, are reliable, and have significant long-term stability. The humidity measurement is resistive, and the temperature measurement is carried out indirectly through an NTC thermistor. The sensor can capture temperatures from 0 to 50 degrees Celsius with $\pm$ 2 degree accuracy and humidity from 20% to 80% with 5% accuracy. The sensors are connected to a high-performance 8-bit microcontroller, which offers excellent quality, quick response times, and is cost-efficient. The DHT11 is convenient for this project as it is easy to program. Its C++ libraries are freely available on the Internet. It can measure two types of data (temperature and humidity).

Other components that we used were a GPS Module (Neo-6 M, Position accuracy: 2 m and better with good satellite signals) and an infrared camera (Noir V2 8Mpx) (Fig. 3). The first allows the detection of the exact location of a victim. Its characteristics are: Velocity Accuracy: 0.1 m/s, Maximum Velocity: 500 m/s, Heading Accuracy: 0.5 degrees while moving. The second gives a real-time visual of the area. This type of camera does not require external light, and its principal characteristics are Optical size 1/6 inch, Resolution 640 $\times$ 480 VGA, Onboard regulator; only a single and 3.3 V supply is needed. Finally, we used L298N Dual H Bridge DC Motors Driver for the movement of the wheels. Their

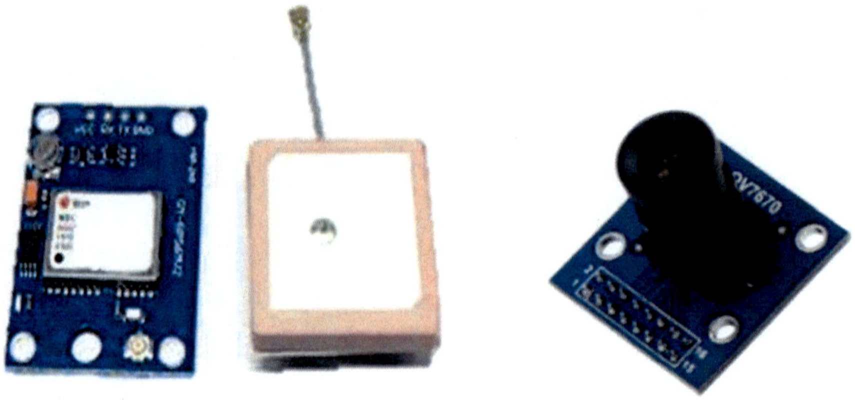

Fig. 3 From left to right (Neo-6 M & Noir V2 8Mpx, GPS module). (Source: https://udvabony.com/product/ublox-neo-6m-gps-module/)

characteristics are: Input Voltage: 3.2 V ~ 40Vdc, Power Supply: DC 5–35 V, Peak current: 2A, operating current range: 0 ~ 36 mA. The power supply of the Rescue Bot is an 11.1 V, three cells, 2200 mAh, rechargeable battery.

The data will be sent via a Wi-Fi signal through an ESP8266 ESP-13 Wi-Fi Web Server Shield device whose characteristics are as follows: 802.11 b/g/n wireless standards, TCP/IP protocol stack, one socket, Supports standard TCP/UDP Server and Client. This makes communication easy over a local network which does not require functioning Internet access. It also supports serial port base rates of 1200/2400/4800/9600/19200/38400/57600/74800/115200 bps, allows continuous transmission operation with a current of ≈70 mA (200 mA max), temperatures of −40 °C ~ + 125 °C and humidity from 10% to 90% non-condensing, which makes it ideal for rescue tasks in comparison to other wireless devices or Wi-Fi integrated Arduino cards such as ESP32, ESP8266, among others.

The design needs to be light. Because of that, we use an Arduino Mega with a large number of pines instead of protoboards. In any case, all components must be able to work and draw power simultaneously. The battery has a discharge capability of 2200 mA/h, and if connected in series, all parts would utilize at a maximum of 180.5 mA/h (15*4 + 2.5 + 10 + 20 + 36 + 70).

As shown in Fig. 4, Rescue Bot is an automatic light car with big, versatile wheels for all-terrain travel. The arrangement of components is also light and ordered to carry out its function without being hindered by design. The full dimensions of the prototype are 15 × 13 × 17 cm.

Finally, data collection and management are done through the IoT Cayenne system.[2] On this online and desktop-based platform, a rescuer can monitor all the information collected by Rescue Bot's sensors during the operation. As shown in

[2] Available on: www.thethingsnetwork.org

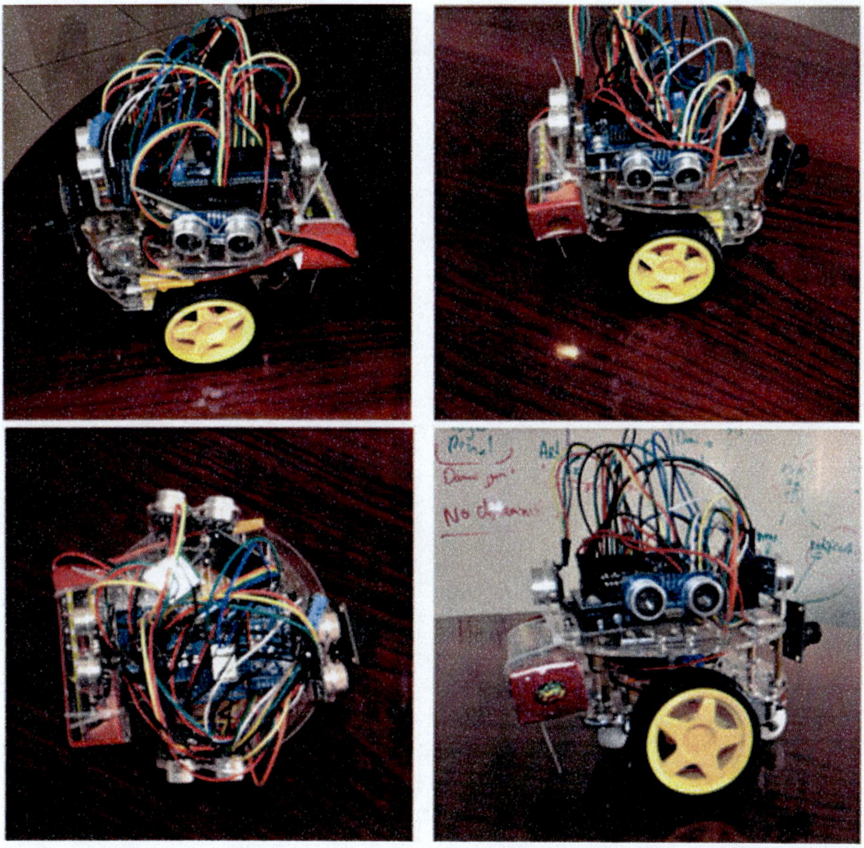

Fig. 4 Rescue Bot prototype; side and top-down views. (Source: Own elaboration)

Fig. 5, the Cayenne platform offers real-time readings of the Arduino's serial monitor.

The code for this project can be found at https://github.com/PilarHidalgo/Rescue-bot.git

The cost of the individual components is broken down in Table 1:

5 Discussion

The Rescue Bot has multiple advantages over other search methods during rescue operations. While the prototype itself may not carry out any physical rescue actions, it significantly assists rescue activities without putting at risk the lives of humans and animals often used for this purpose. Despite this, there are several opportunities to

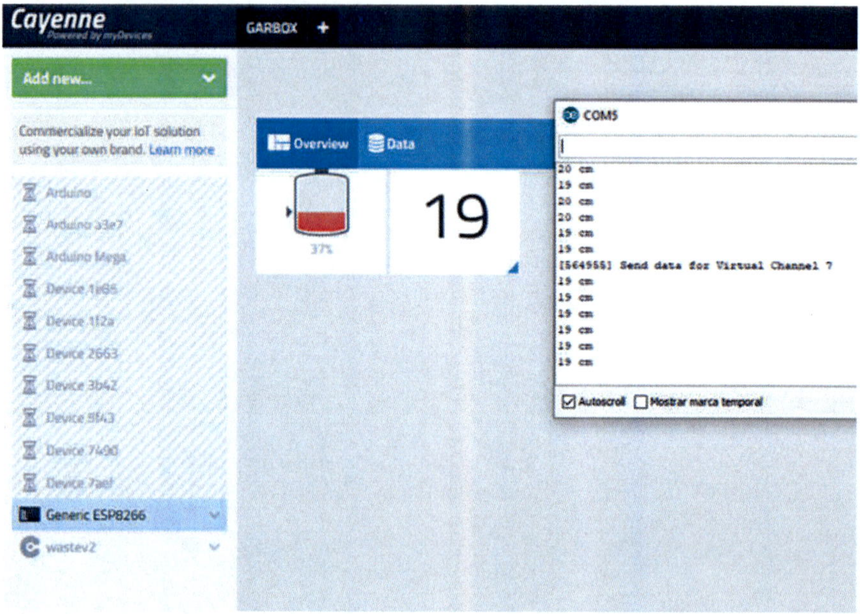

Fig. 5 Screen capture of Cayenne and serial Arduino reading of the proximity sensor HC-SR04

Table 1 Device costs

Component	USD Cost	Amount	Total
Ultrasound Sensor HC-SR04	$2.50	4	$10.0
Temperature and Humidity Sensor DHT11	$3.20	1	$3.20
ESP8266 Wi-Fi Module	$18.52	1	$18.52
Nighttime Camera Noir V2 8MP	$42.10	1	$42.10
GPS GY-NEO6MV2	$3.10	1	$3.10
2200 mAh Battery	$22.85	1	$22.85
Arduino Mega 2560 Ch340 + USB Cable	$14.30	1	$14.30
Wheels Kit (Chassis, DC Motors)	$3.30	1	$3.30
L298 Driver	$3.20	1	$3.20
Cables Kit	$1.00	1	$1.00
Total			**$121.57**

improve the design that we detected during the development of this project, which we will mention below.

The use of the ultrasonic sensor allows the rescuer to avoid obstacles that may arise along the way. However, as stated in Sect. 4, these sensors are limited in their detection range between 2 and 400 cm. They have an angular field of view of 30 degrees, which does not cover the full 360 degrees around the robot across four sensors. The DHT11 humidity and temperature sensors are very sensitive to gas leaks or fires. The GPS helps the rescue process through accurate reporting of latitude and longitude data that will be used to find out where one of the victims of the disaster is in real-time.

In addition to this, to improve the detection and recovery of disappeared persons during a natural disaster, a facial recognition system could be added that indicates who and how many people the Rescue Bot finds in a disaster area.

On the other hand, Rescue Bot can be operated without depending on the Internet as it is possible to connect it to any local network. It can move on any terrain and is very useful for monitoring purposes. The light and compact size of the robot are two of the advantages of this prototype: more opportunities to access small or unsafe places during rescue operations.

From the point of view of the achieved competency in social responsibility, the students successfully developed and improved the skills obtained in class and in the lab to solve a real problem for the community. We recommend including this strategy in other assignments in the program to build up thus better the interest and importance of social responsibility in scientific projects.

6 Conclusions and Future Research

We have designed and constructed a rescue robot with several functions, including reducing human risk in dangerous environments after a natural or anthropogenic disaster. We have presented a low-cost prototype that could be a potential complementary tool for rescuers in this work. It can explore areas that are not easily accessible to humans and measure environmental parameters in that area, such as temperature and humidity. In addition, a camera module allows rescuers to obtain real-time images of the remote location. Control of all these units is easy with the use of the Arduino MEGA board.

An effort that implies the construction and development of projects of this kind builds and strengthens soft skills and motivates to think about problems that affect others outside of the university sphere. The result is a positive achievement of a final objective that goes beyond empathy: professional social responsibility.

Acknowledgements We thank the Universidad del Pacífico for their support and for allowing us to use their resources and laboratory spaces during this project's timeline.

References

1. INEI. Anuario de Estadísticas Ambientales. Perú, https://www.inei.gob.pe/media/MenuRecursivo/publicaciones_digitales/Est/Lib1342/cap06.pdf, 2015.
2. University Law 30220, CHAPTER XIII Responsabilidad Social Universitaria pp. 56, http://www.minedu.gob.pe/reforma-universitaria/pdf/ley_universitaria.pdf

3. Kohlbrecher, S., Meyer, J., Graber, T., Petersen, K., Klingauf, U., and von Stryk, O.: Hector open source modules for autonomous mapping and navigation with rescue robots. In Robot Soccer World Cup, 624–631 (2013).
4. Bernard, M., Kondak, K., Maza, I., & Ollero, A.: Autonomous transportation and deployment with aerial robots for search and rescue missions. Journal of Field Robotics, 28(6), pp. 914–931. (2011).
5. Dixit, D. S. K., and Dhayagonde, M. S.: Design and implementation of e-surveillance robot for video monitoring and living body detection. International Journal of Scientific and Research Publications, 4(4), pp. 2250–3153. (2014).
6. Papoutsidakis, M., Kalovrektis, K., Drosos, C., & Stamoulis, G.: Design of an Autonomous Robotic Vehicle for Area Mapping and Remote Monitoring. Int. J. Comput. Appl, 167(12), pp. 36–41. (2017).
7. Câmara D.: Cavalry to the rescue: Drones fleet to help rescuers operations over disasters scenarios, 2014 IEEE Conference on Antenna Measurements & Applications (CAMA), pp. 1–4. (2014). https://doi.org/10.1109/CAMA.2014.7003421.
8. Sunehra, D., Bano, A., & Yandrathi, S.: Remote monitoring and control of a mobile robot system with obstacle avoidance capability. In IEEE 2015 International Conference on Advances in Computing, Communications and Informatics (ICACCI), pp. 1803–1809. (2015).
9. Eguchi, A.: Robotics as a learning tool for educational transformation. In Proceeding of 4th international workshop teaching robotics, teaching with robotics & 5th international conference robotics in education Padova, Italy (2014).
10. Araújo, A., Portugal, D., Couceiro, M. S., & Rocha, R. P.: Integrating Arduino-based educational mobile robots in ROS. Journal of Intelligent & Robotic Systems 77(2), 281–298 (2015).
11. Osborne, R. B., Thomas, A. J., & Forbes, J. R.: Teaching with robots: a service-learning approach to mentor training. In Proceedings of the 41st ACM technical symposium on Computer science education, pp. 172–176. (2010).
12. Calderon, J. M., Rojas, E. R., Rodriguez, S., Baez, H. R., & Lopez, J. A.: A Robot soccer team as a strategy to develop educational iniciatives. In Latin American and Caribbean Conference for Engineering and Technology, Panama City, Panama (2012).
13. Izu, C., & Luna, A.: Using IF-AT cards to Engage Students in Reading and Understanding CS Course Materials. In 2020 IEEE World Conference on Engineering Education (EDUNINE), pp. 1–6. (2020).
14. Metoyer. S. K., Miller S. T., Mount J., and Westmoreland S. L: Examples from the trenches: student learning in the sciences using team-based learning. Journal of College Science Teaching, vol. 43, no. 5, pp. 40–47, 2014.
15. Hodges L. C., Anderson E. C., Carpenter T. S., Cui L., Gierasch T. M., Leupen S., Nanes K. M., and Wagner C. R.: Using reading quizzes in stem classes the what, why, and how. Journal of College Science Teaching, vol. 45, no. 1, pp. 49–55. (2015).
16. Heiner C. E., Banet A. I., and Wieman C.: Preparing students for class: How to get 80% of students reading the textbook before class. American Journal of Physics, vol. 82, no. 10, pp. 989–996. (2014).
17. Slepkov A. D.: Integrated testlets and the immediate feedback assessment technique. American Journal of Physics, vol. 81, no. 10, pp. 782–791. (2013). [Online]. Available: https://doi.org/10.1119/1.4820241
18. Robots de rescate para grandes catástrofes 2013, ABC-Ciencia, https://www.abc.es/ciencia/20130618/abci-robot-espanol-rescate-catastrofe-201306171617.html

Part IV
Freight Logistics and Distribution

Radio Frequency Identification and Rapid Response Code as Portable and Traceable Logistics Management Devices

Douglas Markonne, Annibal Scavarda, Gláucya Lima Daú, Purna Prasad Chapagai, and Mohammad Aljarrah

Abstract The costs in the health area with patients cared for and admitted to health institutions is a matter of debate that has been carried out by academic leaders and researchers around the world; those concerned with an economic generation of sustainable industrial products and services. Radio frequency identification technology helps in tracking and controlling stocks and logistics systems, in order to reduce losses and reduce operating losses. The study aims to analyze the use of radiofrequency identification and rapid response code in medical portables, in order to improve the logistical management in healthcare. The research was developed through a literature review, using the Web of Science database. A total of 259 texts were retrieved based on the following search equation: "(Hospital and radio frequency identification device)". Of these, 239 abstracts were selected for reading, 20 were excluded for not meeting the objectives, leaving 219, of which 48 were read. The top ten articles were selected for analysis. Research period was between February and August 2021. In the results, the authors of this study present a table with the selected articles in order to organize the findings regarding the applicability and usefulness of radiofrequency identification. The discussion showed that there are ways and strategies for using technology in the supply chain in various health contexts. It is concluded that the adoption of radiofrequency identification in

D. Markonne (✉)
Nursing and Bio Science Graduated Program, Federal University of the State of Rio de Janeiro, Rio de Janeiro, RJ, Brazil

A. Scavarda
Production Engineering School, Federal University of the State of Rio de Janeiro, Rio de Janeiro, RJ, Brazil

G. L. Daú
Federal University of the State of Rio de Janeiro, Rio de Janeiro, RJ, Brazil

P. P. Chapagai
Department of Sustainable Development, Royal University of Bhutan, College of Natural Resources, Punakha, Bhutan

M. Aljarrah
Industrial Engineering, The Hashemite University-College of Engineering, Zarqa, Jordan

J. Vargas Florez et al. (eds.), *Production and Operations Management*, Springer Proceedings in Mathematics & Statistics 391,
https://doi.org/10.1007/978-3-031-06862-1_31

logistics provide promising health outcomes, although there are still some surmountable barriers that need further studies.

Keywords Logistic models · Hospital equipment and supplies · Radio frequency identification device

1 Introduction

In the health area, the interest of academia and industry in studying the applicability of radio frequency identification technology (RFID) to monitor and track portable medical devices, fixed assets and people has been observed; in order to find beneficial logistical management solutions.

Traditionally, activities related to logistics management aim to provide healthcare professionals with supplies or equipment available to provide technical support, as well as people and the organization. RFID technology comes to promote the automation of systems and logistics in order to raise sustainable management models in health.

According to Caredda et al. [1], radio frequency identification (RFID) is a wireless technology that is being widely used in the implementation of labels on products in order to improve the coordination of the entire supply chain, bringing advantages to actors and supply networks.

Healthcare industry and services supply chains, which use RFID to track their fixed and mobile assets, require adequate flows of products and devices. These requirements must be contemplated in the plans and actions of public and private health systems, to which they intend to offer quality resources and capabilities. However, it is known that these attribute requirements are not fully met due to technological and managerial situations; complexity and technical competitiveness, organizational resistance and obstacles in data transmission through environmental barriers are mentioned.

Given these problem situations, the study in question, a literature review, seeks to investigate how RFID technology can help in inventory control and logistics in order to reduce losses and costs in health, in order to find knowledge gaps that can fill any eventual questions.

The study by Araujo et al. [2] indicates that logistics represents up to 45% of a hospital's operating budget. Inductively, it is believed that the significant percentage value may express similar patterns throughout the entire supply chain of the healthcare network, composed of several care units, suppliers and organizations located in different places. For this reason, the study becomes relevant insofar as it intends to discuss the different applicability and contributions of RFID in healthcare logistics.

The motivation for the applicability and usefulness of the RIFID technology in the health area is justified by the fact that in a hospital unit, for example, in an emergency, the control of the stock of supplies, medicines, equipment and portable medical devices is complex due to the very dynamics of the sector. Thus, it is believed that technology could be useful and provide administrative, technical and financial solutions for health institutions.

2 Radio Frequency Identification Technology – RFID: Concepts and Fundamentals

A study with a historical approach carried out by Ting et al. [3] states that RFID technology emerged right after World War II. Wireless labeling technology has made it possible to identify products and enable their consumption by public, private and industrial organizations for commercial and scientific purposes. Depending on how the tags are powered (internally or externally by energy) they are typified as active or passive for use in marking and communicating materials or people.

The information system is integrated with RFID technology and organized in multilayers, composed and structured by computer hardware and software. The first layer is the identification of objects by RFID tags (tags), which automatically capture marking data. This information system allows healthcare labels to be frequently used in cold chain logistics management and in places where people need real-time information. In addition, it is considered a financially sustainable technological acquisition for being low-cost and presenting excellent performance [4].

The authors also affirm the existence of types of RFID readers (fixed, portable and mounted) that follow their technical specifications based on requirements, environmental barriers and specific purposes. The registered data, by RFID readers, are transferred to the management layer communication network; where the second coat of the data processing and distribution system, analysis and information conversion is constituted to improve the decision-making process.

In the hospital environment, the RFID system architecture is fully scalable to meet different deficiencies of medical and institutional services. Mention is made of ultra-high frequency (UHF) RFID, a technological advance of RFID; a selected technology for Near-Field Communication (NFC) near-field communication. In particular, the use of UHF RFID technology in healthcare translates to some advantages: higher frequency power, lower cost of hardware infrastructure, easier reading and can be monitored and tracked by portable medical devices (tablets and smartphones), which are considered mobile assets [5].

In addition, there are contributions of the RFID system by causing changes in the work routine of professionals in the adoption of new work protocols where there is the application of technology in order to minimize errors and help doctors and nurses to monitor care and management. In the following section the methodology of this study will be explained.

3 Methodology

This is an exploratory study through a literature review. In the thoughts of Whittemore et al. [6], an integrative literature review is a broad method that allows the inclusion of theoretical literature for analysis and, as a scientific parameter, allows the realization of abstracts of published studies, based on general deductions regarding a particular area.

The bibliographic survey took place in the Web of Science database, the searches were based on a mathematical algorithm expression where a specific code was used, the TS descriptor (the symbol represents that the search was to find keywords in the following places on the platform: topic, title, abstract and keywords). Thus, the search equation for this study was "TS = (Hospital and radiofrequency identification device)" where 259 texts were found. The delimited inclusion criteria for pre-selection of studies were: complete and electronically available articles, peer-reviewed, in English, published from 2016 to 2021 and contemplating the proposed object with updated articles. Editorials, letters to the editor, works published in event proceedings, incomplete texts, articles written in other languages and duplicates were excluded.

Content analysis is based on Bardin [7] method, which initially proclaims a pre-reading analysis of selected texts, which are characterized in a floating and intuitive way of exploring the material. Thus, 239 abstracts were included for floating reading, of which 10 were excluded because they were conference proceedings and ten from the year 2015. Making 219, in which 48 articles were selected for full reading, where the ten most relevant articles remained, from according to criticality assessments by the authors, who met the scope of the work and composed the analysis synthesis. The study period took place from February to August 2021.

The selection process of the findings was performed by reading the abstracts in full with the help of the Rayyan software (text manager) and the plenitude of the articles by the Medley software (citation and bibliography manager) so that they were for the final selection ten articles that met the criteria guided by the PRISMA flowchart.

4 Results (Table 1)

Table 1 Synthesis of articles on the applicability of RFID technology in the logistics of portable medical devices

No.	Title	Authors	Application	Utility
01	RFID technology for blood tracking: an experimental approach to comparing different devices	V. Caredda, PF Orrú, G. Romagnoli, A. Volpi, F. Zedda	Track the blood supply chain	Enables performance assessment of devices, improves patient safety, assesses technical and logistical feasibility, promotes blood supply chain reengineering processes
02	RFID medical equipment tracking system based on a technical service location [10]	Meng-Hsiun Tsai, Chiu-Shu Pan, Chi-Wei Wang, Jui-Ming Chen, Cheng-Bang kuo.	Track medical equipment	Inventory control provides medical staff with location and inventory information for medical devices and supplies
03	Dual mode RFID antenna design for inventory management and IV: fluid level warning system	Ssu-Han Ting, Chih-Kuang Wu, Ching-Hsing Luo	Inventory management and intravenous fluid drip	Provides a method for pharmaceutical and inventory management, reduces nursing workload
04	Evaluation of surgical instruments with radiofrequency tag identification in the operating room [11]	Kazuhiko Yamashita, Kaori Kusuda, Yoshitomo Ito, Masaru Komino, Kiyohito Tanaka, Satoru Kurokawa, Michitaka Ameya, Daiji Eba, Ken Masamune, Yoshihiro Muragaki, Yuji Ohta, Chugo Rinoie, Kenji Yamada, Yoshiki Sawa	Surgical instrument tracking system	Warns surgeons about the retention of instruments in the body cavity and monitors their use, reduces the risk of breakage, instrument counting method
05	Improving surgical instrument logistics processes: the case of RFID technology	Afrooz Moatari-Kazerouni, Ygal Bendavid	Surgical instrument tracking system	Improves tracking and location of surgical instruments. It reduces expenses with rentals, purchases, time spent tracking and the amount of missing instruments
06	Where is my infusion pump? Dynamic utilization network to improve hospital equipment fleet management [9]	Diego A Martinez, Jiarui Cai, Jimi B Oke, Andrew S Jarrell, Felipe Feijoo, Jeffrey Appelbaum, Eili Klein, Sean Barnes, Scott R Levin	Manages the equipment fleet between hospital units	Improves inventory management of drug/diet and active infusion pumps. Safe balancing system for hospital equipment

(continued)

Table 1 (continued)

No.	Title	Authors	Application	Utility
07	Implementation and evaluation of an RFID smart cabinet to improve traceability and efficient consumption of high cost drugs and supplies in a large hospital	Maria Del Carmen León, Elisa Gómez-Inhiesto, Maria Teresa Acaiturri-Ayesta	Tracking system for high value cardiac surgical products	Improves information on inventory management, saves time in the supply chain, capable of monitoring the tracking and consumption of surgical products
08	The value of radiofrequency identification in the quality of blood transfusion chain management in an academic hospital setting	Linda W Dusseljee-Peute, Remko Van der Togt, Bas Jansen, Monique W Jaspers	Manages the blood transfusion logistics chain	Logistic and temperature data generation. Evaluates real-time location, time, temperature, transfusion quality data. Promotes process redesign and evaluation of storage, transport and distribution
09	RFID Application in the Management of Medical Asset and Patient Operations: A Technology, Organizational and Environmental Perspective (TOE) on Key Enablers and Impediments [8]	Mohamed Aboelmaged, Ghraib Hashem	Medical Asset Management	It generates technological, organizational and environmental impacts. Promoting technical, organizational, environmental competitive, technical complexity and environmental uncertainty advantages. They provide robust implications for healthcare managers and RFID providers
10	RFID Technology for Management and Tracking: eHealth Applications	Yuri Álvarez López, Jacqueline Franssen, Guillermo Álvarez Narciandi, Janet Pagnozzi, Ignacio González-Pinto Arrillaga, Fernando Las-Heras Andrés	System for tracking and managing assets in hospitals	Applied to electronic health devices; improves screening of patients, drugs and medical assets. Improves asset efficiency and security. Provides additional logistics details, increases the level of process automation

5 Discussion

The analysis of the selected articles allowed revealing and being presented in the discussion some paths that the radiofrequency identification technology is able to contemplate in the area of health logistics. Despite the many apparent benefits, the adoption of RFID in the supply chain encounters significant and surmountable barriers; which will also be briefly discussed in this part of the text.

There are countless applications and utility of solutions that RFID offers for the field of production and health services. The study by Dusseljee-Peute et al. [8] addresses the management, tracking and inventory of the supply chain of intravenous fluids and blood transfusion logistics in the hospital environment. It was contacted that RFID enables the evaluation of portable medical devices, improves patient safety care activities, allows the evaluation of the logistics and technique used; enables processes to redesign supply systems, storage, transport and distribution of products and inputs, provides strategies for the management and inventory of medications; evaluates in real time the creation of logistic data of temperature, location and drip quality of the blood-concentrate bags.

In addition, it reduces the workload of managerial and care nursing, regarding the monitoring of administrative information and patients' blood transfusions. With this, it is believed that care can be prioritized in exclusive functions of human dedication; that does not have the preeminent need for support of RFID technology to carry out tasks.

Another use found for RFID in healthcare logistics, especially hospital, was mentioned by Martinez et al. [9] on the management and tracking of medical equipment and drug infusion pumps and enteral and parenteral diet nutrition. It was shown that RFID technology improves the management of fixed and mobile infusion devices, dietary materials and stock monitoring, usage information and secure balancing of medical equipment.

RFID tags are used in surgical centers and centers for sterilization materials, as shown in the statements by Araújo et al. [2] when they state that the technology can be applied in tracking systems for surgical instruments and high-value products of cardiac surgeries. In addition, they monitor the use of instruments and swabs, warning physicians through visual or audible alerts of possible forgetfulness within the body of patients undergoing surgery; facilitates the counting of instruments in monitoring their locations, minimizes the percentage of lost and broken instruments, reduces expenses with unnecessary purchases, promotes savings in time and consumption of products in the supply chain between hospital sectors.

Research developed by López et al. [5] on healthcare logistics, RFID and its impacts on the management and tracking of medical assets, proved that the technology has effects on work processes and promotes environmental competitiveness, improves patient and device follow-up care digitals, provides detailed and additional information to healthcare managers and technology providers, reduces technical risks and organizational uncertainties, increases the level of automation of asset safety processes and protocols.

6 Conclusion

In this literature review carried out after data processing, 10 articles listed on how radiofrequency technology – RFID and rapid response code are applied and used in the logistic management of portable medical devices in health institutions were analyzed.

Although studies still present limitations of results regarding the financial return related to the adoption of the technology in healthcare, some authors mention that RFID contributes to significantly reduce the costs of the supply network and has been shown to have great impacts on the hospital supply chain.

The technology has the power to locate portable medical and nursing objects and mobile medical equipment; ensuring quick access, effective control and improved maintenance. These advantages promote the reduction of time spent searching for technical instruments, freeing staff to treat and care for patients.

In managing the inventory of materials, supplies, medicines, diets and intravenous fluids, RFID technology provides options for implementing new methods of organizing the hospital warehouse and pharmacy. It introduces alarm systems to work protocols that warn of low levels of fluids administered to patients and the decrease in the inventory of stored products.

There are research initiatives that show promising results about the performance of RFID technology and how it contributes to process reengineering; with the intention of reducing clinical risks, improving the quality of services, advancing patient safety standards, progressing in the tracking and control of high-cost healthcare tools.

Other studies demonstrate that tests carried out with RFID technology, in the field of health communication and logistics, have positive effects for people in reducing the risk of contamination by contact-borne diseases, reducing errors in hospitals due to human prescribing mistakes, requisition and administration of medications.

The implications found in this study through the interpretation of the results reveal that RFID has pragmatic effects on the economics of health systems and institutions. However, it would be useful in future research to investigate the use of technology in people management and patient care management; in order to discover other elements of different contexts of RFID application health.

References

1. Caredda, V., Orrù, P. F., Romagnoli, G., Volpi, A., & Zedda, F. (2016). RFID technology for blood tracking: An experimental approach for benchmarking different devices. *International Journal of RF Technologies*, *7*(4), 209–228.
2. del Carmen León-Araujo, M., Gómez-Inhiesto, E., & Acaiturri-Ayesta, M. T. (2019). Implementation and Evaluation of a RFID Smart Cabinet to Improve Traceability and the Efficient Consumption of High Cost Medical Supplies in a Large Hospital. *Journal of medical systems*, *43*(6), 178.

3. Ting, S. H., Wu, C. K., & Luo, C. H. (2017). Design of dual mode RFID antenna for inventory management and IV fluid level warning system. *International Journal of Antennas and Propagation, 2017.*
4. Moatari-Kazerouni, A., & Bendavid, Y. (2017). Improving logistics processes of surgical instruments: case of RFID technology. *Business Process Management Journal.*
5. Álvarez López, Y., Franssen, J., Álvarez Narciandi, G., Pagnozzi, J., González-Pinto Arrillaga, I., & Las-Heras Andrés, F. (2018). RFID technology for management and tracking: E-health applications. *Sensors, 18*(8), 2663.
6. Whittemore, R., & Knafl, K. (2005). The integrative review: updated methodology. *Journal of advanced nursing, 52*(5), 546–553.
7. BARDIN, L. Content Analysis. Publisher: Editions 70. 2011
8. Dusseljee-Peute, L. W., Van der Togt, R., Jansen, B., & Jaspers, M. W. (2019). The Value of Radio Frequency Identification in Quality Management of the Blood Transfusion Chain in an Academic Hospital Setting. *JMIR medical informatics, 7*(3), e9510.
9. Martinez, D. A., Cai, J., Oke, J. B., Jarrell, A. S., Feijoo, F., Appelbaum, J., ... & AHRQ Patient Safety Learning Laboratory Program and the CDC MinD Healthcare Program. (2020). Where is my infusion pump? Harnessing network dynamics for improved hospital equipment fleet management. *Journal of the American Medical Informatics Association, 27*(6), 884–892.
10. Tsai, M. H., Pan, C. S., Wang, C. W., Chen, J. M., & Kuo, C. B. (2019). RFID medical equipment tracking system based on a location-based service technique. *Journal of Medical and Biological Engineering, 39*(1), 163–169.
11. Yamashita, K., Kusuda, K., Ito, Y., Komino, M., Tanaka, K., Kurokawa, S., ... & Sawa, Y. (2018). Evaluation of surgical instruments with radiofrequency identification tags in the operating room. *Surgical innovation, 25*(4), 374–379.

Urban Road Network Resilience Assessment on Freight Logistics by Simulating Disruptive Events

Leonardo Flores-González, Jorge Vargas Florez, Lorena Monteza-Valdivia, Alexia Cáceres-Cansaya, Javier García-Salinas, and Luciano Silva-Alarco

Abstract The assessment of resilience in port road networks under disruptive events is a key issue related to urban logistics. This paper addresses an original simulation method to evaluate resilience using macro and micro simulation based on stochastic theory. The results provide insight into the resilience index of the network.

This paper specifies the most influential network links around the area of influence produced by a logistics transport avenue in Lima-Peru. A function that includes redundancy and robustness of the system as a performance measure is proposed to measure resilience.

Keywords Freight logistic · Urban shape resilience · Macro and micro simulation · Stochastic theory · Metrics · Dry ports

1 Introduction

1.1 Sea Port as Hub Logistic

Dry ports represent logistics infrastructure specific to maritime trade that is of great importance for the efficiency of logistics operations, especially in areas that do not have an appropriate road development or a logistics activity zone close to the port. For this reason, dry ports or temporary warehouses are considered as the lungs of the ports, as they allow to decongest the accumulation of traffic from and to the port.

L. Flores-González (✉)
National University of Engineering, Lima, Peru
e-mail: flores.leonardo@pucp.edu.pe

J. Vargas Florez · L. Monteza-Valdivia · A. Cáceres-Cansaya · J. García-Salinas · L. Silva-Alarco
Pontifical Catholic University of Peru, Lima, Peru
e-mail: jorge.vargas@pucp.edu.pe; lorena.montezav@pucp.pe; alexia.caceres@pucp.pe; jegarcia@pucp.pe; lsilva@pucp.pe

J. Vargas Florez et al. (eds.), *Production and Operations Management*, Springer Proceedings in Mathematics & Statistics 391,
https://doi.org/10.1007/978-3-031-06862-1_32

Furthermore, this allows cargo to be stored for longer periods of time than in a port, since in the latter the constant rotation of cargo is more important.

There is little research on dry ports from a supply chain perspective [1]; however, dry port integration improves costs, responsiveness, safety, security, safety, sustainability, resilience, and innovation outcomes [2]. Costs are reduced for the following reasons: synergies generated in the supply chain, reduced transportation, and reduction of port fees risks caused by delayed container pick-up.

2 Literature Review

2.1 Resilience

From an ecological point of view, resilience is the property of adapting and improving its functions after the occurrence of a disruption that changes the initial conditions of the environment [3]. It can also be defined as the measure of resistance of a system to perturbations and the speed with which the system recovers its functions before the perturbation [4].

Other concepts include the ability of a system to preserve its functions when subjected to disruptive events [5]. It is also defined as the ability of a system to absorb changes and reorganize itself while undergoing such changes and to maintain the same functions, structure, and identity after the occurrence of a disruption [6].

From the above, two types of concepts can be distinguished; one, the ecological concept that seeks adaptation and improvement; and the other, the engineering concept that seeks to maintain properties. Both concepts are contingent on the occurrence of a disruption and the capability to predict changes in systems with complex topologies. Transport road network topology refers to the connectivity among its nodes. Therefore, logistics and transport systems, due to their complexity, need to be studied from a resilient engineering point of view.

In the article [7] resilience means the capability to prepare for and adapt to changing conditions and to resist and recover quickly from disruptions and associate it with quantification metrics. Resilience includes the capability to resist and recover from disruptions of the types of deliberate attacks, accidents, naturally occurring threats or incidents. The resilience of a system function can be measured in terms of the persistence of a corresponding functional performance under uncertainty in the face of disruptions.

A suitable concept for measuring road network resilience between dry ports and specific port is given by [5], defines resilience as the capability of a system (with the help of immediate recovery activities) to meet transport demand within a limited period (recovery time) when faced with network disruptions.

Within the context of resilience and dry ports it is important to answer the following question: how can urban road resilience be calculated in the context of an inland port container transport network [IPCTN]? This question was addressed by [5]. This study contributes to fill the knowledge gap concerning resilience analysis in IPCTN

Table 1 Paper with key word

Paper	Freight logistic	Urban shape resilience	Macro & micro simulation	Stochastic theory	Metrics	Dry ports
Gao Y, Wang J (2020) [8]	✔	✔				
Khaslavskaya A, Roso V (2019) [1]		✔				✔
Sun W, Bocchini P, Davison B (2018) [6]		✔		✔	✔	
Jung H, Do M (2018) [9]		✔	✔			
Calvert S, Snelder M (2017) [10]		✔		✔		
Chen H, Cullinane K, Liu N (2017) [5]	✔	✔		✔	✔	✔
Ganin A, et al. (2017) [4]		✔			✔	
Crainic T, Dell'Olmo P, Ricciardi N, Sgalambro A (2015) [11]	✔					✔

and the quantitative measurement of a resilient transport network with dry ports. However, disruptive situations that affect urban road logistics and that occur daily in a road network similar to the port of Callao are not considered, for this reason it is proposed to measure resilience for mixed networks.

The calculation of resilience considering all the agents of the urban road network, is an integrative approach for calculating it and contributes to the management of network performance.

Table 1 summarizes 8 papers of urban road network [1, 4–6, 8–11] published on or after 2015. Each column indicates if the paper addresses a keyword from the present paper. From Table 1 it can be concluded that there are not recent research on resilience in mixed networks with macro and micro simulation.

2.2 *Network Cost*

The cost of the network is obtained according to [12], which is a robustness measure. The travel time of the network link is the independent variable and the cost is dependent variable, it is calculated according to the following formula in the influence area A:

$$C = \sum_{a \in A} x_a t_a \tag{1}$$

Where x_a is the flow in link a, t_a is the travel time in link a, C is the base cost and C_i is the cost of the network without link i. The formula (1) is the sum of the link road network travel times in the influence area.

2.3 Resilience Metrics

It should be noted that [7] considers that associating the appropriate metric to resilience leads to massive economic savings since system recovery is included.

There are several proposals to measure the resilience of a system, notably [13–16], in which the attributes that should be considered in each system are weighted. It is observed that the attributes of robustness, redundancy and speed of recovery are common to various types of systems. This indicates that resilience metrics should measure these attributes or be designed for a model that considers them.

A traditional way of calculating the resilience index [13] is the use of the resilience triangle concept, however, there are other criteria such as [14] which considers the following attributes to quantify resilience: robustness, redundancy, resource capacity and recovery speed [15]. It measures the resilience of a network as the percentage of damaged links versus network throughput and the percentage of damaged nodes compared to network throughput. Tierney and Bruneau [16] Based on the observation that resilient systems reduce the probability of failure and improve recovery, they conclude that resilience can be measured by the functionality of a system after a disruptive event, during the time it takes to return to the initial level of functionality.

The recommended metric to measure resilience in logistics roads networks would be [7], which considers three fundamental elements: demand, network topology and network resilience. Demand can be defined as "number of trips" for most networks.

Demand is a measure of how much of a commodity is transported from the origin to destination through the network. The performance requirements to satisfy demand are expressed in terms of the following three concepts:

I. Performance: number of trips completed per unit of time within the area of influence.
II. Resilience: measure of system performance as a function of cost.
III. Travel time: Time taken to travel between an origin and a destination.

The output variable to be considered is the travel time in the system under study; the independent variables are travel demand, network shape, vehicular flow, and finally the resiliency index which is a function of all the variables mentioned above. The scheme proposed for the study follows the guidelines of [11] and those proposed by this article, see Fig. 1.

The dependent variable to be considered in a numerical simulation model, which helps to calculate the resilience index, is the travel time in the links of the system under study. The independent variables are travel demand, network shape, and

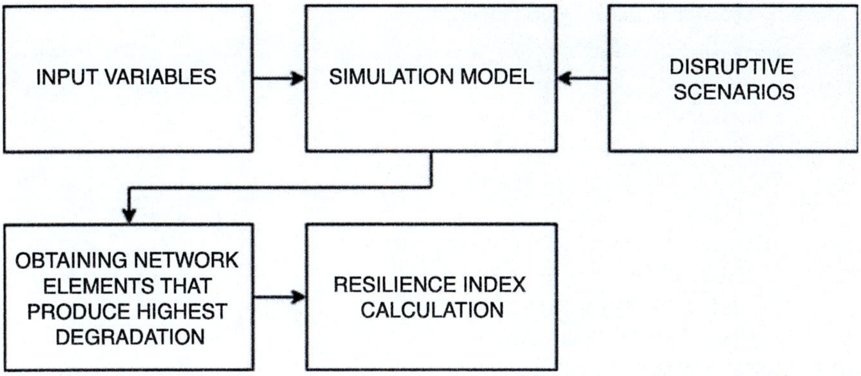

Fig. 1 Network resilience measurement proposal

disruptive scenarios. Thus, the scheme proposed for the present paper follows the guidelines of [7, 15], see Fig. 1.

To measure the average efficiency ε of an urban road network, Nagurmey proposes the formula (2) [17].

$$\varepsilon = \frac{\sum_{i \in M} \frac{1}{t_i}}{n} \tag{2}$$

Where t_i is the travel time on the path i, n is the total number of trips in a given time interval. In this formula the travel time is inversely proportional to the efficiency ratio ε, a disadvantage of (2) is that it does not measure the efficiency of a network link, so in this paper the authors propose to determine the most important link with the help of the Perron-Frrobenius theorem and the adjacency matrix of the influence area.

3 Problem Statement

The transportation system within the supply chain is one of the most important economic activities. Transportation costs represent between one third and two thirds of the total logistics costs, so it is key to be able to analyze them in depth as it could mean significant savings for companies. This requires supply chain managers to have a thorough understanding of how a transportation system works and its cost structure.

Transportation plays a connective role between companies and the end consumer and its proper planning is a responsibility that directly affects inventories, production, and customer service. Thus, management must focus on minimizing costs and maximizing service levels simultaneously. Over the years, the development of

Table 2 Export volume mobilized in TEUs [18]

#	Terminal	%	TEUs
1	DP WORLD CALLAO SRL	16.73%	74,695
2	APM TERMINALS CALLAO SA	15.63%	69,783
3	UNDECLARED	13.23%	59,090
4	RANSA COMERCIAL SA TERMINAL	11.22%	50,121
5	NEPTUNIA SA	8.94%	39,941
6	TERMINALES PORTUARIOS PERUANOS SAC	6.81%	30,416
7	FARGOLINE SA	4.41%	19,707
8	INVERSIONES MARITIMAS UNIVERSALES PERU SA	3.88%	17,343
9	LOGISTICA INTEGRAL CALLAO SA	3.77%	16,826
10	IMPALA TERMINALS PERU SAC	3.67%	16,388
11	CONTRANS SAC	2.92%	13,022
12	APM TERMINALS INLAND SERVICES SA	2.81%	12,560
13	LOGISTIC INDUSTRY & MINING SA - LOGISMINSA	2.12%	9454
14	UNIMAR SA	1.68%	7504
15	SAKJ DEPOT SAC	1.44%	6415
16	VILLAS OQUENDO SA	0.66%	2963
17	TRABAJOS MARITIMOS SA	0.05%	234
18	EMPRESA NACIONAL DE PUERTOS SA	0.02%	91
19	ALMACENES MUNDO SA	0.00%	8
20	TERMINAL: 4459	0.00%	6
	Grand Total	100.00%	446,567

consolidation facilities and distribution centers has, in part, improved transportation services and overall logistics management performance.

To explain the proposal of this research, data related to the volume of cargo moved through temporary warehouses in Callao, located mainly on Argentina and Gambetta Avenues, are presented. Table 2 shows that more than 50% of the export volume in TEUs is mobilized through the extra-port warehouses or dry ports during 2019. The first three positions in Table 2 represent cargo not moved through dry ports (intra-port warehouses and "undeclared" advance shipments) and the remaining is a volume transported from a dry port to the container port of Callao. Due to the amount of merchandise that is mobilized, it is necessary to mitigate the negative effects of urban and cargo transportation on the road network leading to the port of Callao, which implies having a measure of the performance of the urban network in this sector of Lima.

Figure 2 shows a map with the location of the port of Callao, the Jorge Chávez international airport and Gambetta, Argentina and Faucett Avenues.

In Lima, there is no logistics road corridor and in the Callao port area, urban transport and logistics transport circulate on the same roads, so it is necessary to define a methodology for calculating the resilience index for mixed networks; therefore, this paper proposes to answer the following question: How to measure

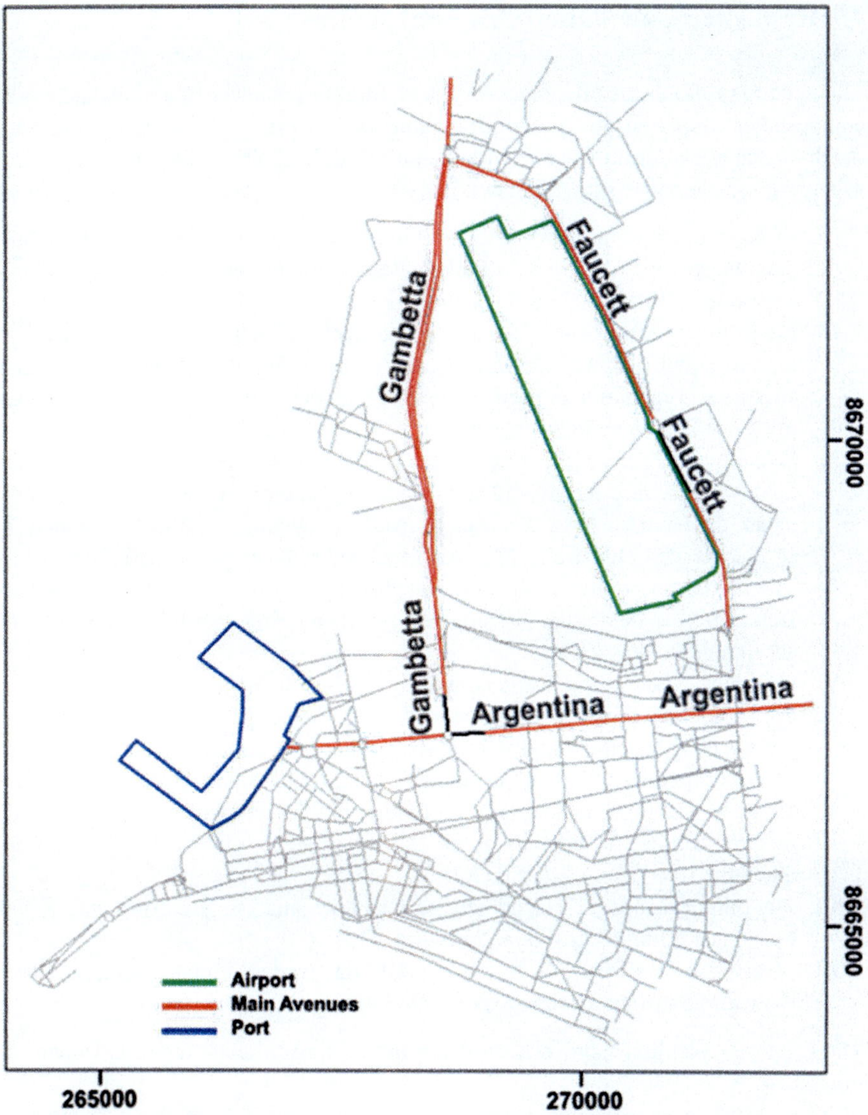

Fig. 2 Map of Callao port

the resilience of the urban and logistics transport road network located in the vicinity of the port of Callao under ordinary disruptive events?

4 Resilience Calculation Proposal

This section proposes the procedure for calculating the resilience of a mixed network composed of urban transport and logistic transport. The urban road network connects the dry ports of Gambetta Avenue with the port of Callao. The sequence of activities to measure resilience is shown below:

I. Build the entire network connectivity, select the default internal model parameters and compare the simulated result with the field measurement data.
II. Calibrate the macro model by vehicle counts.
III. Determine the influence area of the network segment under study with a macro model (in this case, Gambetta Avenue). For this purpose, the four-step transport model is used. Consider that the entrances and exits of the influence area are additional centroids.
IV. Selecting an efficiency metric for the model (MOE) and a micro simulation model to measure the overall performance of the proposed network; in this paper, the Krauss model is chosen to reach the dynamic equilibrium, which is combined with Dijkstra's algorithm to consider the effect of disruptions.
V. The model proposed in IV is calibrated to minimize the difference between simulated and field data. A calibrated model requires that this difference be acceptable.
VI. Introduce disruptions to generate disruptive scenarios by removing or decreasing the capacity of the network link. The importance of the network links is established by adjacency matrix, where the dominant eigenvector has all its components greater than zero and the level of importance of each link is assigned according to the numerical value of each component. This form of hierarchization of network links is supported by the Perron-Frobenius theorem.
VII. The output data to be considered for resilience analysis are flows and travel times; according to the literature review [19].
VIII. Measure the resilience of the system with the proposed metric, e.g., measure the variation in performance over time intervals.

The proposal calculation process for determining the resilience index is shown in Fig. 3.

The resilience of the whole system considering the interrelation of logistic transport and urban transport is calculated with the formula (3) proposed by the authors of this paper, for each scenario a, which is determined as a cost function. The difference with the formula proposal in [5] is that the interrelationship with mixed urban transport is considered.

Formula (3) implicitly considers redundancy, since network links are eliminated in the model to generate new scenarios and calculate their impact on system resilience.

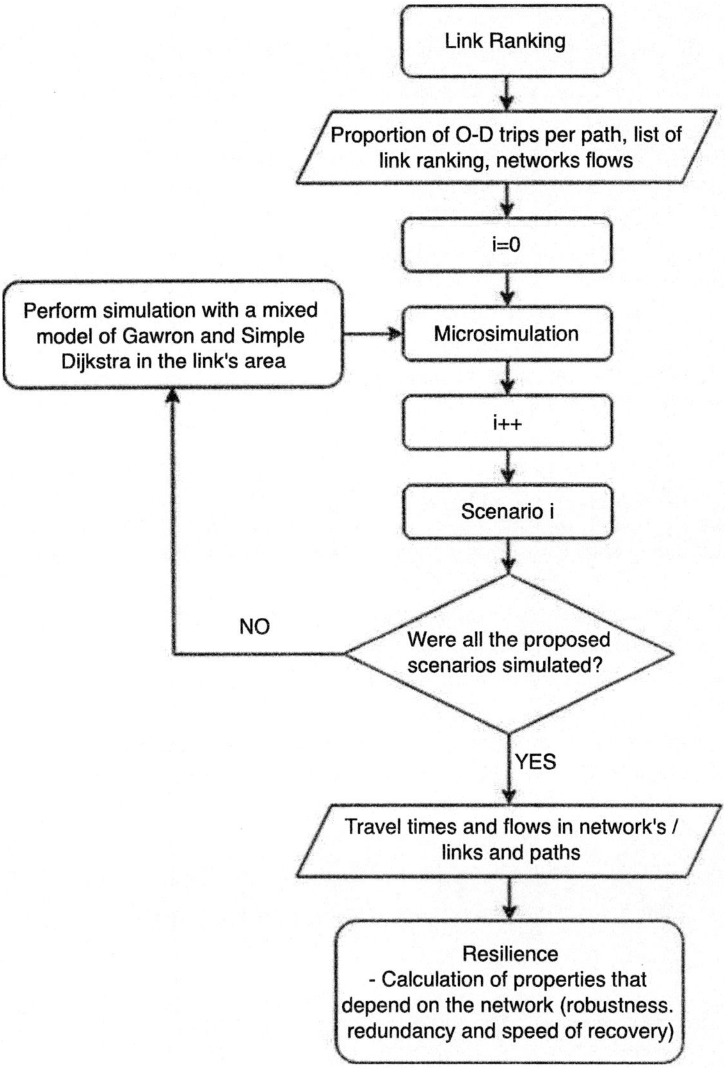

Fig. 3 Resilience calculation process

$$R(a) = 1 + \frac{C - C_a}{C} \tag{3}$$

4.1 Model Calibration

In the area of influence of Gambetta Avenue, a macro simulation analysis is performed to determine the routes used in the trips made within the area in question. The model is initially obtained from the Urban Transportation model of Lima and Callao, and then calibrated with traffic counts made within the area of influence. However, its use is exclusively to determine the area of influence of the avenue, the rest of the traffic model is performed with a micro simulation model based on following model and lane change theory as proposed in the SUMO v1.9.0 program.

The program uses internal parameters that serve to produce simulation results of the proposed traffic system and these are compared with the results of traffic count data. When the difference of simulated and field measured results according to a calibration variable is acceptable, a calibrated model is available, and the newly obtained parameters can be used in the simulation.

The use of a calibration process requires consideration of the recommendations given in [20], which are listed below:

I. Pre-model: selection of an efficiency metric (MOE) and a micro simulation model.
II. Initial model: consists of building the network, selecting the default parameters of the model and comparing the simulated results with the field data.
III. Calibration: consists of verifying that the difference between the simulated results and the field data is acceptable, it is considered acceptable if it is within the objectives of the paper, otherwise, selecting the internal parameters of the model that most influence the simulation results by means of a sensitivity analysis (regression or experimental factorial). Optimization algorithms (e.g., genetic algorithm) are used to minimize the difference between simulated results and field data.

Figure 4 shows schematically stage III of the calibration process:

5 Case: Road Network in Callao, Peru

The inland transport of containerized cargo in the port of Callao faces complexities that have not been resolved due to there are various factors that affect the problem, such as social conflicts, limited link capacity, and recurrent incidents, among other events, which in sum prevent the optimization of logistical transport to the detriment of competitiveness at the national level. The occurrence of these factors is evidenced by long queues on the immediate access roads to the ports, long waiting times for carriers, logistics cost overruns, dissatisfaction of importers and/or exporters, and in the worst cases leading to the devaluation of the cargo or even legal abandonment.

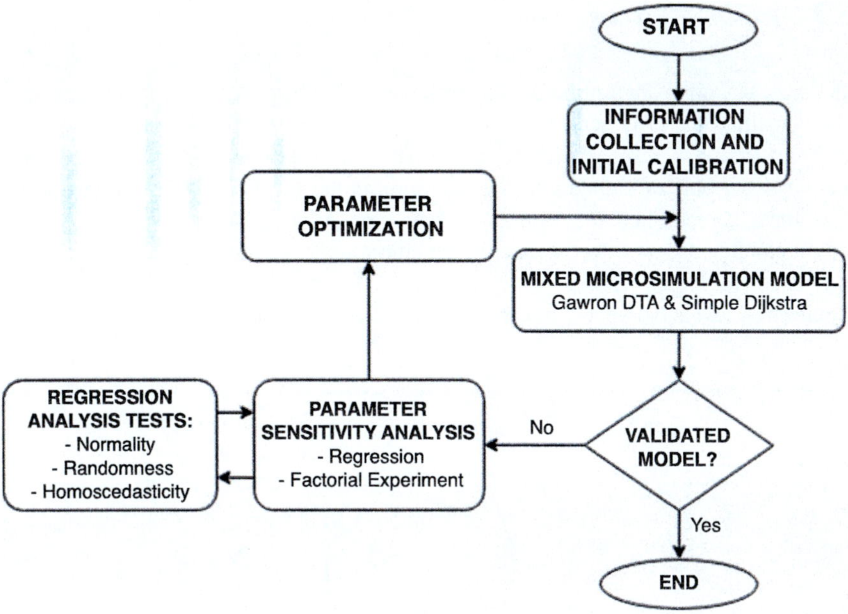

Fig. 4 Calibration process

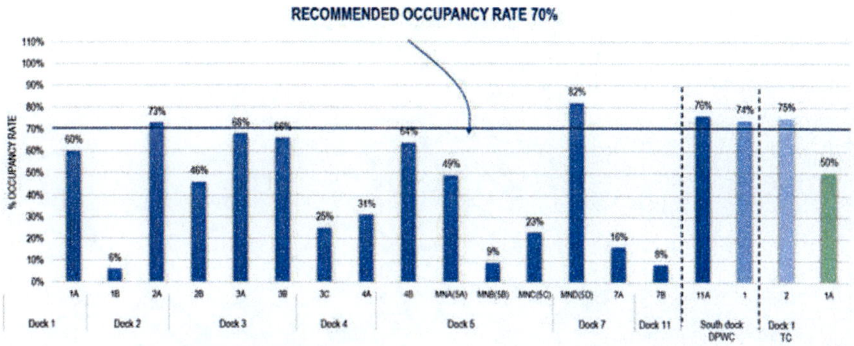

Fig. 5 Pier occupancy ratios – Callao [18]

The occupancy of the Callao North Pier 5D and Callao South Pier shown in Fig. 5 exceed the limits suggested by the port authorities demonstrating the limited capacity to satisfy market demand, further compromising the containerized cargo transport flow.

5.1 Influence Area

In Lima, dry ports are around the Argentina and Gambetta Avenues, both located in the constitutional province of Callao, this province has the highest concentration of dry ports in the country. The most visible effect of the problem of logistics transport between the port of Callao and the dry ports is the level of traffic congestion on Gambetta Avenue and the section of Argentina Avenue closest to the port. This effect can be observed from a macro simulation model of travel supply and demand based on the Urban Transport Master Plan of Lima and Callao [21]. The current congestion levels between 7 and 9 am can be visualized in Fig. 6.

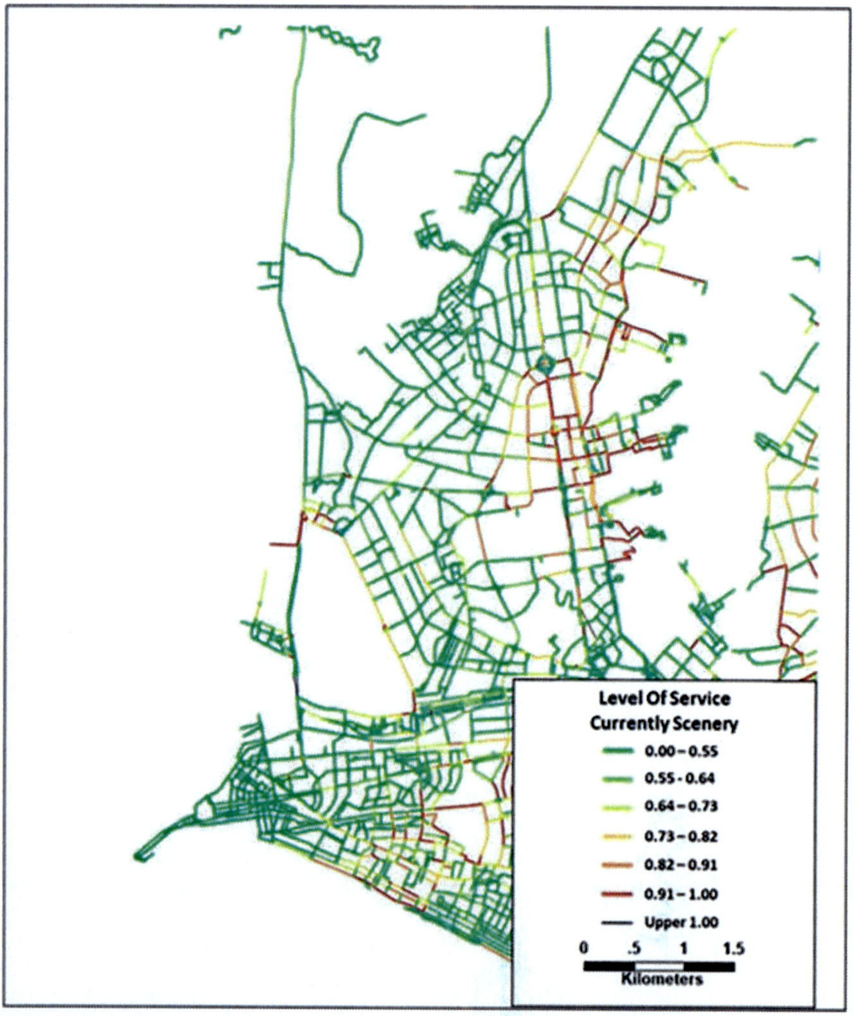

Fig. 6 Influence area, initial estate [21]

In order to evaluate the performance of the road network under study, disruptive conditions that cannot be predicted must be added. Therefore, it is important to consider the resilience of the system, which becomes a determining factor in the design or redesign of the network. To determine how Gambetta Avenue influences the urban transport system of Lima, it is removed from the system, which means that the macro simulation model no longer presents Gambetta as a segment of the road network.

Figure 7 shows that the Jorge Chávez International Airport and the urban area between Argentina, Faucett and Colonial Avenues are affected. The aforementioned

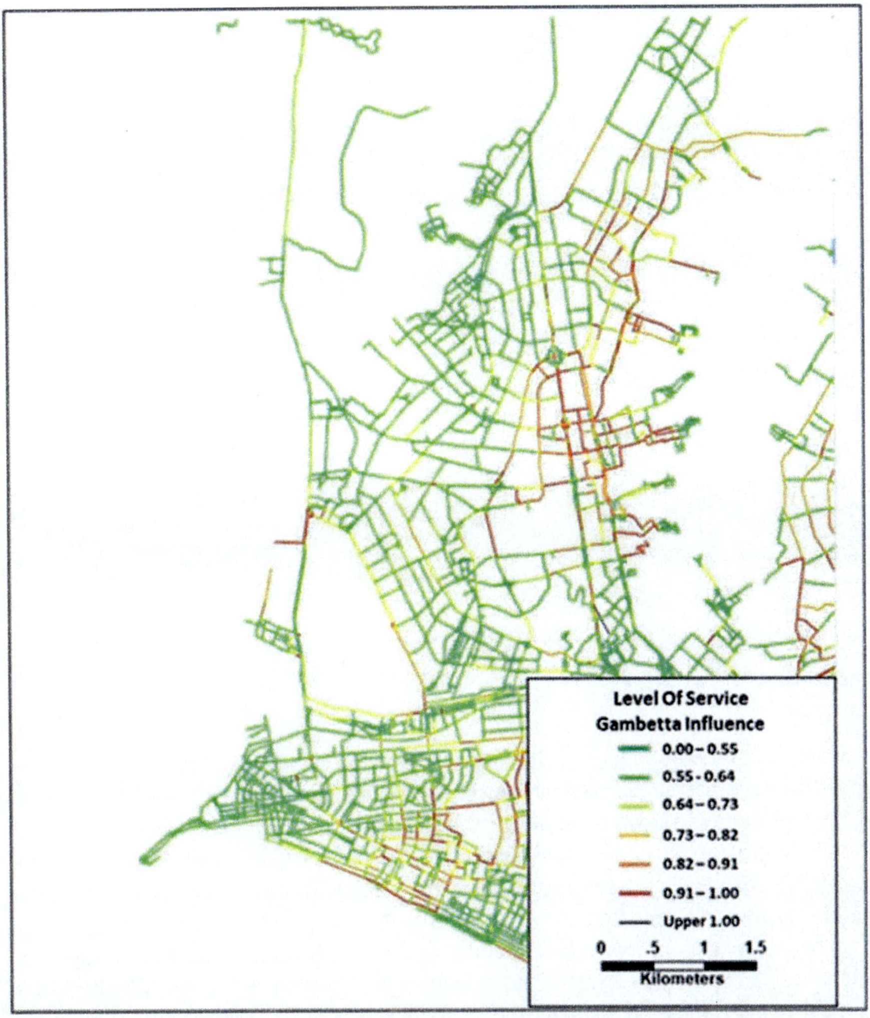

Fig. 7 Influence area without Gambetta Avenue [21]

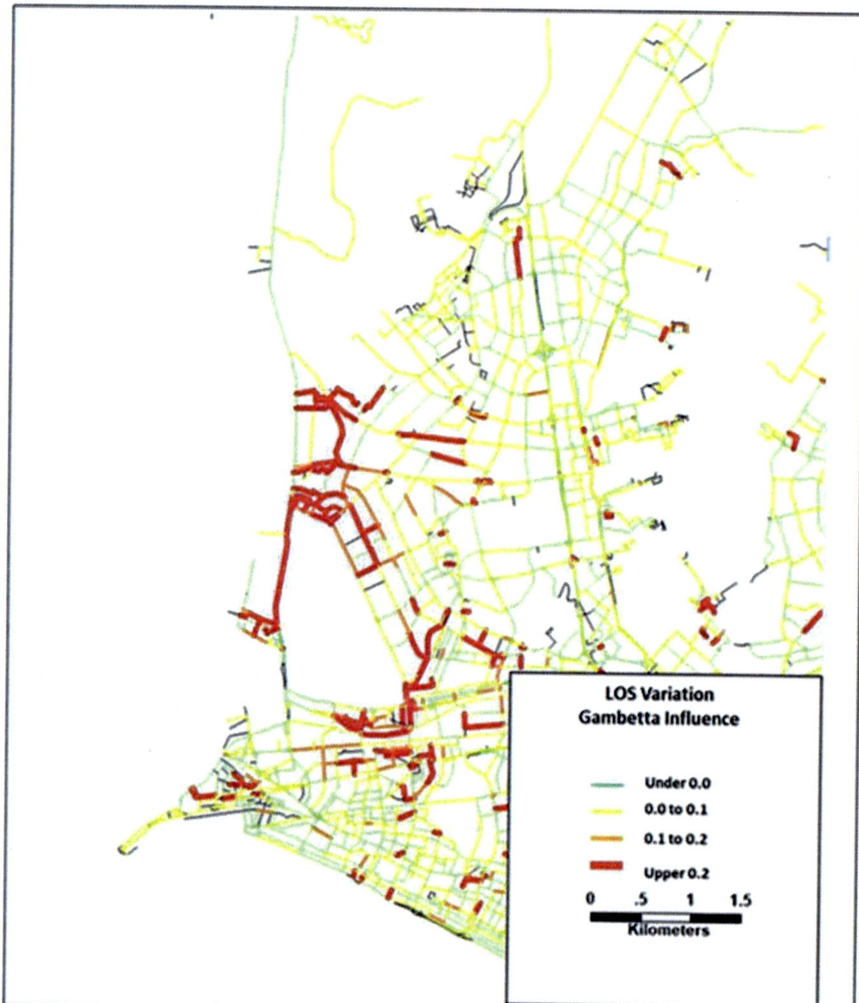

Fig. 8 Cost differences [21]

avenues form the boundary of the influence area of Gambetta. Figure 7 also shows that Gambetta has a localized influence mainly in Callao.

Figure 8 shows the cost difference among the scenarios of Figs. 6 and 7. It can be concluded that exist urban roads leading to the port of Callao with cost increases, however, other avenues are not affected, the links with cost increase will form the influence area of Gambetta.

The logistics mobility system included in the Gambetta area must have improvements in its capacity to adapt or recover from unforeseen situations, which clearly indicates that it must be resilient to disruptive events, so it is imperative to research how to measure resilience.

5.2 Influence Area Model

The network model considers an influence area where there is a high density of variation in service levels due to the absence of Gambetta Avenue. The influence area is connected and of adequate size to facilitate the calculation of flows and travel times. The trips are made between the centroids of the traffic analysis zones (TAZ) as shown in Fig. 9.

Within the micro simulation model, a parallel logistic transport network is considered, for this purpose, origin-destination matrices of the aforementioned networks are considered. The matrices were obtained from the macro model of the urban transport master plan of Lima and Callao [21].

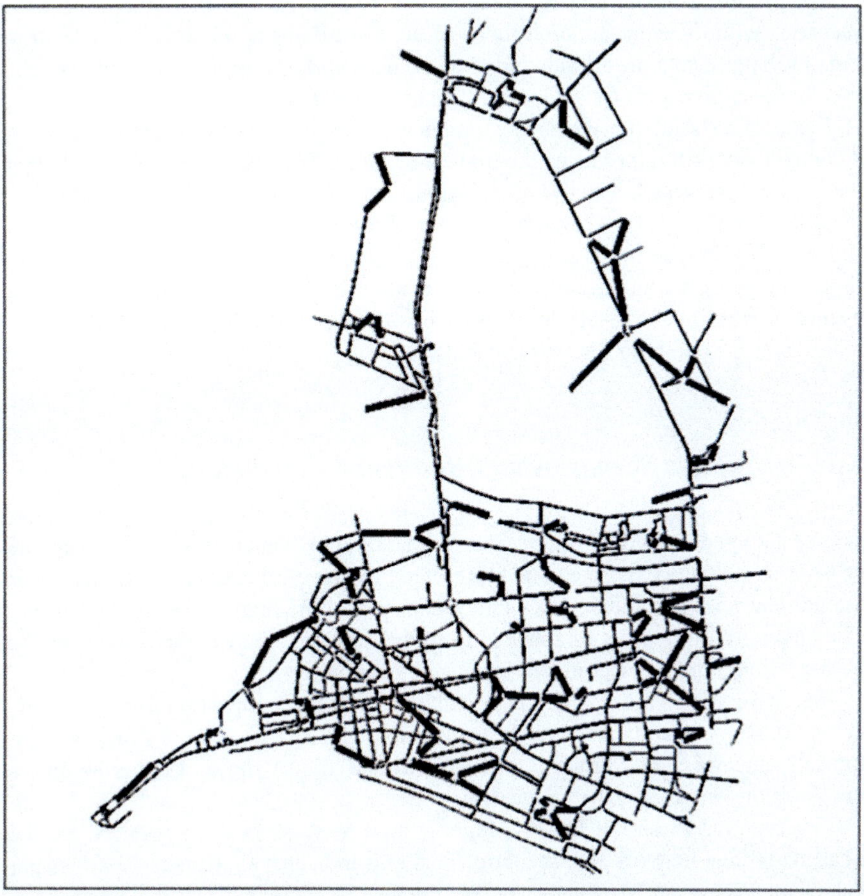

Fig. 9 Graphic representation of the model to be used

The Krauss model with 5 parameters [22] is used; with this number of parameters, its calibration is not complicated according to the SUMO user's guide; this model is sufficient to achieve the objectives proposed in this paper.

In this paper, the model is considered to be calibrated if the difference between the flows of the proposal Krauss – Dijkstra model and the flows of the macro model of the Lima and Callao urban transportation master plan is minimal.

In Fig. 9, the thick lines represent the centroids of the TAZs which are the centers where trips are generated and attracted, i.e., they are the origins and destinations of the trips. The system presents 4 types of trips: internal, inside-out, outside-in and trips crossing the influence area. The hierarchy of the links is calculated with the adjacency matrix as shown in Fig. 10. The numbers at each node are the components of the eigenvector of the adjacency matrix where a high numerical value indicates that it is an important node (Perron nodes). With the help of these numerical values, the most important network link is determined, which is located on Faucett Avenue as shown in Fig. 10.

The proposed model is used to determine the resilience of the system for three scenarios with different duration times of the disruptions at the locations shown in Fig. 11. The results are in qualitative agreement with the numerical values of the dominant eigenvector of the adjacency matrix (Perron vector).

Figure 13 shows the resilience triangles for each scenario where the Faucett Avenue network link produces the highest demand. The above results indicate that the network should be analyzed to assess the impact of an ordinary one-hour disruptive event, it is concluded that Faucett Avenue is the most affected. Figure 14 shows the difference in network costs for the proposed scenarios. It is also seen in Figs. 13 and 14 that the network exhibits a slow recovery from ordinary disruptive events; therefore it can be concluded that the network should be designed or redesigned for one-hour disruptive events.

5.3 *Disruptive Scenarios and Resilience Calculation*

This section generates disruptive scenarios that refer to events that occur daily on the urban road network. These can be represented by the closure of a link located near the port entrance and the network link that produces the highest cost in the system. The hierarchy of nodes is established from the numerical value associated with the network link, which is a component of the Perron vector.

Figure 14 shows that Gambetta Avenue has a greater impact on the cost of the network than Argentina Avenue. Transportation flow near the port is slow, as cargo trucks queue for a long time. Figures 12 (a) and 12 (c) show that the disruptive problems on Gambetta affect Faucett Avenue.

Finally, Fig. 15 shows that the highest total network cost is produced by the Faucett Avenue network link scenario. This cost indicates an average time loss per vehicle of 25 min, and the Nagurney performance measure indicates that the average time to traverse the system is 110 min.

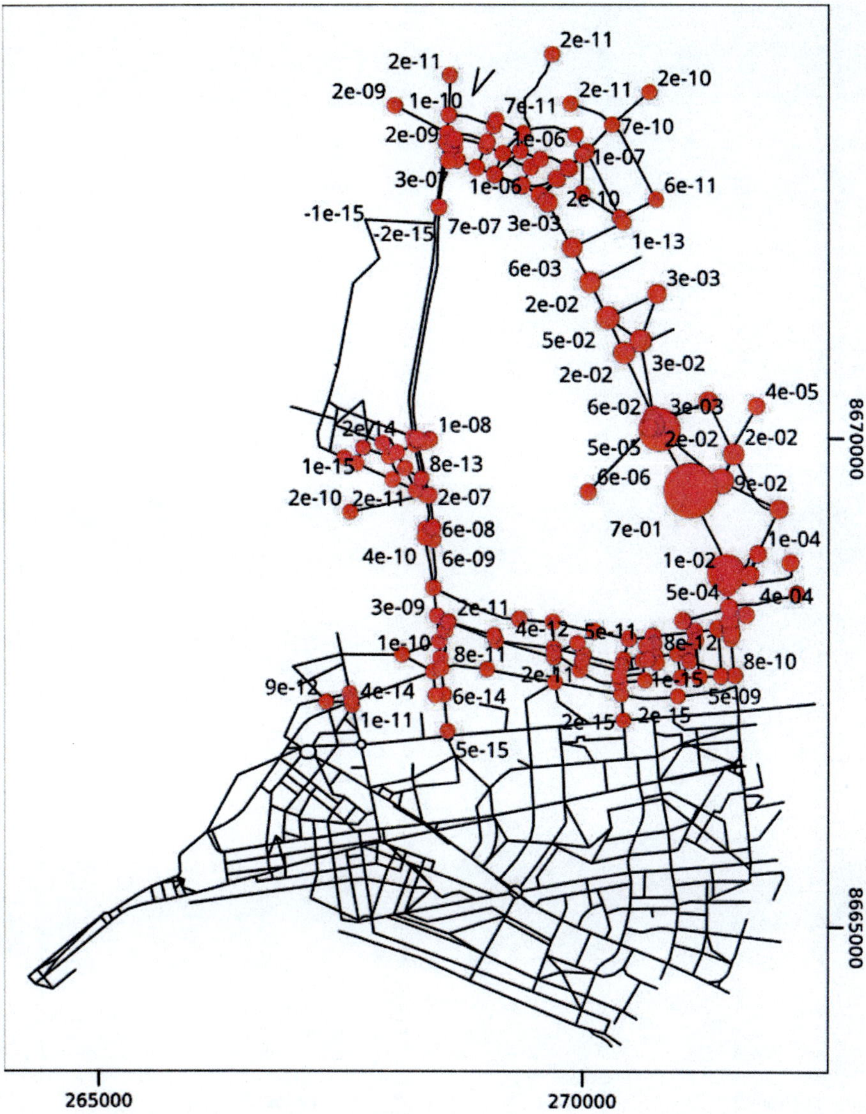

Fig. 10 Importance of each node in the network

6 Discussion

In the case of logistics transportation, it is important to consider its interaction with public and private transportation because this is what really happens. This interaction necessarily increases the possibility of disruptive events occurring due to the

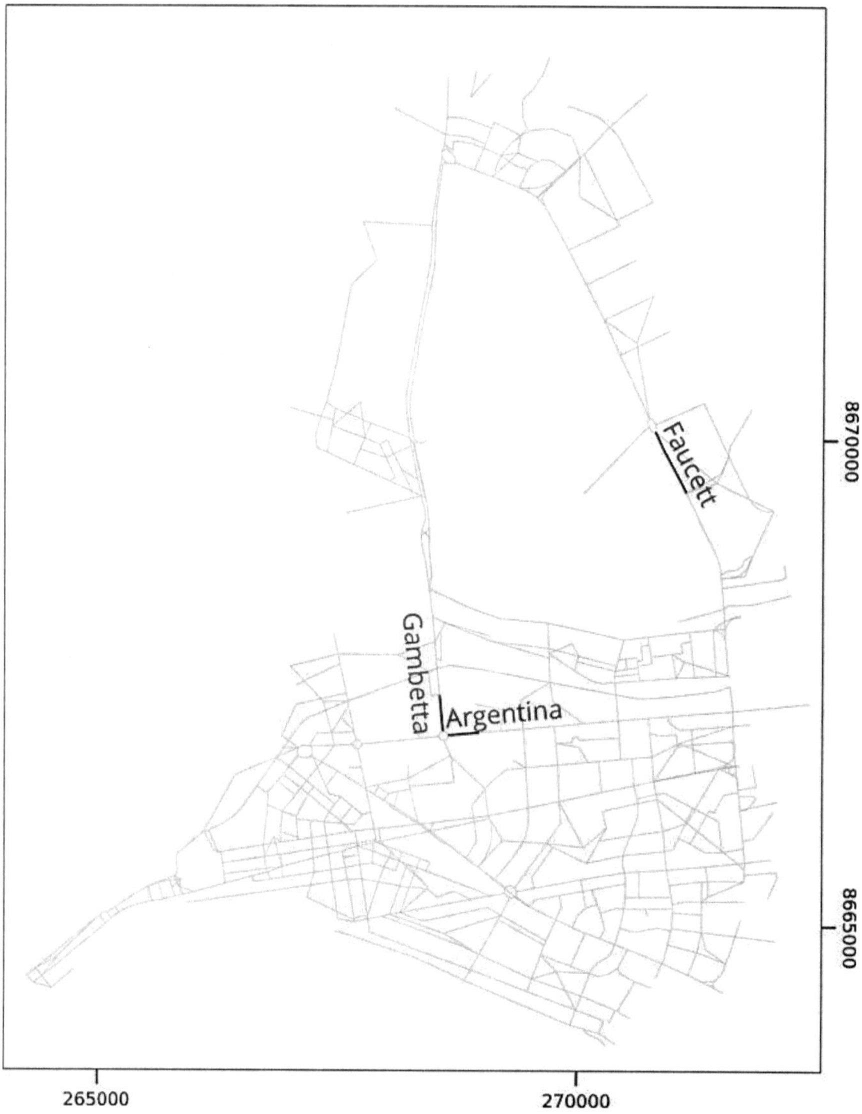

Fig. 11 Location of links that present disruption

participation of all the agents of the system under study. From this it is possible to design an integrated model that includes the location of a logistics corridor.

It is important to measure the performance of a system; however, it must have a metric that not only serves to verify whether the objectives are achieved, but can also indicate its resilience during a disruption. In a transport network, it is important to choose resilience attributes and not leave out the system topology. Not considering

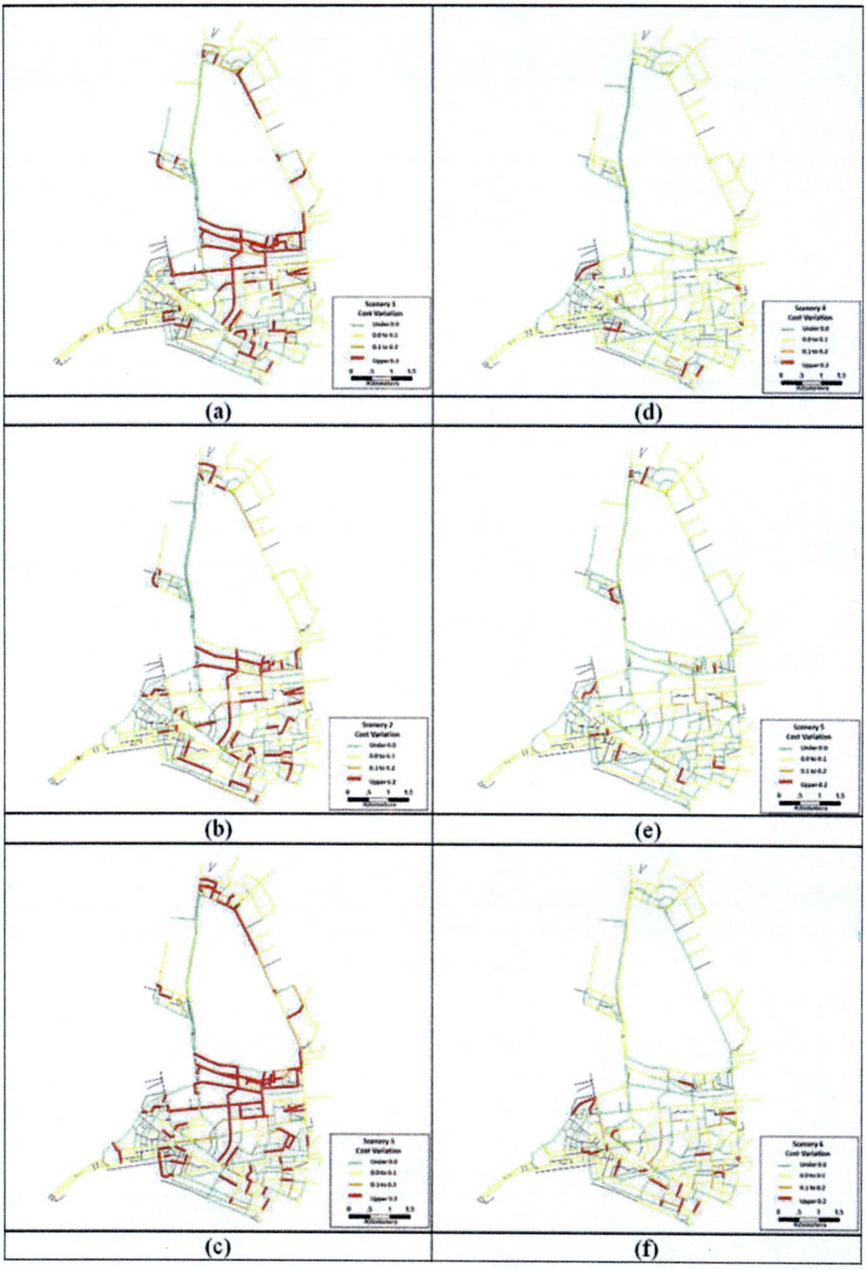

Fig. 12 Disruptive scenarios produced by the elimination of the link between Gambetta Avenue and the Centenario roundabout in the **(a)** south-north direction, **(b)** north-south direction, and **(c)** both directions; and disruptive scenarios produced by the elimination of the link between Avenida Argentina and the Centenario roundabout in the **(d)** west-east direction, **(e)** east-west direction, and **(f)** both directions

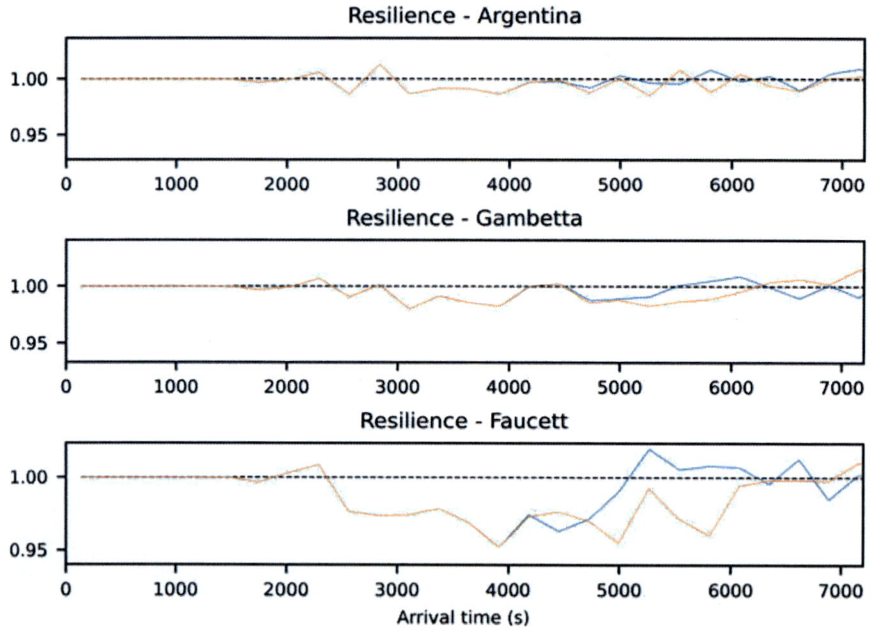

Fig. 13 Network resilience triangles for different scenarios

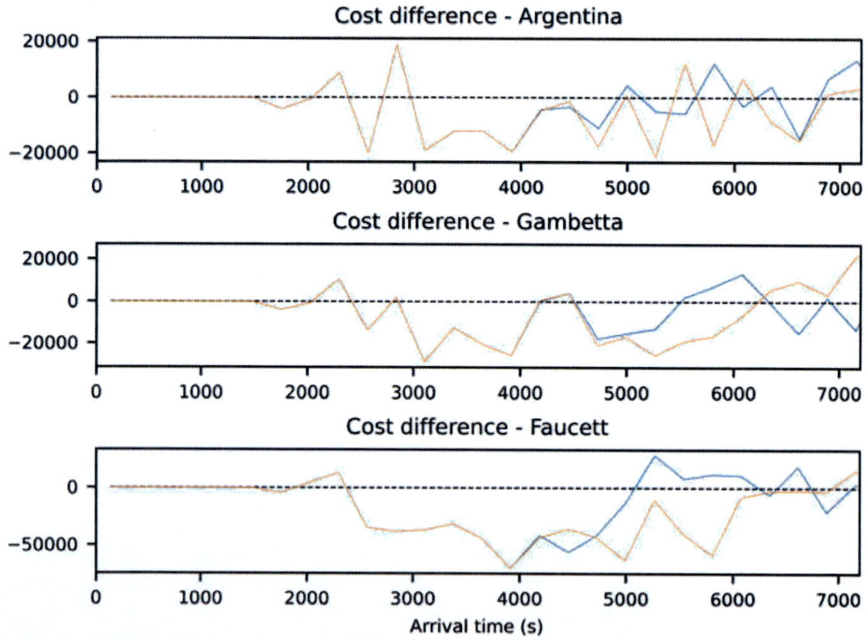

Fig. 14 Network cost difference for different scenarios

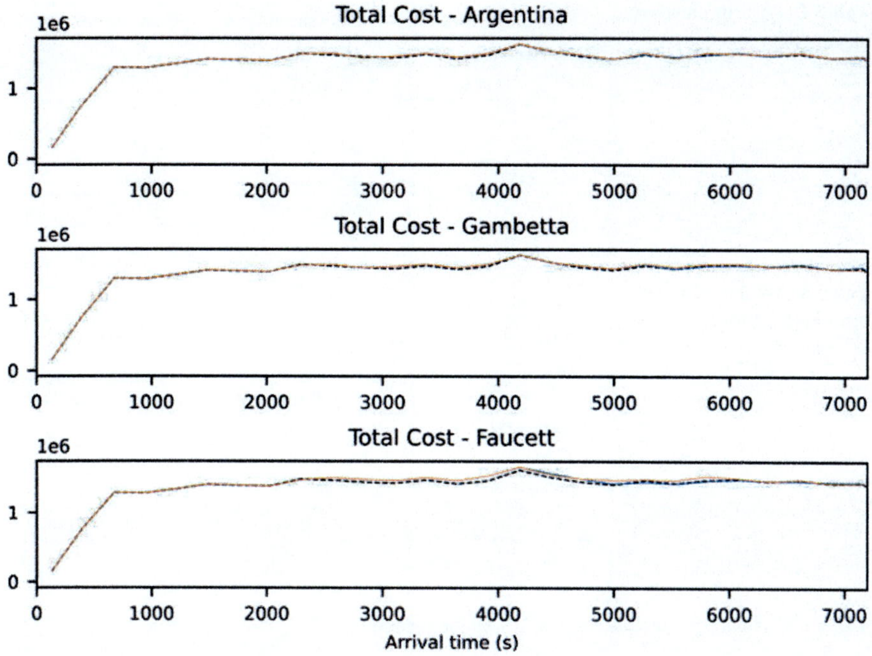

Fig. 15 Total cost of the system

Table 3 Average resilience due to disruption location

Disruption location	Disruption time	Average resilience
Argentina	40 min	0.00215
	60 min	0.00160
Gambetta	40 min	0.00060
	60 min	0.00517
Faucett	40 min	0.00923
	60 min	0.01090

topology means evaluating the performance of the network as if its shape were not relevant in the calculation of the system attributes. Robustness, redundancy and recovery speed are the independent variables of the system cost function, which implies considering resilience as a function of the system cost function.

Table 3 allows us to conclude that the location and duration of the disruptive event increases the numerical value of the resilience triangle, which means that, if the failure occurs in an important network link and is not resolved quickly, the system costs increase. This cost difference can be seen in Table 4 where the cost of the network produced by a disruptive event on a network link is subtracted from the cost of the reference scenario.

The results in Table 5 agree with the results observed in Fig. 15. The highest system costs occur at the Perron nodes with the highest numerical value, so it is

Table 4 Cost difference due to disruption location

Disruption location	Disruption time	Cost difference (s)
La Argentina	40 min	−62623
	60 min	−116915
Gambetta	40 min	−102914
	60 min	−142861
Faucett	40 min	−354910
	60 min	−529442

Table 5 Total cost due to disruption location

Disruption location	Disruption time	Total cost (s)
Base		39048941
Argentina	40 min	39111564
	60 min	39165856
Gambetta	40 min	39151855
	60 min	39191802
Faucett	40 min	39403851
	60 min	39578383

Table 6 Network performance due to disruption location

Disruption location	Disruption time	Network Performance
Base		0.007147819
Argentina	40 min	0.002981062
	60 min	0.00297693
Gambetta	40 min	0.002977994
	60 min	0.002974959
Faucett	40 min	0.002958949
	60 min	0.002945901

sufficient to have the dominant eigenvector of the adjacency matrix to plan actions at the most important locations in the network. The adjacency matrix should be constructed with the costs of each link of the network, and if it were a function of time it would help in the management of urban logistics. This is possible with the proposed model.

The numerical values of the system performance index according to Nagurney, shown in Table 6 are qualitatively consistent with the results shown in Fig. 10 and the resiliency index values in Table 3. These results indicate that the resiliency index of the proposal can be used to measure network performance and hierarchically rank network links.

7 Conclusions

The Callao dry dock scenario can be studied with a mixed model of macro simulation and micro simulation of traffic as shown in Figs. 8, 9 and 10.

Logistics transport influences urban road transport, as shown in Figs. 12 and 8, which implies that decisions made by the port management directly influence the development of urban mobility in Callao.

The importance of logistic transport is evidenced by the volume of TEUs transported from the port of Callao to the dry ports as shown in Table 2.

The proposed methodology is adequate to evaluate the resilience of logistics transportation since the macro simulation theory, combined with the micro simulation theory, allows the study of dry ports similar to the port of Callao in different scenarios where the performance of the network can be calculated through the proposed resilience index as a function of time.

The study of resilience contributes to increase the knowledge gap of the interrelation between logistics transportation and urban networks.

8 Recomendations

An agent-based model is recommended to calculate the resilience of a system; in addition, travel demand should be obtained from ITS intelligent transportation systems [23]; and, finally, the interaction of demand with the model should be calculated in real time.

It would be convenient to conduct a study considering the probability of occurrence of disruptive events and thus be able to simulate what happens daily in a system near a port, which in this case turned out to be Callao.

References

1. Khaslavskaya A, Roso V (2019) Outcome-driven supply chain perspective on dry ports. Sustainability https://doi.org/10.3390/su11051492
2. Crainic T, Dell'Olmo P, Ricciardi N, Sgalambro A (2015) Modeling dry-port-based freight distribution planning. Transportation Research Part C: Emerging Technologies 55:518-534. https://doi.org/10.1016/j.trc.2015.03.026.
3. Galbusera L, Azzini I, Jonkeren O, Giannopoulos G (2016) Inoperability Input-Output Modeling: Inventory Optimization and Resilience Estimation during Critical Events. ASCE-ASME Journal of Risk and Uncertainty in Engineering Systems Part A: Civil Engineering. https://doi.org/10.1061/AJRUA6.0000861
4. Ganin A, Kitsak A, Marchese M, Keisler D, Seager J, Linkov I (2017) Resilience and efficiency in transportation networks. Science Advances. https://doi.org/10.1126/sciadv.1701079

5. Chen Hong, Cullinane K, Liu N (2017) Developing a Model for Measuring the Resilience of a Port-Hinterland Container Transportation Network. Transportation Research Part E: Logistics and Transportation Review 97:282–301. https://doi.org/10.1016/j.tre.2016.10.008
6. Sun W, Bocchini P, Davison B (2018) Resilience metrics and measurement methods for transportation infrastructure: the state of the art. Sustainable and Resilient Infrastructure 5: 168-199 https://doi.org/10.1080/23789689.2018.1448663
7. Ayyub B (2014) Systems Resilience for Multihazard Environments: Definition, Metrics, and Valuation for Decision Making. Risk Analysis 34:340–355. https://doi.org/10.1111/risa.12093
8. Gao Y, Wang J (2020) A resilience assessment framework for urban transportation systems. International Journal of Production Research 59:2177–2192. https://doi.org/10.1080/00207543.2020.1847339
9. Jung H, Do M (2018) Enhancing Road Network Resilience by Considering the Performance Loss and Asset Value. MDPI Sustainability 10:4188. https://doi.org/10.3390/su10114188
10. Calvert S, Snelder M (2017) A Methodology for Road Traffic Resilience Analysis and Review of Related Concepts. Transportmetrica A: Transport Science 14:130–154. https://doi.org/10.1080/23249935.2017.1363315
11. Crainic T, Dell'Olmo P, Ricciardi N, Sgalambro A (2015) Modeling dry-port-based freight distribution planning. Transportation Research Part C 55:518–534. https://doi.org/10.1016/j.trc.2015.03.026
12. Scott D, Novak D, Aultman L, Guo F (2006) Network Robustness Index: A new method for identifying critical links and evaluating the performance of transportation networks. Journal of Transport Geography 14:215–227. https://doi.org/10.1016/j.jtrangeo.2005.10.003
13. Attoh-Okine N, Cooper A, Mensah S (2009) Formulation of resilience index of urban infrastructure using belief functions. IEEE Systems Journal 3:147–153. https://doi.org/10.1109/JSYST.2009.2019148
14. Bruneau M, Reinhorn A (2007) Exploring the concept of seismic resilience for acute care facilities. Earthquake Spectra 23:41–62. https://doi.org/10.1193/1.2431396
15. Garbin D, Shortle J (2007) Measuring Resilience in Network-Based Infrastructures. In: McCarthy JA (ed). Critical Thinking: Moving from Infrastructure Protection to Infrastructure Resiliency. Fairfax, Virginia, pp. 73–86.
16. Tierney K, Bruneau M (2007) Conceptualized and measuring resilience. TR News 250:14–17.
17. Nagurney A, Qiang Q (2007) A Network Efficiency Measure for Congested Networks. Europhysics Letters 79:1–5. https://doi.org/10.1209/0295-5075/79/38005
18. Ministerio de Transportes y Comunicaciones – Autoridad Portuaria Nacional (2019) Tasa de ocupación en los muelles de los terminales concesionados de uso público. Report RO 03, MTC-APN, Lima.
19. Otkóvic I, Tollazi T, Sraml M (2013) Calibration of microsimulation traffic model using neural network approach. Expert Systems with Applications 40:5965–5974. https://doi.org/10.1016/j.eswa.2013.05.003
20. Maheshwari P, Batthacharyya K, Maitra B, Boltze M (2020) A methodology for calibration of traffic micro-simulator for urban heterogeneous traffic operations. Journal of traffic and transportation engineering (English Edition) 7:507–519. https://doi.org/10.1016/j.jtte.2018.06.007
21. Japan International Cooperation Agency (2013) Encuesta de recolección de información básica del transporte urbano en el área metropolitana de Lima y Callao. MTC, Lima.
22. Gawron C (1998) An Iterative Algorithm to Determine the Dynamic User Equilibrium in a Traffic Simulation Model. International Journal of Modern Physics C 9:393–407. https://doi.org/10.1142/S0129183198000303
23. Ganin A, Mersky A, Jin A, Kitsak M, Keisler J, Linkov I (2019) Resilience in Intelligent Transportation Systems (ITS). Transportation Research Part C: Emerging Technologies 100: 318–329. https://doi.org/10.1016/j.trc.2019.01.014

Intelligent Route Planning for Effective Police Patrolling in a Peruvian District

Bradith Zevallos, Alessandro Huamán, Luis Polanco, Jonatán Rojas, and César Corrales

Abstract Citizen security directly influences life quality and is a great source of concern in Latin America. A scheduling and vehicle routing model is proposed for the allocation and routing of police resources, increasing visits in locations with a higher crime rate, and strategic standby locations based on real data.

Keywords Security · Route planning · Criminality · Resource allocation · Mathematical optimization

1 Introduction

The improvement project is located in the district of La Victoria in the metropolitan city of Lima, Peru, where it seeks to increase the presence of the patrol corps with the current availability of resources. District limits, critical points, and points of interest are defined, which are areas of historical criminal activity, high mobilization, or public areas such as squares, parks, etc. For the allocation of resources, which in this case are the patrol corps personnel, a linear programming model is proposed, imposing constraints on the work hours of resources, critical days of the week, and time windows. The final deliverable of the project is a set of patrol routes that ensures police presence at the study site while minimizing travel distance.

B. Zevallos · A. Huamán · L. Polanco · J. Rojas (✉)
Pontificia Universidad Católica del Perú, Lima, Peru

Grupo de Investigación de Operaciones Aplicadas-GIOPA, Lima, Peru
e-mail: bradith.zevallos@pucp.pe; alessandro.huaman@pucp.edu.pe; luis.polanco@pucp.pe; jrojasp@pucp.pe

C. Corrales
Pontificia Universidad Católica del Perú, Lima, Peru
e-mail: ccorral@pucp.edu.pe

J. Vargas Florez et al. (eds.), *Production and Operations Management*, Springer Proceedings in Mathematics & Statistics 391,
https://doi.org/10.1007/978-3-031-06862-1_33

2 Current Situation

La Victoria district is part of Lima Metropolitana, which has an extension of 8.74 km^2 and a total of 166, 657 citizens. It is divided into 43 zones, as shown in Fig. 1 [1].

According to criminal statistics calculated from 2017 to 2019, presented by the district administration, it is evident that the biggest problem it faces is citizen insecurity (crimes against private property).

Therefore, there is a latent need to improve the surveillance and routing system of personnel to ensure an adequate reach (Table 1).

Fig. 1 Zones distribution in La Victoria district [1]

Table 1 Crime typology statistics from 2017 to 2019 [1]

Year	Patrimony	Life, physical harm and health	Public safety	Freedom	Other	Aggregate value
2017	5667	694	666	211	98	7336
2018	5454	697	661	301	87	7200
2019	2321	398	191	114	77	3101

Table 2 Current district resources per location [1]

Resource	General administration	Police station 1	Police station 2	Police station 3
Officers	231	108	108	84
Wagon	22	11	6	6
Motorcycles	36	3	2	2
Bicycles	16	0	0	0

Fig. 2 (**a**) Points of interest distribution, (**b**) historical crime distribution

The current administration has the following resources to ensure district security (Table 2):

Citizens who live or interact with places with low security are those who, according to studies, report low levels of well-being. In the same way, people who show confidence in the police force are those who feel less fear of being victims of crime. The study of [3] concludes that places with high violence are key points to improve the perception of the police, alongside spaces that facilitate exposure of citizens to patrols, such as parks, hospitals, schools, etc. For this reason, a set of points of interest were defined to ensure police presence in these high-exposure locations (Fig. 2; Table 3).

Table 3 Statistics of points of interest

	Parks	Malls	Markets	Medical centers	Academic establishments
Quantity	57	2	2	2	9
Percentage (%)	79.2	2.8	2.8	2.8	12,5

3 Contribution

A resource-scheduling model is proposed, which aims to maximize a fictitious objective function that forces a greater allocation of security agents on Fridays and Saturdays, in shifts from 11:00 am to 11:00 pm. In addition, a security agent routing model is proposed using a VRP model formulation. In this way, we seek to reach adequate coverage in the most dangerous areas at night and during the day, increasing the perception of safety by the population while visiting the previously mentioned points of interest. Initially, the first resource allocation model is presented and then the vehicle routing model.

3.1 Security Agent Scheduling Model

The following sets are defined, containing different values whose meaning is explained in Table 4:

- I: Agent travel mode
- J: Agent entry work shift (4 h each)
- K: Day of the week an agent starts its shift

3.2 Security Agent Scheduling Model – Decision Variables

The decision variable is defined as follows:

X_{ijk}: Number of agents of type i that starts its shift in turn j in day k.

An agent that starts its shift must also work in the next immediate shift, for 6 days a week. It is allowed a single day of rest in the week. In Fig. 3 a pair of examples is shown to further understand these implications, where the entry position is colored in green, the X_{ijk} variable is present in the shifts and days an agent is actively working and the rest slots are colored in red.

Table 4 Meaning of set values (each column represents one value)

Set description	Set	1	2	3	4	5	6	7
Agent travel mode	I	Foot	Motorcycle	Wagon	–	–	–	–
Agent entry work shift	J	3:00–7:00	7:00–11:00	11:00–15:00	15:00–19:00	19:00–23:00	23:00–3:00	–
Day of the week an agent starts its shift	K	Monday	Tuesday	Wednesday	Thursday	Friday	Saturday	Sunday

$i = 1$		$K = 1$						
	x_{1jk}	1	2	3	4	5	6	7
$J = 1$	1	x_{111}	x_{112}	x_{113}	x_{114}	x_{115}	x_{116}	x_{117}
	2	x_{121}	x_{122}	x_{123}	x_{124}	x_{125}	x_{126}	x_{127}
	3							
	4							
	5							
	6							

Fig. 3 Example of X_{ijk} presence along shifts and days

3.3 Security Agent Scheduling Model – Objective Function

The objective function is defined as follows:

$$Max\ Z = \sum\nolimits_{ijk} X_{ijk} * G1_j * G2_k \tag{1}$$

Where $G1_j$ and $G2_k$ are factors that drive a higher allocation of resources in certain shifts and days of the week:

- $G1_j$: 1.5 if j = 3, 4 or 5. Else, it takes the value of 1.
- $G2_k$: 2 if j = 5 or 6. Else, it takes the value of 1.

3.4 Security Agent Scheduling Model – Constraints

Two constraints are needed to ensure a minimum and maximum required quantity of agents:

Constraint 1 The number of agents of a given travel mode, located in each shift of each day of the week, must be greater than a previously defined minimum number (Q_i), which depends solely on the travel mode *i* of the agent. Hereby, it is ensured that each shift is adequately covered with personnel. To achieve this requirement on shift j of day *k*, it is needed to take into account agents X_{ijk} that start their shifts in *j* and (*j*-1), workdays *k*, (*k*-1), (*k*-2), (*k*-3), (*k*-4), and (*k*-5), with a work coverage of 8 continuous hours and the working range of 6 days a week. Next, the representation of the constraint for the wagon agent ($i = 3$) for shift 1 ($j = 1$) on Monday ($k = 1$) is shown. The same logic is applied for each combination *j* and *k* (Fig. 4):

$$\sum\nolimits_{k=1..7} X_{36k} - X_{361} + \sum\nolimits_{k=1..7} X_{31k} - X_{312} \geq Q_3 \tag{2}$$

$i = 3$ x_{3jk}		K						
J	Description	1	2	3	4	5	6	7
1	03 - 07	x_{311}	x_{312}	x_{313}	x_{314}	x_{315}	x_{316}	x_{317}
2	07 - 11							
3	11 - 15							
4	15 - 19							
5	19 - 23							
6	23 - 03	x_{361}	x_{362}	x_{363}	x_{364}	x_{365}	x_{366}	x_{367}

Fig. 4 Example of constraint 1 evaluation for $i = 3, j = 1$, and $k = 1$

Table 5 Minimum quantities available per agent

Agent type	I	Minimum amount	Available
Foot	1	20	136
Motorcycle	2	8	47
Wagon	3	12	48

Constraint 2 The number of agents in each category, located in each shift of each day of the week must be less than the maximum number of agents available, which depends only on category i of the agent:

$$\sum\nolimits_{jk} X_{ijk} \leq C_i \tag{3}$$

The data corresponding to the minimum and the available number of agents per shift are as follows (Table 5):

3.5 *Agent Routing Model*

The CVRP model presented below follows the logic of a variant proposed by Toth and Vigo [2].

3.6 *Agent Routing Model – Decision Variables*

The model generates m routes, where each route is carried out by a single agent. On each route, the agent visits each location associated with the route only once. A total of n locations are visited only once by a different route. In our case, La Victoria district has $n = 186$ and $m = 23$. As the district has three police stations, vehicles depart from these locations as depots. For ease of planning, the district is divided into three different zones, as shown in Fig. 5, where the number of vehicles, the number

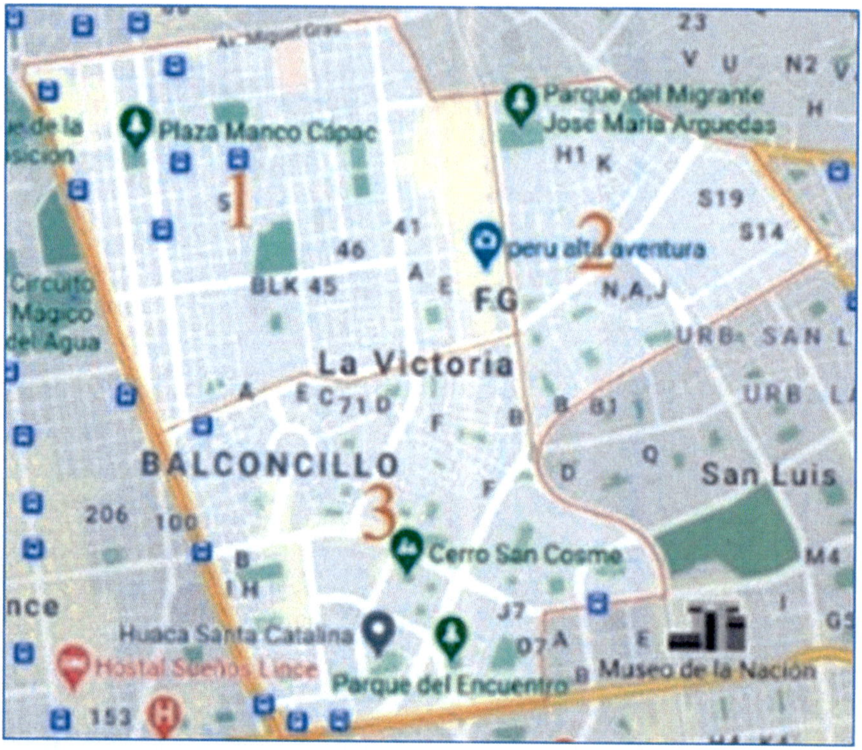

Fig. 5 Zones in La Victoria

Table 6 Zone's characteristics

Zone	Depot	*n*	*m*	*C*
1	Police station 1: PNP LA VICTORIA	56	11	7
2	Police station 2: PNP APOLO	60	6	10
3	Police station 3: PNP SAN COSME	70	6	13

of locations to visit, and the maximum number of visited locations (C) per route depends on each area's available resources, as shown in Table 6.

The agents start their route from the assigned police station, indexed by 0, carry out their routes, and return to the station.

The indices of the model are presented below, which depend on the zone as previously described:

- m: Index of the route traveled by an agent.
- n: Index of the location to visit
- $i = 0,\ldots,$n: Index of the location from which the agent departs.
- $j = 0,\ldots,$n: Index of the location to which the agent arrives.

The decision variables of the model are presented below:

- X_{ij}:Binary variable that indicates that the agent starts from location i and travels to location j.
- u_i:Integer support variable, associated with each location indexed by a $i >= 1$. It allows the satisfaction of the subtour elimination constraint.

3.7 Agent Routing Model – Objective Function

The objective is to find the optimal trajectories that each route should follow, such that the total distance traveled by the agents is minimized:

$$\text{Min Z} = \sum_i \sum_j d_{ij} * X_{ij} \tag{4}$$

Where d_{ij} is the distance necessary to travel the linear section between locations i and j. It comes from a matrix of Euclidean distances between pairs of locations, using the latitude and longitude coordinates. Given that these distances are merely an approximation of the real distances to travel, it is of interest to have the sequence of locations to visit on each route. With this sequence, routes can be generated that, when traveling through streets, will have different distances. When $i = j$, the value of d_{ij} is infinite, causing the model to avoid inserting recurring visits.

3.8 Agent Routing Model – Constraints

First, the constraints that limit the number of entrances and exits to each location are defined. In Eqs. 5 and 6, it is indicated that the number of visits to the police station (indexed by 0) must be equal to the number of routes, while in Eqs. 7 and 8 it is indicated that the rest of the locations must be visited only one once by a single route.

$$\sum_i X_{i0} = m \tag{5}$$

$$\sum_j X_{0j} = m \tag{6}$$

$$\sum_i X_{ij} = 1, \forall j \in \{1, \ldots, \text{n}\} \tag{7}$$

$$\sum_j X_{ij} = 1, \forall i \in \{1, \ldots, \text{n}\} \tag{8}$$

Finally, the capacity constraint and elimination of subtours is defined, following the formulation of Miller-Tucker-Zemlin [2]:

$$u_i - u_j + X_{ij} * C <= C - 1 \forall i, j \in \{1, \ldots, \mathrm{n}\}, i \neq j \tag{9}$$

4 Results

The allocation model gives, as a result, the number of agents of each travel type *i*, that begin its shift *j* in day *k* (X_{ijk}), which is presented in Fig. 6.

The implemented routing model provides a series of routes that cover the locations with the highest criminal incidence in La Victoria. The visit sequences of each route by each zone are presented in Fig. 7.

5 Conclusions

The proposed models systematically cover two important needs of the current police system, referring to the management of police resources, ensuring the availability of security agents at the most critical hours and days, and establishing standard routes to be followed during patrol work, ensuring coverage of critical locations. The model

Fig. 6 Scheduled number of agents by table type, shift, and day

Agent type	1		2			3			
Day	5	6	1	5	6	1	2	5	6
Shift									
1	20	0	1	7	0	4	4	4	0
2	16	20	0	0	7	0	0	0	4
3	20	0	0	8	1	4	4	4	0
4	0	20	0	0	7	0	0	0	4
5	20	0	0	8	1	4	4	4	0
6	0	20	0	0	7	0	0	0	4

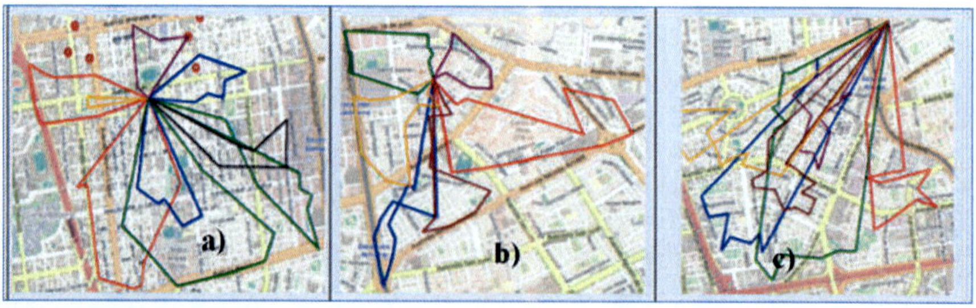

Fig. 7 (**a**) Travel routes by zone 1, (**b**) Travel routes by zone 1, (**c**) Travel routes by zone 1

for assigning human resources to shifts provides a flexible tool that can be enriched by new procedures determined by the police station, optimizing new situations easily. The routing model generates a set of standard routes corresponding to each of the sectors, which cover the most critical locations in the district, considering data from 2017–2019. The assignment of agents to each route can be done with total freedom, at the discretion of the police station, without losing the ability to adapt as it can be updated according to variations in the criminal dynamics of the district.

References

1. Municipalidad Distrital de la Victoria: Plan de acción distrital ciudadana – 2020, CODISE – La Victoria, https://www.munilavictoria.gob.pe/files/codisec/2020/plan-de-accion/PADSC%20LA%20VICTORIA%202020%20FINAL%20PDF.pdf, (2020).
2. P. Toth, D. Vigo. The Vehicle Routing Problem, Philadelphia: Society for Industrial and Applied Mathematics (2002).
3. R. Zahnow, L. Mazerolle, J. Corcoran. Living near violence: How proximity to violence shapes perceptions of police effectiveness and confidence in police. Journal of Environmental Psychology, no 53, pp. 138–148. https://doi.org/10.1016/j.jenvp.2017.07.007 (2017)

UrbanPy: A Library to Download, Process and Visualize High Resolution Urban Data to Support Transportation and Urban Planning Decisions

Andres Regal, Claudio Ortega, Antonio Vazquez Brust, Michelle Rodriguez, and Patricio Zambrano-Barragan

Abstract Research demonstrates that the large-scale access and analysis to data can help cities plan for more inclusive and efficient growth. However, policy makers and researchers around the world still lack sufficient access to granular and dynamic sources of data – a challenge that has come to prominence as cities attempt to respond to the local impacts of the COVID-19 pandemic. To address this challenge, this paper presents UrbanPy, a new open-source library that makes the automated collection, processing, and visualization of spatial urban data simple and consistent for cities. UrbanPy, first developed as a tool to help Latin American cities design rapid responses to COVID-19, presents six innovative capabilities for researchers and practitioners focused on data collection, processing and visualization. To illustrate the potential applications of UrbanPy, this paper presents a case study from Lima, Peru, where the library helped municipal leaders with their COVID-19 response by identifying vulnerable populations, creating food accessibility metrics, and optimizing the location of food supply facilities.

Keywords Urban data · Transportation · Urban planning decisions

1 Introduction

Cities today are home to close to 60% of the world's population and are collectively responsible for an increasingly larger share of economic output (UN-Habitat 2016). Given current trends, by the end of the XXI century 85% of people will live in cities, primarily in developing countries in Africa, Asia, and Latin America (OECD.

A. Regal (✉) · C. Ortega · M. Rodriguez
Universidad del Pacifico, Lima, Peru
e-mail: a.regalludowieg@up.edu.pe; c.ortegaariza@up.edu.pe; rodrigezm@up.edu.pe

A. Vazquez Brust · P. Zambrano-Barragan
Inter-American Development Bank, Washington DC, USA
e-mail: pzambrano@iadb.org

J. Vargas Florez et al. (eds.), *Production and Operations Management*, Springer Proceedings in Mathematics & Statistics 391,
https://doi.org/10.1007/978-3-031-06862-1_34

Publishing 2015). Rapid urban growth in LAC is the result of agglomeration benefits: people want to live in places that offer opportunities for a better life, while firms locate where demand and talent lie, thus providing conditions for human development through the effective delivery of basic services and the establishment of social and political organizations to solve difficult human problems (Glaeser 1999, Duranton and Puga 2014). However, unplanned urban growth can also present considerable challenges, particularly unequal access to quality infrastructure services and negative environmental impacts amidst sprawling urban expansion. Overall, these characteristics make cities "*the example par excellence of complex systems: emergent, far from equilibrium, requiring enormous energies to maintain themselves, displaying patterns of inequality spawned through agglomeration and intense competition for space, and saturated flow systems that use capacity in what appear to be barely sustainable but paradoxically resilient networks*" (Batty 2008).

While rapid urban growth, with all its challenges and opportunities, has defined global development since at least the mid-twentieth century, there is another, equally transformational dual phenomenon: digitalization and rapid technological change. Consider that, between 2016 to 2017, 300 million people started using the internet (DOMO 2020), increasingly via mobile devices (GSM 2018). Much of this growth has been driven by developing regions, including Latin America, where over the last three decades, the proportion of people using the internet jumped from 0% to 56% (World Bank 2020). Yet, much like urbanization, digitalization and technological change present challenges, ranging from privacy and security breaches, to overreliance on potentially biased data processing tools, and to persistent conditions of inequality in terms of access. The question is whether both urban growth and digitalization can benefit citizens.

Altogether, the possibilities new technologies present to urban development are concrete and likely to grow substantially. Recent reports show that the application of existing data-driven solutions could improve quality of life indicators by 10–30% in areas such as mobility, security, healthcare, energy, water, waste, civic engagement, economic development and housing (Woetzel et al. 2018). Dynamic analysis of urban data is thus essential to the proper management of cities and, when committed to principles of inclusion and open data, can improve citizen participation and policy outcomes (Townsend and Zambrano-Barragan 2020).

How can growing cities enable data-driven governance? Contemporary urban systems already generate tremendous amounts of spatial information – through mobile and transportation GPS trackers, geolocated data on land use and cadaster, and broader internet of things (IOT) – which can, in turn provide critical insights to improve living conditions. At present, there are proprietary and open-source tools for technical spatial analysis (ArcGIS and QGIS, respectively), as well as more user-focused platforms for navigation and location services in the city (Google Maps, and the crowdsourced open platform, OpenStreetMap). However, from the perspective of policy makers and city leaders, proprietary tools can be prohibitive in terms of cost and technical capacity, while open source solutions lack spatial resolution and functionality.

In this context, several efforts have been made to produce an open-source set of processing and visualitazion tools. Overpass API, OSMnx (Boeing 2017) and the Nominatim API aim to facilitate the gathering and processing of road network data, while Geopandas (GeoPandas developers 2020) abstracts spatial operations with points, polylines, and polygons. However, the workflow for managing several spatial data sets with different levels of aggregation, with disparate coordinate systems and sources, becomes increasingly more difficult as more layers of data are added to the analysis.

To address these limitations and to support data-driven urban governance, this paper presents UrbanPy, a library to download, process and visualize high resolution urban data. This open source tool facilitates best-available open data collection (using sources like OpenStreetMap and Humanitarian Data Exchange), while supporting standard spatial aggregation of data sets using regular polygons (i.e. grids or hexagons), as well as fast processing of spatial aggregations into these regular polygons. Ultimately, UrbanPy enables the visualization of these disparate data sources – and the combined layers and statistics built from them – thus strengthening decision support systems for key policy actors such as municipal authorities. Working with shortest path calculation, population density layers (grouped by age brackets), and Points of Interest (POI) data, UrbanPy can provide a detailed assessment of access to key services, such as food and health, and offers an optimization-based approach to improve this accessibility.

This paper is structured as follows. Section 2 provides a functional analysis and a comparison with similar tools. Section 3 presents a case study of the application of UrbanPy in Lima, Peru, as part of municipal efforts to address the impacts of the COVID-19 pandemic. Finally, Sect. 4 discusses the implications of UrbanPy and its future direction.

2 Main Functionality and Modules

To address the challenges that arise from working with large and diverse spatial urban data sets, UrbanPy provides a tool for the collection of data, aggregation into spatial units, and processing of road network statistics within these units consistent, automatable, and easy to use primarily for urban and transportation planning purposes.

UrbanPy was originally conceived as part of a collaboration between the Municipality of Lima, the Inter-American Development Bank, and the Universidad del Pac'ıfico, as a way to improve policy responses to the COVID-19 pandemic in Lima, Peru. The first 'test' application of UrbanPy in this city shows that, beyond urban and transport planning, the tool can also be applied for health and social protection policy decisions (see Sect. 3 below).

UrbanPy is an open-source Python package that downloads administrative boundary geometries, street networks from OpenStreetMap, and population data from Humanitarian Data Exchange. It wraps graph download and graph statistic computation via OSMnx (Boeing 2017), spatial operations with R-tree and Geopandas (GeoPandas developers 2020) and computes distance matrices and travel times using different service providers (open-source and proprietary, as preferred by users).

This library enables users to easily construct, project, visualize, and analyze several layers of data by constructing a city's or neighborhood's regular polygon segmentation, as well as downloading walking, driving, or biking networks within these polygons. As mentioned before, UrbanPy was built using Python because of its easy of use and the growing strength of its spatial data analysis libraries stack. UrbanPy is built on top of Python's NetworkX, Matplotlib, Geopandas, OSMnx and H3 libraries for rich spatial analysis capabilities, fast spatial queries and operations with R-tree indexing and crisp visualizations. This section discusses UrbanPy's primary features in order.

2.1 *Downloading Spatial Data Sets*

As described above, UrbanPy provides an interface to download different spatial datasets. In its current release,[1] UrbanPy provides the following methods for data download: *download.nominatim osm*, *download.hdx dataset*, *download.hdx fb population*, *download.overpass_pois*, and *download.osmnx graph*. As their names indicate, the Nominatim and Overpass APIs are supported, as well as graph download using OSMnx and population data sets using Humanitarian Data Exchange (HDX).

For population layers, using HDX, Facebook high resolution population estimates may be downloaded and turned into DataFrame structure using *download.hdx fb population*. This representation allows the user to harness both query operations (filtering, grouping, sorting) and spatial operations (intersection, spatial filtering, etc). It is important to note that HDX data can be downloaded with a specific function for Facebook data, but any resource that is stored on HDX as a compressed CSV (or just a CSV) can be downloaded using *download.hdx dataset*.

The second layer that is supported is the road network graph. UrbanPy provides three main ways to download these graph: by point (a lat, lon coordinate with a buffer radius), by place name (a string address) or based on a polygon bounding box using *download.osmnx graph*. The latter allows for road network coefficients (*e.g.* betweenness centrality, page rank) to be computed for specific geographical partitions.

The third information layer, which enables local filters using spatial operations, is *download.nominatim osm*. It provides an interface with the Nominatim API by providing a query string (*i.e.* Los Angeles, California) and an optional expected

[1]Source code may be found at https://github.com/EL-BID/urbanpy, as well as through the IDB's Code for Development platform, code.iadb.org

position (the polygon is not always located as the first result). This function queries the API and converts the polygon result into a GeoDataFrame, with EPSG 4326 as its coordinate reference system.

Finally, UrbanPy provides an interface with the Overpass API through the *download.overpass_pois* function. A set of queries for food supply, healthcare facilities, education centers, and financial establishments are supported *a priori* with the automatic construction of a DataFrame based on the result. Aside from these pre-built queries, a custom query string may be supplied, where the corresponding JSON response is returned for the user to process into the format they prefer. This function can also create GeoDataFrames for points and polygons (including linestrings) for spatial filters to be performed later on.

2.2 *Constructing Hexagons While Providing Spatial Aggregation and Down-Sampling*

To analyze a city in standardized spatial units, as mentioned before, several approaches can be followed. UrbanPy provides an interface with Uber's H3 hexagon partitioning system (Uber 2020). H3 was chosen to provide a regular polygon partition of a city that allows higher resolution in urban areas and lower resolution in peripheral areas, accelerating distance matrix computation and spatial aggregation procedures. Given a polygon representing a city's administrative boundaries, UrbanPy is able to partition the area to an input resolution (see H3's documentation[2] for detail into resolution, hexagon area and number of unique indexes).

Once a base hexagon layer is created, UrbanPy can be used to aggregate data from lower resolutions (points and polylines) into hexagons or perform overlay and spatial operations to downsample data from a higher resolution (administrative boundaries, census zones). These merging and aggregation operations are flexible – both the aggregation function (sum, min, max, mean, mode) and the spatial operation (intersection, union, overlay) are user-determined. This enables different metrics to be aggregated in different ways, reducing code footprint and delivering consistent results.

2.3 *Constructing Distance and Travel Time Matrices*

Within the routing module, UrbanPy provides support for several routing engines. If the user has docker and wget already installed, UrbanPy provides an interface to download, install and process a road network using OSM and OpenSource Routing

[2]https://h3geo.org/docs/core-library/restable

Machine (OSRM). With this processing, OSRM is capable of handling distance matrix and point-to-point path computations.

If a service provider is preferred, UrbanPy provides interfaces with the Google Maps Distance Matrix and OpenRoute Service for distance matrix computation (with a user provided API key). Note that NetworkX is also used for graph shortest path finding, whether it is used to compute the number of nodes in a path or the total length.

To compute these matrices, only the geometries are needed (in EPSG 4326). Latitude and longitude pairs are also accepted, as well as data structures other than DataFrames such as lists of (lat, lon) tuples. This added flexibility lets the user read coordinate data from different sources, merge them into a single structure and compute both distance and travel time matrices with the service of choice, letting UrbanPy handle the abstraction into API specific patterns.

3 Case Study: Lima, Peru

Lima, Peru's capital and largest city, concentrates 30% of the country's population (11 million inhabitants, INEI (2018)). Once the COVID-19 pandemic hit the country, the national government introduced a strict quarantine and a curfew to 'flatten the epidemic curve' and mitigate the long-term impact on the economy. During this mandatory quarantine, guaranteeing a high level of accessibility to food supplies (e.g. markets, nanostores, convenience stores, and similar facilities) without risking vulnerable populations, such as the elderly, became a significant challenge. The need to optimize service delivery was a common challenge among cities in Latin America, and, as a result, the IDB and the UP launched a collaboration to design digital tools to support municipal authorities. UrbanPy was thus conceived as a tool to satisfy the need for rapid access to accurate and useful information for local authorities and support data-driven decision-making. The tool was developed through an iterative solution design process, in which the core development team and non-technical decision-makers from Lima's municipality where involved.

In this section, the process of sourcing data, converting it to spatial units and applying an optimization model to allocate itinerary markets will be presented as a use case for UrbanPy. The first stage in this initiative was to source different layers of data. To estimate demand an population flow, HDX was chosen due to the high resolution of the data collected from Facebook. These points have been filtered to Lima's administrative boundaries but due to their high resolution, they need to be aggregated for comparable analysis.

Since each record in this data set if of point geometry, a spatial unit to aggregate the data and produce relevant visualizations is needed. Thus, Lima's administrative boundaries were split into resolution 8 hexagons. Figure 1 presents the initial HDX data aggregated to hexagon level. This representation is particularly useful to aggregate other sources of data (i.e. socio-economic variables, where the median

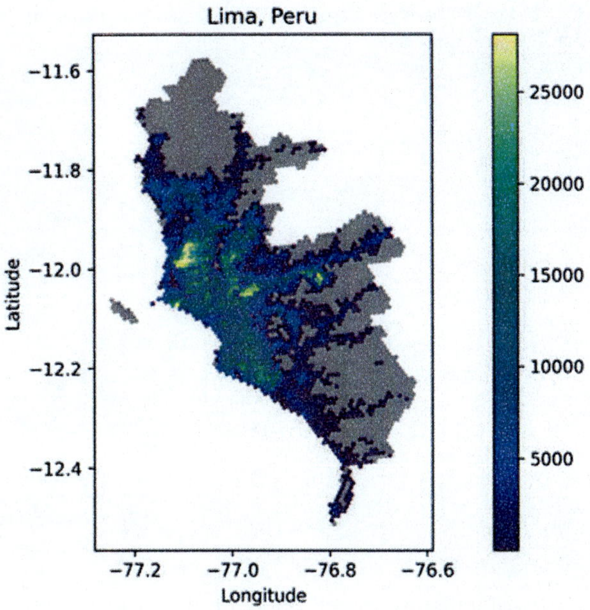

Fig. 1 Hexagon population density layer (for 2020)

income level can be represented as the percentage of city blocks within the hexagon that perceive the median income).

Using these layers and UrbanPy's POI download functions to obtain food facilities within the city, food accessibility metrics can be measured using the estimated travel time from each spatial unit to the facilities. Then, it was determined that the current food facilities were not capable of successfully reaching the most densely population zones in the city. Also, in Lima, traditional markets are especially critical agglomeration points, which leads to a potential increase in the spread of COVID-19 among vendors and customers. In this context, parks and sport fields arise as potential relief spaces to allocate temporal food facilities. These facilities, commonly referred to as itinerary markets, may help to improve the food access and to reduce the infection rate among the daily customers in current markets and supermarkets.

A Facility Location Model was proposed to obtain the optimal itinerary markets locations. Results were delivered to different departments within the Municipality to support real-time decision making. The data acquisition and modelling process is described as follows: First, a query to the Overpass API was made to get all the candidate itinerary markets (parks, sport fields, etc). These initial results are concatenated with the national market census, while filtering to a specific district to reduce the search space for the model (as to make it computationally viable). Then, a distance matrix can be computed to serve as input to the optimization model. In Fig. 2, the filter to San Juan de Lurigancho (Lima's most densely populated district) is presented. This district will be the focus of this case study given its high levels of population and available markets in the candidate set.

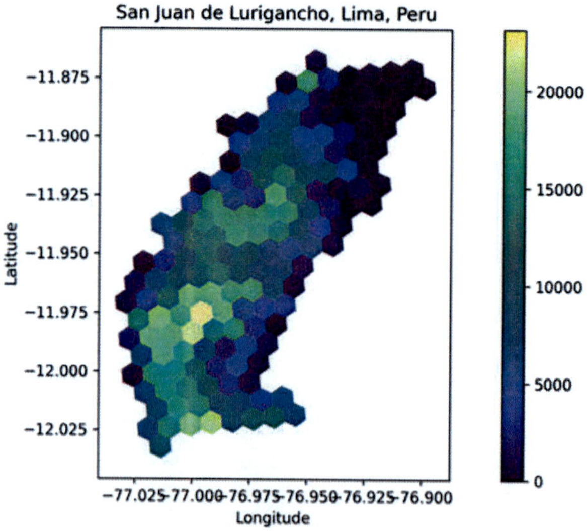

Fig. 2 Population layer for San Juan de Lurigancho, Lima, Peru

In particular, a modification of the p-median problem was applied, adding capacity constraints to the facilities and forcing the optimizer to keep certain facilities active.

The p-median problem uses two main decision variables: x_{ij} representing the percentage of the demand from customer j assigned to facility i, and y_i representing the activation of the i-th facility. As its parameters, d_j represents the demand of customer j, c_{ij} represents the cost of travelling to facility i by customer j. Finally, the number of facilities to activate must be specified using p.

$$\min \sum_{i=1}^{n} \sum_{j=1}^{m} d_j c_{ij} x_{ij} \tag{1}$$

$$\sum_{i=1}^{n} x_{ij} = 1 \forall j \in \quad \text{Customer} \tag{2}$$

$$\sum_{i=1}^{n} y_1 = p \tag{3}$$

$$\sum_{j=1}^{m} d_j x_{ij} \leq c_1 \forall i \quad \text{Facilities} \tag{4}$$

$$x_{ij} \leq y_i \forall i \in \quad \text{Facilities} \quad \forall i \in \quad \text{Customers} \tag{5}$$

$$x_{ij}, y_i \in \{0, 1\} \forall i \in \quad \text{Facilities} \quad \forall i \in \quad \text{Customers} \tag{6}$$

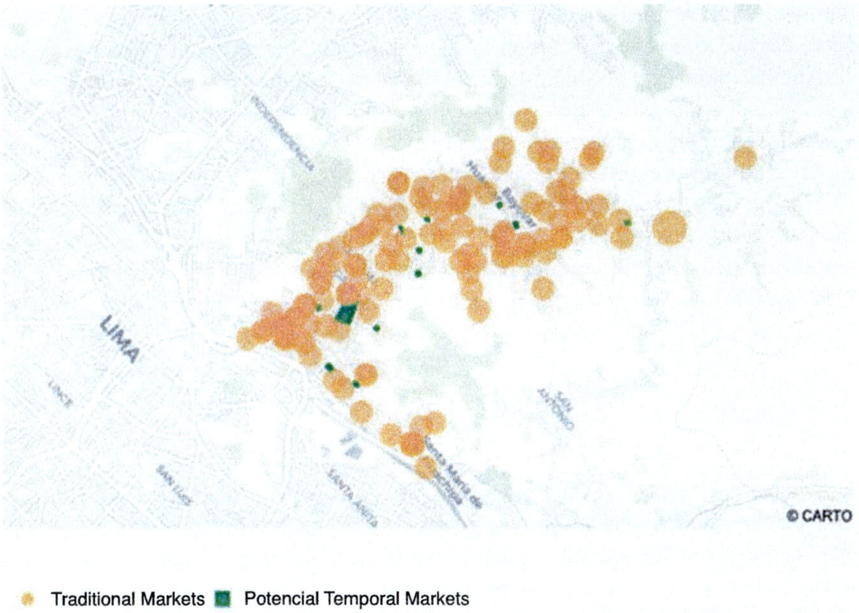

Fig. 3 Optimization results, green points and polygons represent the selected facilities and orange circles represent the active markets

$$0 \leq c_{ij} \leq 1 \forall i \in \quad \text{Facilities} \quad \forall i \in \quad \text{Customers} \tag{7}$$

$$d_j \geq 0 \forall j \in \quad \text{Customers} \tag{8}$$

Equation 1 presents the objective function: minimizing the cost of assigning x_{ij}% of the demand to the active facility. In Eq. 2 an equality constraint is enforced to ensure that for each customer, the sum percentages equals 1 (conserving the total demand for each customer). Using Eq. 3 the total number of activated facilities is constrained to p. Constraint 3 (Eq. 4) sets the capacity constraints. Since the demand is expressed in people attending a market, summing $d_j x_{ij}$ and limiting it to c_i is adequate to express this restriction. Eq. 5 links both decision variables while the last three constraints (Eqs. 6, 7 and 8) set the decision variable bounds and the cost and demand non-negativity constraints.

Finally, the results of this optimization[3] can be visualized in Fig. 3 (see the library's examples folder[4] for detail on how the data was obtained and transformed). The result of minimizing the total amount of people assigned to each candidate facility yielded some interesting insights. First, adding a constraint that ensures all

[3] A web-based version of these results is available at https://www.observatoriodelima.org/mercados_itinerantes.

[4] https://github.com/EL-BID/urbanpy/tree/master/notebooks

wholesale and retail markets remain active, ensures that all candidate activations serve the purpose of redistribution the demand of the highest population hexagons.

Another point is that a couple of football pitches and parks were selected deep inside the district, which, though located within hard access zones, still represent a valuable addition to guarantee that at-risk population does not engage in long distance travel to acquire supplies for their quarantine.

Finally, these results encourage a large scale location of itinerary markets. Even though an optimal solution may be computationally challenging (there are about 13,000 candidate facilities and 3200 hexagons), an approximate solution using lagrange relaxation or genetic algorithms can be viable for a public policy approach.

4 Discussion

In this paper, UrbanPy was presented as a library to download, process and visualize high resolution urban data. It provides support for population, administrative boundaries and POI data download, regular polygon partitioning of administrative boundaries, spatial operations for aggregation and down-sampling, routing APIs, and visualization functions.

Within the case study, UrbanPy showed its critical role in providing data to the optimization model in the context of the COVID-19 pandemic. This was done by using all the available modules to construct the candidate and customer sets for the FLP as well as the cost matrix and the demand vector. This application also showed UrbanPy's seamless integration into a Pandas/Geopandas workflow, such as incorporating data from external sources to complement the optimization's constraints.

UrbanPy can still accommodate additional features. First, UrbanPy is currently shallow in down-sampling methods (such as k-ring smoothing). When merging data from sources larger than the hexagon's resolution, the approach for merging and downsampling (how data is assigned to each hexagon) is critical for coherent and consistent analysis. Within the plotting functions, currently UrbanPy has support for Plotly choropleth maps, while the standard hexagon output is compatible with Deckgl and Pydeck. To produce some of the figures shown in this paper, Geopandas and Pandas were harnessed for their plotting features, which by design is arguably the cleanest way to produce static plots. Still, there are avenues of support for other web-based visualizations such as Bokeh, D3, among others.

Finally, is important to note that, as mentioned, UrbanPy stemmed from a collaboration between the IDB, Universidad del Pacifico and the Municipality of Lima, which allowed for both an iterative design process and a quick and rapid application of the tool. For the tool to truly enable data-driven governance, in the context of COVID-19 as well as given the twin challenges of urbanization and rapid technological change, it is essential that UrbanPy be scaled up and tested in additional cities, among a similar diverse set of stakeholders, and for new policy questions, all while maintaining a balance between a user-friendly experience, high-performing code, and rapid-deployment experience.

References

Batty, M. (2008), 'The size, scale, and shape of cities', *science* 319(5864), 769–771.

Boeing, G. (2017), 'Osmnx: New methods for acquiring, constructing, analyzing, and visualizing complex street networks', *Computers, Environment and Urban Systems* 65, 126–139.

DOMO (2020), 'Data never sleeps 5.0', Available at https://web-assets.domo.com/blog/wp-content/uploads/2017/07/17_domo_data\-never\-sleeps\-5\-01.png. Accessed 10 July 2020.

Duranton, G. & Puga, D. (2014), The growth of cities, *in* 'Handbook of economic growth', Vol. 2, Elsevier, pp. 781–853.

GeoPandas developers (2020), 'Geopandas'. URL: https://geopandas.org

Glaeser, E. L. (1999), 'Learning in cities', *Journal of urban Economics* 46(2), 254–277. GSM (2018), 'The mobile economy 2018', *London: GSM Association.*

INEI (2018), 'Peru: Crecimiento y distribuci´ on de la poblaci´ on, 2017. primeros resul-´ tados'.

OECD. Publishing (2015), *The metropolitan century: Understanding urbanisation and its consequences*, OECD Publishing.

Townsend, A. & Zambrano-Barragan, P. (2020), 'Big urban data: A strategic guide for cities', *Washington, DC: Inter-American Development Bank.*

Uber (2020), 'H3: A hexagonal hierarchical geospatial indexing system'. URL: https://github.com/uber/H3

UN-Habitat (2016), 'World cities report', *Nairobi: UN-Habitat.*

Woetzel, J., Remes, J., Boland, B., Lv, K., Sinha, S., Strube, G., Means, J., Law, J., Cadena, A. & Von der Tann, V. (2018). 'Smart cities: Digital solutions for a more livable future', *McKinsey Global Institute: New York, NY, USA* pp. 1–152.

World Bank (2020), 'Databank – world development indicators', Available at https://databank.worldbank.org/source/world-development-indicators. Accessed 10 July 2020

Delivery Bay Location and Dimensioning for City Logistics Uses: An Interactive Modelling Approach

Andrés Regal-Ludowieg, Jesus Gonzalez-Feliu, and Michelle Rodríguez

Abstract This paper aims to present an interactive modelling and solving method for delivery bay location in Miraflores, Lima and its generalization for city logistics uses. An MIP model has been built following a co-construction approach involving researchers, experts and the municipality, then solved via Branch-and-Cut, is proposed. Several scenarios are analysed. Finally, generalizations are proposed.

Keywords City logistics · Delivery bay design · FTG · Facility location · Interactive modelling

1 Introduction

The design and location of urban delivery bay systems is an important issue in city planning. Indeed, public space, mainly for parking, is a rare resource, but loading and unloading operations, crucial for the city functioning, need to use that space. The lack of availability of such space leads to an under-optimization of delivery systems, which impacts also the entire urban transport system: search of parking space generates traffic (Magniol et al. 2018), double line parking increase congestion, and then a general cost increase but also pollution and global warming (Figliozzi and Tipagornwong 2017), etc.

For those reasons, anticipating needs and issues or urban logistics, and taking into account the various stakeholders involved, are crucial for designing and planning city logistics solutions (Gonzalez-Feliu 2018), from which delivery areas networks are an important component. Thus, the aim of this paper is to propose an interactive,

A. Regal-Ludowieg (✉) · M. Rodríguez
Facultad de Ingenieria, Universidad del Pacifico, Lima, Peru
e-mail: a.regalludowieg@up.edu.pe; rodriguez_m@up.edu.pe

J. Gonzalez-Feliu
Centre de Recherche en Innovation et Intelligence Managériales,
Excelia Business School, La Rochelle, France
e-mail: gonzalezfeliuj@excelia-group.com

J. Vargas Florez et al. (eds.), *Production and Operations Management*, Springer Proceedings in Mathematics & Statistics 391,
https://doi.org/10.1007/978-3-031-06862-1_35

systems thinking planning methodology to plan delivery area networks in a metropolitan district (Miraflores, Lima, Peru) and analyze the implications of its generalization. First, the related work and background on the field is examined. Then, the methodological framework, based on an active research-practice collaboration, is presented. It proposed an interactive problem solving methodology, that needs to be related to a practice field and involves various stakeholders. The main results of the methodology and both research and practice implications are finally presented and discussed.

2 Related Work

The design, location and management of urban delivery bays has lead an increasing attention in scientific literature over the years. Indeed, before the twenty-first century, works dealt mainly with personal mobility, and freight parking were either considered as marginal contributions to urban traffic or resulted on exploratory works about how they impacted passenger mobility (Habib 1981; Falcocchio et al. 1995). Only a small number of works tries either to analyze public policies in terms of freight transport parking (Topp 1991) or to model goods transport parking choices (Van Der Goot 1982). With the increasing interest of both scientific and practice communities in city logistics, the first decade of the twenty-first century sees a big interest on freight transport policy, but the question of delivery spaces remains marginal, although it starts to increase. It is after 2015 that we start to see a constructed body of works on the question, organized around 2 main subjects.

The first is that of analyzing the impacts of delivery space use, and double lines, on overall traffic, but not only as a negative impact for personal mobility. Indeed, those works focus on how a good use of public space for delivery purposes increase the performance of freight transport sector in urban areas, and contributes to an improvement of overall traffic. Main methods are related to simulation, such as queue theory (Figliozzi and Tipagornwong 2017), traffic micro-simulation (Magniol et al. 2018) or agent-based simulation (Kladeftiras and Antoniou 2013), or analytical models (Simoni and Claudel 2018), among others.

The second is that of designing and planning urban delivery bays. In that category, most works aim to develop methodologies (Aiura and Taniguchi 2005; Comi et al. 2017; Roca-Riu et al. 2017; Lopez et al. 2021; McDonald and Yuan 2021) for designing such spaces, or tools for planning and managing the operations and the use of such spaces.

In the design of such spaces, the location of areas is a major issue (Lopez et al. 2019). However, literature shows works dealing with a pure application of location problems which does not interact with pracitionners' needs and aims. To allow a higher continuity of proposed solutions, a more interactive vision of problem solving is preferred (Ackoff 1994) since it allows to take into account the systemic and specific issues of the modeled reality and rely on both scientific rigor and capability to respond to practitioner's needs. The present work is issued from an applied research project with the Municipality of Miraflores, Lima (Peru) and, needing to address the needs of city planners and at the same time improving city distribution operations of private stakeholders, interactive planning seems to be the most suitable approach to solve the problem addressed by the municipality. But to do that, a cyclic methodology needed to be deployed. It will be described in next section.

Methodology

The proposed methodology is based on the interactive planning (Ackoff, 1970; Ackoff 2016) and the optimization and problem solving cycle (Gonzalez-Feliu 2019) frameworks. They are based on the same principle, that of iteratively, and interactively, finding both the best representation of a decision problem and a solution to it (Ackoff 1977).

To tackle the delivery bay location as an application in Miraflores (Lima, Peru), the following methodology was deployed. First, exchanges with the Municipality on the aims an needs of the delivery bay location program were carried out. In parallel, a territorial characterization, starting from Regal et al. (2019) was made, examining data aggregations other than square grids. Exchanges with practitioners and a literature analysis on delivery practices in urban area where also carried out to complete the problem characterization. Finally, for the socioeconomic characteristics of the territory, a hexagonal division was chosen.

After that, a field quantification, mainly in terms of demand, was made, starting from the deployment of Freight Trip Generation (FTG) models (Gonzalez-Feliu and Sanchez-Diaz 2019; Holgu'ın-Veras et al. 2019) completing the socioeconomic characterization of the territory already made. After examining the different types of combinatorial optimization problems, a first decision problem was formulated. The core decision problem addressed by this methodology is where each delivery bay must be located. This is a traditional facility location problem (FLP), where different formulations such as set covering, p-median or set partitioning are well known in the literature (Daskin 2011). To approach this problem, some input data is needed: a set of candidate facilities (j), a set of customers (i), the demand at each customer (d), a cost matrix representing the cost of serving a customer from a facility (c_{ij}), and depending on the formulation, a coverage range.

Different formulations were considered. After three iterations of exchanges with various stakeholders (having the municipality as the central one), a standard formulation of the set covering problem was implemented (Daskin 2011). Given that the implementation cost of all delivery bays is the same, the set covering formulation can be represented mathematically as follows:

$$min \sum_{i=1}^{n} X_j \tag{1}$$

st.

$$\sum_{j=1}^{n} a_{ij} X_j \geq 1 \quad \forall i \in Customers \tag{2}$$

$$Xj \in \{0, 1\} \tag{3}$$

Where X_j is a binary variable representing which facility is active, and a_{ij} is a binary matrix that has values of 1 when the demand of customer i can be served within the cover range of facility j. This coverage range is not set as a single expected value by the municipality. To test how the solution varies across different coverage ranges, the same formulation was solved several times for values between 1 km and 5 km (Miraflores' diameter). All models are implemented using the Coin-or Branch and Cut (CBC) solver.

3 Results and Discussion

To solve the FLP, in collaboration with the Miraflores municipality, a database of all commercial establishments was collected using the operation licenses. With these, an FTG survey was used to estimate the demand as the number of delivery trips in a week. Figure 1 presents the spatial distribution of FTG rates, which has a mean of 175.84 trips per week and a standard deviation of 446.09.

With the estimated rates, the municipality developed a candidate set of potential delivery bay locations (see Fig. 2) that consists of 39 potential bays. The location of these bays is based on the FTG rates, commercial density and population density, prioritizing areas that perceive the highest influx of freight trips while having a set of bays in residential or tourist areas. This set of candidate locations is used as the X binary variable in the set covering formulation.

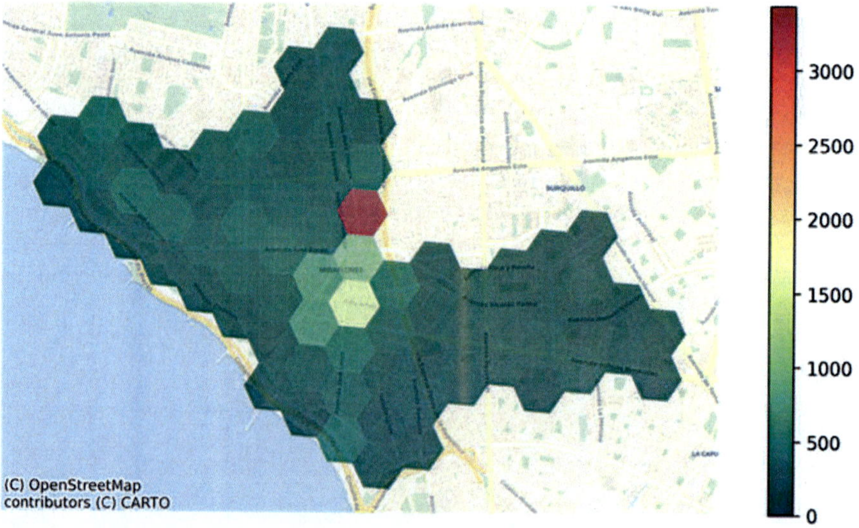

Fig. 1 Spatial FTG rate distribution within Miraflores

Fig. 2 Candidate bay locations

To estimate the customer availability matrix a_{ij}, a threshold distance was used to determine which delivery bays can serve a given customer. Since all bays have a geospatial component, the distance to the customers can be computed using the haversine distance as follows:

$$d = 2r \arcsin\left(\sqrt{\sin^2\left(\frac{\varphi_2 - \varphi_1}{2}\right) + \cos(\varphi_1)\cos(\varphi_2)\sin^2\left(\frac{\lambda_2 - \lambda_1}{2}\right)}\right) \quad (4)$$

Where r represents the earth's radius, ϕ the longitude of either point and λ the latitude. It is worth noting that for accurate computation, the input lat-lon pairs must be in radians, not sexagesimal degrees. Using this metric allows the distance computation to consider the earth's curvature and adjust it accordingly to have a better distance estimate.

With the bay-customer distance matrix, determining a_{ij} for a given threshold is as simple as evaluating which values of the distance matrix are less than or equal to the threshold. The quantitative result of changing this threshold, or coverage range, and its effect on the total number of facilities implemented is shown in Fig. 3.

As mentioned in the previous section, values for the coverage range were tested in the range between 1 km and 5 km. Even so, through our experimentation, any coverage size below 1.1 km results in an unfeasible model. This is, in part, due to the candidate set location, which would make the coverage of all customers under lower radii impossible, but it also reflects an important balance to the municipality's objective: the trade-off between bay location and pickup travel distance. Since walking 1.1 km is within the acceptable range set by the municipality, only this solution was acceptable from our experiments.

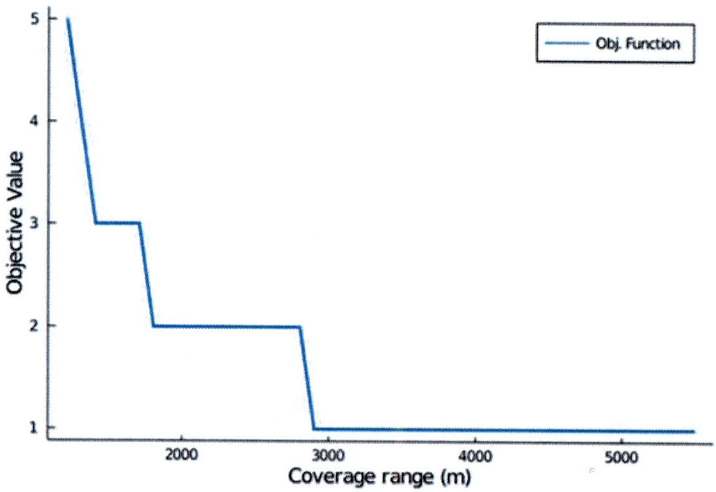

Fig. 3 Number of facilities activated given the coverage range

4 Conclusions and Future Research

In this paper, an optimization framework based on interactive modelling for delivery bay location was presented. A practical problem of locating a set of delivery bays in a metropolitan district of Lima (Peru) was solved, and an optimization model defined, solved, adjusted and re-proposed in order to obtain a solution able to respond to the various stakeholders' needs. Results show that data-driven and interactive approaches can help in identifying decision problems and their consequent solutions that both optimize and satisfy the stakeholders involved in practice. Through interactive planning and spatial analysis, the iteration of possible solutions to a facility location problem yielded a set of bay locations that balanced the coverage range of a bay and the acceptable walking distance by customers. Further developments see an evolution of the method to continue improving the decision problem and the adjusting of the consequent model.

Bibliography

R. Ackoff. Systems, messes and interactive planning. In E. Trist, F. Emery, and M. Hugh, editors, *The Social Engagement of Social Science, a Tavistock Anthology, Volume 3*, pages 417–438. University of Pennsylvania Press, 2016. R. L. Ackoff. *A concept of corporate planning*. Wiley, 1970.

R. L. Ackoff. Optimization+ objectivity = optout. *European Journal of Operational Research*, 1(1):1–7, 1977.

R. L. Ackoff. Systems thinking and thinking systems. *System Dynamics Review*, 10(2): 175–178, 1994.

N. Aiura and E. Taniguchi. Planning on-street loading-unloading spaces considering the behaviour of pickup-delivery vehicles. *Journal of the Eastern Asia Society for Transportation Studies*, 6: 2963–2974, 2005.

A. Comi, B. Buttarazzi, M. M. Schiraldi, R. Innarella, M. Varisco, and L. Rosati. Dynaload: a simulation framework for planning, managing and controlling urban delivery bays. *Transportation research procedia*, 22:335–344, 2017.

M. S. Daskin. *Network and discrete location: models, algorithms, and applications*. John Wiley & Sons, 2011.

J. Falcocchio, J. Darsin, and E. Prassas. An inquiry on the traffic congestion impacts of parking and pricing policies in the manhattan cbd. Technical report, City College of New York, University Transport Research Center, 1995.

M. Figliozzi and C. Tipagornwong. Impact of last mile parking availability on commercial vehicle costs and operations. *Supply Chain Forum: An International Journal*, 18(2):60–68, 2017.

J. Gonzalez-Feliu. *Sustainable Urban Logistics: Planning and Evaluation*. John Wiley & Sons, London, UK, 2018.

J. Gonzalez-Feliu. *Logistics and Transport Modeling in Urban Goods Movement*. IGI Global, Hershey, PA, 2019.

J. Gonzalez-Feliu and I. Sanchez-Diaz. The influence of aggregation level and category construction on estimation quality for freight trip generation models. *Transportation Research Part E: Logistics and Transportation Review*, 121: 134–148, 2019.

P. A. Habib. Curbside pickup and delivery operations and arterial traffic impacts. Technical report, Polytechnic Institute of New York, 1981.

J. Holguín-Veras, D. Ramirez-Rios, T. Encarnación, J. Gonzalez Feliu, E. Caspersen, C. Rivera-González, C. González-Calderón, and R. da Silva Lima. Metropolitan economies and the generation of freight and service activity: An international perspective. In M. Browne, S. Behrends, J. Woxenius, G. Giuliano, and J. Holguín-Veras, editors, *Urban Logistics. Management, Policy and Innovation in a Rapidly Changing Environment*, pages 19–51. Kogan Page, London, UK, 2019.

M. Kladeftiras and C. Antoniou. Simulation-based assessment of double-parking impacts on traffic and environmental conditions. *Transportation research record*, 2390(1):121–130, 2013.

C. Lopez, C.-L. Zhao, S. Magniol, N. Chiabaut, and L. Leclercq. Microscopic simulation of cruising for parking of trucks as a measure to manage freight loading zone. *Sustainability*, 11(5):1276, 2019.

C. Lopez, O. Rifki, and N. Chiabaut. Optimal freight loading zones: a graphtheoretic approach. In *9th Symposium of the European Association for Research in Transportation (hEART 2020)*, 2021.

S. Magniol, C. Lopez, J. Gonzalez-Feliu, N. Chiabaut, and L. Leclercq. The searching time to measure the freight loading zone accessibility using microscopic traffic simulation. In *Proceedings of the 7th International Conference on Information Systems, Logistics and Supply Chain, ILS 2018, Lyon, 8-11 July*, pages 406–414, 2018.

N. McDonald and Q. Yuan. Freight loading space provision: Evidence from the us. *Journal of Urban Planning and Development*, 147(2):04021015, 2021.

A. Regal, J. Gonzalez-Feliu, M. Rodriguez, and M. Juganaru-Mathieu. Defining urban logistics profile zones in south american metropolis by combining functional and spatial clustering techniques. In *2019 International Conference on Control, Automation and Diagnosis (ICCAD)*, pages 1–6. IEEE, 2019.

M. Roca-Riu, J. Cao, I. Dakic, and M. Menendez. Designing dynamic delivery parking spots in urban areas to reduce traffic disruptions. *Journal of Advanced Transportation*, 2017, 2017.

M. D. Simoni and C. G. Claudel. A simulation framework for modeling urban freight operations impacts on traffic networks. *Simulation Modelling Practice and Theory*, 86:36–54, 2018.

H. H. Topp. Parking policies in large cities in germany. *Transportation*, 18(1): 3–21, 1991.

D. Van Der Goot. A model to describe the choice of parking places. *Transportation Research Part A: General*, 16(2):109–115, 1982.

Instant Deliveries: A Latin America Overview

Leise Kelli Oliveira, Julio Castillo, Mario Chong, and Jackeline Murillo

Abstract This paper contributes with an overview of Brazil and Peru instant deliveries according to new forms of distributing different products focus on urban freight describing the customer profile and digital platforms consumers use. Customers currently demand new ways to receive or deliver products; thus, through mobile applications (app's), customers have tools to improve interaction with commerce or solve your necessities. Although technology helps improve quality of life, not knowing the current situation can be harmful in traffic and could increase safety issues. For these reasons, this research performed a product analysis, consumers demand and instant deliveries impact. A dataset of 558 consumers in Brazil and Peru were collected. This project proposes public policies and guidelines for regulating this emergence delivery services.

Keywords Instant delivery · Last-mile delivery · Digital market · Food delivery

1 Introduction

Internet access and the growing use of smartphones have created a new market. Currently, 139.1 and 24 million people have Internet access in Brazil and Peru, respectively (BlackSip 2019). Internet access supports an environment in which e-commerce has opportunities for increasing. Revenue from e-commerce in the year

L. K. Oliveira
Universidade Federal de Minas Gerais, Belo Horizonte, Brasil
e-mail: leise@etg.ufmg.br

J. Castillo
Faculty of Engineering, Universidad del Pacífico, Lima, Perú
e-mail: j.castillomorales@alum.up.edu.pe

M. Chong (✉) · J. Murillo
Universidad del Valle, Cali, Colombia
e-mail: m.chong@up.edu.pe; jackeline.murillo@correounivalle.edu.co

J. Vargas Florez et al. (eds.), *Production and Operations Management*, Springer Proceedings in Mathematics & Statistics 391,
https://doi.org/10.1007/978-3-031-06862-1_36

2018 in Peru was USD 791 million (BlackSip 2019), and in Brazil, it was USD 14.7 billion (Statista 2019).

Tech products help daily lives in terms of productivity, education system, purchasing capacity and healthy lifestyles. M-commerce, which allows shop anytime, anywhere, has gained more and more consumers in this context. Digital applications are one of the fastest ways to get desired products in a short time. The delivery generated by a digital application with a time window of 2 h is called instant deliveries (Dablanc et al. 2017). Understanding the instant delivery market and the challenges of this activity are still a challenge for researchers in the field.

In this way, this paper presents an overview of the instant deliveries in Peru and Brazil. A web-based survey was carried out to identify the type of products, behaviour,s and frequency of orders made using digital apps. The research questions intended to provide a picture of instant deliveries in two Latin-American cities.

- What is the customers profile for instant delivery?
- What are the leading companies placed in the instant delivery market?
- What are the challenges for regulating instant delivery?

2 Conceptual Framework

Instant delivery is defined as services that provide on-demand delivery within two hours – by either private individuals, independent contractors, or employees – by connecting shippers, couriers and consumers via a digital platform (Dablanc et al. 2017). The main challenges in this topic are focused on analyzing the delivery frequency and complexity of logistic activities.

In addition, M-commerce is defined as any transaction involving the transfer of ownership or rights to use goods and services, which is initiated or completed by using mobile access to computer-mediated networks with the help of an electronic device (Chib, M-Commerce 2011). Regarding the mobile transaction, the customer could access information about different products in any place with a mobile device. M-commerce offers value as products tracking, specific location according to your location, instant connectivity to take orders, and flexibility.

European Union Institutions have noticed the importance of last-mile deliveries related to online sales (Monganti et al. 2014). One reason for the customers to use applications is finding parking slots close to favourite stores. Parking is a significant issue in large urban areas, one that aggravates congestion and increases vehicles miles travelled (Holguín-Veras et al. 2015).

Although research has already been carried out about online shopping behaviours and their impact on delivery activities in urban areas, the literature about instant delivery services is still scarce, reported by Dablanc et al. (2017). The deliveries services (Rougès and Montreuil 2014; Shaheen et al. 2016), sharing economy (Shaheen et al. 2016; Briffaz and Darvey 2016), and job conditions (Dablanc et al. 2017; Dablanc and Saidi 2017) are the most cited topics in the literature.

Previous research was developed between December 2016 and March 2018 applying a survey in 100 bike couriers in Paris (Dablanc et al. 2017), with 35 questions about worker's situations, facts of the job, including characteristics of the platform, revenue, and the perception of the job including benefices and challenges. The principal findings were that the scooter increased by 16%, the use of a motor vehicle, including a scooter, requires a freight transport licence. Finally, delivery workers live outside Paris. Main findings show issues between safety and urban mobility as weakness, threat. Industries of low cost are increasing with a business model that is constantly adapting. The cities policies will need to adjust to minimize traffic, urban mobility, and sustainability.

With a similar approach, Oliveira et al. (2019) analyzed the profile of commercial motorcycle couriers in Belo Horizonte (Brazil). The results indicated the digital platforms increase the number of deliveries performed by these couriers in Belo Horizonte.

3 Methods and Procedures

A web-based survey was carried out to collect data about the behaviour of instant delivery consumers. The first version of the survey was conducted in Brazil in November 2018. Recently, the questionnaire was revised, and a survey was conducted in Peru. Table 1 presents the most relevant themes obtained from the questionnaire. A set of 578 responses were collected through an online survey, a distribution of 86% and 14% of the responses correspond to Brazil and Peru customers, respectively.

A descriptive statistics analysis was performed for understanding patterns regarding variables related to instant delivery services. Also, a linear regression model was developed for estimating the instant deliveries per week. The chi-squared test evaluated variables dependence between instant delivery orders, delivery fee and orders cost. The results are presented in the following section.

Table 1 Information to identify customers profiles of instant delivery services

Theme	Variables	Response type
Delivery information	Request instant delivery	Binary
	Products and frequency	Categorical (6 classes)
	Number of instant deliveries	Continuous
Delivery charge	Digital platforms	Categorical (13 classes)
	Maximum delivery charge	Continuous
	Maximum delivery charge by foot or bike	Continuous
	Value of product for no delivery charge	Continuous
Customer profile	Age	Categorical (5 classes)
	Income	Categorical (5 classes)

Regarding Chi-Square statistic, it is mainly used to evaluate tests of independence when using a cross of two categorical variables simultaneously. In this case, the independent variables are delivery fee and order cost, while the dependent variable is the number of orders. The Test of Independence assesses whether an association exists between the two variables by comparing the observed pattern of responses in the cells to the pattern expected if the variables were genuinely independent of each other.

$$x^2 = \sum \frac{(fo - fe)^2}{fe}$$

Where:

fo = the observed frequency (the observed counts in the cells)
fe = the expected frequency if NO relationship existed between the variables

Finally, the Chi-Square statistic is based on the difference between what is observed in the data and what would be expected if there was truly no relationship between the variables.

4 Results and Discussion

Respondents and instant delivery customers profiles are presented in Table 2. The profile of respondents using apps to order products by instant deliveries is different in both countries. In the case of Brazil, customers' profile is a person between 25 to 34 years old with a medium-high revenue level, which allows buying several categories of products. In contrast, Peruvian customers are mostly 24 years old,

Table 2 Profile of respondents (*only instant delivery customers)

Profile of respondents		Brazil	Peru
Total of respondents		478	100
Number of instant deliveries customers		244 (52%)	90 (90%)
Profile of instant delivery customers*			
Gender	Female	47%	50%
	Male	53%	50%
Age	15–24 years old	26%	42%
	25–34 years old	42%	27%
	35–49 years old	24%	29%
	> 50 years old	9%	2%
Income	Until $ 920	17%	29%
	$ 921 to $ 2,300	41%	33%
	$ 2,301 to $ 4,580	30%	23%
	> $ 4,581	12%	15%

Table 3 Use of digital platforms for ordering instant deliveries

Digital platforms	Brazil	Peru
Ifood	84.40%	0.00%
Uber eats	20.10%	13.93%
Rappi	18.90%	38.78%
Glovo	8.20%	26.53%
Social networks and WhatsApp	28.30%	5.44%
Digital apps from the establishment	30.30%	13.61%
Others	25.10%	30.20%

Table 4 Order delivery frequency by product type

Type of product	Country	Once by week	Once by month	Rarely	Never
Prepared food	Brazil	38%	49%	3%	0%
	Peru	51%	32%	14%	3%
Drugs	Brazil	4%	15%	28%	43%
	Peru	9%	11%	23%	57%
Home gas cylinder	Brazil	0%	15%	17%	67%
	Peru	5%	12%	10%	73%
Water gallon	Brazil	5%	9%	11%	76%
	Peru	5%	3%	9%	83%
Other beverages	Brazil	4%	14%	23%	60%
	Peru	9%	18%	26%	47%
Other products	Brazil	5%	13%	25%	58%
	Peru	15%	8%	18%	59%

low revenue level, and only buy a few products. As the sample is not representative, the data were analyzed individually for each country. The results cannot be evaluated in terms of order of importance but terms of the phenomenon's effect.

The respondents reported the use of different digital platforms for ordering goods. Table 3 presents the main platforms used by the country. Firstly, Rappi and Glovo are the principal deliveries company in Peru due to the variety of products and services offered. In addition, digital apps from their establishments represent a good share as the delivery cost is included in the final price of the order. In Brazil, the main digital platform is the Brazilian company, Ifood. Glovo started operations in Brazil in 2018 and stopped it one year later due to market competitiveness. The number of digital apps from establishments has decreased yearly due to the increment of the market from Ifood and Uber eats.

Analyzing the type of products and the frequency by which the products are ordered using apps, Table 4 shows that 71% of Peruvian consumers prefer prepared food orders two times a week. In all cases, considering the type of products related to prepared food, the vehicle's characteristics must be small. In Brazil, M-commerce is increasing daily, and prepared food is an emerging market for instant deliveries. Still, the frequency of products ordered by digital platforms presents some differences between Brazil and Peru.

Table 5 Average number of orders by month

Profile		Brazil		Peru	
		General orders	Food orders	General orders	Food orders
Gender	Female	1.25	1.56	1.99	2.00
	Male	1.13	1.75	2.00	2.05
Income	Until $ 920	0.90	1.63	1.49	1.61
	$ 921 to $ 2,300	1.25	1.50	2.21	2.35
	$ 2,301 to $ 4,580	1.15	1.90	1.61	1.65
	> $ 4,581	1.28	1.70	2.48	2.60
Age	15–24 years old	1.45	1.63	1.74	1.86
	25–34 years old	1.13	1.63	2.38	2.56
	35–49 years old	1.15	2.00	3.48	3.88
	> 50 years old	0.90	1.35	0	0

Table 6 Instant delivery order rate

Country	Independent variable	Coefficient	t-test	p-value	R^2
Brazil	Delivery fee	0.04706	4.553	7.4e-06	0.058
	Order cost	0.055236	11.89	<2e-16	0.296
Peru	Delivery fee	0.8958	7.124	4.33e-10	0.391
	Order cost	0.12894	7.139	4.06e-10	0.392

Table 7 Chi-squared test

Country	Dependent variable	Independent variable	Chi^2	df	p-value
Brazil	Number of orders	Delivery fee	544.92	285	0.00
		Order cost	487.1	315	0.00
Peru		Delivery fee	263.38	228	0.05
		Order cost	234.29	266	0.92

Table 5 presents the number of orders (general and prepared food) by month in each country, ddifferentiated by some profile characteristics. Orders rates in Peru are higher than in Brazil for both general and food orders. On average, males request more orders than females. The income of respondents that order products in Peru are higher than in Brazil. Also, the respondents between 35 and 39 years old order more frequently than other age range.

Results in Table 6 indicated that delivery fees contribute more than order costs to the number of instant delivery orders in Peru. The inverse situation occurs in Brazil. Both independent variables contribute positively to the number of instant delivery orders.

Table 7 presents the results of the chi-squared test to evaluate the dependence of attributes since the coefficient of determination (R2) has a low value. Results indicated the independence of attributes, except the by number of orders and order cost in Peru, showing that consumers order products by digital platforms. The delivery fee and order cost do not interfere with the number of weekly orders.

Finally, all results conclude there is a different profile of the customer behaviour of Brazil and Peru. The number of orders is influenced by delivery fees and order value. However, the number of orders, delivery fee, and order cost are independent variables. This result indicates that users of digital platforms made it by convenience. In this way, it is necessary to create public policies to regulate the instant delivery service. Otherwise, this activity may be making work marginalized by society, despite its inherent importance for this new market.

5 Conclusions and Future Research

Customer profile that uses apps in Brazil and Peru is different concerning the type of products that request. Brazil customers use the application to coordinate deliveries of different products, while Peru customers focus on food deliveries and groceries. Furthermore, both countries have several restaurants and grocery stores that are innovating and using apps that allow new forms of attending to customers to get a competitive advantage. In this contest, opportunities to increase instant delivery orders considering the easy access to app's via smartphones and the high quantity of restaurants and stores are real.

Recently, research showed the importance of analysing urban freight and emphasising the type of products consumers request, the scheduled delivery, including if they are done at peak hours and the quantities of transactions that consumers perform per week. Considering the high frequency of deliveries orders, we need to propose policies on Infrastructure and traffic management.

To restrict the use of multi-use lanes and propose the creation of lanes exclusively for bike transit or particular vehicles help us allow more efficiency on transport. In South America, m-commerce is increasing with orders of orders up to 3 times by week, so the use of exclusive truck lanes dedicated to deliveries will improve instant delivery.

Policies of vehicle size and weight restrictions are significant to create and upgrade intermodal infrastructure terminals and improve on/off street parking and loading for delivery products. According to the product type, the delivery unit must be adequate, allowing conserving the product with quality and safety.

In the future, we have plans to continue with the survey project to collect more data, including more countries, to understand more consumer behaviour regarding instant deliveries and develop comparisons, including a question based on the immediate need on order. This question can help us understand the necessity of urgent orders and implement various forms of off-hour delivery policies or pick-up delivery in alternate locations, such as lockers.

References

Blacksip, El e-commerce en Perú, Lima, June. (2019).

Briffaz, M., Darvey, C. Crowd-Shipping in Geneva – Exploratory and Descriptive Study of Crowd-Shipping. Master Thesis, University of Gothenburg, Sweden. (2016).

Chib, S. M-Commerce, Himalaya publishing house, New Delhi, India. (2011).

Dablanc, L., Morganti, E., Arvidsson, N., Woxenius, J.; Browne, M.; Saidi, N. The rise of on-demand 'Instant Deliveries' in European cities. Supply Chain Forum: An International Journal, 4, p. 203–2017. https://doi.org/10.1080/16258312.2017.1375375 (2017).

Dablanc, L., Saidi, N. Instant Deliveries: New Jobs and New Questions for Cities. Presented at Royal Geographical Society, London, United Kingdom, August 30. (2017).

Holguín-Veras, J., Jaller, M., Wang, X. Large urban freight traffic generators: Opportunities for city logistics initiatives. The journal of transport and land use Vol 8 pp 51–67. (2015).

Monganti, E., Dablanc, L., Blanquart, C., Lenz, B., Seidel, S. The impact of e-commerce on final deliveries: alternative parcel delivery services in France and Germany. Presented at International Scientific Conference on Mobility and Transport Sustainable Mobility in Metropolitan Regions (2014).

Oliveira, L. K., Oliveira, I. K., Nascimento, C. O. L., Cordeiro, C. H.O. L. Análise econométrica do perfil do motociclista profissional e a influência na produtividade das entregas de mercadorias e na segurança viária: um estudo para Belo Horizonte. In 33 Congresso Nacional de Ensino e Pesquisa em Transportes, Balneario Camboriu, 10–14 november 2019. (2019).

Rougès, J.-F., Montreuil, B. Crowdsourcing Delivery: New Interconnected Business Models to Reinvent Delivery." Presentation at the 1st International Physical Internet Conference, Quebec City, Canada. (2014).

Shaheen, S., Cohen, A., Zohdy, I. Shared Mobility: Current Practices and Guiding Principles. FHWA report. FHWA-HOP-16-022. Washington. (2016).

Statista. E-Commerce in Latin America. (2019).

Bundling Strategy Through Text Mining Tools

Miguel Rodriguez, Carlos Trujillo, Jonatán Rojas, Alessandro Huaman, Luigi Flores, Andrea Caballero, and Adrian Mendoza

Abstract Nowadays, there is a growing tendency to order food through apps for delivery, which are based on a free market system in which sporadic jobs are obtained and known as "gig economy" [1]. In Peru, the growth of that market was at a rate of 30% in 2019 [2] and, currently, these platforms have about 14,000 registered dealers with more than 1600 associated stores and an average of 20 daily orders served by each distributor [3]. This research seeks to improve the revenue of a food delivery company, by stimulating targeted partnerships with local businesses. A text mining analysis and a bag-of-words model were developed to get a bundling strategy that helps encourage consumers to purchase this combination, selecting those that have the biggest lift.

Keywords Bundling strategy · Food delivery applications · Text mining · Bag-of-words model · Turnover increase

1 Introduction

Food delivery applications (FDAs) have emerged as "new delivery" players since their first appearance in 2013. They allow users to contrast menus, post reviews, and order from a wide range of restaurants with ease [4]. Once users sign up, it is proved that 80 percent never or rarely left for another application, which leads to a strong winner-take-all dynamic. In addition, they manage their distribution network through the outsourcing of drivers whose sporadic job is known as part of a market system called "gig economy". These platforms are retributed by restaurants with a fixed margin of each order, while customers pay a flat fee.

M. Rodriguez · C. Trujillo · J. Rojas (✉) · A. Huaman · L. Flores · A. Caballero · A. Mendoza
Pontificia Universidad Católica del Perú, Lima, Perú
e-mail: miguel.rodrigueza@pucp.edu.pe; carlos.trujillod@pucp.pe; jrojasp@pucp.pe; alessandro.huaman@pucp.pe; luigi.flores@pucp.pe; andrea.caballero@pucp.edu.pe; adrian.mendoza@pucp.edu.pe

J. Vargas Florez et al. (eds.), *Production and Operations Management*, Springer Proceedings in Mathematics & Statistics 391,
https://doi.org/10.1007/978-3-031-06862-1_37

New-delivery platforms personalize the experience by storing significant customer data. Information quality has been found to positively influence continuous use intention due to its connection with performance expectancy [5]. Thus, providing accurate and reliable information in a detailed manner via an adequate app design will lead consumers to perceive delivery apps as useful.

Currently, there are more than a dozen FDAs in Latin America and they are all seeking a slice of market share. Consumer preferences and digital behavior differ greatly across the region as, for instance, 20% of the orders are paid in cash so that companies ought to adapt to local payment preferences. Also, to scale their operations effectively, there are many barriers to overcome like the lack of address standardization and general business information in less-developed areas [6].

Despite that, situation in Lima, Peru looks promising as sales have reached a yearly value of 12.15 million dollars and 61 percent of customers spend more than 15 dollars per use [7]. FDA's have evolved into "anything" delivery applications and that has generated many opportunities to take advantage of the data gathered by their CRM system, which is a technology used for managing relationships with their current customer portfolio and potential acquisition. Therefore, bundling, the practice of including two or more products within a separate product package [8], could be applied as part of product strategy for an increase in their turnover.

The main objective of this research is to examine the feasibility of mixed bundling to understand its dynamic effects with complementary products, so consumers would be able to obtain additional utility from having both items. Consequently, natural language processing (NLP) and text mining tools are proper techniques to be used so that statistical analysis can be carried out.

2 Literature Review

The state of the art and the conceptual definitions of the main tools that will support this study will be detailed:

2.1 *Natural Language Processing (NLP)*

Natural Language Processing is a field of Artificial Intelligence that gives ma-chines the ability to read, understand, and derive meaning from human languages [9]. The main techniques used to fulfill its tasks are related to text normalization through syntactic and semantic analysis to transform it into a uniform sequence. They can be described as follows:

Bag-of-Words It is a representation of text that describes the occurrence of words within a document after its tokenization, which is executed by separating it into pieces [10]. Therefore, any information about the order or structure of words in the

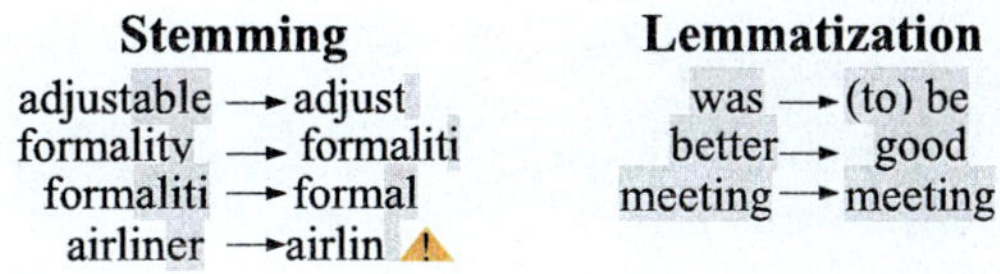

Fig. 1 Difference between Lemmatization and Stemming

document is not considered and this approach shows us the histogram of the words within the text as they are treated as a feature on their own [11]. Nevertheless, it may imply some disadvantages because stop-words add noise to the results.

Regular Expressions (REGEX) A Regular Expression is an instruction that describes a search pattern that can be used to identify what and how to match or replace patterns inside a set of strings efficiently [12]. They are commonly used for punctuation marks, stop-words, and whitespaces removal.

Lemmatization and Stemming The goal of both techniques is to reduce inflectional forms and sometimes derivationally related forms of a word to a common base form. Lemmatization aims to remove inflectional endings only and to return the base or dictionary form of a word, which is known as the lemma, by running a full morphological analysis. On the other hand, Stemming refers to a heuristic process that chops off the ends of words in search of achieving this goal correctly most of the time and often includes the removal of derivational affixes [13]. In order to illustrate the difference between them, Fig. 1 is displayed.

2.2 *Text Mining*

Text mining is a variation in the field of data mining that refers generally to the process of extracting interesting and non-trivial information and knowledge from unstructured text. It is also known as Knowledge-Discovery in Text (KDT) and it is believed to have a high commercial potential value as most information (over 80%) is stored as text. Thus, it is deeply rooted in NLP, but it also relies on methods from statistics, machine learning, reasoning, information extraction, knowledge management, and others for its discovery process [14]. The main tools that will be applied in this research are the following:

TF-IDF (Term Frequency-Inverse Document Frequency) It is a statistical measure that evaluates how relevant a word is to a document in a collection of documents. Its calculation is done by multiplying two metrics: how many times a word appears in a document and the inverse document frequency as the logarithm of the total number of documents divided by the number of documents that contain the word [15]. On that account, if the word is common, IDF would approach zero. The TF-IDF score for the word t in document d from the document set D is calculated as follows:

$$tfidf\ (t, d, D) = \log(1 + freq(t, d)) * \log(N/count(d\epsilon D : t\epsilon d))$$

This method is useful for extracting keywords from the text as the score straightforwardly reflects each word's importance.

Topic Modelling Topic modeling can be described as a method for finding a group of words (known as topics) from a collection of documents that best represents the underlying information in the collection [16]. The key to this summarization is to reduce the length and detail of a document while retaining its main points and overall meaning. To achieve that, a widely used technique is the Latent Dirichlet Allocation (LDA) model, which was developed by Prof. David M. Blei in 2003. LDA model says that each word in each document comes from a topic and the topic is selected from a per-document distribution over topics. Then, the probability of a word given document (assuming conditional independence) equals:

$$p(d) = \sum\nolimits_{t \in T} p(w \mid t) p(t \mid d) \tag{2}$$

p (t|d) = *probability distribution* of *topics in documents*
p (w|t) = *probability distribution* of *words in topics*

This can be thought similar to a matrix factorization, where the distribution of words in a document is decomposed into two matrices. To maximize the likelihood of the data given these two matrices, an algorithm called Gibb's sampling is used to identify the correct weights. The result obtained due to its application is a conditional probability equation that is expressed like this for a word w in document d that belongs to topic k:

$$p(z_{d,n}, w, \alpha, \lambda) = \frac{n_{d,k} + \alpha_k}{\sum n_i d,\ i + \alpha_i} \frac{v_{k,w_{d,n}} + \lambda_{w_{d,n}}}{\sum_i v_{k,i} + \lambda_{w_{d,n}}} \tag{3}$$

nd,k = *number* of *times document d uses topic k*
vk,w = *number* of *times topic k uses given word*
αk = *Dirichlet* parameter for *document to topic distribution*
λ*w* = *Dirichlet* parameter for *topic to word distribution*

For each word, a vector of probabilities will be created to explain how likely this word belongs to each topic. Consequently, the algorithm randomly picks any of the topic and assigns them to a word. These steps are repeated for all other words as well. Intuitively, a topic with the highest conditional probability should be selected.

3 Case Study

Along this section, the employed methodology for the bundling strategy analysis will be set forth.

3.1 Data Extraction

As it was mentioned before, it was possible to obtain a data set from the CRM system owned by the leading FDA in Lima, Peru. Due to its volume and semi-structured organization because of customers' freedom to write anything they desire, a pre-processing phase was needed so that the capability of using the text mining tools, as they were previously defined, was ensured. The main features included were the following: date, order description, pick-up point address, delivery point address, and revenue per purchase. In addition, it had over 150,000 instances and a cyclical pattern could be further analyzed as the orders were gathered during the Peruvian summer season, between January and March.

3.2 Text Pre-processing

At first, it was necessary to apply several techniques before the modeling stage. Tokenizing was executed for separating words into pieces.

A stable and scalable data analysis pipeline was necessary for the iterative creation and improvement of our solution. For this purpose, we relied on the Python [17] programming language and Stanford NLP package [18], which contains useful functions and object classes that allow the rapid application of common NLP processing steps. The final text preprocessing pipeline is composed of 3 steps:

Undesired Character Cleaning White spaces and punctuation marks were removed from every row of our dataset. The application of textual pattern recognition for character location and replacement was made possible by regular expression packages inside the Python ecosystem.

Text Normalization Applied for scalable conversion of Spanish words as language primitives and reduction of inflectional and derived forms. For this purpose, we applied tokenization and lemmatizing capabilities of the Stanford NLP package to every row of our dataset. Lemmatization was made possible as this package leverages the power of pretrained Recurrent Neural Networks in Spanish corpus. Words that could not be accurately lemmatized were further preprocessed with stemming.

Tokenization Executed to extract features from the text as unigrams (unique words) and bigrams (adjacent word pairs).

Fig. 2 Word cloud model

The word cloud model shown in Fig. 2 was executed over the preprocessed dataset to highlight popular products and assess the quality of normalized tokens. It is shown that the applied preprocessing steps assure the adequate generation of meaningful tokens for further NLP modeling.

3.3 Modeling

A bag-of-words model was developed to be aware of the occurrence of each token in the dataset as part of a statistical measure of its relevance. That measure was complemented by the application of TF-IDF in search of a reliable ranking as it also considers their inter-document uniqueness inside the dataset. Therefore, keywords could be extracted preliminarily and important tokens for each topic were highlighted. Later, topic modeling was carried out through the Latent Dirichlet Allocation (LDA) algorithm [19] and it was considered that, at least, 10 topics would be able to explain the whole sample. Obtained results are is in Fig. 2.

In Fig. 3, the 10 topics generated after applying topic modeling techniques are shown. There are 2 topics highly differentiated from the others which are topics 1 and 2. Among the remaining groups, topic 5 is slightly separated from the rest. The remaining topics are very closed to each other, therefore a bundle resulting from them could not be accurate enough. Given these results, the next step is analyzing the most frequent and relevant terms in topics 1, 2, and 5 because these would be more prone to target groups with certain preferences.

The distribution of items in the orders of each group is shown in Fig. 4, where the light blue bars represent the token popularity across all orders and the red bar

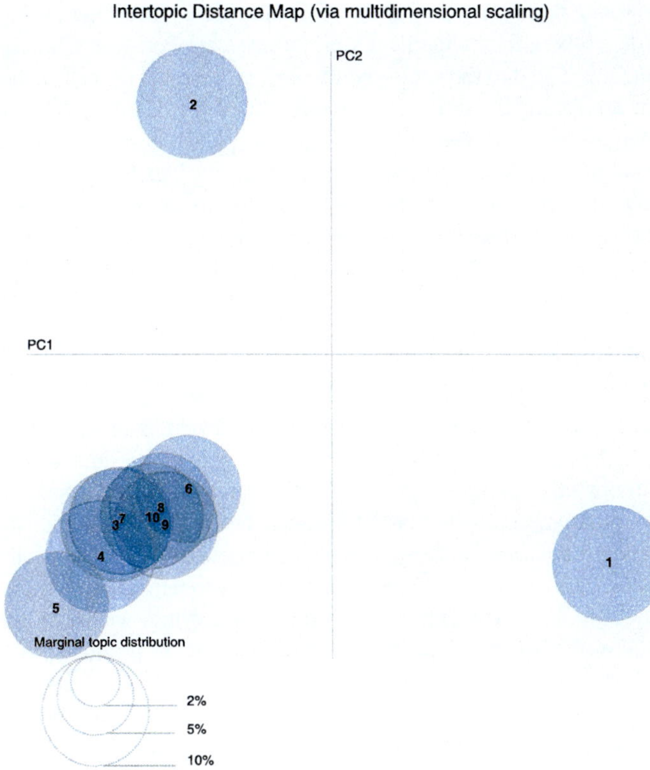

Fig. 3 Topic representation from terms in orders from the delivery app

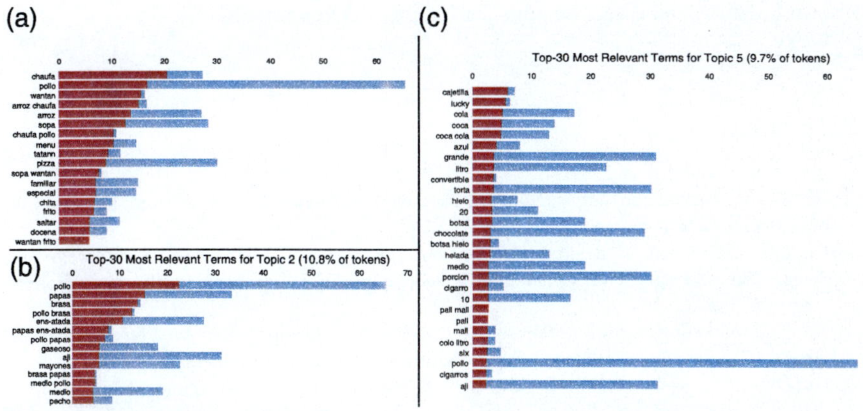

Fig. 4 Top 30 most relevant terms for the topic of interest 1, 2, and 5

indicates the token popularity only in the topic specified. According to Fig. 4a, topic 1 gathers orders where the main client's preferences are Peruvian Chinese food.

Several dishes related to those offered by those restaurants as well as their various presentations are among the most relevant terms for this first group. The next group, topic 2, shown in Fig. 4b, terms related to a specific presentation of "pollo a la brasa" (Peruvian rotisserie chicken) and their complements such as French fries, salads, or sodas. Finally, topic 5 in Fig. 4c, indicates the predominance of cigarettes and sodas, items that can be ordered by people in a social meeting or small gathering.

4 Conclusions

Thanks to this study the company can improve its revenues through an intelligent partnership with local businesses for the satisfaction of product demands. It is mandatory to prioritize partnerships with local Peruvian Chinese restaurants names "chifas" and popular Peruvian roasted chicken restaurants that offer "pollo a la brasa", a typical Peruvian food. Both of them represent 20% of the total demand by the channel. The company is losing money if customers request these kinds of products through the "anything" app button, specifying whatever they desire because if there was a specialized button for these specific products, the company could take 30% of the total billing. An additional benefit is the assurance of the quality and timeframe of the delivery of these popular products, meaning an increase in a competitive advantage for the company. The next step would be the geographic localization of large unsatisfied demand volumes for "chifas" and "pollo a la brasa", along with the timestamp analysis for the revelation of seasonalities. With these approaches, the proactive search of local businesses that can satisfy customers could grant immediate and direct benefits for business development.

References

1. Iglesias, M. What is the "gig economy"?. Retrieved on August 28th, 2019 from: https://www.bbva.com/en/whatis-gig-economy/ (2018).
2. Diario La República. Mercado de delivery por aplicativo creció 30% en Lima. Retrieved on August 28th, 2019 from: https://larepublica.pe/economia/1233400-mercadodelivery-aplicativo-crecio-30-lima/ (2019).
3. Arellano Marketing. Apps de delivery: Una lucha a toda velocidad. Retrieved on August 28th, 2019 from: https://www.arellano.pe/apps-delivery-una-lucha-toda-velocidad/ (2019).
4. Hirchberg, C., Rajko, A., Schumacher, T. & Wrulich, M. The changing market for food delivery. Retrieved on August 28th, 2019 from: https://www.mckinsey.com/industries/technology-media-and-telecommunications/our-insights/the-changing-market-for-food-delivery (2016).
5. Lee, S., Sung, H. & Jeon, H. Determinants of Continuous Intention on Food Delivery Apps: Extending UTAUT2 with Information Quality. Retrieved on August 28th, 2019 from: https://www.mdpi.com/2071-1050/11/11/3141/pdf (2019).

6. Michener, G. The biggest food delivery apps are moving into Latin America. Retrieved on August 28th, 2019 from: https://www.idgconnect.com/opinion/1503702/biggest-food-delivery-apps-moving-latinamerica (2019).
7. Peru Retail. Perú: Ventas de comida por delivery mueven alrededor de S/400 millones al año. Retrieved on August 28th, 2019 from: https://www.peru-retail.com/peruventas-delivery-mueven-alrededor-de-s-400-millones/ (2019).
8. Derdenger, T. & Kumar, V. The Dynamic Effects of Bundling as a Product Strategy. Retrieved on August 28th, 2019 from: https://www.hbs.edu/faculty/Publication%20Files/dynamicbundling_2013_06_20_13def9fb-1904-479a8a37-fa64eff9663a.pdf (2013).
9. Rabby, S. What is NLP & What Do NLP Scientists Do?. Retrieved on August 28th, 2019 from: https://towardsdatascience.com/whatnlpscientistsdo-905aa987c5c0 (2019).
10. Cambridge University Press. Tokenization. Retrieved on August 28th, 2019 from: https://nlp.stanford.edu/IRbook/html/htmledition/tokenization-1.html (2008).
11. Brownlee, J. A Gentle Introduction to the Bag-of-Words Model. Retrieved on August 28th, 2019 from: https://machinelearningmastery.com/gentle-introductionbag-words-model/ (2019).
12. Van den Rul, C. [NLP] Basics: Understanding Regular Expressions. Retrieved on August 28th, 2019 from: https://towardsdatascience.com/nlp-basics-understanding-regular-expressions-fc7c7746bc70 (2019).
13. Cambridge University Press. Stemming and Lemmatization. Retrieved on August 28th, 2019 from: https://nlp.stanford.edu/IR-book/html/htmledition/stemming-andlemmatization-1.html (2008).
14. Lehal, G. A Survey of Text Mining Techniques and Applications. Retrieved on August 28th, 2019 from: https://www.researchgate.net/publication/42802926_A_Survey_of_Text_Mining_Techniques_and_Applications (2009).
15. Stecanella, B. What is TF-IDF?. Retrieved on August 28th, 2019 from: https://monkeylearn.com/blog/whatis-tf-idf/ (2019).
16. Tomar, A. Topic modeling using Latent Dirichlet Allocation (LDA) and Gibbs Sampling explained. Retrieved on August 28th, 2019 from: https://medium.com/analytics-vidhya/topic-modeling-using-lda-and-gibbs-sampling-explained-49d49b3d1045 (2018).
17. Python Software Foundation. Python Language Reference, version 2.7. Retrieved on Au-gust 28th, 2019 from: http://www.python.org (2019).
18. Qi, P., Dozat, T., Zhang, Y. & Manning, C. Universal Dependency Parsing from Scratch In Proceedings of the CoNLL 2018 Shared Task: Multilingual Parsing from Raw Text to Universal Dependencies, pp. 160–170. Retrieved on August 28th, 2019 from: https://nlp.stanford.edu/pubs/qi2018universal.pdf (2018).
19. Sievert, C. & Shirley, K. LDAvis: A method for visualizing and interpreting topics. Retrieved on August 28th, 2019 from: https://nlp.stanford.edu/events/illvi2014/papers/sievert-illvi2014.pdf (2014).

Machine Learning Applied to Last Mile Operations: Applying Machine Learning Models for Stops Classification in Urban Logistics

Bernardo Puente-Mejia, Carlos Suárez-Núñez, David Calahorrano, Martin Gavilanes, and Daniel Masaquiza

Abstract This study analyzes urban freight GPS data to achieve more efficient public policies and better last-mile operations planning. The main objective is to obtain the time stops of each urban freight vehicle and classify the stop as traffic, delivery, or rest. After classifying each stop using data mining algorithms and applying machine learning models such as K-Means and HDBSCAN, the study demonstrates that it is possible to apply machine learning models capable of grouping GPS stops by similar features. The clusters obtained are relevant inputs to future research and potential applications, such as in public policy decision making or routing optimization allowing private and public entities to optimize urban logistics and last-mile operations.

Keywords Machine learning · Urban logistics · Last mile operations · Geospatial data analysis · K-means · HDBSCAN · Cluster

1 First Section

According to the United Nations, nowadays, more than half of the world's population lives or works in urban areas of cities worldwide, and predictions show that it will reach a top of 70% in the following 30 years [1]. Thus, the efficient movement of daily consumption products such as groceries, industrial supplies, and other staples of modern life is critical to reach competitiveness and develop urban economies [2].

GPS data has been used to observe patterns, identify possible solutions and aim for different and better road systems in traffic and transportation management [4]. Likewise, GPS data has gained popularity when measuring freight performance,

B. Puente-Mejia (✉) · C. Suárez-Núñez · D. Calahorrano · M. Gavilanes · D. Masaquiza
Universidad San Francisco de Quito USFQ, Colegio de Ciencias e Ingenierías, Departamento de Ingeniería Industrial and Instituto de Innovación en Productividad y Logística CATENA-USFQ, Quito, Ecuador
e-mail: bpuentem@usfq.edu.ec; csuarez@usfq.edu.ec

J. Vargas Florez et al. (eds.), *Production and Operations Management*, Springer Proceedings in Mathematics & Statistics 391,
https://doi.org/10.1007/978-3-031-06862-1_38

as it presents advantages over traditional measuring tools such as traffic cameras, loop detectors, and others [5]. However, GPS data has limitations, like spatial inaccuracy caused by urban canyons or misleading information caused by signal loss [5]. Furthermore, [6] states that access to large GPS databases could benefit emerging countries where data collection is rarely common.

The increase in logistics activities and the lack of geospatial data has created the necessity of finding better solutions for both public and private sectors [7]. Ecuador is one of the countries developing a logistics observatory [8] as it has found the need to create different entities to deal with all logistics aspects caused by urban population increase, creating an unsustainable situation for city vehicular mobility [9]. At the same time, various typical sets of policies have been applied, hoping to mitigate and control mobility congestion [10]. However, these policies, including vehicle access and mobility restrictions, and other regulations, are not often wholly evaluated before implementation [11]. Consequently, inefficient mobility inside urban areas is created, and commercial strategies are affected [12]. These consequences include the need to develop sustainable public mobility plans and other different commercial strategy for companies [1]. Under these circumstances, an initiative to collect private freight truck GPS information has been established in Quito by an academic institution. These efforts have allowed to collect GPS data of around 100 urban freight vehicles for a period of over a year, with a total of approximately 4.5 million records per month available for this study. The data was downloaded monthly by means of an automated Python code which access the vehicle GPS data repository. In addition, the data obtained has 12 common features; the most important used in our study are date and hour, longitude, latitude, velocity, and car label.

This research proposes applying machine learning models able to identify and classify each vehicle stop from GPS urban freight data. Second, presenting results that showcase the different stops classification as an important input for future potential applications like public mobility policy and give valuable insights for private companies and help them modify commercial strategies that will make urban logistics more efficient [13]. Finally, the analysis of the stops obtained with other calculations such as variation through the week, schedule congestion over time or geographic area, and others will demonstrate the potential value of geospatial data analytics for different diagnoses of urban logistics and planning, which is important from the perspective of a city for public policy [12].

2 Related Works

2.1 Global Positioning Systems

Since the creation of Global Positioning System (GPS) in the 70s, this device has been used for many applications including transportation [14]. GPS devices record information such as latitude, longitude, speed, heading, altitude, engine status, etc. In addition, GPS allows different settings, such as record frequency and accuracy

[5]. Data acquisition changed exponentially with the availability of the GPS devices allowing to measure traffic data as records in sequences of discrete positioning signals transmitted by the cars' GPS [4]. On the other hand, drawbacks of collecting GPS data may be: the uncertainty and bias generated by imprecise and inaccurate instruments; GPS signals could output less accurate information due to the erroneous calibration or high-rise buildings or structures, especially in big cities [15]. Despite this, GPS data opens an opportunity for extensive analysis between geospatial and other data sources [16].

2.2 *Vehicle Stops Identification*

Even though a massive volume of GPS data collected is available, many companies use GPS devices just for vehicle monitoring [4] and the use of geospatial data to obtain valuable metrics and generate different insights is limited [11].

From freight trucks and its GPS data, [12] developed an algorithm by analyzing the difference between consecutive GPS location points to determine if a vehicle stopped [12]. The study proposes a parameter of 240 s as a correct threshold to flag a group of records as a stop and the geographic distance between locations of a vehicle at consecutive geospatial points of 6 m. Similarly, [6] uses a Support Vector Machine (SVM) model to identify delivery and non-delivery stops applied to GPS datasets in New York. Yang uses stop duration and distance from a stop to the city's center as classification features. It was noticed that the delivery time has a variation between 10 and 20 min. Whereas [3] propose an algorithm able to identify delivery stops if and only if trucks' dwell time exceeds 180 s. Equally important, small truck movements were not reported because it was considered a GPS fluctuation when a vehicle idles. Consequently, it was decided that the record was considered a stop only if the distance between two consecutive data points was less than 20 m [3].

2.3 *K-Means and Hierarchical Density Based Spatial Clustering Application with Noise HDBSCAN*

Clustering of unlabeled data, such as the classification of vehicle stops, can be performed with a clustering algorithm such as K-means. The model works efficiently with flat geometry variables, and it allows for large scalability. K-means algorithm has the main objective of finding optimal solutions between the record and the clustering by associating a number of records to a specific cluster and grouping them between others placed within the established Euclidean distance [17].

The Hierarchical Density-based spatial clustering of applications with noise (HDBSCAN) model was developed as a significant update to DBSCAN (Density-based spatial clustering of applications with noise) model [25]. HDBSCAN's main objective is to create a flat cluster by grouping points from a distance from a centroid to the border of the cluster [19]. The points will be grouped in the cluster if and only if they share common features, if not, those will automatically fall in another cluster where its features are shared in common [20].

3 Methodology

A hybrid combination between [21] Methodology and CRISP-DM methodology is established for the algorithm development. According to [21], the first step is to define every general tool that could be applied for the algorithm development. For this study, the tools found and classified were the geospatial datasets, coding literature and software for data cleaning and code construction.

In the other hand, CRISP-DM methodology analyzes the use of data mining techniques in order to build a machine learning model for a specific purpose [22]. This methodology groups every task to be done into six concatenated phases: Business Understanding, Data Understanding, Data Preparation, Modelling, Model Evaluation and Deployment. As a result, the following methodology shown in Fig. 1 is proposed to achieve the objective of the study.

3.1 Exploratory Data Analysis

The first step is to understand and analyze the different characteristics of the GPS data-sets obtained using data analytics tools [23]. Then, the Exploratory Data Analysis (EDA) is started, for the present study five geospatial data-sets are studied. Each data-set belongs to a specific month from September 2019 to January 2020 giving us a total of 25,475,082 records.

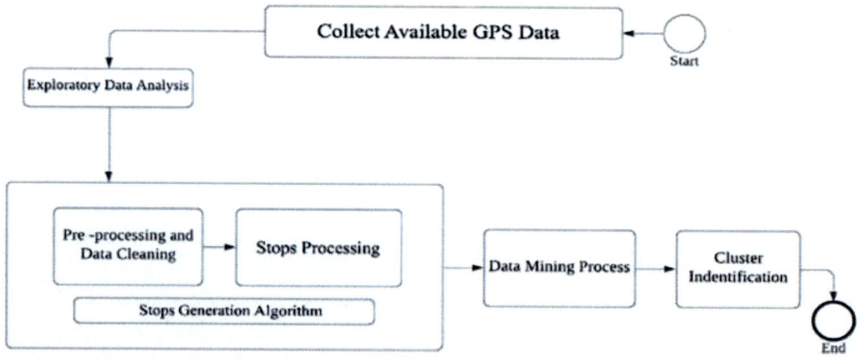

Fig. 1 Proposed methodology for the study

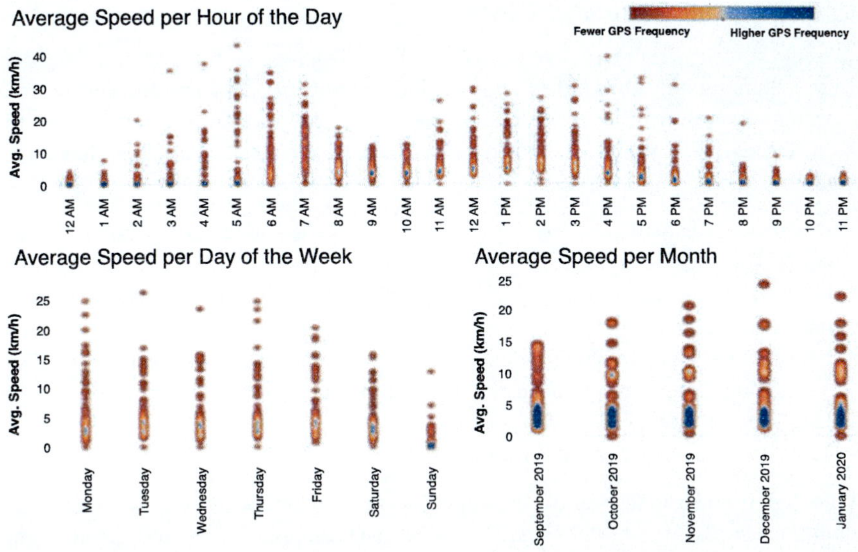

Fig. 2 Analysis of the key variable [Speed]

The 25 million records show an average standard deviation of 3462 records per week. In the different GPS data-sets, a record is registered on average, every 12 s with a maximum of 14.7 s between 6 pm and 6 am, and a minimum of 8.3 s between 6 am and 10 am. The next step in the algorithm development is the analysis of the key variable and other target variables, as follows [1]:

- Vehicle: Car_ID (Total of 83 vehicles in the fleet).
- Date-Hour: Date and hour of record collected.
- Latitude: Degrees from 0° of the equator to 90°N (+90°) of the North Pole or 90°S (−90°) of the South Pole.
- Longitude: Degrees from 0° meridian to 180°E (+180°) to the east and 180°W (−180°) to the west.
- Vehicle Velocity: Velocity of the vehicle [km/h]

The aggregation of average velocities by Hour of the Day, Day of the Week and Day of the Month was done. Then, the analysis of the key variable gives as a result that between 7 pm and 5 am most of the variables are zero, which is expected since there are no deliveries outside working hours. Also, the average velocity per vehicle decreases in the most concurred hours of the day. This daily behavior is similar through all the studied months as shown in Fig. 2.

3.2 Pre-processing and Cleaning

Before defining the algorithm and model development, data pre-processing and cleaning steps are needed in order to guarantee that only useful information gets to the next stage [1]. Therefore, the following group of concatenated steps has been applied:

1. Delete records with missing data since it is labeled as a GPS' signal lost or variability.
2. Drop records with negative velocity and with latitude and longitude values out of range.
3. Records that have a velocity value less than or equal to 14 km/h are assumed as 0 km/h and considered as a short vehicle stop [6].
4. Duplicate data records are dropped so a unique record is kept.
5. Velocity values are rounded to the nearest integer.

After the pre-processing and cleaning procedure was completed, we identified multiple records with GPS' signal lost. This was identified specially at midnight when a record was recorded with a velocity greater than zero whereas previous and following records showed a velocity of zero. Thus, these records were labeled as zero km/hr. Finally, by the Tukey interquartile range method allowed to detect and eliminate outliers for the upper boundary [24].

Once completed, the following step was to transform the velocity variable data type from a float variable to an integer variable to make the algorithm processing more efficient [18]. To have a smoother method for this variable, each vehicle was processed independently by creating a single data set for every vehicle.

3.3 GPS Stops Generation Algorithm

The first step towards vehicle stop classification is processing raw GPS data points into stops. Since GPS raw files only provide information related to location and velocity, stops need to be obtained from the underlying behavior of trucks. The proposed algorithm follows a 5-step process in which vehicle stops data points are separated and consolidated into a different data set.

1. The first correction step is to measure the distance between GPS data points. If the distance between two points is less or equals than 20 m, then the velocity is corrected to zero and the position (latitude and longitude) of the record is updated to the previous record's position [26]).
2. All data-set records are sorted from oldest to newest. Then the algorithm evaluates the vehicle's velocity feature. When the velocity of the record (n) and the record ($n + 1$) is zero, for the record ($n + 1$) the algorithm starts the variable K that calculates the time difference between the timestamp of the record ($n + 1$) and the record (n).

3. The algorithm generates the variable L and accumulates the time difference between records as long as the velocity remains 0 kilometers per hour.
4. Once the velocity of the record ($n > =2$) is greater than zero the algorithm generates the variable M in which a vehicle stop label is assigned to the previous record. It is important to notice that the record with the stop label also has the variable L which accumulated the time the vehicle remained stopped.
5. After each stop label generated, the variables K and L reset to zero and the loop runs for all the remaining points in the data set.

After algorithm 1 run through all the GPS data points for each vehicle independently, only the records with a stop label are kept. The new dataset has the following features: Vehicle ID, timestamp when the stop started, timestamp when the stop ended, stop position (Latitude and Longitude), and total time the vehicle remained stopped. These new features will be needed as the input for the stop classification model.

Algorithm 1: GPS Stops Generation

```
Require: Sort all points from oldest to newest
  for All n in Data-Set do                          ▷ GPS noise
correction
     if Distance between points n, n + 1 ≤ 20m then
      Speed_{n + 1} = 0
      (lat, long)_{n + 1} = (lat, long)_n
    end if
Next n
end for
for All n in Data-Set do                          ▷ Stops Generation
L = 0                               ▷ Stop time accumulating variable
  if Speed_n and Speed_{n + 1} = 0 then
    while Speed_n and Speed_{n + 1} = 0 do
      Register current timestamp as stop start
       k = 0
       k = GPS time between points n, n + 1
       L = L + k
       Next n
    end while
    Register current timestamp as stop end
    Register label M as "Stop Event"
  end if
Next n
end for
```

3.4 Vehicle Stops Classification Model: K-Means

According to [12], one of the most efficient models for cluster classification of vehicle stops based on stop time is the non-supervised clustering K-Means model.

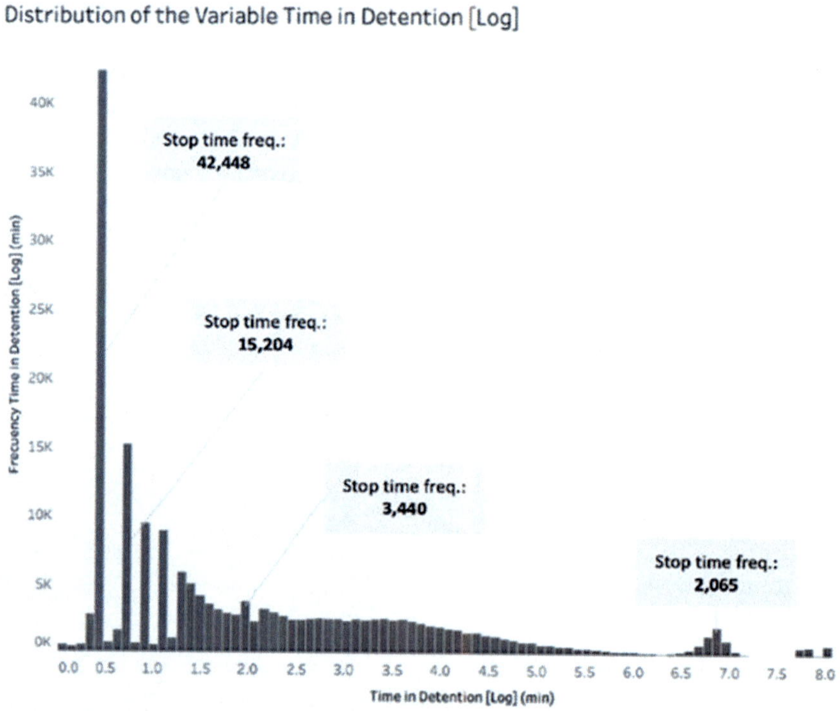

Fig. 3 Distribution of the variable time in stop logarithmic transformation

Before the model is built, the analysis of the variable time stopped is needed. This analysis shows that the variable is heavily skewed to the left. However, it is not unexpected since a great number of stops are short and recurrent as normal behavior in urban traffic. According to [17], if K-Means model is used, the input variables must be continuous and tend to a normal distribution. To deal with the skewness problem [18], recommend a base 10 logarithm transformation to correct normality into the variable. The new variable distribution is shown in Fig. 3. In order to calibrate K-Means model, the hyperparameter K which determines the number of clusters was tested with different values. Looking forward to obtaining an automatically data flow for the model deployment, the Silhouette Score tool was used to obtain the optimal value of K setting a range between 3 and 6 clusters to optimize processing efficiency, [27]. The clustering process run independently for each vehicle considering that each vehicle has a different route and that it remains usually constant.

3.5 *Vehicle Stops Classification Model – HDBSCAN*

After running the K-Means model, every vehicle's stops are classified within a cluster considering only the time stopped. The next step is to run a new clustering process, this time considering the stop's geo-localization. This way stops with similar time length and happening in recurrent positions will be effectively clustered together [28]. The HDBSCAN model uses only the stop's geo-localization as a variable to find optimal clusters grouped by position. Considering [5], recommendation this model was independently applied for each cluster and vehicle's stop dataset.

4 Results and Discussion

After running the proposed algorithm, it was found that the average quantity of stops per vehicle was 2429 per month. The first model implemented was the K-Means model, which allowed identifying the optimal number of clusters based on the variable time in stop [min]. Figure 4 shows the obtained results where clusters are presented as a function of the stop duration.

After obtaining the first results for the K-means clustering, HDBSCAN algorithm was run to obtain the spatial clustering of K-means results as shown in Table 1. Cluster 1 reports an average of 1330 min, 268 total stops, and 22 HDBSCAN clusters. These results indicate that this stops cluster resembles long-duration

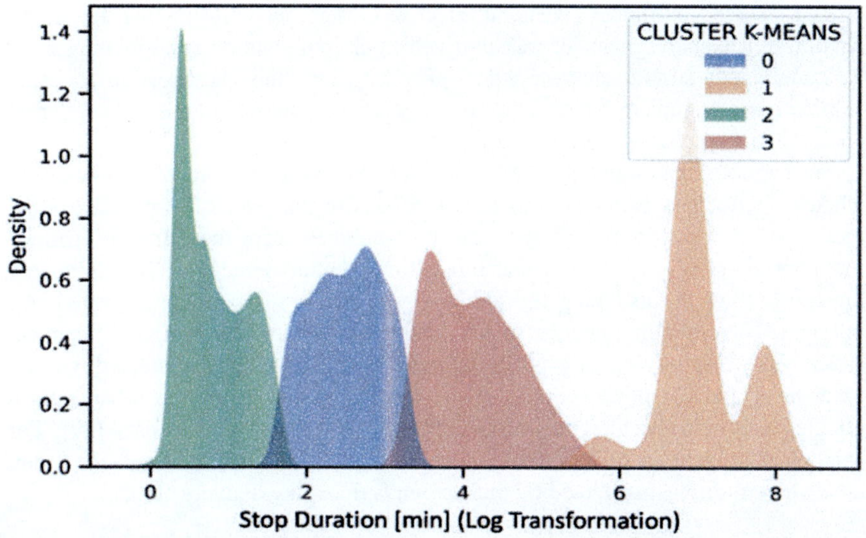

Fig. 4 K-Means algorithm results

Table 1 Clustering results summary

K-means cluster	Avg. stop time [min]	Stops	HDBSCAN clusters
0	13	519	119
1	1330	268	22
2	2	626	141
3	75	430	102

stops. Cluster 2 shows an average of 2 min, 626 total detentions, and 141 HDBSCAN clusters, suggesting short stops in traffic. Cluster 0 shows an average of 13 min stop, 519 total stops, and 119 HDBSCAN clusters, resembling short delivery detention. Finally, cluster 3 reports an average of 75 minutes, 430 total stops, and 102 HDBSCAN clusters, resembling long delivery stops. In general, it can be seen that for the stops in transit with a minimal average time in stop, the number of stops and HDBSCAN clusters is higher. On the other hand, the number of stops and HDBSCAN clusters for long-duration stops is much lower. This makes sense as long-duration stops are less recurrent and usually occur at specific locations, while in transit stops are much more recurrent and occur at different locations.

5 Conclusions

This study shows that the proposed stop identification algorithm gives reliable data regarding vehicle stop time since it does not require additional parameters other than the GPS raw data. The algorithm works on the given data and uses fixed parameters based on statistics and previous GPS data studies.

The HDBSCAN model generated effective clusters when discussing the identification of geospatial points of traffic or vehicular congregation and grouping them by common or similar characteristics, providing the stop classification for each vehicle dataset studied. These allowed us to divide the detentions into different groups like traffic stops, delivery stops, and resting stops.

Furthermore, the model is as robust as it can be to receive, process, and analyze whatever set of geospatial datasets only if it has the characteristics specified in the model. Also, the stop classification clusters obtained from the different artificial intelligence models could be used as inputs for potential applications for optimizing logistic routing by analyzing the points where traffic stops are congregated. An optimized distribution logistics system yields a relevant competitive advantage; either better responsiveness capacity or more efficiency could be attained. On the other hand, public entities could use the model's results to design urban freight policies and infrastructure by improved public policy inputs and data [29]. For example, the creation of specific loading bays for urban freight inside the urban area where traffic gets affected by unforeseen actions like delivery stops.

References

1. Laranjeiro, P. F., Merchán, D., Godoy, L. A., Giannotti, M., Yoshizaki, H. T., Winkenbach, M., Cunha, C. B: Using GPS data to explore speed patterns and temporal fluctuations in urban logistics: The case of São Paulo, Brazil. Journal of Transport Geography, 76, 114–129. (2019) https://doi.org/10.1016/j.jtrangeo.2019.03.003
2. Hess, S., Quddus, M., Rieser-Schüssler, N., Daly, A.: Developing advanced route choice models for heavy goods vehicles using GPS data. Transportation Research Part E: Logistics and Transportation Review, 77, 29–44. (2015) https://doi.org/10.1016/j.tre.2015.01.010
3. Sharman, B., Roorda, M.: Analysis of Freight Global Positioning System Data Clustering Approach for Identifying Trip Destinations. Transportation Research Record Journal of the Transportation Research Board, 83–91. (2011). https://doi.org/10.3141/2246-11
4. Spaccapietra, S., Parent, C., Spinsanti, L.: Trajectories and Their Representations. Mobility Data, 3-22. (2013). https://doi.org/10.1017/cbo9781139128926.002
5. Du, J., Aultman-Hall, L.: Increasing the accuracy of trip rate information from passive multi-day GPS travel datasets: Automatic trip end identification issues. Transportation Research Part A: Policy and Practice, 41(3), 220–232. (2007). https://doi.org/10.1016/j.tra.2006.05.001
6. Yang, X., Sun, Z., Ban, X. J., Holguín-Veras, J.: Urban Freight Delivery Stop Identification with GPS Data. Transportation Research Record: Journal of the Transportation Research Board, 2411(1), 55–61. (2014). https://doi.org/10.3141/2411-07
7. Khan, A. M.: Reducing Traffic Density: The Experience of Hong Kong and Singapore. Journal of Urban Technology, 8(1), 69–87. (2001). https://doi.org/10.1080/10630730120052181
8. Banco Interamericano de Desarrollo. Observatorio Regional de Transporte de Carga y Logística. Departamento de Transporte y Medio Ambiente, 8–17. Nota Ťecnica IDB-NT-508. (2013) https://publications.iadb.org/publications/spanish/document/Observatorio-regional-de-transporte-de-carga-y-logADstica. pdf
9. Municipio del Distrito Metropolitano de Quito: Diagńostico De La Movilidad En El Distrito Metropolitano De Quito Para El Plan Metropolitano De Desarrollo Territorial (Pmot), 5–10. (2014) http://gobiernoabierto.quito.gob.ec/wp-content/uploads/documentos/pdf/diagnosticomovilidad.pdf
10. Puente-Mejia, B., Orellana-Rojas, C., Suárez-Núñez, C.: Data-driven Solutions for Evaluating and Planning Last Mile Operations in Latin America: A Methodological Approach Focused on Quito, Ecuador. Supply Chain Management and Logistics in Emerging Markets. "Data-driven Solutions for Evaluating and Planning Last Mile Operations in Latin America: A Methodological Approach Focused in Quito, Ecuador", Yoshizaki, H.T.Y., Mejía Argueta, C. and Mattos, M.G. (Ed.) Supply Chain Management and Logistics in Emerging Markets, Emerald Publishing Limited, Bingley, pp. 107–129. (2020). https://doi.org/10.1108/978-1-83909-331-920201006
11. Puente-Mejia, B., Palacios-Argüello, L., Suárez-Núñez, C. Gonzalez-Feliu, J.: Freight trip generation modeling and data collection processes in Latin American cities. Modeling framework for Quito and generalization issues. Transportation Research Part A: Policy and Practice, Volume 132, 226–241. (2020). https://doi.org/10.1016/j.tra.2019.10.013
12. Greaves, S., Figliozzi, M.: Collecting Commercial Vehicle Tour Data with Passive Global Positioning System Technology Issues and Potential Applications. Transportation Research Record: Journal of the Transportation Research Board, 158–165. (2008). https://doi.org/10.3141/2049-19
13. Romano, A., Tkanori, S,. Hong, M., Jeong, K., Jing, P., Ben-Akiva, M.: Exploring Algorithms for Revealing Freight Vehicle Tours, Tour Types, and Tour Chain Types from GPS Vehicle Traces and Stop Activity Data. Journal of Big Data Analytics in Transportation. 175–177. (2019). https://doi.org/10.1007/s42421-019-00011-x
14. McNeff, J.: The global positioning system. IEEE Transactions on Microwave Theory and Techniques, 50, 645–652. (2002), https://doi.org/10.1109/22.989949

15. Zhao, X., Cheng, X., Zhou, J., Xu, Z., Dey, N., Ashour, A., Satapathy, S.: Advanced Topological Map Matching Algorithm Based on D–S Theory. Arabian Journal for Science and Engineering, 43, pages 3863–3874. (2018). https://doi.org/10.1007/s13369-017-2569-0
16. Hariharan, R., Toyama, K.: Project Lachesis: Parsing and Modeling Location Histories. (2004) Retrieved October 18, 2020. https://doi.org/10.1007/978-3-540-30231-58
17. Morissette, L., Chartier, S.: The k-means clustering technique: General considerations and implementation in Mathematica. Tutorials in Quantitative Methods for Psychology, 9(1), 15–24. (2013). https://doi.org/10.20982/tqmp.09.1.p015
18. Beauchamp, J. J., Olson, J. S.: Corrections for Bias in Regression Estimates After Logarithmic Transformation. Ecology, 54(6), 1403-1407. (1973). https://doi.org/10.2307/1934208
19. Malzer, C., Baum, M.: A Hybrid Approach To Hierarchical Density-based Cluster Selection. 2020 IEEE International Conference on Multisensor Fusion and Integration for Intelligent Systems (MFI). 223–228. (2019). https://doi.org/10.1109/MFI49285.2020.9235263
20. Rahman, M. F., Liu, W., Suhaim, S. B., Thirumuruganathan, S., Zhang, N., Das, G.: Density Based Clustering over Location Based Services. 2017 IEEE 33rd Inter- national Conference on Data Engineering (ICDE). (2017). https://doi.org/10.1109/icde.2017.103
21. Comendador, J., Lopez-Lambas, M. E., Monz'on, A.: A GPS Analysis for Urban Freight Distribution. Procedia – Social and Behavioral Sciences, 39, 521–533. (2012). https://doi.org/10.1016/j.sbspro.2012.03.127
22. Wowczko, I.: A Case Study of Evaluating Job Readiness with Data Mining Tools and CRISP-DM Methodology. International Journal for Infonomics, 8(3), 1066–1070. (2015). https://doi.org/10.20533/iji.1742.4712.2015.0126
23. Yu, C. H.: Exploratory data analysis in the context of data mining and resampling. International Journal of Psychological Research, 3(1), 9–22. (2010). https://doi.org/10.21500/20112084.819
24. Conagin, A., Barbin, D., Demetrio, C. G.: Modifications for the tukey test procedure and evaluation of the power and efficiency of multiple comparison procedures. Scientia Agricola, 65(4), 428–432. (2008). https://doi.org/10.1590/s0103-90162008000400016
25. McInnes, L., Healy, J., Astels, S.: hdbscan: Hierarchical density based clustering. The Journal of Open Source Software, 2(11), 205. (2017). https://doi.org/10.21105/joss.00205
26. Zhao, W., Mccormack, E., Dailey, D. J., Scharnhorst, E.: Using Truck Probe GPS Data to Identify and Rank Roadway Bottlenecks. Journal of Transportation Engineering, 139(1), 1–7. (2013) https://doi.org/10.1061/(ASCE)TE.1943-5436.0000444
27. Wang, J., Chai, R., Xue, X.: The Effects of Stop-and-go Wave on the Immediate Follower and Change in Driver Characteristics. Procedia Engineering, 137, 289–298. (2016). https://doi.org/10.1016/j.proeng.2016.01.261
28. Birant, D., Kut, A.: ST-DBSCAN: An algorithm for clustering spatial–temporal data. Data & Knowledge Engineering, 60(1), 208–221. (2007). https://doi.org/10.1016/j.datak.2006.01.013
29. Ayala- Jaramillo, M., Celleri-Maldonado, J., Camacho, C., Puente-Mejia, B.: Optimization of Distribution of Humanitarian Aid: Cotopaxi Volcano. Natural Hazards Review, 23(2), 05022001. (2022). https://doi.org/10.1061/(ASCE)NH.1527-6996.0000549

Comparison of Nanostore Supply Chain Strategies in Urban Areas: The Case of Ica, Peru

Mariana Moyano, Julio Castillo, Mario Chong, and Christopher Mejía

Abstract In Peru, the traditional channel dominates with a 70% of market share; however, the modern channel and other organized retail chains (e.g., convenience stores) have been growing in recent years in the retail landscape. Moreover, the COVID-19 pandemic represented a new challenge for retailers due to the mobilization restrictions that affected consumption. Thus, the present work aims to describe the main characteristics of two nanostores from Ica, Peru, and capture the activities that compose their supply chain strategy. The applied methodology was adapted to the retail environment from a widely known procedure. We also supplemented it by adding a semi-structured interview based on qualitative semi-open questions to characterize processes, context and performance metrics. Finally, we added a 26-item questionnaire to score good business practices of the studied nanostores. This information allows us to visualize how these nanostores have adapted their business practices and operations after the pandemic and understand how nanostore owners manage their business.

Keywords Retail · Mom and pop stores · Business practices · Strategy · COVID-19 crisis

1 Introduction

The population in developing countries is rapidly growing in metropolitan areas; thus, huge cities are starting to dominate the landscape (Fransoo et al. 2017). In 2020, there were 1934 metropolises with more than 300,000 inhabitants, accounting for approximately 60% of the world's urban population (United Nations 2020).

M. Moyano · J. Castillo · M. Chong (✉)
Facultad de Ingeniería, Universidad del Pacifico, Lima, Peru
e-mail: m.chong@up.edu.pe

C. Mejía
Massachusetts Institute of Technology, Cambridge, MA, USA

J. Vargas Florez et al. (eds.), *Production and Operations Management*, Springer Proceedings in Mathematics & Statistics 391,
https://doi.org/10.1007/978-3-031-06862-1_39

Moreover, according to the United Nations (2018) statistics, Latin America and the Caribbean are the most significant proportion of the population concentrated in areas with over 10 million inhabitants. For instance, Peru is the seventh most populated country in the region with 32.6 million inhabitants and, globally, the eighth most densely populated country with an average of 25 inhabitants per square kilometer (Instituto Nacional de Estadistica e Informatica – INEI 2020).

Furthermore, in Lima, the capital of Peru, 80% of the population lives in urban areas (INEI 2017). The urban population outnumbers the rural population in almost all the other Peruvian states. Nonetheless, there's a marked difference in the case of the province of Ica, where the total population is 850,765 inhabitants. This represents 93% of the urban population and only around 7% of the rural sector (INEI 2017). This imbalance also exists in other Latin-American cities, has implications regarding the management of resources and brings new challenges in the value chain and product management (Mejia-Argueta et al. 2015). Finally, it is relevant to consider that the Peruvian retail landscape is highly fragmented and shows a predominant market share of the traditional channel over the modern one with 70% (Peru Retail 2018) (BBVA Research 2018). Small, family-owned retailers play a fundamental role. We will call this channel *nanostores* to refer to these informal retailers and mom-and-pop stores (Fransoo et al. 2017).

Given this high portion of nano and micro businesses, in particular retailers, the arrival of the COVID-19 crisis affected them considerably with a dramatic reduction of consumption and shortage of cash due to lockdowns and other safety protocols. According to a survey developed by McKinsey & Company in late 2020, most Peruvian consumers experienced reduced income and savings; moreover, purchase intent remained negative across most product categories except for groceries, household supplies and personal care products (McKinsey and Company 2020). On the side, shopkeepers faced challenges like constant exposure to the virus and reduced working hours, which resulted in low productivity and low income (*United Nations Development Programme – Peru* 2021).

Given all these dynamics in highly urbanized cities or provinces like Ica, Peru across emerging market economies, it is relevant to study how nanostores adapted their operations to prevail productively and survive. Thus, this study aims to compare the supply chain strategy and business practices of two nanostores located in Ica, Peru, under the 2020–2021 pandemic. The methodology applied to this research is based on Perez-Franco's (2010) study for capturing the supply chain strategy. In addition, a semi-structured interview with nanostores' owners and a 26-item questionnaire proposed by McKenzie and Woodruff (2017) was applied to measure the business practices of nanostores. We found that the two nanostores from Ica, Peru, that we analyzed in detail are highly demand-driven. Still, each one responds differently to create an effective strategy to serve their patronage differently. There is a high focus on marketing and record-keeping strategies in family-owned retailers, while inventory management and financial performance are not tracked in detail. Finally, we found that during the pandemic, nanostores' owners started using technology for order generation and as alternative payment methods and offering home delivery to respond to the lockdowns brought by the pandemic.

The following section presents an overview of the literature on nanostore supply chains, best practices and business models. Section 3 presents the methodological framework we use to perform this study. Section 4 discusses the main highlights of applying the methodology in two nanostores. Finally, Sect. 5 presents conclusions and future work.

2 Literature Review

In the last decade, public policy and brand strategies have focused on taking advantage of nanostores' potential to grow as an economic and social force in every community, particularly in the most vulnerable areas. There are several characterization approaches that researchers have used to understand the evolution of consumers in emerging markets. For example, Boulaksil and Belkora (2017) conducted an empirical study in Moroccan cities using surveys to compare the strategies of two types of distribution: van sales vs presales to determine which one was more effective for manufacturers and distributors. Likewise, Narayan and Chandra (2015) used a questionnaire to understand key factors that influence consumer purchase to offer suggestions to organized retailers in India. A similar case was introduced by Zameer and Mukherjee (2011), who characterized the value that consumers look for at Indian nanostores via factor analysis.

In addition, there are three case studies developed in Latin America that pointed out some characteristics of these businesses, which allows them to prevail in the retail landscape despite the entry of the modern channel and other retail formats in emerging economies. For instance, D'Andrea et al. (2006) carried out a study in Argentina, Brazil, Chile, Colombia, Costa Rica, and Mexico. They applied store checks and in-depth interviews to shopkeepers to build a business model assessment to analyze the value proposition. Likewise, Rangel-Espinosa et al. (2020) identified opportunities. They made suggestions to increase nanostores' survival based on their research findings at different socioeconomic levels in Mexico City, using data collection instruments like observation, surveys and interviews. Furthermore, Castillo et al. (2019) applied the methodology of the square kilometer in last-mile deliveries. They analyzed the logistics profile of an area of the city of Lima, Peru, intending to contribute with updated information that could help public policymakers to create parking and urban distribution regulations.

McKenzie and Woodruff (2017) developed a study in Bangladesh, Chile, Ghana, Kenya, Mexico, Nigeria, and Sri Lanka; where they administered a survey consisting of 26 questions to characterize business practices in marketing and stock-keeping record-keeping and financial planning for micro and small firms. They applied statistical tools such as regressions to find out that the index of business practices has a high correlation with enterprise heterogeneity in the cross-sectional research. This means that firms with better business practices also have higher sales, profits, labor productivity and total factor productivity.

Other studies like Salinas et al. (2018) highlighted the importance of logistics and nanostores in combating food malnutrition. They surveyed end consumers in the Mexico City metropolitan area, processed the data and created a causal loop diagram to understand interactions and dynamics among variables.

Lastly, a few approaches consider the COVID-19 crisis context; however, they do it from a qualitative perspective. For instance, Kumar et al. (2020) compared and discussed different operations and supply chain perspectives to propose strategies in a post-pandemic era. While Sillo-Acuna (2021) summarized the circumstances in which the Peruvian nanostores were situated when the COVID-19 restrictions were established, it also accentuated the importance of resilience and entrepreneurship that shopkeepers have and helped them to come out quickly with new strategies to deal with the uncertainty.

In summary, there are only some studies in the literature body documenting from a quantitative albeit still empirical perspective the practices of the fragmented retail landscape in highly urbanized areas in developing countries. Second, there is still scarce documentation about how bottom-of-the-pyramid consumers and businesses serving them adopted strategies during the pandemic to survive in the pandemic. Finally, there is still a scarce link of all business practices to the supply chain strategy of the fragmented retail landscape in countries with high dependence on the nanostores channel. This study aims to address these gaps and contribute to state of the art with expanding knowledge by applying standard data collection tools and strategic analysis techniques.

3 Methods and Procedures

This study's methodology is broken down into five phases (Perez-Franco et al. 2011) to capture the essence of the supply chain strategy and business practices. However, considering nanostores characteristics, some adaptations were performed, such as the aggrupation of phases with similarities. Next, the steps followed in this research will be presented.

3.1 Phase 1: Scope of the Project and Selection of Respondents

Escamilla-González-Aragón et al. (2020) argued that getting reliable information from nanostore shopkeepers is usually a big challenge. Thus, a key component to ensure getting valuable data is to build trust or get access to respondents who are committed to the study. Based on these challenges and to make this study more practical, a direct sampling by convenience was performed in two nanostores: "Comercial Mireli" and "Bodega Pierina's". Even though this is not a representative sample, this study can be used as a first-order approach to build empirical understanding on the mindset used by shopkeepers to make decisions and to extend descriptive literature of nanostores in highly urbanized areas such as Ica, Peru. Nanostores' owners selected for this study helped get access to detailed information than in previous experiences was limited based also on the authors' experience.

3.2 Phase 2: Conduct Qualitative Interviews and Identification of Areas of Activity

The semi-structured interviews were carried out with the owners of selected nanostores. Both were conducted on August 10th, 2021 and had 1–1.5 h each. Each interview consisted of two parts. The first one focused on documenting qualitative semi-open questions related to working hours, assortment, top products, logistics operations and the impact of the COVID-19 crisis on their business performance (see Appendix #1). The second one consisted of applying the index of business practices developed by McKenzie and Woodruff (2017) to give a score to each nanostore depending on their practices in marketing, stock-keeping, record-keeping and financial planning.

After documenting and analyzing the recordings of interviews, the broad areas of activity were identified. In this case, both nanostores sell first essential products; however, the way they sell them shows a different consumption pattern. Thus, this phase let us identify activities and convert them into a hierarchical summary.

3.3 Phase 3: Translate Each Hierarchical Summary into a Partial Map and Get Validation of the Nanostore Owner

For this phase, Perez-Franco (2010) indicates that each identified area with its hierarchical summary needs to be converted into a pyramidal conceptual map, which helps visualize the interaction among their components. After this step was done, the validation with each nanostore owner was held a week after phase 2 was conducted to confirm the collected information from the interviews. In our case, the retailers requested no modifications.

3.4 Phase 4: Combine the Partial Maps of Strongly Related Areas, Simplify and Create an Abstract of the Nominal Business Strategy

To reduce the complexity of the resulting partial maps, it was necessary to simplify them by using the layer of activities and the layer of supporting means. However, on the way to identifying the core strategy, we did not have access to any written documentation because the selected nanostore owners declared that they do not regularly keep any formal record of their operations. However, from the provided information in the interviews, we made an effort to identify the business strategy to determine a more detailed idea of how their business works.

3.5 Phase 5: Assemble the Functional Strategy Map

The resulting functional strategy map features two distinct halves. The left side shows a conceptualization of the nominal strategy, and the right side shows a conceptualization of the executed strategy. This phase allows us to visualize how the core strategy is followed by specific activities that each nanostore applies daily in their operations. This phase also let us identify opportunities, misalignments, goal achievement and resource allocation.

The following section presents the results of applying the proposed methodological framework documented in this section.

4 Results

This section discusses the main findings of our research. We use the five-stage approach presented in the previous section to facilitate the interpretation.

4.1 Phase 1: Description of the Selected Nanostores

When describing the main characteristics of a nanostore, authors like Fransoo et al. (2017) and Escamilla-González-Aragón et al. (2020) detailed that there are different sizes of these businesses. Some could be as small as 15 to 40 m^2 of store surface. Additionally, Rangel-Espinosa et al. (2020) extended this characterization by describing them as small, independent, family-owned retailers that do not belong to any franchise and commonly sell consumer packaged goods (CPG) and other essential goods. For this case, both selected nanostores meet this theoretical description in terms of total area, and they are family-owned businesses. However, a few

Table 1 Main characteristics of each nanostore

	"Comercial Mireli"	"Pierina's"
Total Area	8 m^2	20 m^2
Location	Inside a local market	Part of a house
Years of operation	Founded in 1997 (24 years)	Founded in 1992 (29 years)
Offered product categories	Bakery and sweets Eggs, dairy and coffee Oil, rice, pasta and legumes Canned food Juices and non-alcoholic beverages Cosmetics and personal care Home cleaning	Sweets and snacks Dairy and coffee Oil, rice and pasta Juices and alcoholic beverages Copies and photocopies
Accepted payment methods	Only cash payments	Cash and digital wallet *Yape*

Source: Developed by the authors

characteristics distinguish one from another, like the accepted payment methods, the product categories that they offer, and where they are located.

As expected, these differences may change the type of patronage they serve, the processes they perform to supply, keep and offer those products at the nanostores, the physical, information and cash flow shopkeepers manage, as well as their autonomy to make decisions and take actions to respond to uncertainty. In other words, these features change the degree of agility, adaptability and alignment shopkeepers will show to respond to changing conditions due to the pandemic or any other potential disruption or shock in the nanostore supply chain (Table 1).

4.2 *Phase 2: Comparison of Results from Questionnaires*

During the interview with Iris Gavilan, owner of "Comercial Mireli", she pointed out that the main reason why she chose to focus on selling essential products was because of experience. When she started her business, she only offered fruits and vegetables; however, as these foods may spoil quickly, she decided to move on to selling more non-perishable products. Nonetheless, this statement focuses on the assortment she offers but not on the service that nanostores play in a community. Likewise, during the interview with Carmen Ramos, owner of "Pierina's", she indicated that her business started as a telephone station for the community. But with the advancement of technology and the massification of cellphones, her business had to move on to selling essential products. Still, she started selling alcoholic beverages and services like internet booths, copies, photocopies and mobile top-up services. Thus, "Pierina's" offers products and services.

Additionally, from the business practices index, the score obtained by "Comercial Mireli" and "Pierina's" was 14 and 12 out of 26 business practices, respectively.

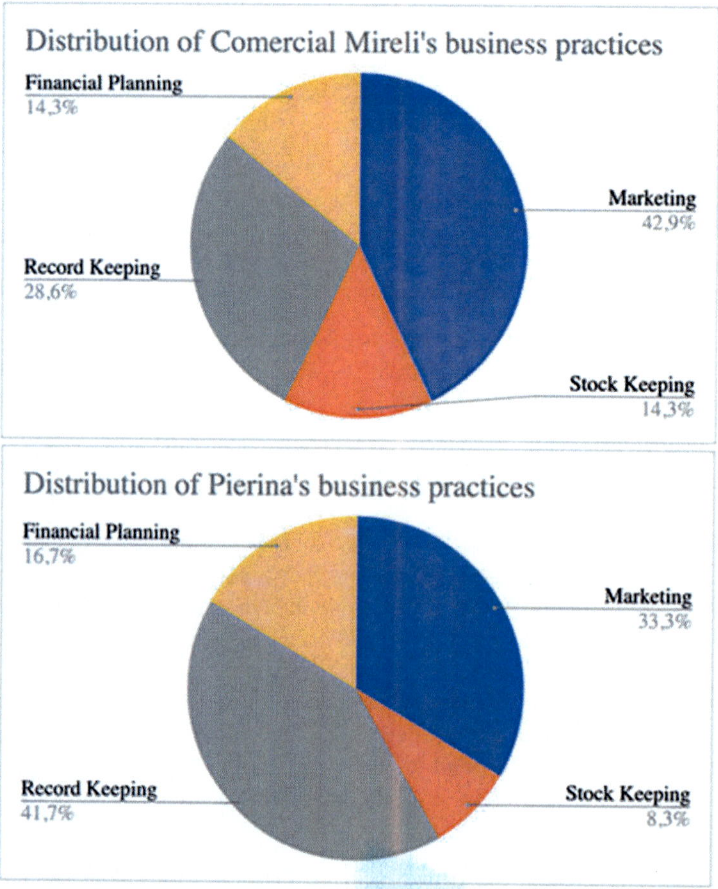

Source: Developed by authors with data provided by the shopkeepers

Fig. 1 Comparison of business practices at nanostores in Ica, Peru. (Source: Developed by authors with data provided by the shopkeepers)

This translates into a proportion of 0.5385 and 0.4615 for the index of each one, respectively (see Appendix #3). Figure 1 shows that "Comercial Mireli" focuses its efforts on marketing business practices (42.9%) followed by record-keeping practices, while "Pierina's" does the opposite, focusing on record-keeping business practices (41.7%) followed by marketing practices. However, both nanostores have a low proportion of their business practices dedicated to stock keeping and financial planning, which in the case of a larger store may represent a considerable opportunity to handle inventory properly.

This finding also shows that both owners need to improve their control and performance in financial and inventory management. Finally, it also shows why the owners decided to sell non-perishable products to concentrate their efforts on promotion, supply and getting in close contact with their consumers; while not concentrating that much on diversifying their suppliers or ensuring the quality of perishable products, which will require higher skills on inventory management.

4.3 *Phase 3: Comparison of the Hierarchical Summary of Each Nanostore*

Both nanostores provide the same areas of activity, such as offering essential products to their patronage, giving clients friendly personal care while offering convenience and immediate attention. Nevertheless, they have different activities that support each area depending on their business model strategy (Tables 2 and 3).

It is important to note that before and after the COVID-19 crisis, shopkeepers declared that the areas of activity remained the same for both nanostores. However, the activities that support them changed and adapted to new customer requirements. For example, in the case of "Comercial Mireli", the owner added the home delivery option based on the mobilization restrictions due to the pandemic. After the circulation restrictions were lifted, she continued offering this service because many clients demanded it. In the case of "Pierina's", the owner started to accept payments

Table 2 Hierarchical summary of nanostore "Comercial Mireli"

Ensure access to essential goods to clients from the neighbourhood
The shopkeeper possesses a storage room where she keeps products in stock.
She brings new products that clients request regularly.
She designed the layout to maintain products in good condition.
Offer clients a high-quality customer service
The owner advises clients to choose what to cook that day based on the availability of products and taste preferences.
She recommends clients choose a product that aligns with their necessities.
Regular clients have the owner's cellphone number to order in advance or to ask for specific products.
Bring convenience and fast attention to consumers
In-store attention hours go from 7 am to 3 pm.
The owner provides home delivery of products for regular consumers.
Speed up the attention and order preparation by also receiving orders via WhatsApp.

Source: Developed by authors with data provided by the shopkeepers

Table 3 Hierarchical summary of nanostore "Pierina's"

Ensure access to essential goods to clients from the neighbourhood
The owner disinfects all the products displayed on the shelves
She also brings new products that clients request regularly
Offer clients a high-quality customer service
The owner has a place outside the nanostore where clients can hang out or sit together
Bring convenience and fast attention to consumers
In-store attention hours go from 8 am to 12 pm.
She accepts payments in cash and with digital wallets

Source: Developed by authors with data provided by the shopkeepers with the digital wallet *Yape* to avoid direct contact with money and facilitate her customers' other payment methods. She continues using this payment method.

4.4 Phase 4: Comparison of the Nominal Business Strategy of Each Nanostore

These two different business models show the importance of adapting quickly to demand shocks, the main focus of both businesses, according to Fig. 1. The resulting partial maps (see Appendix #4 for further detail) helped to support what the nanostore owners suggested as their core strategy. There is a marked difference between "Comercial Mireli" and "Pierina's" from a strategic perspective. While the owner of "Comercial Mireli" focuses more on the prices and delivery, the owner of "Pierina's" does it on customer service.

The shopkeeper of "Comercial Mireli" created an agile way to ensure her clients keep receiving their products at their doorsteps given the commodities, food-related and essential products she sells physically, by phone or via WhatsApp. She is also interested in providing a wider assortment. On the other hand, the owner of "Pierina's" decided to work on extending her working hours, building trust in the community by disinfecting the products and getting access to more new clients via digital payment methods. Her strategy mainly focuses on capitalizing on those SKUs and clients with the highest potential.

A summary of the aforementioned nominal strategies is provided in Fig. 2.

4.5 Phase 5: Comparison of the Functional Strategy Map of Each Nanostore

The Functional Strategy Map consists of the nominal strategy presented in the previous point and the executed strategy, which includes daily, operational activities

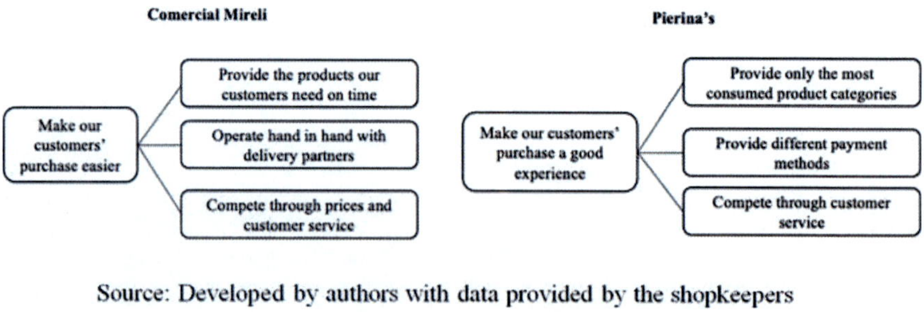

Fig. 2 A mapped nominal strategy for each nanostore. (Source: Developed by authors with data provided by the shopkeepers)

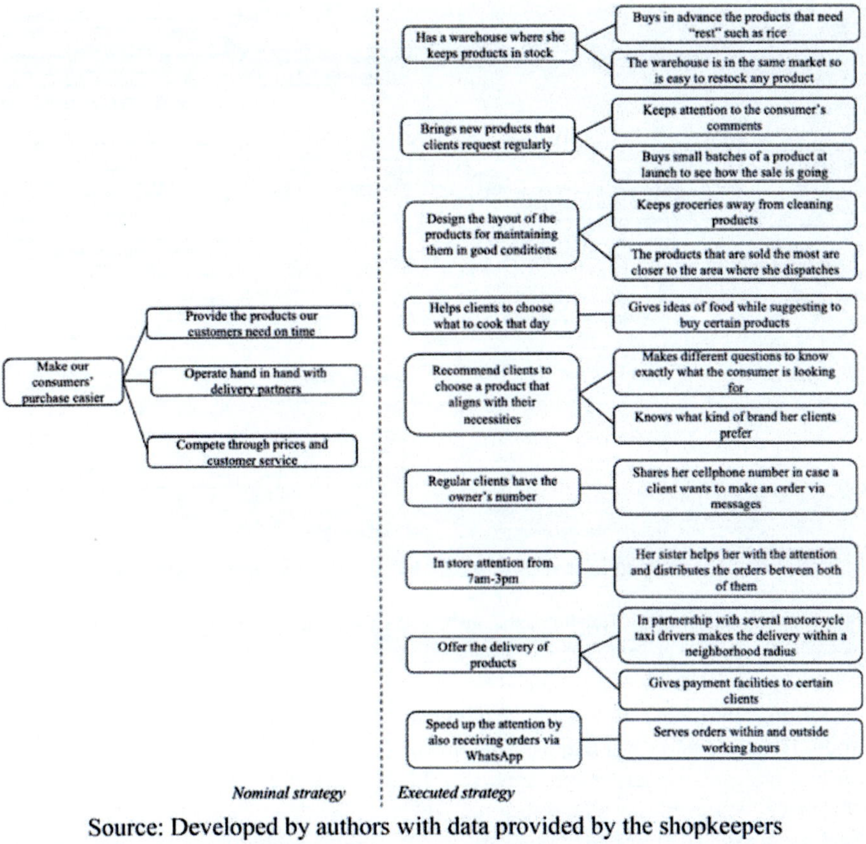

Fig. 3 Functional Strategy Map for nanostore "Comercial Mireli". (Source: Developed by authors with data provided by the shopkeepers)

performed by the shopkeepers to take advantage of the nominal strategy. This phase lets us visualize the bigger picture of operations corresponding to the core strategy and identify alignment between the strategic and operational levels, agility, etc. (Fig. 3)

The nominal strategy of "Comercial Mireli" is well-aligned with the executed strategy. Daily activities are consistent with the owner and her helpers' tactics to respond to consumers' needs. Unlike the other nanostores, "Comercial Mireli" takes advantage of its storage room to replenish fast-moving goods quickly. The owner is risk-averse but has a closer relationship with her regular patrons, even providing informal credit to some of them. She differentiates by having an interesting delivery method and other logistics activities that support goals, tactics and operations (Fig. 4).

Finally, a nominal strategy of "Pierina's" is more straightforward but more focused on providing an excellent consumer experience through an executed strategy based on differentiation and visible safety protocols. Due to the absence of the

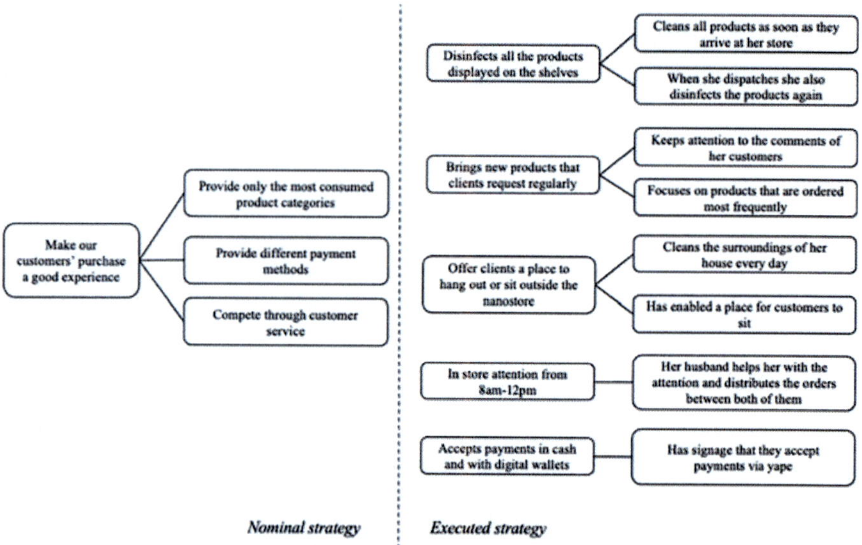

Source: Developed by the authors with data provided by the shopkeepers

Fig. 4 Functional Strategy Map for nanostore "Pierina's". (Source: Developed by the authors with data provided by the shopkeepers)

backroom, the owner and her husband ensure the high availability of fast movers through diverse supply channels. They introduce products to their assortment based on highly rated consumer requests, mainly from loyal consumers. The owner of "Pierina's" keeps closer attention to clients and looks for converting her nanostore into a community hub, which supports and is aligned with the main strategy of the nanostore.

5 Conclusions

Nanostores have strong relevance to improving communities and neighbourhoods' local economy, especially the most underserved locations. Salinas et al. (2018) highlighted that its importance consists of being the principal source of food products in affordable units. They also offer convenience, an assortment of items and credit facilities for unbanked citizens. However, nanostores are constantly exposed to external factors such as a rise in product prices, supply shocks and high variation of demand patterns, which affects the cash flows, the sales and the portfolio and inventory management. Moreover, the COVID-19 crisis meant an even more significant challenge, not only for the risk of direct exposure to the virus but also for the restrictions that came with the pandemic. This event marked a milestone on how to address nanostore supply chains and business strategies. It pushed many nanostore owners to move on to adding other value-added activities, like home delivery, accepting payment methods different from cash, as well as differentiating by offering a safe environment to do groceries.

In the present study case, we analyzed the operations of two nanostores to understand how they adapted their operations to requirements related to safety protocols while keep offering convenience. Nevertheless, how these nanostores have implemented these strategies has followed a different approach depending on their core business model. By applying the methodology for capturing the supply chain strategy, we could see if the nanostore had activities that supported them consistently. We found that nanostore's owners can better understand if their operations correspond to their objectives from the proposed methodology. Thus, the improvement of their sales and their better positioning in front of other channels must be directly related to potentiating and maintaining their strengths while working on their weakness.

Through the business practices index, we were able to see that the main strength of "Comercial Mireli" was the one related to marketing business practices while "Pierina's" was the record-keeping practices. On the other hand, the main weakness was the stock keeping and the lack of monitoring of financial performance in both nanostores. It is also important to highlight that results of the business practices index show the average proportion score of both nanostores is 0.5 and, even though this is not a representative measure, it is interesting to visualize it next to the results of McKenzie and Woodruff (2017) for another type of businesses.

Nevertheless, these results cannot be used for cross-country comparisons since the sampling strategy varies by country (McKenzie and Woodruff 2017, 2971). It sheds light on opportunity to keep analyzing nanostore performance by using primary data collection instruments, especially among Peru, Mexico and Chile but using statistically representative samples.

Finally, future research should include a more significant and representative sample to build more general conclusions and specific recommendations in handling inventory and portfolio management, cash and financial management, among others. However, this paper described the main characteristics of two nanostores from Ica, Peru, and captures the areas and activities that compose their supply chain strategy, which provides relevant information on how these nanostores have adapted their business practices and operations to face high-uncertainty contexts in the coming future like the COVID-19 crisis.

Appendices

Appendix #1: Semi-structured Interviews with Shopkeepers

Goal: To get to know basic data about the nanostore operations from a primary source, in addition to characterizing both the interior (supply, assortment and distribution of products in the layout) and the exterior of the premises (neighbourhood, security, close businesses -competitors-, location, among others).

1. How did you start your nanostore?
2. How long have you owned and worked your nanostore?
3. What is the total area of your business?
4. What payment methods are accepted in your store?
5. Do you offer any type of credit to your clients?
6. What products with the highest turnover do you sell in your store?
7. Do you offer promotions or discounts in a given month or season?
8. What were your service hours before the pandemic? How did they vary according to the imposed safety protocols during the pandemic?
9. What are your service hours to serve consumers nowadays?
10. How many people are hired by you? What is their journey (i.e., number of working hours and days)? What activities do they perform?
11. What is the day of the week with the highest influx of consumers (i.e., foot traffic)?
12. What are the hours with the highest influx of consumers (i.e., foot traffic)?
13. Where and how do you buy your products? Who are your main suppliers?
14. What days and hours do your suppliers arrive at your nanostore?
15. Do you have a storage room or backroom? How do you manage your inventory?
16. How do you arrange the products in the nanostores (i.e., layout)? At what time of the day do you replenish the shelves with products?
17. How many brands do you usually offer per product category?
18. Do you offer healthy products? Which products and what brands? Concerning other products, how profitable are their sales?
19. Have you ever brought new products because your customers requested them?
20. What has been the biggest challenge you faced with the COVID-19 crisis?
21. How was the process of adopting the safety protocols presented by the government to your store?
22. What actions did you take to contact your consumers?
23. What problems do you usually have in your business? How do you solve them?

Appendix #2: Index of Business Practices Developed by McKenzie and Woodruff (2017)

Goal: To apply a standardized survey to get data about business practices in the nanostore and compare the results.

The total score is the composite business practice score. It ranges between 0 and 1. The total is the proportion of the following 26 business practices used by the firm:

Marketing Practices (M class practices): Each practice is coded as 1 for each of the following items that the nanostore owner has done in the last three months:

- M1: Visited at least one of its competitor's businesses to see what prices they are charging to similar consumers.

- M2: Visited at least one of its competitor's businesses to see what products they have available for sale.
- M3: Asked existing consumers whether there are any other products they would like the business to sell.
- M4: Talked with at least one former consumer to find out why they stopped buying from the nanostore.
- M5: Asked a supplier about which products are selling well in the industry (i.e., for similar nanostores).
- M6: Attracted consumers with a special offer, discount or promotion.
- M7: Advertised the nanostores or their products in the last SIX months.

Buying and Stock Control Practices (B class practices): Each item is coded as 1 for the following practices:

- B1: Attempted to negotiate with a supplier for a lower price on specific products.
- B2: Compared the prices or quality offered by alternative suppliers or sources of products/services to the current suppliers or vendors
- B3: The business does not run out of stock monthly or more (coded as one of the businesses has no stock at some point).

Costing and Record-Keeping Practices (R class practices): Each item is coded as 1 for the following practices:

- R1: Keeps written business records.
- R2: Records every purchase and sale made by the business.
- R3: Able to use records to see how much cash the business has on hand at any point in time.
- R4: Uses records regularly to know whether sales of a particular product are increasing or decreasing from one month to another.
- R5: Works out the cost to the business of each main product the shopkeeper sells.
- R6: Knows which goods make the most profit per sold item.
- R7: Has a written budget, which states how much is owed each month for rent, electricity, equipment maintenance, transport, advertising, and other indirect costs to business.
- R8: Has records documenting that there exists enough money each month after paying business expenses to repay a loan in the hypothetical situation that this business wants a bank loan.

Financial Planning Practices (F class practices): Each item is coded as 1 for the following practices:

- F1: Review the financial performance of their business and analyze where there are areas for improvement at least once per month.
- F2: Has a target set for sales over the next year.
- F3: Compare their sales achieved to their target at least once per month.

- F4: Has a budget of the likely costs their business will have to face over the next year.
- F5: Has an annual profit and loss statement.
- F6: Has an annual statement of cash flow.
- F7: Has an annual balance sheet.
- F8: Has an annual income/expenditure sheet.

Appendix #3: Detailed Results from the Index of Business Practices

Business practices		Comercial Mireli	Pierina's
Marketing	M1	1	1
	M2	1	0
	M3	1	1
	M4	0	0
	M5	1	1
	M6	1	1
	M7	1	0
Total in marketing		6	4
Stock keeping	B1	1	0
	B2	1	1
	B3	0	0
Total in stock keeping		2	1
Record keeping	R1	1	0
	R2	1	1
	R3	0	0
	R4	0	0
	R5	1	1
	R6	1	1
	R7	0	1
	R8	0	1
Total in record keeping		4	5
Financial planning	F1	0	0
	F2	0	0
	F3	1	0
	F4	1	0
	F5	0	1
	F6	0	0
	F7	0	0
	F8	0	1
Total in financial planning		2	2
Total score		14	12
Proportion		0,5385	0,4615

Source: Developed by authors with data provided by the shopkeepers

Appendix #4: Additional Information

Nanostore 1: "Comercial Mireli"

It is located inside a local market – Mercado Carlos Ramos Loayza at Av. Manuel Santana Chiri, Ica 11001, Peru

Photos taken by the authors

Nanostore 2: " Pierina's"

It is part of a house located at Av. Manuel Santana Chiri, Ica 11001, Peru

Photos taken by the authors

References

BBVA Research. (2018). *Perú Situación retail moderno 2018*. BBVA Research. https://www.bbvaresearch.com/wp-content/uploads/2018/11/Peru-Retail-Moderno.pdf

Boulaksil, Y., & Belkora, M. J. (2017). Distribution Strategies Toward Nanostores in Emerging Markets: The Valencia Case. *Informs Journal*, *47*(6), 473–579. https://doi.org/10.1287/inte.2017.0914

Castillo, J. C., Goicochea, E., Chong, M., & Rodriguez, M. (2019). Inmegacity characterization: trends and realities. *Management Research*, *17*(2), 187–204. https://doi.org/10.1108/MRJIAM-05-2018-0835

D'Andrea, G., Lopez-Aleman, B., & Stengel, A. (2006). Why small retailers endure in Latin America. *International Journal of Retail & Distribution Management*, *34*(9), 661–673. https://doi.org/10.1108/09590550610683184

Escamilla-González-Aragón, R., Fransoo, J.C., Mejía Argueta, C., Velazquez-Martinez J.C. & Cedillo-Campos, M.G. (2020). Nanostores: Emerging Research in Retail Microbusinesses at the Base of the Pyramid, *Academy of Management Global Proceedings*, 161.

Fransoo, J. C., Blanco, E. E., & Mejia-Argueta, C. (2017). *Reaching 50 million nanostores: retail distribution in emerging megacities*. CreateSpace Independent Publishing Platform: Cambridge.

INEI. (2017). *Proyecciones y estimaciones en base al censo 2017*. https://www.ine.cl/estadisticas/sociales/demografia-y-vitales/proyecciones-de-poblacion

INEI. (2020). *Estado de la población peruana 2020*. https://www.inei.gob.pe/media/MenuRecursivo/publicaciones_digitales/Est/Lib1743/Libro.pdf

Kumar, A., Luthra, S., Mangla, S., & Kazançoğlu, Y. (2020). COVID-19 impact on sustainable production and operations management. *Sustainable Operations and Computers*, *1*, 1–7. https://doi.org/10.1016/j.susoc.2020.06.001

McKenzie, D., & Woodruff, C. (2017). Business Practices in Small Firms in Developing Countries. *Management Science*, *63*(9), 2967–2981. https://doi.org/10.3386/w21505

McKinsey & Company. (2020). *Survey: Peruvian consumer sentiment during the coronavirus crisis*. Our Insights – McKinsey & Company. https://www.mckinsey.com/business-functions/marketing-and-sales/our-insights/survey-peruvian-consumer-sentiment-during-the-coronavirus-crisis

Mejia-Argueta, C., Soto, O. C., Gamez, H. M., & Moreno, J. P. (2015). Análisis del tamaño de empaque en la cadena de valor para minimizar costos logísticos: un caso de estudio en Colombia. *Journal of Management and Economics for Iberoamerica, 31*(134). https://doi.org/10.1016/j.estger.2014.06.009

Narayan, G., & Chandra, R. (2015). Factors Affecting the Purchase of Food and Grocery Products from Modern Retail Stores: An Empirical Study. *Journal of Management and Research, 14*, 7. https://www.semanticscholar.org/paper/Factors-Affecting-the-Purchase-of-Food-and-Grocery-Narayan-Chandra/62dc31b8dc2134cf74a22692bba31eb47dedcd6c

Perez-Franco, R. J. (2010). *A Methodology to Capture, Evaluate and Reformulate a Firm's Supply Chain Strategy as a Conceptual System*. https://www.researchgate.net/publication/266044104_A_Methodology_to_Capture_Evaluate_and_Reformulate_a_Firm's_Supply_Chain_Strategy_as_a_Conceptual_System

Perez-Franco, R., Singh, M., & Sheffi, Y. (2011). An approach to evaluate a firm's supply chain strategy as a conceptual system. Center for Transportation and Logistics, Massachusetts Institute of Technology.

Peru Retail. (2018). *Perú: Canal tradicional vs Canal moderno*. Peru Retail: La web de retail y los canales comerciales. https://www.peru-retail.com/peru-canal-tradicional-vs-canal-moderno/

Rangel-Espinosa, M. F., Hernandez-Arreola, J. R., Pale-Jimenez, E., Salinas-Navarro, D. E., & Mejia-Argueta, C. (2020). Increasing Competitiveness of Nanostore Business Models for Different Socioeconomic Levels. *Supply Chain Management and Logistics in Emerging Markets*, 273–298. https://doi.org/10.1108/978-1-83909-331-920201013

Salinas, D., Fransoo, J., Mejia-Argueta, C., & Benitez, V. (2018). *An Overlooked Weapon in the Fight Against Malnutrition: Small Store Logistics*. Food Logistics. https://www.foodlogistics.com/warehousing/blog/21032010/an-overlooked-weapon-in-the-fight-against-malnutrition-small-store-logistics

Sillo-Acuna, E. L. (2021). Emprendimiento y resiliencia : caso de las bodegas de barrio en el Perú durante la pandemia de covid-19. *Desde el Sur, 13*(1), 1–18. https://doi.org/10.21142/DES-1301-2021-0012

United Nations. (2018). *The World's Cities in 2018*. https://www.un.org/en/events/citiesday/assets/pdf/the_worlds_cities_in_2018_data_booklet.pdf

United Nations. (2020). *Global State of Metropolis 2020 – Population Data Booklet*. https://unhabitat.org/global-state-of-metropolis-2020-%E2%80%93-population-data-booklet

United Nations Development Programme – Peru. (2021). *INNOVA TU MERCADO: Emprendedoras lideran la transformación digital de centros de abasto*. PNUD Peru. https://www.pe.undp.org/content/peru/es/home/presscenter/articles/2021/innova-tu-mercado%2D%2Demprendedoras-lideran-la-transformacion-digit.html

Zameer, A., & Mukherjee, D. (2011). Food and Grocery Retail: Patronage Behavior of Indian Urban Consumers. *South Asian Journal of Management, 18*(1), 119–148.

Evaluation of COVID Restrictions in Airport Flight Management Using Discrete Event Simulation

Eduardo Carbajal, François Marmier, and Ivana Rasovska

Abstract The present work uses discrete event simulation to incorporate COVID restrictions into passenger flow service in an international airport. It includes modifications in gate assignment processes, passenger transportation and operations handling during restricted time intervals to the covid pandemic. The impact of this modifications in service indicators such as process time and level of satisfaction (LoS) are presented and discussed focusing on the update of the bus fleet and the gate allocation results.

Keywords Airport Management · Discrete event simulation · Covid-19

1 Introduction

The general closure of air borders for the transit of national passengers and international tourists led the airline industry to a serious crisis as a consequence of the COVID-19 pandemic during the beginning of 2020. It is estimated that during the first eight months of 2020, airlines globally lost revenues in the order of USD 256 billion. Of this, 7% (USD 19 billion) correspond to losses originating in Latin America and the Caribbean. In this region alone, the fall endangered more than five million jobs, in addition to seriously affecting the tourism industry that depends on the arrival of foreign tourists. The pandemic also casts doubt on numerous airport expansion plans and air logistics infrastructures, as projected passenger flows will not recover in the medium term. The aerospace industry sector, both for the repair and certification of aircraft, as well as for the construction of parts and new devices, has also been affected by the crisis in the sector, which has caused the cancellation of

E. Carbajal (✉)
Pontificia Universidad Católica del Perú, San Miguel, Perú
e-mail: ecarbajal@pucp.pe

F. Marmier · I. Rasovska
University of Strasbourg, Strasbourg Cedex, France

J. Vargas Florez et al. (eds.), *Production and Operations Management*, Springer Proceedings in Mathematics & Statistics 391,
https://doi.org/10.1007/978-3-031-06862-1_40

purchase orders for aircraft already under construction. Although there has been a significant recovery in the sector since the beginning of 2021, the flow of passengers has not yet reached the expected levels. This represents a significant challenge in the context of adapting to the new style of the pandemic due to the change in passenger behavior after the crisis of the year 2020 and before the successive waves in different countries that affect the airport operations.

In the next couple of years, airlines face two key uncertainties: (1) the cost of health-related measures and (2) the new protocols necessary for the management of airport operations. An increase in operating costs has been observed since the gradual reactivation of operations, both for airlines and airports, due to additional health and safety requirements (for example, disinfection, properties, modifications in infrastructure and adaptation of specialized signage, necessary additional equipment, temperature controls or viral tests) before they can be passed on to consumers. Furthermore, if applied for air travel, social distancing measures could force a reduction in the passenger load factor (i.e. the number of seats that can be occupied during a flight) of up to 50% in some extreme cases when the pandemic peaks in specific waves.

The other fundamental aspect refers to international travel restrictions, the contraction of economic activity and changes in transport habits by cautious consumers may prevent a return to pre-crisis levels of demand. This is expected even as that confinement measures and domestic travel restrictions are relaxed in many countries. Commercial air traffic is recovering slowly: as of September 2020, the number of flights remains more than 40% below pre-crisis levels worldwide according to IATA.

The combination of negative demand and supply shocks, as well as uncertainty around the medium-term outlook, creates an uncertain outlook for airlines. Through inter-industry links, this uncertainty affects the entire airline industry. Furthermore, the industry continues to be exposed to a possible outbreak of the pandemic, as governments may impose new restrictions on air travel to combat outbreaks or a possible second wave of infections. These changing conditions make it of particular interest to study the immediate consequences of the main effects on the logic of operations in airport management in order to anticipate their effects on the optimization of the usual services related to passenger care at the terminals. This leads to introduce into airport operations the necessity of social distancing in each possible process, especially in those that require in regular conditions crowd congestion or long waiting times.

All thought this numbers will suppose less flights at the start of the pandemic, the number of flights is increasing daily as restrictions of travelling are being lifted. However, the new reality shows that social distancing is a new restriction that will stay in long terms on the airport operations. Social distancing reduces the effective capacity of infrastructure and resources (less sits available in apron buses, fewer gates available to board at the same time if they are close and similar cases.) so as numbers rise this will pose a critical issue to address related to resource sizing as in bus fleet sizing and gate allocation. The reason is because the regular decision parameters in airport processes still need to satisfy international standards (such as those stablish by the International Air Transport Association (IATA)) which focus on LoS and process time.

In order to present this work, Sect. 2 presents literature related to the impact of the COVID-19 pandemic in international airports management. Section 3 presents the problems statement. Section 4 establishes the model proposal for this problem while Sect. 5 presents the detailed case study approached. Finally, in Sect. 6 the results will be presented and analyzed.

2 Literature Review

During the last year there have been some investigations related to the impact of the COVID-19 pandemic on airport operations. This researchs mainly relate to the analysis of the new restrictions imposed by the introduction of social distancing in the processes related to passenger departures and arrivals in airports.

Within these studies, some focus on the strategic direction to optimize costs to reduce the impact of operating cost overruns generated in airport operations to guarantee the necessary conditions to minimize the possibility of contagion, as in (Jong Hae 2021). Other works relate to focusing the analysis on how commercial aviation can play a successful role in maintaining its operations while not playing a role that is detrimental to the spread of the pandemic. In this case, the impact of COVID-19 on operations is analyzed in detail and how to adapt them for the commercial restart stage that occurred during 2021. An example of this is (Xiaoqian 2021) where some important operating conditions are defined within of the flights related to guarantee the spacing in different stages of the process that minimize the possibilities of contagion. Variations of different boarding methodologies are presented and tested used agent based simulation being the reverse pyramid method the most frequently used due to being the most efficient in terms of average process time.

In this sense, the challenge posed in 2020 meant that airports can adapt their operations to guarantee social distancing, since they generate a significant impact on the capacity of different processes related to the departure and arrival of passengers. Thus, related studies have been developed to focus on different processes where these conditions have a direct impact, for example in proposing analytical solutions on how to guarantee the processes of boarding and disembarking passengers to their flights during the pandemic as in (Schultz 2021) where boarding processes using passenger loading bridges(PLBs) or apron buses are studied.

On the other hand, it is important to recognize that in the commercial aviation industry a widely used tool to optimize processes and solve operational problems has been the simulation of discrete events for several years. There are works developed in simulation of different airport processes, taking many case studies in international airports, such as in (Fonseca i Casas 2017) where the simulation of passenger flow is used, or simulation is used to improve passenger flow by reducing the service times as in (Ashish 2021), to other studies focused on other more specific processes that have a direct impact on airport management as in (Shangyao 2021) where the allocation of boarding gates is evaluated using simulation. It is also important to cite examples of simulation cases to analyze the resilience of the terminals just due to the highly changing conditions as it is a common scenario currently as a consequence of COVID-19 as it develops in (Metznera 2019). These studies within the last year add to many other evidence of simulation application cases in airports, which given the context of the pandemic provide the opportunity to adapt existing models to include greater controllable parameter management capabilities to adapt solutions and capacity estimates to the current reality of the pandemic. In this context the problem at hand is to optimize action parameters in the boarding processes, because they suppose the highest risk because of passenger crowding.

3 Problem Statement and Proposal

The COVID-19 pandemic has generated substantial changes in airport operations aimed at establishing conditions to minimize the risk of exposure and contagion during the boarding, flight and onboarding of commercial flights. The main requirement is related to guaranteeing adequate social distancing within the terminals and the aircraft themselves. In particular, with regard to terminals, this means implementing measures to ensure that spacing is guaranteed throughout all the processes involved in the departure and arrival of passengers. Ensuring this spacing is obviously subject to existing infrastructures in most airports, so the operating conditions must be adjusted to be able to meet these new conditions. It is necessary to size the resources involved in various processes in an appropriate way.

Within the scope of the boarding and unboarding process it will focus on some processes that are considered critical. Within these processes the critical ones correspond to the adequate spacing in the waiting rooms and the adequate spacing in the transport to the airplanes when this is not carried out by PLBs but by apron buses.

Within waiting rooms, the infrastructure of the boarding gates commonly places successive gates closely to allow access to consecutive PLBs. Under current conditions, operating with the total gates of the terminals would mean that the allowed capacity is exceeded given the new capacities due to pandemic, so that in many terminals policies must be reassigned to be able to use a lower number of gate than those available or redirect shipments on the first floors of the terminals where they can board by buses. These modifications in the policies for assigning gates should be considered so that the simulations include the process of assigning gates, disabling gates close to the time they are assigned for flight boarding.

The issue of transportation by buses in terminals that have boarding gates on the first floor is also critical, since normally the standard buses in airports in the Latin American region usually operate with a regular capacity of 78–80 passengers, but in pandemic conditions these same buses cannot operate with more than 40 passengers to ensure that a bus of the same dimensions maintains a proper spacing between passengers. Obviously this means that aircraft boardings in many cases need to increase the number of movements of buses to transport passengers to keep the process time under international IATA standards. As many airports normally sized their bus fleet according to normal operating conditions, the need of social distancing imply a recalculation of the fleet estimates. This must also be sync with the gates assignment process described previously.

4 Simulation Model of the Airport Processes

The starting point is the simulation model developed in (Carbajal et al. 2019), which comprises a simulation model on the determination of the bus fleet but which includes the entire process of generation of arrivals and arrivals flights including in its reach the modeling of the waiting areas, allocation of gates and processes of passenger transport to and from the aircraft. Based on this model, two main sub-models need to be enhanced in order to include the modifications of the COVID-19 policies in the model of a port terminal. This includes in the allocation of gates and in the process of loading passengers.

4.1 Gate Assignment Policies

The distancing conditions imply including operational restrictions in the allocation of gates, so the model can be used to be able to obtain, along with the bus fleet sizing, the list of recommended gates per day and per hour according to the schedule of projected scheduled flights updated after the impact of the pandemic. It is proposed not only to define gates as simple resources, but also to convert them into variables so that a logic similar to that developed in (Carbajal et al. 2019) used for remote locations can be used, but now for gates so that a gate assignment matrix is developed that allows blocking some gates while others are assigned (taking the proximity of distance between them as a criterion) so that in the adjoining waiting rooms the flights are distributed ensuring greater availability of space to wait for passengers on the same flight. A small fraction of this binary variable locking logic for gates can be seen in Fig. 1.

4.2 Loading/Unloading Process in Aircrafts

In particular, another sub-process of the model that must be updated is the transfer of passengers to and from airplanes, particularly in cases where buses are used. In the first place, the capacity of the bus fleet is updated to 50% of the initial capacity to reflect the distancing conditions in the transfer. As a second modification, priority is

Fig. 1 Cross blocking matrix for gate assignments

GATES	1	2	3	4	5	6	7	8	9	10
1	1	1	0	0	0	0	0	0	0	0
2	1	1	0	0	0	0	0	0	0	0
3	0	0	1	1	0	0	0	0	0	0
4	0	0	1	1	0	0	0	0	0	0
5	0	0	0	0	1	1	1	0	0	0
6	0	0	0	0	1	1	1	0	0	0
7	0	0	0	0	1	1	1	0	0	0
8	0	0	0	0	0	0	0	1	1	0
9	0	0	0	0	0	0	0	1	1	0
10	0	0	0	0	0	0	0	0	0	1

given to ensuring continuity of care for one flight before starting care for another. It should be noted that the model previously developed optimized the proximity of the bus since in extreme cases two buses were required for full flights of national departure, but currently this can mean up to 4 buses per flight in these cases. The objective will be to rerun the optimization model to determine the required fleet size update under this logic. To achieve the priority assignment to a flight of the necessary buses, an additional attribute is introduced that can generate a compound expression of the priority linked to the proximity of the gate but with greater weight to the assignment of a new bus to a flight that has not yet completed the total number of buses needed to cover the flight.

5 Case Study

The application of the model is developed in an international airport in Peru. In addition to the modifications in the logic described in the previous section, the input data for the arrival and departure of flights is updated based on updated projections for the current year and the following years based on changes in demand based on the effect. Estimated by the COVID-19 pandemic. This involves updating the input files of the simulation model, as well as the new estimated flight occupancy levels to have a more accurate approximation of the number of passengers per flights. The case study is applicable to the issue of gates spacing as well as bus transportation as it has several gates on the first floor that require such type of passenger transportation when the aircraft are in remote position. From the case study, we will seek to obtain the suggested gates programming from the simulation model as an answer and also through the use of simulation-based optimization the updated fleet sizing. These results are presented in the next section.

6 Results and Discussion

The change in passenger behavior after the COVID-19 crisis, travel restrictions and the ensuing economic crisis led to a drastic reduction in demand for airline services. According to the International Air Transport Association (IATA), passenger air transport measured as revenue per passenger per kilometer was down 90% year-on-year in April 2020 and was still down 75% in August. The collapse of economic activity and trade affected cargo operations, which were almost 30% lower year-on-year in April and still nearly 12% lower in August. This can be confirmed in the study case. After running the simulation-based optimization of the model, it was possible to size the bus fleet required annually. Figure 2 comparatively shows the new fleet sizes compared to those previously estimated before the COVID-19 pandemic. The reduction in bus capacity to 50% allows evidence of an increase in fleet size of around 40% on average compared to pre-pandemic estimates. Although a 200% increase could have been expected in theory, in many cases the new projections of flight passengers, as they do not have the same occupancy level and do not always double the number of buses per flight, allow us to obtain a manageable result with respect to the fleet.

On the other hand, given that the fleet size was obtained, the model was run to obtain the recommended allocation of gates per hour and day for the current year. Obtaining these results allows to validate that the adaptability of previously built models can easily include updated logic to obtain applicable results in the context of the needs of the pandemic. It would be relatively and easy to be based on this premise to be able to use simulation in the sizing and programming of airport resources, in other processes not reviewed in this document, such as in other airport processes such as Check-in, security control, migrations. Customs, among others.

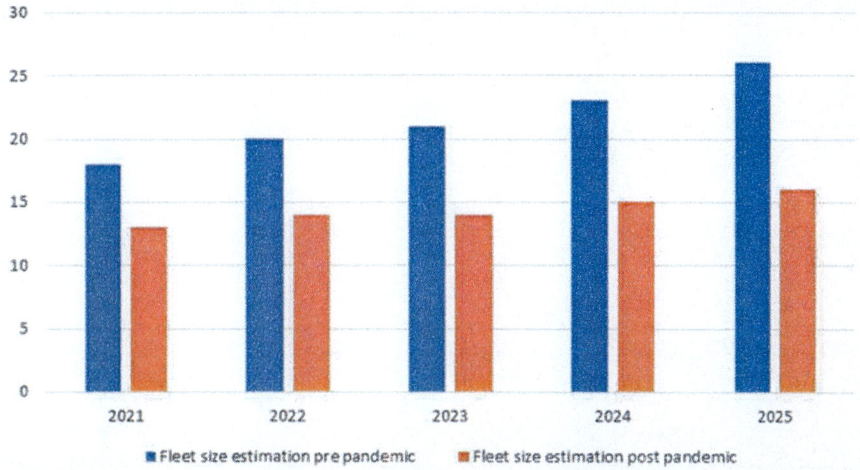

Fig. 2 Fleet size estimation: Comparison between pre and post pandemic

The drop in air traffic is even more pronounced in the case of long-haul flights. In the long term, changes in consumer behavior can generate structural changes in the demand for air transport. While the upturn in domestic flights in China indicates that traffic could return to pre-crisis levels, a permanent drop in demand from pre-crisis levels cannot be excluded, either through modal shifts in trade in goods. Services or, to a lesser extent, by substitution with other modes of transport (for example, high-speed trains). These modifications should be taken into account for future instances of studies in this field since they will generate the need to experiment and enhance simulation models of this operations in order to achieve results adequate to the new conditions in decision making.

References

Ashish, V. 2021. "Agent Based simulation model for improving passenger service time at Bangalore airport". In Case Studies on Transport Policy Volume 8

Carbajal, E., Marmier, f. & Fontanili, F 2019, "Bus Fleet Size Dimensioning In an International Airport Using Discrete Event Simulation," 2019 Winter Simulation Conference (WSC), 2019, pp. 464–475, https://doi.org/10.1109/WSC40007.2019.9004878

Pau Fonseca i Casas 2017. "Passenger flow simulation in a hub airport: An application to the Barcelona International Airport". In Simulation Modelling Practice and Theory Volume 75

Jong Hae, C. 2021. "Changes in airport operating procedures and implications for airport strategies post-COVID-19" In *Journal of Air Transport Management Volume 94*

Metznera, N. 2019. "A comparison of agent-based and discrete event simulation for assessing aiport terminal resilience". In *Transportation Research Procedia Volume 43.*

Schultz, M. 2021 "Analytical approach to solve the problem of aircraft passenger boarding during the coronavirus pandemic" In *Transportation Research Part C: Emerging Technlogies Volume 124.*

Shangyao, Y. 2021 "A simulation framework for evaluating airport gate assignments". In *Transportation Research Part A: Policy and Practice*

Xiaoqian, S. 2021 "COVID-19 pandemir and air transportation:; Successfully navigating the paper hurricane". In *Journal of Air Transport Management Volume 94*

Evaluation of Variables to Determine Cycling Routes in Lima-Perú

Miguel Rodríguez, Jonatán Rojas, Héctor Mattos, Steffano Reyes, Carlos Trujillo, Alexia Cáceres, and Jackeline Alva

Abstract The demand for transport without agglomerations in many cities has increased lately to avoid infections of COVID19. Therefore, this research is focused on a mathematical model for a cycling network minimizing the distances of cycle paths and maximizing the factors that are more important for citizens of Lima using linear programming and analytics.

Keywords Cycling network · Optimization routes · Operations research in urban logistics

1 Introduction

Lima with approximately 9 million inhabitants [1], whose 65% [2] are not satisfied with their city. The main causes of this dissatisfaction are insecurity and public transport with a 75% and 51% frequency, respectively (the survey admitted multiple options). Additionally, Lima is considered by the WHO [3] as the South American capital with the worst air quality because of the higher PM2.5 and PM10 rates and this pollution is mainly due to the automotive park growth. For this reason, it is important to implement less polluting transportation systems in the Peruvian capital.

In recent years, transport specialists and urban planners have promoted urban cycling as an effective and sustainable mode of transport [4]. However, the construction of cycle paths has not increased the proportion of bicycle trips towards workplaces or educational institutions as they have become less frequent [2]. This lessening can be seen in Fig. 1, where there is a clear trend of constant decrease. It is precise to recall that the information of 2016 must-have contained an error in the method of analysis by the entity that made it.

M. Rodríguez · J. Rojas · H. Mattos · S. Reyes · C. Trujillo · A. Cáceres · J. Alva (✉)
Pontificia Universidad Católica del Perú, San Miguel, Peru
e-mail: miguel.rodrigueza@pucp.edu.pe; jrojasp@pucp.pe; hmattos@pucp.pe; s.reyes@pucp.pe; carlos.trujillod@pucp.pe; alexia.caceres@pucp.pe; jalvaz@pucp.pe

J. Vargas Florez et al. (eds.), *Production and Operations Management*, Springer Proceedings in Mathematics & Statistics 391,
https://doi.org/10.1007/978-3-031-06862-1_41

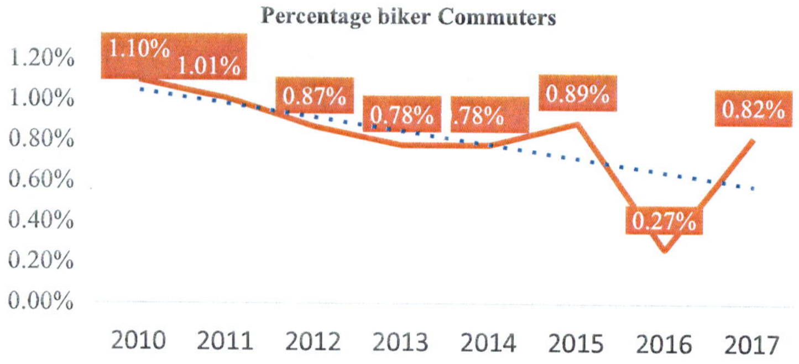

Fig. 1 Percentage biker Commuters

Therefore, it is essential to know the characteristics of how Lima cyclists design their routes to set up an infrastructure according to their needs and, consequently, increase the proportion of bicycle trips to work or study. The choice of routes for cyclists depends not only on the minimization of distance but also on some interconnected factors such as avoiding high positive slopes, reducing high bus flow paths and so on [4–6].

2 State of the Art

Within the first studies on the choice of routes by cyclists, cyclists were asked to draw on a map of the routes they chose [7–9]. These studies revealed that most gave very imprecise information and therefore unreliable one [4].

Nonetheless, thanks to the new GIS and GPS technologies, studies have been better refined, for instance [10] it is possible to identify those cyclists in a US city tend to drive more comfortably in high traffic conditions because vehicles advance at very low speed. A [4] study of the behavior of cyclists in Sao Carlos – Brazil was carried out with a sample of 50 cyclists where it was found that 19% of the trips traveled fewer distances than the minimum possible distance because it is rare for Brazilian cyclists to take routes in the opposite direction of traffic or going byways that are not legally possible.

Also, [11] cyclists from San Francisco were asked to use a mobile application to record their route information and it found out that cyclists prefer bicycle lane routes, and those female cyclists prefer routes with low accumulated slope. In some studies that took place in Canada [12], GPS devices were used to collect information on trips made by cyclists, and it was found that there is a strong relationship between population density and the number of bicycle trips generated.

The research was conducted in the Czech Republic to analyze the circumstances and causes of death of 83 registered cyclists in the regions of Olomouc and Zlín [13] between 2005 and 2013, where it was determined that the driver was guilty in 68.7%

and the cyclist in 31.3%. The cars were involved in 62%, heavy vehicles, including buses, by 31%. Trucks were more dangerous to cyclists than cars at intersections, while cars were in straight sections. The most important pattern that was identified was that motorized vehicles hit a cyclist from behind in a straight section in 22% of cases.

In the Netherlands, a mode change in the direction of sustainable transport was observed through two national cycling policy programs, promoting that people who drive on congestion-prone routes change their transport behavior towards using bicycles [16]. The Inter-American Development Bank indicates that there is a high correlation in Latin America and the Caribbean between the level of adoption of bicycles and an improvement in indicators such as road safety, sales, and equality, as well as in the reduction of theft and greenhouse gas emissions [17].

3 Case Study

Along this section, the employed methodology for the research will be set forth.

3.1 Data from Surveys

A survey was carried out in a Google Form that can be viewed at the following link https://goo.gl/ugv4ha where they were asked to place the Strava URL of the first oneway route to work or study. The first route is a better indicator for each individual than the route back home because it is very important to get to his duties without signs of tiredness or intense sweating as most jobs in Lima or schools do not have showers or there are very few showers so that cyclists can take a bath comfortably.

240 responses were obtained of which only 50 responses included the URL of their routes. Of the 240 people, 73% were male and 27% were female, as can be seen in Table 1. This percentage shows that it is mainly a male mode of transport in a similar percentage to countries such as Australia, United Kingdom, United States, and Canada and that is a very low percentage compared to the female participation of European countries that are about 50% [14].

Table 1 Frequency of trips

Gender		Frequency of trips to work or study per week	
Male	73%	[0,1] No or almost no frequency	24%
Female	27%	[2,3] Moderate frequency	17%
		[4,7] High frequency	59%

3.2 Data from Strava

Data was collected from the Strava app, which is a platform for cyclists to record all the details of their routes. The sample comprehended 100 random records from Lima cyclists of the club "Lima Bike", and it can be visualized in Fig. 2.

3.3 Analysis from Survey Data

The factors analyzed were a number of buses, neighborhood security, pavement quality, street width, and street lighting which are plotted in a boxplot grouped by gender and frequency of weekly use of respondents in Fig. 3.

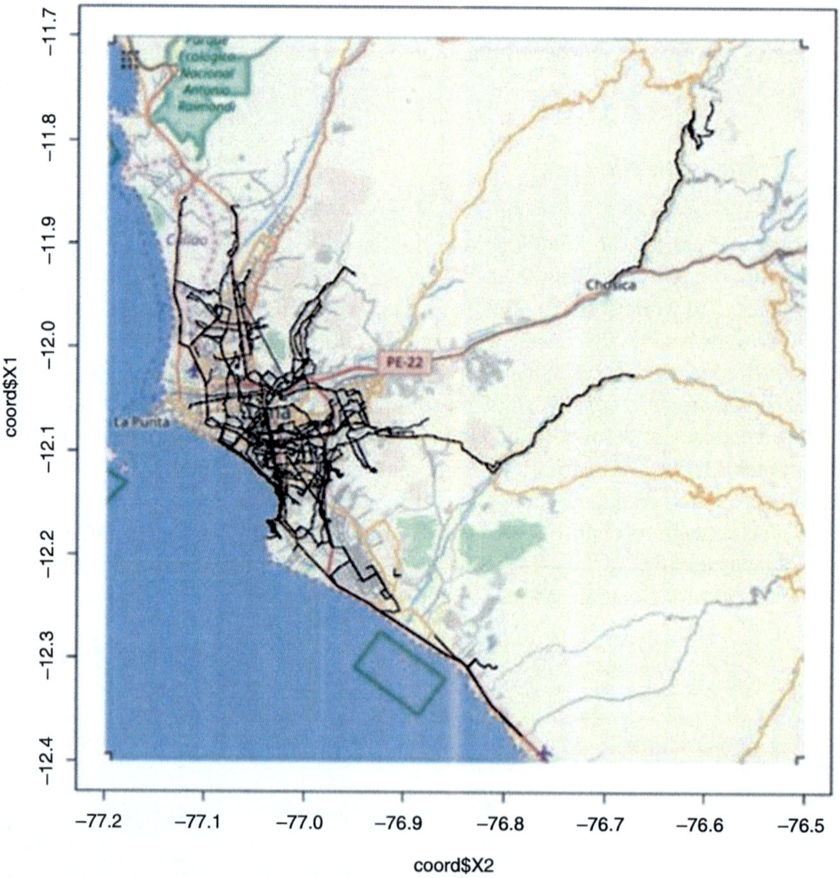

Fig. 2 Data gathered from Strava

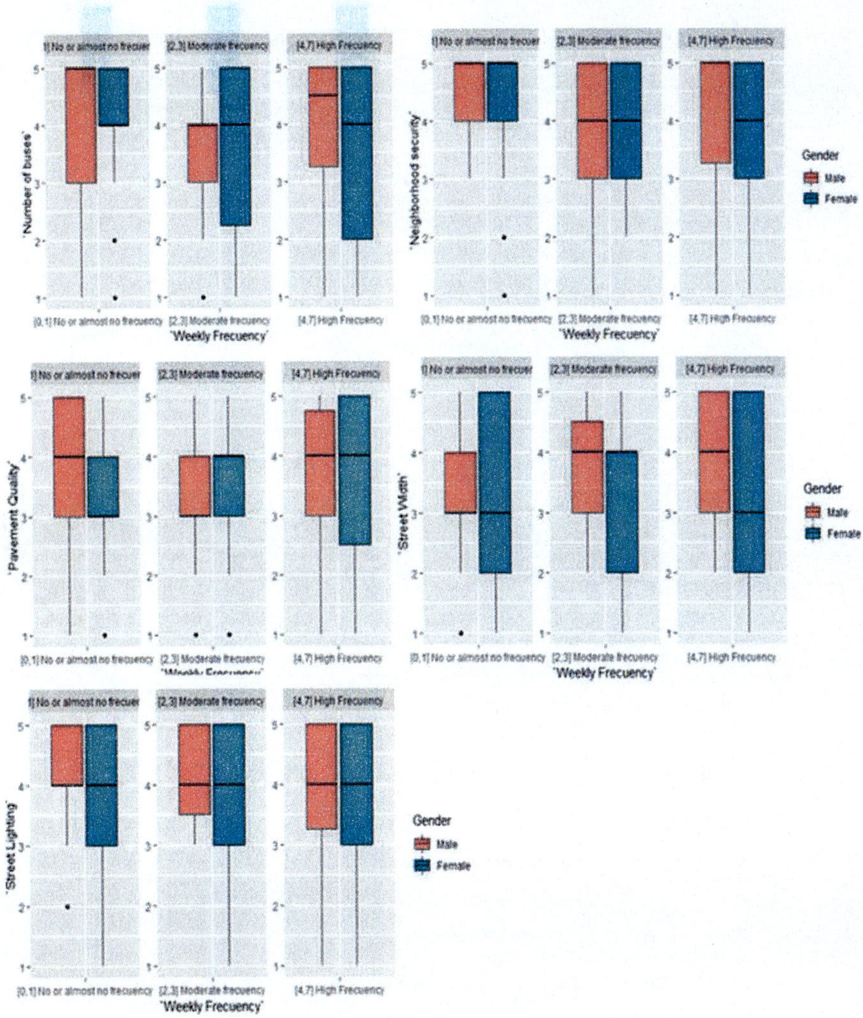

Fig. 3 Boxplot by gender and frequency of weekly use

To identify whether the effects of gender or frequency of bicycle use per week are statistically significant, an Analysis of Variance (ANOVA) is applied to each factor using the statistical software R and the results can be visualized in Fig. 4. It can be observed that there is a difference in the factor of street lighting in the gender grouping, while in the others there is no significant effect of the groupings of gender and frequency of use, which therefore does not deserve a dissection in those parts.

The Duncan method was used to rank the means of the number of buses to have a view of which subgroup has a greater sensitivity of this and the results are shown in Fig. 5.

Fig. 4 Importance of the factors for the choice of route and Duncan analysis

```
> anova(lm(`Number of buses` ~ `Weekly Frecuency` + Gender))
Analysis of Variance Table

Response: Number of buses
                    Df Sum Sq Mean Sq F value  Pr(>F)
`Weekly Frecuency`   2  12.60  6.3007  3.1552 0.04443 *
Gender               1   3.03  3.0260  1.5153 0.21955
Residuals          236 471.27  1.9969
---
Signif. codes:  0 '***' 0.001 '**' 0.01 '*' 0.05 '.' 0.1 ' ' 1
> anova(lm(`Neighborhood security` ~ `Weekly Frecuency` + Gender))
Analysis of Variance Table

Response: Neighborhood security
                    Df  Sum Sq Mean Sq F value Pr(>F)
`Weekly Frecuency`   2   4.783 2.39173  1.9754  0.141
Gender               1   0.637 0.63677  0.5259  0.469
Residuals          236 285.742 1.21077
> anova(lm(`Pavement Quality` ~ `Weekly Frecuency` + Gender))
Analysis of Variance Table

Response: Pavement Quality
                    Df Sum Sq Mean Sq F value Pr(>F)
`Weekly Frecuency`   2   3.10 1.54938  0.8993 0.4082
Gender               1   0.23 0.23295  0.1352 0.7134
Residuals          236 406.60 1.72289
> anova(lm(`Street Width` ~ `Weekly Frecuency` + Gender))
Analysis of Variance Table

Response: Street Width
                    Df Sum Sq Mean Sq F value  Pr(>F)
`Weekly Frecuency`   2   2.80  1.4017  0.8619 0.42370
Gender               1   5.36  5.3578  3.2942 0.07079 .
Residuals          236 383.84  1.6264
---
Signif. codes:  0 '***' 0.001 '**' 0.01 '*' 0.05 '.' 0.1 ' ' 1
> anova(lm(`Street Lighting` ~ `Weekly Frecuency` + Gender))
Analysis of Variance Table

Response: Street Lighting
                    Df Sum Sq Mean Sq F value Pr(>F)
`Weekly Frecuency`   2   2.64  1.3205  0.9246 0.3981
Gender               1   3.49  3.4852  2.4403 0.1196
Residuals          236 337.06  1.4282
```

```
> model <- aov(`Cantidad de buses` ~ Frecuencia_uso_semanal)
> duncan.test(model, "Frecuencia_uso_semanal", group=T, console=T)

Study: model ~ "Frecuencia_uso_semanal"

Duncan's new multiple range test
for Cantidad de buses

Mean Square Error:  2.001242

Frecuencia_uso_semanal,  means

                                  Cantidad.de.buses      std   r Min Max
[0,1] Nula o casi nula frecuencia          4.051724 1.190937  58   1   5
[2,3] Poca frecuencia                      3.512195 1.398606  41   1   5
[4,7] Alta frecuencia                      3.517730 1.500490 141   1   5

Groups according to probability of means differences and alpha level( 0.05 )

Means with the same letter are not significantly different.

                                  Cantidad de buses groups
[0,1] Nula o casi nula frecuencia          4.051724      a
[4,7] Alta frecuencia                      3.517730      b
[2,3] Poca frecuencia                      3.512195      b
```

The agricolae library is used in the statistical software R and it leads to the conclusion that people who rarely drive to their center of studies give greater importance to the number of buses when they choose their routes than people who drive with a moderate or high frequency. The perception of the factors of this subgroup will be preferred because it is more critical to evaluate them with the other factors, therefore the values of the answers of people who ride a little or with a

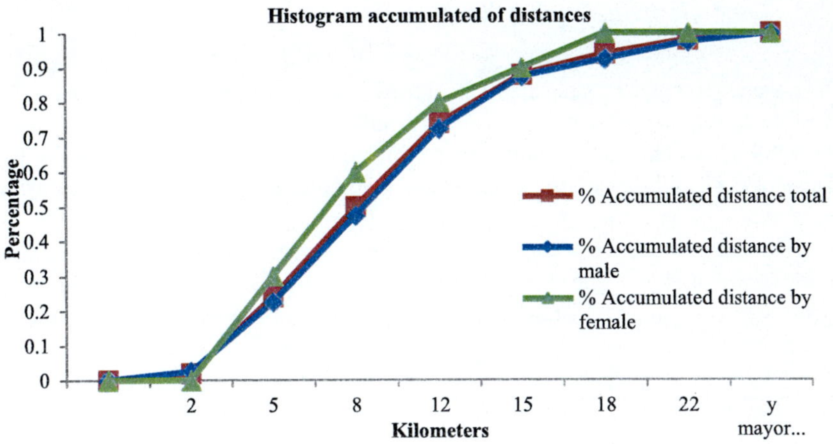

Fig. 5 Histogram accumulated of distances

Table 2 Rank of factors

ID	Factor	Lower limit	Average	Upper limit
F1	Number of buses	3.98	4.05	4.12
F2	Neighborhood Security	3.95	4.09	4.23
F3	Quality of Street	3.35	3.52	3.68
F4	Street Width	3.34	3.50	3.66
F5	Street Light	3.66	3.81	3.96

Table 3 Weights of factors

Factor	Weight	Factor	Weight
F1	0.4	F3	0.04
F2	0.4	F4	0.04
F5	0.13		

high frequency will be replaced by the average value of those who ride with no or almost null frequency.

Afterward, the average of all the factors will be compared through the analysis of their confidence intervals at 95%, as can be seen in Table 2 to evaluate which one we should prioritize over the others. Having the results of Table 2 we rank the factors as F4 = F3 <F5 <F1 = F2, being the factors of the number of buses and security of the neighborhood the most critical for Lima urban cyclists.

Then, the AHP technique was applied to assign weights to the factors that were hierarchized in Table 3, with the criterion of assigning a value of 9 when the difference of importance is large between the two variables of 5 when the difference is medium and 1 when they estimate the same degree of importance. Each value of the paired matrix was divided by the sum of the corresponding row, obtaining the normalized matrix. Finally, the averages per row were calculated to obtain the weights to assign to each factor shown in Table 3 ordered by importance.

3.4 Analysis from Strava Data

The distance traveled of male and female trips to work is shown in Fig. 5, where it is possible to appreciate that approximately 50% of travel distances have a length below 8 km, which is in line with Marketwin's study [15] and the *Plan Maestro de Transporte Urbano de Lima*, which states that Lima residents generally travel an average of 8 kilometers to work.

Also, approximately 90% of travel distances cover less than 15 kilometers to go to work, which shows that if good policies are implemented to promote the use of bicycles as a means of transport, the transport problem in Lima could be solved to a greater extent.

4 Conclusions

Contrasting the map obtained by the routes carried out by Strava in Fig. 2 and the current cycleways map in Lima, it will be seen that the bike paths of Villa María del Triunfo are underutilized and the cyclists from Villa el Salvador have created different routes to the demarcated by their cycleways. We also visualize those districts such as San Juan de Lurigancho that lack cycle paths have a high power because there are cyclists who have created routes to connect their district with others, the same with Callao in the north near Ventanilla. La Molina, which has very short cycle paths of no more than 5 km per segment, is seen to be short because of the connections that its cyclists make from there, even though this district has very few vehicular entrances and has several slopes, unlike most flat territory of Lima.

Although the sample from this study is indeed small, it is much more representative than many other studies on the subject, so it is advisable to enlarge the sample. For instance, data could be directly gathered from Strava Metro so that it would be possible to carry out the bike lanes under the interactive planning of its population. Another limitation of the study is that Lima does not have a statistic of injured and dead cyclists and that is georeferenced.

Areas where bike lanes are noted and high demand for cyclists should be prioritized for the construction of bikeways that would help make their transport much safer and comfortably avoid areas with many buses that meet these bikeways. Many bicycle lanes are built in Lima, but they lack systemic thinking that reproduces the behavior of Lima urban cyclists. It is proved that there are stretches of bike lanes smaller than 8 kilometers, although this is the average distance for which cyclists mobilize. Lima culture tends to believe that urban cycling is not for women but, in this study, it is shown that the physical effort measured by average distance is practically the same as executed by men. Thus, the policies on urban cycling should be more inclusive. Based upon this quantification of the degree of importance of the variables, further research might use these results to propose a bike path network in Lima that has a better representation of the preferences of the bike users.

References

1. Instituto Nacional de Estadística e Informática (INEI). Encuesta Nacional, 2017
2. Lima Como Vamos. VIII Informe de resultados sobre calidad de vida. Lima, 2017
3. OMS. Global Urban Ambient Air Pollution Database, 2016
4. Pereira and Penha. Identification of factors that influence cyclists route choice, 2014
5. Howards and Burns, Cycling to Work in Phoenix: Route Choice, Travel Behavior, and Commuter Characteristics, 2016
6. Broach et al. Where do cyclist ride? A route choice model developed with revealed preference GPS data, 2014
7. Aultman-Hall, L., Hall, F., & Baetz, B. Analysis of bicycle commuter routes using geographic information systems: implications for bicycle planning, 1997
8. Hyodo, T., Suzuki, N., & Takahashi, K. Modeling of bicycle route and destination choice behavior for bicycle road network plan. Transportation Research Record: Journal of the Transportation Research Board, 70–76. 2000.
9. Howard, C., & Burns, E. Cycling to work in Phoenix: route choice, travel behavior, and commuter characteristics. Transportation Research Record: Journal of the Transportation Research Board, 39–46. 2001
10. Casello, J., Rewa, K., & Nour, A. An Analysis of Empirical Evidence of Cyclists' Route Choice and their Implications for Planning. In 91st Annual Meeting of the Transportation Research Board. 2012
11. Hood et al, A GPS-based bicycle route choice model for San Francisco, California. 2011
12. Cassello et al, Modeling Cyclists' Route Choice Based on GPS Data. 2012
13. Bil et al, Fatal cycling accidents in the Czech Republic: factors and causes of death. 2015
14. Sustrans, Inclusive city cycling. Women: reducing the gender gap. 2017
15. Marketwin Estudio del tráfico en Lima. Disponible en: http://marketwin.pe/estudiotrafico/ 2017
16. Bruno, M. & Nikolaev, A. Towards a maintenance-based approach to mode shift: Comparing two cases of Dutch cycling policy using social practice theory. 2020
17. Banco Interamericano de Desarrollo. Ciclo-inclusión en América Latina y el Caribe: Guía para impulsar el uso de la bicicleta. 2015

Optimization Model Applied to the Distribution of Covid-19 Vaccines in Lima and Callao

Juan Diego Guerra Vargas, Julio C. Quezada Rojas, and Eduardo J. Siuse Calixto

Abstract Currently, Peru has started a vaccination plan to face the crisis generated by Covid-19; however, it has not been the most efficient. The current study seeks to reduce vaccine distribution times and costs in Lima and Callao by applying a mathematical optimization model.

Keywords Optimization · Mathematical model · Supply chain · Covid-19

1 Introduction

As of August 01, 2021, the World Health Organization (WHO) has reported 210 million cases of COVID-19 globally; of which 39% has occurred only in America [1]. Particularly in Latin America, seeing the distant possibility of a rapid acquisition of vaccines, we opted for non-pharmaceutical interventions, such as mitigation and suppression, adapted to the capacity of each country [2]. In that area, the capacity of the Peru, although it has been increasing in recent years, continues to be low according to Dargent [3]. In Peru, since the first case was detected on March 6, 2020 until today, more than 2 million cases and more than 197 thousand deaths from this new disease have been confirmed [4]. In this context, vaccines came as the main solution, since they induce immunity against COVID-19, reducing symptoms and the risk of serious health consequences. This immunity generated by vaccines not only helps people to fight against this virus in case of infection, but also reduces the probability of infecting other people as well, therefore, indirectly reducing the transmission of the virus throughout the population [5, 6]. According to Fuentes [7], with the arrival of vaccines, the Peruvian economy is expected to increase from 7% to 12%, motivating people to consume products and services confidently as they are already vaccinated, thus reactivating businesses and promoting new jobs.

J. D. Guerra Vargas (✉) · J. C. Quezada Rojas · E. J. Siuse Calixto
Pontificia Universidad Católica del Perú, San Miguel, Perú
e-mail: jdguerra@pucp.edu.pe; jcquezada@pucp.edu.pe; eduardo.siuse@pucp.pe

J. Vargas Florez et al. (eds.), *Production and Operations Management*, Springer Proceedings in Mathematics & Statistics 391,
https://doi.org/10.1007/978-3-031-06862-1_42

While vaccines remain a rare and coveted commodity in the world, Gooding et al. [8] demonstrated that it is important to improve the supply chain to achieve greater immunization in less time and face up frequently stockouts because of supply chain issues. In that sense, the type, quantity, allocation and distribution of the vaccine are considered the four main components of the vaccine supply chain, Rastegar et al. [9]. It is important to meet demand of vaccines in the period of high request, for that Peru follows a vaccination plan for age ranges and as Han et al. [10] mentioned, the mortality rate can be minimized by prioritizing vaccination for people who work in the health sector, followed by adults over 65 years old; then continue with people between 40 and 65 years old and thus to gradually decrease until the majority of the population is vaccinated.

As a result of improving the allocation and distribution of vaccines, Shakiba et al. [11] presented a mathematical optimization model for the optimal distribution of vaccines, which consists of dividing the population into different subgroups and distributing the vaccine in an equitable and justified manner to each. Recently, Georgiadis et al. [12] developed a mixed integer linear programming (MILP) model for optimal planning of the supply chain of COVID-19 vaccines in order to reduce costs related to the storage and transportation. To face up the stockouts, Sabouhi et al. [13] proposed a multi-objective optimization model which it uses clustering in order to improve the network design and supply in the regions. A similar study by Souto et al. [14] proposed a clustering methodology, and so they minimize costs and increase the flow of products demanded by customers through the installation of intermediate warehouses. Altabeeb et al. [15] carried out a research where she used a problem of generation of routes for trained vehicles (CVRP) with the objective of finding the routes of the vehicles minimizing the distances traveled; she was able to achieve far more promising results compared to other known methods. Further-more, Akhtar et al. [16] also used a CVRP and got the best routes in order to reduce the distances traveled, reduce transport costs and increase the efficiency of meeting demand for each point in their research case.

In this research, the improvement proposal will be presented to save distribution and storage costs in the supply chain of vaccines against COVID-19 for Lima and Callao. The proposal will consist of two stages. The first stage consists of a clustering process for the location and use of secondary warehouses, and the second stage, consists of a mathematical model applied to a vehicle routing problem for the optimal distribution of vaccines from secondary warehouses to vaccination points. Its proposed to determine and select the facilities of the secondary warehouses of the comprehensive health network for each zonal division of Lima and Callao. The health centers, communitary health posts, and the headquarters of the health directions and the vaccination centers will be part of the new proposed supply chain. In addition, logistical aspects such as storage capacity and transport capacity, the cost associated with transport, storage costs and implementation costs will be tackled as well. Lastly, all virtual aspects related to COVID-19 vaccination were considered, such as the monthly demand for vaccines, the segmentation of the population into age ranges, among others. Based on this, the current vaccination plan was analyzed, focusing on real data obtained in the North region of Lima. The objective of the optimization proposal is to demonstrate the percentage of savings associated with costs and time in the distribution and storage of COVID-19 vaccines.

2 Methodology

2.1 Current Procedure

Presently, the vaccination system in Peru begins with the reception of vaccines from abroad, which are stored at the national supply addresses. Then, the vaccine is distributed to the different health directorates that are in charge of delivering the vaccines to the vaccination centers [17] in the areas assigned to them. There are five health directions that divide the capital into five zones: North Lima, East Lima, Central Lima, South Lima and Callao. Each of these is responsible for the distribution and storage of vaccines among the districts assigned to them. Figure 1 represents the vaccine supply chain for each area of Metropolitan Lima.

The current process has compromised the supply to meet the demand of each vaccination post, due to the limited response capacity and unplanned supply routes. According to the vaccination plan issued by the government [18], people from 12 to 39 years old, and those who are 40 years old or more laggards who were unable to vaccinate on the established dates, will be vaccinated between August and December this year. Therefore, the 18.7 million doses will be demanded and must be delivered during the respective periods to each of the vaccination points for each zone.

Figure 2 shows the monthly demand for each area. The demand reflects the population from 12 to 39 years old, information taken from CPI [19] where they estimate the number of people in Peru of this age group. On the other hand, the demand of the population aged 40 and over who had not received the first dose or the second dose was consider. The total population for the current year indicated by the INEI [20] for each district was use and the amount of doses applied in each of these was subtracted according to the government website [21].

With the current vaccination system, there are high costs for supplying vaccination centers. The demand from August to December will be up to 250% more than in the previous months. The cost of transportation is high because spacious units are required to be able to distribute from the Health Direction to all the vaccination

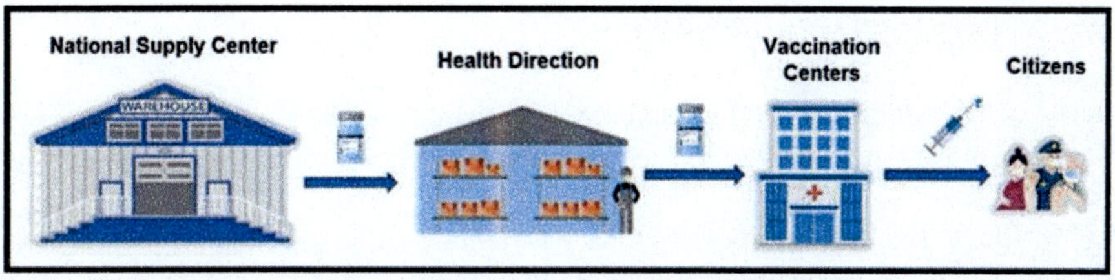

Fig. 1 Current Covid-19 Vaccine Supply Chain

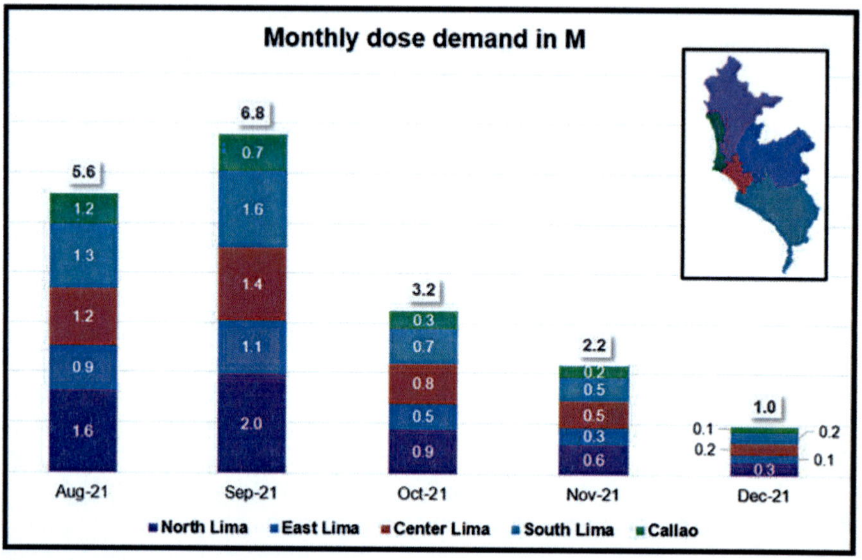

Fig. 2 Monthly dose demand

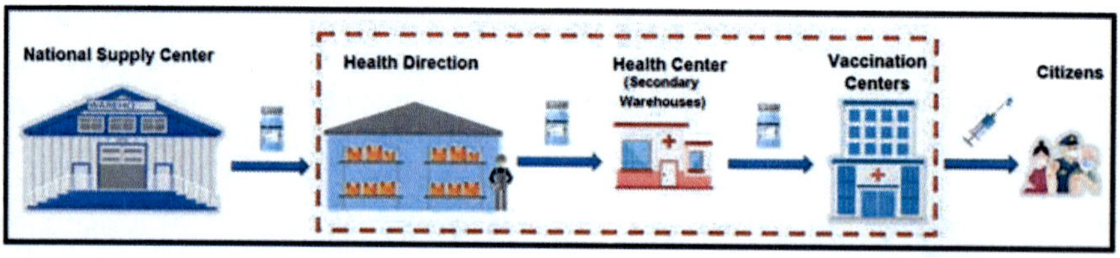

Fig. 3 Proposed supply chain for Covid-19 vaccines

points in each area, which leads to traveling for long stretches incurring cost overruns for not accelerating the rotation rate of the vaccinations and the cost of rescheduling trips to vaccination centers. The vaccination points demand a daily or inter-day supply and when evaluating a greater future demand there is a need to improve the vaccine distribution system in Metropolitan Lima.

2.2 *Improvement Proposal*

Based on the contributions of the referenced research and the analysis of the current situation, it was determined to include secondary warehouses in the supply chain, see Fig. 3.

For the optimal selection of secondary warehouses, the vaccination points were chosen to be in different areas of Metropolitan Lima, in order to find a geographic center. By applying the elbow method, performed in Python, it was possible to

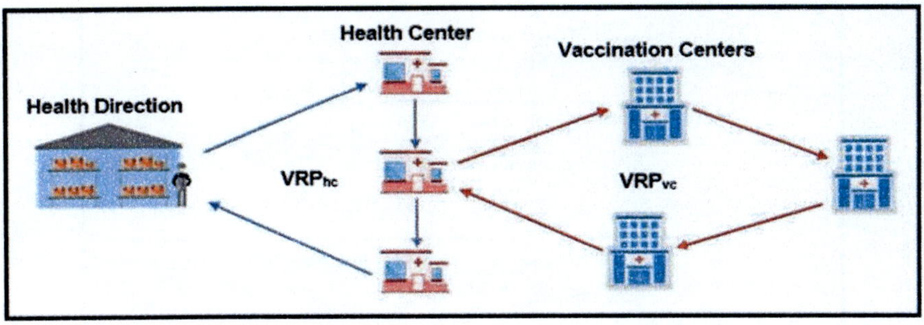

Fig. 4 Generic Net for CVRP application

determine the number of clusters needed for each zone and thus the number of centroids required. The health center or health post with the shortest distance from the centroid was chosen as the secondary warehouse. However, due to the dispersion of the vaccination points, there were some isolated cases from the rest of the points that were assigned to other clusters in order to determine the distribution network. As a result, two distribution networks are obtained: The main distribution network in which the Health Direction distributes to the secondary warehouses and the distribution network of the secondary warehouses to the vaccination points belong to the same cluster. The second stage focused on finding the optimal route to be followed by the vehicle. As shown in Fig. 4, there will be a route that the vehicle must follow through each secondary warehouse used and thus the minimum distribution cost will be obtained.

Currently, the cost of storing vaccines in health centers (c2) is much lower than the cost of storing in health directions (c1), because the spaces proposed are not being used and that these centers provide care outside the facilities to avoid using closed spaces when caring for patients and thus prevent contagion of Covid-19.

$$c1 * t1 - (c1 * t2 + c2 * t3) = Savings$$

On the other hand, the cost of storing a vaccine in the health directions (c1) multiplied by the time it remains in it (t1) gives as a result the total cost of storing a vaccine. In contrast, in the improvement proposal, the storage time in the health direction (t2) will decrease with the help of the secondary warehouses, in which they will be stored for the required time (t3). To find the total cost of storage in the improvement proposal, the storage cost in the health directions (c1) will be multiplied by the time the vaccines will remain in these (t2) and the result will be added with the storage cost in the secondary warehouse (c2) multiplied by the time it will remain in it (t3), thus producing considerable cost savings, as shown in Fig. 5.

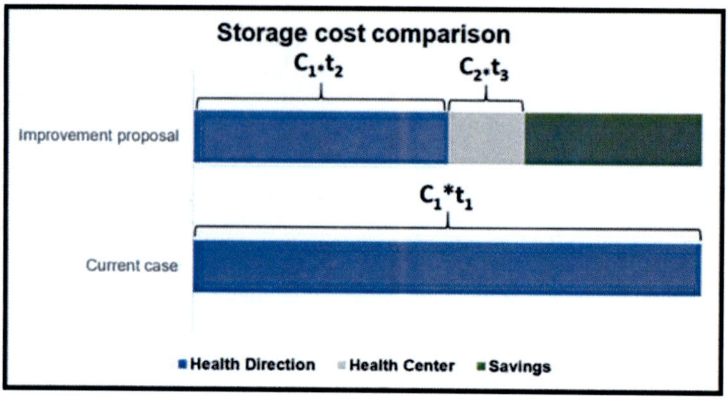

Fig. 5 Storage cost comparison

2.3 *Mathematical Model*

To optimize the use of the resources available to distribute vaccines, an optimization model is proposed in aims of minimizing costs of distribution and meet the demand of each vaccination center.

Below is each one of the parameters used in the optimization model:

- *xij* is the binary decision variable defined to indicate if the vehicle traverses an arc (i, j)
- n is the number of vaccination centers
- N is set of vaccination centers, with $N = \{1, 2, \ldots, n\}$
- V is set of nodes, with $V = \{0\} \cup N$
- A is set of arcs, with $A = \{(i, j) \in V2 : i \neq j\}$
- *cij* is cost of travel over arc $(i, j) \in A$
- Q is the vehicle capacity
- *qi* is the amount of vaccine cooler that has to be delivered to vaccination centers $i \in N$

Based on the previously mentioned, the problem is presented by the following optimization model:

$$\text{Min} \sum_{i,\, j \in A} C_{ij} x_{ij} \tag{1}$$

Subject to:

$$\sum\nolimits_{j \in V\, j \neq i} x_{ij} = 1 \; i \in N \tag{2}$$

$$\sum_{I \in V\, i \neq j} x_{ij} = 1\, j \in N \tag{3}$$

$$if\ x_{ij} = 1 \rightarrow u_i + q_j = u_j\, i, j \in A : j \neq 0, i \neq 0 \tag{4}$$

$$q_i \leq u_i \leq Q\, i \in N \tag{5}$$

$$x_{ij} \in \{0, 1\}\, i, j \in A \tag{6}$$

The objective function (1) minimizes the total costs for the routes chosen to meet the demand of the vaccination centers. The constraints (2) and (3) are the indegree and outdegree constraints for vaccination centers. Constraints (4) are the route continuity and sub-tour elimination constraints, ensuring that the solution contains no sub-tour disconnected from the depot, and that the vehicle load is a step function that accumulates the demand attended by the vaccination centers that are on the vehicle's route. The constraints given in (5) are capacity bounding constraints that restrict the upper and lower bounds of *ui* and constraints (6) are obligatory constraints.

3 Results

The results obtained by coding the elbow method in Python made it possible to determine the geographic locations of the secondary warehouses for each cluster. In this way, the cost of storage in each of these warehouses was calculated by using the estimated cost of these spaces and the amount of square meters that would be used to store the vaccines that will be distributed to each vaccination center.

The mathematical model was coded in CPLEX and supplemented with imported Python tools. Then, the cost of each arch between the nodes was estimated based on the average special cargo transportation rate, in dollars. Initially, the data from each vaccination point grouped in their respective clusters were evaluated, each one supplied by a secondary warehouse. Subsequently, the data from each secondary warehouse in each area was evaluated and the main headquarters of each Health Direction was taken as a reservoir. Both cases provided the costs associated with transportation and storage for each of the zones. In Callao, Central Lima, East Lima, North Lima and South Lima there were savings of 7.12%, 22.5%, 25.75%, 29.15% and 14.76% respectively. Figure 6 shows the monthly supply cost structure for Metropolitan Lima, showing the savings in dollars. Figure 7 shows the comparison of the current and proposed scenario for Central Lima, one of the areas that faces the greatest difficulty in terms of traffic and the time it takes to distribute the vaccines.

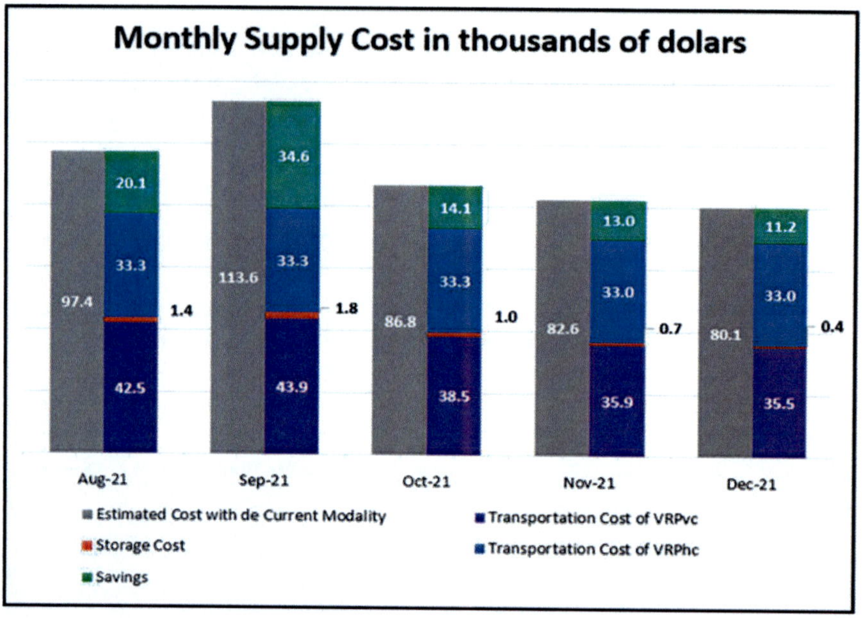

Fig. 6 Generic Net for CVRP application

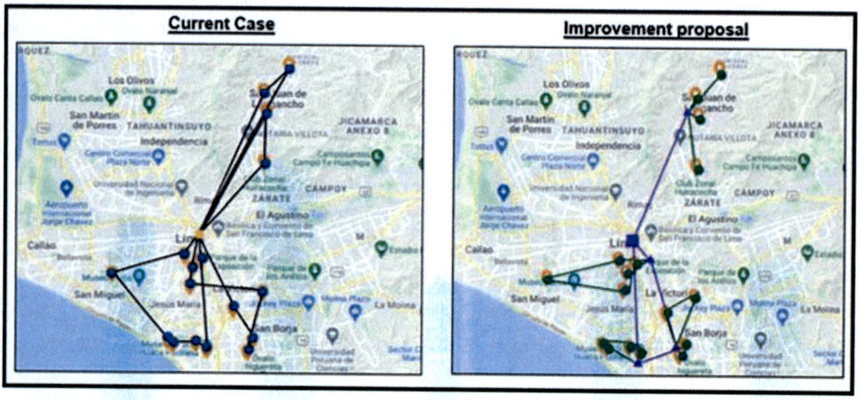

Fig. 7 Routing comparison in Central Lima

4 Conclusions

Based on the analysis, it can be concluded that the implementation of secondary warehouses in the Covid-19 vaccine supply chain in Metropolitan Lima increases response capacity and reduces costs associated with transportation and storage. This is the result of applying the elbow method to group vaccination sites and locate secondary warehouses in strategic geographic points; in addition to the application of a vehicle routing problem model with limited capacity (CVRP) to determine the optimal route from each health direction to the final vaccination point of each cluster. The application of these two stages are operationally and economically feasible; However, it is important to note that if there are future variations due to the shortage of vaccines, the operational efficiency in the supply chain will have to be reorganized and re-evaluated. For future research, it would be relevant to analyze the variants of the proposed model in the face of atypical events and even develop a simulation model to control operational times and foresee a variation in time demands for the vaccination centers or a reduction in the number of vaccination centers.

References

1. World Health Organization: WHO Coronavirus (COVID-19) Dashboard, https://covid19.who.int/table, last accessed 2021/08/01.
2. Ferguson NM, Impact of non-pharmaceutical interventions (NPIs) to reduce COVID-19 mortality and healthcare demand. Imperial College COVID-19 Response Team 2020, https://www.imperial.ac.uk/media/imperial-college/medicine/sph/ide/gida-fellow-ships/Imperial-College-COVID19-NPI-modelling-16-03-2020.pdf
3. Dargent E, Feldmann AE, Luna JP. Greater State Capacity, Lesser Stateness: Lessons from the Peruvian Commodity Boom. Politics & Society. 2017;45(1):3–34.
4. MINSA Peru: Sala Situacional COVID-19 Perú, https://covid19.minsa.gob.pe/sala_situacional.asp, last accessed 2021/07/31.
5. Sadarangani M, Abu Raya B, Conway JM, et al. Importance of COVID-19 vaccine efficacy in older age groups. Vaccine. 2021;39(15):2020–2023.
6. World Health Organization: Enfermedad por el coronavirus (COVID-19): Vacunass, https://covid19.who.int/table, last accessed 2021/08/01.
7. Cesar Fuentes: La vacuna llega al Perú: ¿Cuánto mejorarán las perspectivas económicas?, https://www.esan.edu.pe/conexion/actualidad/2021/02/15/la-vacuna-llega-al-peru-cuanto-mejoraran-las-perspectivas-economicas/, last accessed 2021/07/28.
8. Gooding E, Spiliotopoulou E, Yadav P. Impact of vaccine stockouts on immunization coverage in Nigeria. Vaccine. 2019 37(35):5104–5110.
9. Rastegar M, Tavana M, Meraj A, Mina H. An inventory-location optimization model for equitable influenza vaccine distribution in developing countries during the COVID-19 pandemic. Vaccine. 2021;39(3):495–504.
10. Han, S., Cai, J., Yang, J. et al. Time-varying optimization of COVID-19 vaccine prioritization in the context of limited vaccination capacity. Nat Commun 12, 4673 (2021).
11. Shakiba E., Osman Y. Optimal influenza vaccine distribution with equity. European Journal of Operational Research. 2020;283(2):714–725.
12. Georgiadis GP, Georgiadis MC. Optimal planning of the COVID-19 vaccine supply chain. Vaccine. 2021;39(37):5302–5312.
13. Sabouhi F. et al. An optimization approach for sustainable and resilient supply chain design with regional considerations. Computers & Industrial Engineering. 2021; ed.159
14. Souto G. et al. A hybrid matheuristic for the Two-Stage Capacitated Facility Location problem. Expert Systems with Applications. 2021;ed.185.
15. Altabeeb A.M. et al. Solving capacitated vehicle routing problem using cooperative firefly algorithm. Applied Soft Computing. 2021; ed.108.
16. Akhtar M, Hannan MA, Begum RA, Basri H, Scavino E. Backtracking search algorithm in CVRP models for efficient solid waste collection and route optimization. Waste Manag. 2017 Mar;61:117–128.

17. Government of Peru: Consultar centros de vacunación contra la COVID-19, https://www.gob.pe/13334, last accessed 2021/08/01.
18. MINSA Peru: Vacunar es Vida. Digital edition, July 2021 p.93. Available in: http://bvs.minsa.gob.pe/local/MINSA/5486.pdf, last accessed 2021/07/26.
19. CPI Peru: Perú población 2019. Market report 2019 p.9. Available in: http://www.cpi.pe/images/upload/paginaweb/archivo/26/mr_poblacional_peru_201905.pdf
20. INEI Peru: Estimaciones y Proyecciones de Población por Departamento, Provincia y Distrito, 2018–2020. INEI January 2020 p.55–100. Available in: https://www.inei.gob.pe/media/MenuRecursivo/publicaciones_digitales/Est/Lib1715/libro.pdf
21. MINSA Peru: Vacunación COVID-19 Perú, https://www.minsa.gob.pe/reunis/data/vacunas-covid19.asp, last accessed 2021/08/01.

COVID-19 Impacts and Mitigation Strategies on Food Supply Chains: A Survey to the Brazilian Context

Brenda Cardoso, Luiza Cunha, Adriana Leiras, Hugo Tsugunobu Yoshida Yoshizaki, Paulo Gonçalves, Irineu de Brito Junior, and Frederico Pedroso

Abstract The COVID-19 pandemic has significantly affected Food Supply Chains (FSCs), causing impacts that interfere with the normal flow of operations. This paper analyzes the COVID-19 impacts on FSCs and the main strategies to minimize their adverse effects. The research adopts a scoping review to identify impacts and strategies and an empirical study consisting of a survey with professionals operating in Brazilian SCs to discuss and validate the results. Impacts are categorized into final consumer, internal environment, and external environment. The policies are also divided into COVID-19 prevention policies and FSCs operational policies. The survey presents 45 complete responses to identify the major impacts and implementation of mitigation policies in FSCs during the COVID-19 pandemic. The main impacts are increased demand, increased operating costs, and price instability. The main mitigation policies are isolation of employees with suspicious cases (COVID-19 prevention) and strong and focused leadership (Operational policy). The results provide a first validation of the research, and the authors suggest as future research the development of studies deepening other types of supply chains.

Keywords Disaster impact · Food Supply Chains · Survey

B. Cardoso (✉) · L. Cunha · A. Leiras
Pontifical Catholic University of Rio de Janeiro, Rio de Janeiro, Brazil

H. T. Y. Yoshizaki
Sao Paulo University, Sao Paulo, Brazil

P. Gonçalves
Università della Svizzera Italiana, Lugano, Switzerland

I. de Brito Junior
Sao Paulo State University, Sao Jose dos Campos, Sao Paulo, Brazil

F. Pedroso
World Bank, Washington, DC, USA

J. Vargas Florez et al. (eds.), *Production and Operations Management*, Springer Proceedings in Mathematics & Statistics 391,
https://doi.org/10.1007/978-3-031-06862-1_43

1 Introduction

Supply Chains (SCs) may be dynamically unstable, facing disruptions in production systems that result from variations in consumption due to delays in the adaptation of production and inventory capacities and decision feedback throughout the entire chain [1]. The occurrence of disasters (natural and human-made) is one of the factors that can interrupt supply chain flows [2]. The increase in frequency and intensity of disasters is a significant concern by SC practitioners, as it disrupts the performance of SC in any nation [3]. Disasters are increasingly focusing practitioners' attention on the resilience of Food Supply Chain (FSC) [4]. FSCs refer to all processes in food reaching the final consumer, including cultivation, processing, distribution, sales, and consumption [4].

Recently, the COVID-19 pandemic has severely impacted FSCs [5–8]. During the COVID-19 pandemic, breakdowns or bottlenecks in any part of the FSC have impacted other components up and down the chain [9]. Impacts on FSC involve labor scarcity, price increases, transport blockages, limited market access due to quarantine measures, and production interruptions.

One of the challenges in managing the food sector during a crisis or disaster is sustaining a robust food safety system and implementing the appropriate strategies to correlate the needs and requirements of the overall supply chain, including consumers, producers, distributors, economic environment, and other SC links [10].

To minimize the negative effects of disasters, mitigation policies must be devised to address the turbulent markets to maintain business continuity [11]. In the face of the current pandemic, companies generally seek to preserve the sustainability of their businesses by implementing an appropriate set of organizational policies [12]. Furthermore, stakeholders such as government and international organizations need to continue offering essential services to the population part of the interdependent critical infrastructure systems such as transportation, energy, water, and food distribution [13].

In this context, this paper analyzes the impacts of the COVID-19 pandemic on FSCs and the related mitigation policies in Brazil. We aim to answer the following research questions: (i) What are the main impacts caused by the COVID-19 pandemic on the FSC? (ii) What strategies are used to mitigate the impacts caused by the COVID-19 pandemic?

After this introductory section, Sect. 2 presents the theoretical background. Section 3 brings the research methodology. Next, Sect. 4 shows the main results. Finally, Sect. 5 presents the conclusions, limitations, and future research.

2 Theoretical Background

An SC comprises all the stages involved, directly or indirectly, in fulfilling customer demand. Thus, an SC is defined "as an integrated system in which various business entities such as suppliers, manufacturers, industrial customers, distributors, and retailers work together to address issues of both material flow and information flow" [3]. The successful delivery of goods and services from the manufacturer to the consumer requires a flexible and adaptable supply chain that meets global demands. However, due to the interlinkage of SCs across the world, SC disruptions in one country may affect the supply of goods and services in another, regardless of their geographical position [3]. In this way, global supply chains are interconnected in their sensitivity to disasters.

FSCs refer to all processes in food reaching the final consumer [4]. FSCs need to operate in a relatively delicate balance between consumption, production, and inventory once the SCs imbalance of activities and materials impacts the overall SCs [9]. During the COVID-19 pandemic FSCs faced significant challenges, such as panic buying, staffing shortages, and inventory volatility. We categorize these impacts into three types: final consumer, internal environment, and external environment.

Instability in demand for food impacts the final consumer [4]. The population is forced to modify their habits and, consequently, their buying behavior in response to a disaster. When a disaster such as the COVID-19 pandemic strikes, people become cautious about their spending, buying less often but in larger volumes, preferring small grocery stores, and purchasing food online to avoid agglomerations [6]. Consequently, instability in per capita consumption was also an observed impact. Another impact is transitory sales growth due to panic buying [14]. Just after the COVID-19 pandemic was declared, there was a rush to supermarkets, shelves being quickly emptied, and people shopping to stock up on products at home [15]. It can also be mentioned the growth of in-house consumption. With the closing of restaurants and cafeterias and imposed restrictions, people had movement limitations. Therefore, the digital/online market grown-up in importance and e-commerce reached millions of new customers [16].

Labor shortages, impacts on company's productivity and lead-time, and unavailability of materials are examples of impacts on the internal environment during the COVID-19 pandemic. Staffing shortages (not enough workers) have become a significant problem in several countries. Restrictive movement measures (e.g., lockdown) caused labor shortages, which affect activities in the FSC, harvesting, processing, and distribution [17]. Sales instability and inventory instability are impacts directly related to the instability in demand. The COVID-19 pandemic has also affected the agricultural sector, disrupting the supply (e.g., fertilizers, seeds), thus affecting production systems and supply chains [18]. Consequently, companies in all sectors are prone to financial losses and increases in operating costs, adverse effects for business growth and productivity, delays, and interruptions for the delivery of products. Price instability also has a significant impact on FSCs. At the beginning of the COVID-19 pandemic, the increase in prices was immediately noticed in both essential and in-demand products [16]. There were also interruptions in infrastructure and transportation, which needed to be readjusted to adapt to the new reality.

The relationship with other links in the supply chain stands out as an impact related to the external environment, involving negotiation with suppliers to change options [19], information sharing among stakeholders [20], and instability of partnerships [21]. Also, instability of imports and exports [22] caused by closed borders, social isolation, and other prevention measures interfere with people and materials' movement [16].

In the face of disasters, companies seek to maintain the sustainability of their businesses by implementing an appropriate set of organizational policies [12]. In this context, mitigation policies are extensively covered in the literature. This paper divides into COVID-9 prevention policies and operational supply chain policies.

Implementation of remote work [23], use of personal protective equipment [24], development and monitoring of proto-safety collars [19] are examples of COVID-19 prevention policies.

Operational policies are essential to ensure the effective functioning of activities throughout the FSC. Effective partnerships and integration [25]; synchronization of strategic processes [26]; information and communication availability [27] transparency to all decision-makers [28]; strong, focused country leadership [25]; absenteeism monitoring [29]; government financial assistance [30] are examples of operational policies.

3 Methodology

The research applies a survey to validate the main impacts and policies found in the literature. We follow the steps of [31] for survey research: (i) relationship with theory, (ii) design, (iii) data collection and analysis, and (iv) final report.

In the relationship with the theory, the research considers the findings obtained in the theoretical foundation. The unit of analysis is the COVID-19 and Brazilian FSCs. The design stage focused on practitioners of Brazilian SCs. The survey instrument is a structured questionnaire to identify COVID-19 pandemic impacts on FSCs and mitigation policies. The questionnaire consists of two main parts. The first part addressed the respondents' general information such as years of experience and company. The second part consisted of closed questions related to the impacts of the COVID-19 pandemic and mitigation policies in FSCs. Besides, the questionnaire has open-ended questions for additional comments about the topics.

A pilot test with 5 supply chain experts verified the questionnaire's understanding and reviewed necessary improvements. We used the SurveyMonkey platform for data collection. The survey obtained 45 complete responses after excluding incomplete responses from the analysis.

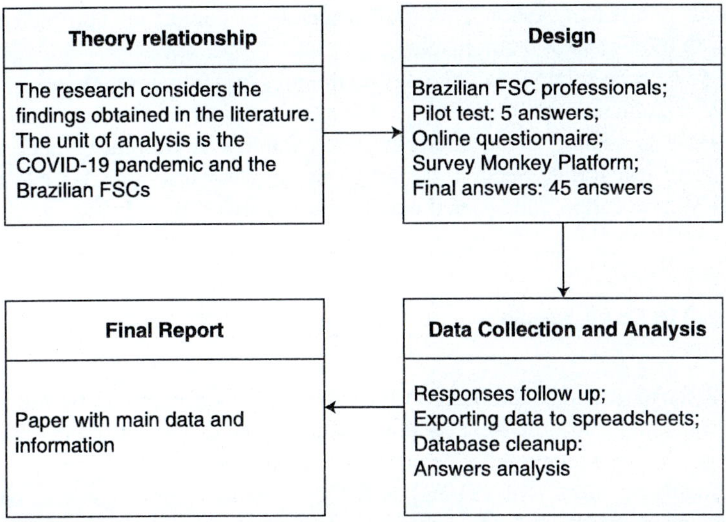

Fig. 1 Survey steps

Table 1 Respondent's profile

Working time	%	Company size	%
0–1 year	9%	Up to 19 employees	13%
1–3 years	27%	From 20 to 99 employees	11%
3–5 years	16%	From 100 to 499 employees	16%
More than 5 years	49%	500 or more employees	60%

Note: Company size classification is based on [32]

In the analysis stage, the results are presented in tables and graphs to explore the critical issues from the perspective of experts. Also, the results consider the convergent validity of the data [31] between the answers and the information found in the literature. The last step, the final report, is presented in this paper.

Figure 1 summarizes the survey steps.

4 Results and Discussions

The sample of the questionnaire respondents include directors, support and management analysts, operation managers, logistics assistant, supply managers, transportation managers, and other positions that deal with supply chain operations.

Table 1 presents the respondent's profile. Most respondents (49%) have worked in their companies for more than 5 years, which allows for greater and better knowledge about activities before and during the COVID-19 pandemic. A significant number of respondents (60%) operate in companies considered to be large, with more than 500 employees.

The following sections highlight the findings regarding the impacts of the COVID-19 pandemic on the FSCs and mitigation policies. Section 4.1 presents the impacts according to the final consumer, internal environment, and external environment categories. Section 4.2 offers the implemented policies considering COVID-19 prevention policies and operational supply chain policies.

4.1 *COVID-19 Impacts*

Figure 2 shows the major impacts of the final consumer identified by the respondents. The main consumer-related impacts are: increased demand, digital/online market growth, and increased sales due to panic buying.

The increase in demand/consumption for food resulted from increased consumer spending on essential goods and from the implementation of emergency income that increased (or avoid a sharp drop) the consumption of food in Brazil [33].

The increase in demand and the growth in sales due to panic buying are intertwined. Panic buying behavior typically originates from customers buying more than usual. This type of consumer behavior is most often observed in periods of crisis and disruptive events, such as natural disasters and public and personal health emergencies [15]. During the COVID-19 pandemic, the media partially drove this trend, often showing photos of empty shelves and warning of food shortages [34].

The digital/online market is a trend resulting from the COVID-19 pandemic [35]. Services offered at physical points have changed to digital services leveraged delivery services throughout Brazil [35]. Brazilian e-commerce experienced the most significant growth in the last 20 years, with a significant increase in sales revenue [36].

An increase in production costs, unavailability of raw or processed materials, and instability of final goods inventory highlight the impacts on organizations' internal environment (Fig. 3).

Increased production costs in the food and beverage industry led to higher consumer prices. During the COVID-19 pandemic, it is estimated that food

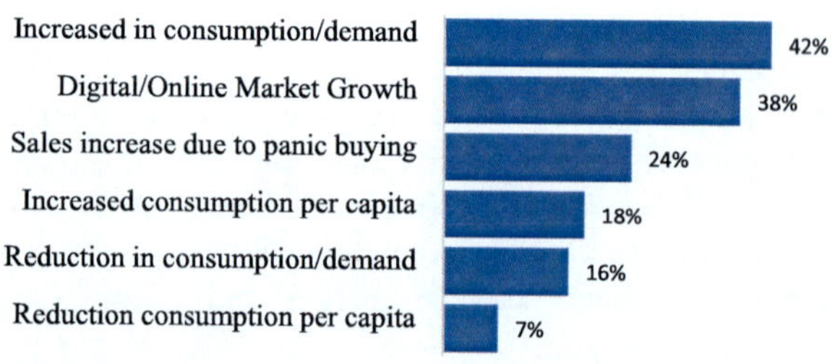

Fig. 2 COVID-19 impacts on FSCs (final consumer)

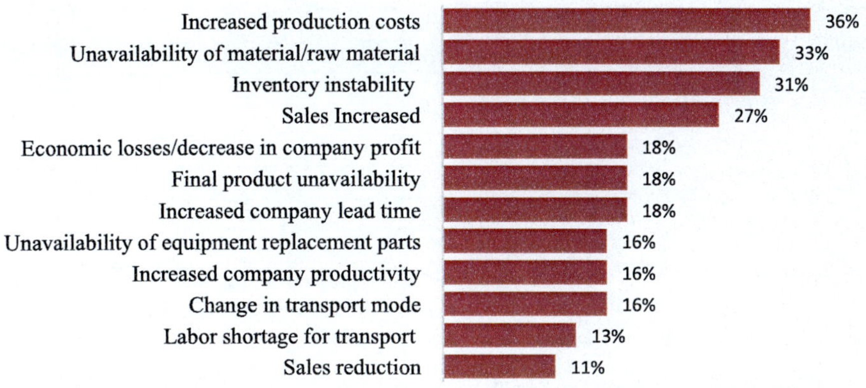

Fig. 3 COVID-19 impacts on FSCs (internal environment)

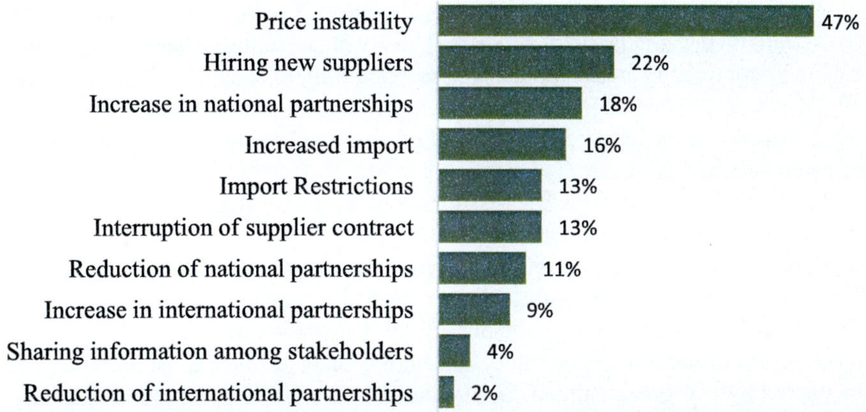

Fig. 4 COVID-19 impacts on FSCs (external environment)

production costs increased by 4.8% in 2020, raising the price of products to the final consumer up to 2.5% [37].

The unavailability of raw or processed materials also affected the food price and its supply chain in general. The COVID-19 pandemic impacted the logistical structure and made it difficult to obtain some inputs or raw materials needed for production [38].

The inventory instability resulted from the fluctuation of supply and demand, emerging from organizations difficulty operating in crisis and emergency scenarios [39]. Such instability required more frequent analysis of inventory policies to balance forecast demand against the current capacity of suppliers [21].

Figure 4 illustrates the impacts related to the organizations' external environment. The main impacts are price instability and hiring new suppliers.

During the COVID-19 pandemic, the increase in prices was immediately observed for essential and high-demand products [16]. In Brazil, in July 2020,

Table 2 Ranking of most affected areas/departments

Area/Department	%
Human Resources	38%
Production	33%
Purchases	33%
Transport	24%
Financial	22%
Storage	20%
Sales	4%

Note: Responses' sum is greater than 100% since each respondent could mark more than one area/department

food consumed at home increased by 0.14%. In August 2020, household food inflation accelerated to 1.15% [40]. Trade frictions, common during the COVID-19 pandemic, also contribute to a global increase in food prices [41].

The hiring of new suppliers is another highlighted point. Due to governments restrictions on the circulation of goods [42], many organizations adapted their supply chains by contracting local suppliers and reconfiguring their supply chains as "short supply chains". Short supply chains play a key role in strengthening the resilience of food networks, contributing to the stability of access to food in periods of crisis such as pandemics and epidemics [43].

The most affected areas or business functions were also analyzed (Table 2). 38% of experts identified human resources as the most affected area. This finding can be explained by the difficulties related to human resources management during the COVID-19 pandemic. The shortage of workers had a negative impact on FSCs. Self-isolation and social distancing guidelines discontinued the usual flow of labor across the industry. In addition to poor physical health, human resource constraints can also be caused by psychological and social factors such as feelings of sadness, apathy, fear, uncertainty, frustration, anxiety, insomnia, [44] all of which were prevalent during the COVID-19 pandemic.

Although it does not appear to emphasise the impacts of the organization's internal environment, it can be said that human resource constraints had a significant impact on the FSC. Lack of staff (insufficient workers) has become a significant problem in several countries. The unavailability of the staffing is perceived since workers often need to stay at home and are not available to carry out their activities due to policies of social distance and even infection by the virus [29, 45, 46].

Information regarding the companies' installed capacity, dismissals, hiring of employees, and interruption of activities are also analyzed.

Regarding the installed capacity of companies, 42% of respondents say that there was an adjustment to improve activities. More specifically, 31% indicate an increase in installed capacity, while the others claim a reduction in installed capacity. This result is directly related to the increase in demand (impact mentioned in Sect. 4.1) for some products. This represents significant changes for companies' operations, as there is a need to fulfil all orders. Related to the workforce, 9% of respondents indicated that employees were laid off, while 29% indicated that new employees were hired. Hiring new employees is also related to the increase in demand, as there is a need for available labor to meet the demand. In 2020, 37.7 thousand companies were operating in the food sector, with 1.68 million direct and formal jobs [47]. About the interruption of activities, 22% of respondents claim to have suffered some interruption for different periods (5 days, 10 days, 30 days). The interruption of activities influenced the implementation of home office in companies to prevent the virus from spreading among workers, especially workers at risk.

4.2 Mitigation Policies

Mitigation policies are divided into COVID-19 prevention and operational policies for business continuity.

Table 3 presents the policies identified to minimize virus transmission during FSC activities. The main policies implemented in the companies are the isolation of employees at risk, the use of personal protective equipment, the creation of security protocols, and the implementation of remote work.

Prevention measures are essential to prevent the spread of the virus both in the workplace and in any other environment. In this way, governments and health agencies work together to guide workers and managers in combating the COVID-19 pandemic. The measures presented in Table 3 are, in summary, part of protocols and guidelines for employers and workers developed by the government of Brazil. The protocols, guides, and plans present the main actions to fight the pandemic: surveillance, laboratory support, infection control measures, medical care, pharmaceutical care, health surveillance, risk communication, and management [48].

Among the actions published by the government of Brazil, there are key questions that also appear in the answers to the questionnaire, such as issuing instructions on the proper use of protective equipment; encourage the monitoring of the

Table 3 COVID-19 prevention policies

%	Policies
62%	Isolation of employees with suspicious cases
58%	Use of masks/face shields
53%	Creation of security protocols against COVID-19
51%	Remote Work Deployment
49%	Social distancing
47%	Investment in protective equipment against COVID-19
47%	Temperature monitoring at the entrance of the company
47%	Removal of employees at risk
44%	Operational plans with contingency
38%	The strict control of security protocols
38%	Use of additional PPE (e.g., gloves and aprons)
33%	Isolation of the group directly in contact with suspected cases

Table 4 FSC operational policies

%	Policies
38%	Strong and focused leadership
31%	Absenteeism monitoring
31%	Strategic/security stocks
24%	Preventive actions due to the international scenario
24%	Availability of information and communication
20%	Centralized decision making
20%	Information transparency
20%	Technologies for sharing information
18%	Partnerships and integration with chain members
16%	Company investment for production adaptation
13%	Benchmarks with other companies/countries
13%	Process synchronization
11%	Change of product mix
11%	Union negotiation for shift change
9%	Company investment in transport activities
7%	Change in import/export strategies
4%	Extension of payment of taxes
4%	Adherence to government funding programs
4%	Simplification of hiring requirements

execution of state/municipal contingency plans for COVID-19 by local managers; support the dissemination of materials developed by the technical area (protocols, manuals, guides, technical notes).

Table 4 presents the policies identified to maintain FSC activities during the COVID-19 pandemic. Table 4 highlights as main policies implemented strong and focused leadership, absenteeism monitoring, and strategic/security stocks.

Strong and focused leadership is a policy indicated by 38% of respondents. High-impact strategies within and outside a country's system are enabled by strong, country-focused leadership, funding, and effective partnerships [25]. Strong, focused leadership approaches are needed, especially in low- and middle-income countries [25].

Absenteeism monitoring is indicated as a mitigation policy by 31% of respondents. Due to human resource constraints during the COVID-19 pandemic, there is a need to proactively develop strategies to address absenteeism and other potential threats to the supply chain [29]. Policymakers need to monitor and plan for absenteeism, exploring ways to recruit and train more workers for critical parts of the supply chain while working with industry to anticipate, identify and resolve new issues as they arise.

The implementation of security stocks is indicated by 31% of respondents. Supply chain responsiveness is a crucial dimension of resilience, and future efforts to increase resilience through strategic inventory management plans and flexible procurement strategies are essential [46]. Flexible procurement strategies involve local suppliers with faster response, low costs, and stimulating the local economy [49].

The respondents also indicated other policies as preventive actions due to the international scenario, availability of information and communication, centralized decision making, information transparency, and technologies for sharing information.

5 Conclusions

This research sought to identify the impacts of the COVID-19 pandemic on the FSC in Brazil and determine related mitigation policies. We analyzed 45 responses from experts working in the Brazil SCs to assess the FSC scenario during the COVID-19 pandemic based on a survey.

The results reveal that the pandemic impacted final consumers and organizations' internal and external environments. In particular, impacts to final consumers include an increase in demand/consumption and growth of the online market). Internally, organizations experienced increased production costs and unavailability of raw materials. Externally, organizations faced price instability and the need for hiring new suppliers. The mitigation policies identified were categorized as COVID-19 impact alleviation (e.g., isolation of suspected cases and use of protective equipment), and operationalization of SC activities included strong and focused leadership, absenteeism monitoring, and strategic inventories.

This study is relevant given the challenges decision-makers face with the occurrence of the COVID-19 pandemic. Identifying impacts and policies is a preparation strategy for future interruptions in supply chains. By analyzing this area of research, our study contributes insights for academics and practitioners on the impacts of disasters on food chains and appropriate policies to mitigate them. Given the greater frequency of disasters, developing more resilient supply chains represents a future necessity.

Future research could perform a comparative analysis of FSC in different countries, considering their particularities and individualities. In addition, future research could also analyze the impact of the pandemic on the supply chains of other industries (e.g., automobile, vaccines) to assess their differences and similarities.

Acknowledgments The authors acknowledge the support of Coordination for the Improvement of Higher Education Personnel (CAPES) [88887.373163/2019-00; 88887.492890/2020-00 – Finance Code 001]; The World Bank [Doc.000208 I UPIs: 569776; 569582; 485892; 569585; 569579]; National Council for Scientific and Technological Development (CNPq) [308084/2019-5, 313687/2019-6], and Foundation for Support of Research in the State of Rio de Janeiro (FAPERJ) [E-26/201.384/2021].

References

1. Thiel, D., Vo, T. L. H., & Hovelaque, V. Forecasts impacts on sanitary risk during a crisis: a case study. The International Journal of Logistics Management 25(2),358–378 (2014).
2. Singh, S., Kumar, R., Panchal, R., & Tiwari, M. K. Impact of COVID-19 on logistics systems and disruptions in food supply chain. International Journal of Production Research 1–16 (2020).
3. Chari, F., & Ngcamu, B. S. An assessment of the impact of disaster risks on dairy supply chain performance in Zimbabwe. Cogent Engineering 4(1), 1409389 (2017).
4. Siche, R. What is the impact of COVID-19 disease on agriculture?. Scientia Agropecuaria 11(1), 3–6 (2020).
5. Montenegro, L. D., & Young, M. N. Operational Challenges in the Food Industry and Supply Chain during the COVID-19 Pandemic: A Literature Review. In 2020 7th International Conference on Frontiers of Industrial Engineering (ICFIE) (pp. 1–5). IEEE, 2020.
6. Rejeb, A., Rejeb, K., & Keogh, J. G. COVID-19 and the food chain? Impacts and future research trends. LogForum 16(4) 2020.
7. Kumar, A., Mangla, S. K., Kumar, P., & Song, M. Mitigate risks in perishable food supply chains: Learning from COVID-19. Technological Forecasting and Social Change 166, 120643 (2021).
8. Cardoso, B., Cunha, L., Leiras, A., Gonçalves, P., Yoshizaki, H., de Brito Junior, I., & Pedroso, F. Causal Impacts of Epidemics and Pandemics on Food Supply Chains: A Systematic Review. Sustainability 13(17), 9799 (2021).
9. Borsellino, V.; Kaliji, S.A.; Schimmenti, E. COVID-19 Drives Consumer Behaviour and Agro-Food Markets towards Healthier and More Sustainable Patterns. Sustainability 12 (2020).
10. Farcas, A. C., Galanakis, C. M., Socaciu, C., Pop, O. L., Tibulca, D., Paucean, A., ... & Socaci, S. A. Food Security during the Pandemic and the Importance of the Bioeconomy in the New Era. Sustainability 13(1), 150 (2021)
11. Kurniawan, R., & Zailani, S. Supply chain vulnerability and mitigation strategy of the manufacturing firms in Indonesia: manager's perspectives. International Business Management 4(3), 116–123 (2010).
12. Silva, M. E., Pereira, S. C., & Gold, S. The response of the Brazilian cashew nut supply chain to natural disasters: A practice-based view. Journal of Cleaner Production 204, 660–671 (2018).
13. Nozhati, S., Rosenheim, N., Ellingwood, B. R., Mahmoud, H., & Perez, M. Probabilistic framework for evaluating food security of households in the aftermath of a disaster. Structure and Infrastructure Engineering 15(8), 1060–1074 (2019).
14. Yoshizaki, H.T.Y., de Brito Junior, I., Hino, C.M., Aguiar, L.L. and Pinheiro, M.C.R. Relationship between Panic Buying and Per Capita Income during COVID-19, Sustainability 12(23) 9968 (2020).
15. Lins, S., & Aquino, S. Development and initial psychometric properties of a panic buying scale during COVID-19 pandemic. Heliyon 6(9) (2020).
16. de Souza Costa, A., da Silva Fôro, G. S., & de Lima Vieira, J. COVID-19 e as cadeias de suprimentos: uma revisão bibliográfica dos principais impactos no Brasil Revista Vianna Sapiens 11(2), 28–28 (2020).
17. Chitrakar, B., Zhang, M., & Bhandari, B. Improvement strategies of food supply chain through novel food processing technologies during COVID-19 pandemic, Food Control 125 (2021).
18. Magar, D. B. T., Pun, S., Pandit, R., & Rola-Rubzen, M. F. Pathways for building resilience to COVID-19 pandemic and revitalizing the Nepalese agriculture sector. Agricultural Systems 187, 103022 (2021).
19. Telukdarie, A., Munsamy, M., & Mohlala, P. Analysis of the Impact of COVID-19 on the Food and Beverages Manufacturing Sector. Sustainability 12(22), 9331 (2020)
20. Schätter, F., Hansen, O., Wiens, M., & Schultmann, F. A decision support methodology for a disaster-caused business continuity management. Decision Support Systems 118, 10–20 (2019).

21. Xu, Z., Elomri, A., Kerbache, L., & El Omri, A. (2020). Impacts of COVID-19 on global supply chains: facts and perspectives. IEEE Engineering Management Review 48(3), 153–166 (2020).
22. Pulighe, G., & Lupia, F. Food first: COVID-19 outbreak and cities lockdown a booster for a wider vision on urban agriculture. Sustainability 12(12), 5012 (2020).
23. Knorr, D., & Khoo, C. S. H. COVID-19 and Food: Challenges and Research Needs. Frontiers in Nutrition 7 (2020).
24. Aday, S., & Aday, M. S. Impact of COVID-19 on the food supply chain. Food Quality and Safety 4(4), 167–180 (2020).
25. Akseer, N., Kandru, G., Keats, E. C., & Bhutta, Z. A. COVID-19 pandemic and mitigation strategies: implications for maternal and child health and nutrition. The American journal of clinical nutrition 00, 1–6 2020.
26. Sharma, R., Shishodia, A., Kamble, S., Gunasekaran, A., & Belhadi, A. Agriculture supply chain risks and COVID-19: mitigation strategies and implications for the practitioners. International Journal of Logistics Research and Applications 1–27 (2020).
27. Sarà, G., Mangano, M. C., Berlino, M., Corbari, L., Lucchese, M., Milisenda, G., ... & Helmuth, B. The synergistic impacts of anthropogenic stressors and COVID-19 on aquaculture: A current global perspective. Reviews in Fisheries Science & Aquaculture 1–13 (2021)
28. Münzberg, T., Berbner, U., Comes, T., Friedrich, H., Groß, W., Pfohl, H. C., & Schultmann, F. Decision support for critical infrastructure disruptions: An integrated approach to secure food supply. In ISCRAM (2013).
29. Gray, R. S. Agriculture, transportation, and the COVID-19 crisis. Canadian Journal of Agricultural Economics/Revue canadienne d'agroeconomie 68(2), 239–243 (2020)
30. Fan, S. Agriculture, food and nutrition security under Covid-19: lessons from China. Review of Agrarian Studies 10 (2020).
31. Forza, C. Survey research in operations management: a process-based perspective. International journal of operations & production management 22(2) 152–194 (2002).
32. SEBRAE-NA/Dieese. Anuário do trabalho na micro e pequena empresa, available at:https://www.sebrae.com.br/Sebrae/Portal%20Sebrae/UFs/SP/Pesquisas/MPE_conceito_empregados.pdf, last accessed 2021/08/15.
33. Baccarin, J. G., & de Oliveira, J. A. Inflação de alimentos no Brasil em período da pandemia da Covid 19, Continuidade e Mudanças. Segurança Alimentar e Nutricional 28 (2021).
34. Boyacı-Gündüz, C. P., Ibrahim, S. A., Wei, O. C., & Galanakis, C. M. Transformation of the Food Sector: Security and Resilience during the COVID-19 Pandemic. Foods 10(3), 497 (2021).
35. De Assunçâo, M. V. D., Medeiros, M., Trump, L. N. R., Paiva, I. V. L., & Paes, C. D. S. Resilience of the Brazilian supply chains due to the impacts of Covid-19. Holos 36(5), 1–20 (2020).
36. EXAME, Vendas online no Brasil crescem 47% no 1° semestre, maior alta em 20 anos, available at: https://exame.com/pme/e-commerce-brasil-cresce-47-primeiro-semestre-alta-20-anos/, last accessed 2021/08/15.
37. ABIA. Alta de custos de produção na indústria de alimentos e bebidas impacta nos preços ao consumidor, available at: https://www.abia.org.br/releases/alta-de-custos-de-producao-na-industria-de-alimentos-e-bebidas-impacta-nos-precos-ao-consumidor, last accessed 2021/08/15.
38. BBC News. Pandemia gera escassez de matéria-prima e faz preços subirem no Brasil, available at: https://www.bbc.com/portuguese/brasil-55048624, last accessed 2021/08/15.
39. Hansen, O., Friedrich, H., & Transchel, S. An inventory management approximation for estimating aggregated regional food stock levels. International Journal of Production Research 1–17 (2019).
40. CNA. Preços dos alimentos sobem menos que a inflação em Julho/2020, available at: https://www.cnabrasil.org.br/boletins-tecnicos/precos-dos-alimentos-sobem-menos-que-a-inflacao-em-julho-2020, last accessed 2021/08/15.

41. Elleby, C., Domínguez, I. P., Adenauer, M., & Genovese, G. Impacts of the COVID-19 pandemic on the global agricultural markets. Environmental and Resource Economics 76(4), 1067–1079 (2020)
42. Butu, A., Brumă, I. S., Tanasă, L., Rodino, S., Dinu Vasiliu, C., Doboş, S., & Butu, M. The impact of covid-19 crisis upon the consumer buying behavior of fresh vegetables directly from local producers. Case study: The quarantined area of suceava county, Romania. International journal of environmental research and public health, 17(15), 5485 (2020).
43. Cavallo, C., Sacchi, G., & Carfora, V. Resilience effects in food consumption behaviour at the time of Covid-19: perspectives from Italy. Heliyon 6(12) (2020).
44. Clemente-Suárez, V.J., Dalamitros, A.A., Beltran-Velasco, A.I., Mielgo-Ayuso, J. and Tornero Aguilera, J.F. Social and psychophysiological consequences of the COVID-19 pandemic: An extensive literature review. Frontiers in Psychology 11, 3077 (2020).
45. Kumar, S. Planning for avian flu disruptions on global operations: a DMAIC case study. International journal of health care quality assurance (2012).
46. Hobbs, J. E. Food supply chains during the COVID-19 pandemic. Canadian Journal of Agricultural Economics/Revue canadienne d'agroeconomie (2020).
47. ABIA (Associação Brasileira da Indústria de Alimentos.) Números do setor. available at: https://www.abia.org.br/downloads/Infograficoanual2020v5.pdf, last accessed 2021/08/28.
48. Ministerio da Saúde Brasil. Plano de Contingência Nacional para Infecção Humana pelo novo Coronavírus COVID-19, available at: https://portalarquivos2.saude.gov.br/images/pdf/2020/fevereiro/13/plano-contingencia-coronavirus-COVID19.pdf, last accessed 2021/08/20.
49. Lamenza, A. A. S.; Fontainha, T. C. ; Leiras, A. Purchasing strategies for relief items in humanitarian operations. Journal of Humanitarian Logistics and Supply Chain Management 9, 151–171 (2019).